MEMOIRS OF THE SOCIETY FOR ENDOCRINOLOGY
NO. 18

HORMONES AND
THE ENVIRONMENT

MEMOIRS OF THE SOCIETY FOR ENDOCRINOLOGY

NO. 18

HORMONES AND THE ENVIRONMENT

PROCEEDINGS OF A SYMPOSIUM
HELD AT THE UNIVERSITY OF SHEFFIELD
ON 2 TO 5 SEPTEMBER 1969

EDITED ON BEHALF OF
THE SOCIETY FOR ENDOCRINOLOGY BY

G. K. BENSON & J. G. PHILLIPS

CAMBRIDGE
AT THE UNIVERSITY PRESS
1970

Published by the Syndics of the Cambridge University Press
Bentley House, 200 Euston Road, London, N.W.1
American Branch: 32 East 57th Street, New York, N.Y.10022

© The Society for Endocrinology 1970

Library of Congress Catalogue Card Number: 71–114601

Standard Book Number: 0 521 07846 6

Printed in Great Britain
at the University Printing House, Cambridge
(Brooke Crutchley, University Printer)

PREFACE

It is logical to seek elucidation of the endocrinological mechanisms in man through a study of other species, principally the common laboratory animals. In recent years, however, our views have been broadened and extended through a growing interest in a whole variety of non-mammalian species, and from this approach has emerged the discipline of comparative endocrinology.

There is also a growing awareness of the importance of the environment in terms of endocrine adaptation, and evidence is available for interaction between the environment and the organism throughout both the vertebrates and invertebrates.

The opportunity to discuss this aspect of endocrinology through the gathering together of a group of interested endocrinologists was made possible by the interest of the Society for Endocrinology through its sponsoring of the meeting. The occasion was provided by the invitation of the University of Sheffield to hold the meeting there.

The programme was devised to afford the opportunity for as wide a coverage as possible, bearing in mind the central theme, the environment. From the meeting emerges this Memoir which records the deliberations of a large number of participants drawn from many countries.

In arranging the Symposium we owe much to the support received from members of the organizing committee, and for valuable help in running the meeting we are indebted to a task-force of academic staff, technicians and graduate students from the Department of Zoology in Sheffield, too numerous to list here, whose enthusiasm and efficiency we warmly acknowledge.

The Society is greatly indebted to the following for support in the form of financial aid or hospitality:

> The University of Sheffield
> The Mayor and Corporation of the City of Sheffield
> The Wellcome Trust
> The Ciba Foundation
> Schering Chemicals Ltd.

Our speakers deserve a special word of thanks for their carefully prepared and stimulating contributions, and our chairmen for their disciplined yet courteous control of the scientific sessions.

The Cambridge University Press proved a constant source of advice in the preparation of this Memoir, and their guidance and patience is gratefully acknowledged.

G. K. BENSON

J. G. PHILLIPS

ORGANIZING COMMITTEE

F. J. EBLING — *Chairman*

G. K. BENSON
J. G. PHILLIPS — *Programme Secretaries*

J. N. BALL
I. W. HENDERSON — *Local Secretaries*

D. BELLAMY

K. C. HIGHNAM

CONTENTS

INTRODUCTION

The present volume contains the proceedings of the Symposium on Hormones and Environment held in the University of Sheffield on the occasion of the opening of an imposing new building for the Department of Zoology, headed by Professor Ian Chester Jones. The meeting was in his honour.

First, I must express my appreciation at having been invited to serve as President of the Symposium. I am well aware that many of those who gathered from near and far had greater expertise and familiarity than I with the type of problems that were considered. What I could not forgo was the opportunity to share in this tribute to Professor Chester Jones.

Although most of the participants might be identified as comparative endocrinologists, the emphasis was not on comparative endocrinology in the modern sense but on endocrine aspects of experimental biology in the classical sense. It will be evident that the papers presented covered a wide range of topics but through them all ran a common theme, the influence of environment. It should not come as a surprise to learn, as we did, that the different species studied exhibit tremendous variations in their endocrine responses to environmental stimuli. Nevertheless, there is among them some similarity of hormonal action and cautious generalizations are beginning to materialize.

Most of the papers dealt with species other than the common laboratory animals. In fact, many of them were concerned with what are often referred to, though with questionable justification, as the lowly inhabitants. In a sense, this quest for knowledge at the older end of the phylogenetic scale seemed refreshingly pure. One could not come away from this Symposium without a healthy respect for the adaptational endocrinology of those species that live in naked contact with nature. That which was expounded seemed clearly for the sake of understanding their biology and for that alone. Questions of applicability were not raised. Indeed, such would have been out of keeping with the serious intent to gain new insights to the relevant biological background of that most vaunted of earth's inhabitants, man himself.

The calibre of the work presented and the quality of the presentations themselves were uniformly excellent. Every paper evoked searching questions by attentive participants. Spirited discussions were commonplace and often carried over into the hall ways and dining-room, but most vociferously, to the bar lounge. This was a lively and stimulating meeting and one that showed the benefits of able and careful planning. Much credit is due to the Society for Endocrinology for the arrangement, and particularly, of course, to the organizing committee. It is my expectation that the resulting volume will greatly intensify the exploration of this new frontier in experimental biology.

ROY O. GREEP
(*Conference President*)

LIST OF PARTICIPANTS

PRESIDENT

GREEP, R. O.
  Laboratory for Human Reproduction and Reproductive Biology, Harvard Medical
  School, 25 Shattuck Street, Boston, Massachusetts, 02115, U.S.A.

CHAIRMAN OF CONFERENCE

CHESTER JONES, I.
  Department of Zoology, The University, Sheffield, England.

CHAIRMEN

BERN, H. A.
  Department of Zoology, University of California, Berkeley, California, 94720, U.S.A.

DODD, J. M.
  Department of Zoology, University College of North Wales, Bangor, Wales.

FOLLEY, S. J.
  Department of Physiology, National Institute for Research in Dairying,
  Shinfield, Reading, England.

HIGHNAM, K. C.
  Department of Zoology, The University, Sheffield, England.

LEATHEM, J. H.
  Bureau of Biological Research, Rutgers—The State University, New Brunswick,
  New Jersey, 08903, U.S.A.

VAN OORDT, P. G. W. J.
  Department of Zoology, The University, Utrecht, Netherlands.

SPEAKERS* AND CO-AUTHORS

BALL, J. N.
  Department of Zoology, The University, Sheffield, England.

*BARTH, R. H.
  Department of Zoology, University of Texas at Austin, Austin, Texas, 78712, U.S.A.

*BELLAMY, D.
  Department of Zoology, University College of South Wales and Monmouthshire,
  Cardiff, Wales.

BLAIR-WEST, J. R.
  Department of Physiology, University of Melbourne, Howard Florey Laboratories,
  Parkville, Victoria 3052, Australia.

BRADLEY, J. S.
  Department of Biological Sciences, University of California, Santa Barbara,
  California, 93106, U.S.A.

*CHAN, D. K. O.
  Department of Zoology, The University of Hong Kong, Hong Kong.

CHAN, MO YIN.
  Department of Biological Sciences, University of California, Santa Barbara,
  California, 93106, U.S.A.

CHIU, K. W.
  Department of Biology, Chung Chi College, Chinese University of Hong Kong,
  Shatin, Hong Kong.

COGHLAN, J. P.
  Department of Physiology, Howard Florey Laboratories, University of Melbourne,
  Parkville, Victoria, 3052, Australia.

[xiii]

*DANTZLER, W. H.
  Department of Physiology, College of Medicine, University of Arizona, Tucson,
  Arizona, 85721, U.S.A.

DENTON, D. A.
  Department of Physiology, Howard Florey Laboratories, University of Melbourne,
  Parkville, Victoria, 3052, Australia.

*EBLING, F. J.
  Department of Zoology, The University, Sheffield, England.

*ETKIN, W.
  Department of Anatomy, Albert Einstein College of Medicine, Yeshiva University,
  Eastchester Road and Morris Park Avenue, Bronx, N.Y. 10461, U.S.A.

FOLLETT, B. K.
  Department of Zoology, The University, Leeds, England.

HALE, PATRICIA A.
  Department of Zoology, The University, Sheffield, England.

*HELLER, H.
  Department of Pharmacology, The Medical School, University Walk, Bristol,
  England.

HENDERSON, I. W.
  Department of Zoology, The University, Sheffield, England.

HIGHNAM, K. C.
  Department of Zoology. The University, Sheffield, England.

HILL, L.
  Department of Zoology, The University, Sheffield, England.

*HOLMES, W. N.
  Department of Biological Sciences, University of California, Santa Barbara,
  California, 93106, U.S.A.

KNAGGS, G. S.
  Department of Physiology, National Institute for Research in Dairying, Shinfield,
  Reading, England.

*LEDERIS, K.
  Department of Pharmacology, Division of Pharmacology and Therapeutics,
  Faculty of Medicine, University of Calgary, Calgary 44, Alberta, Canada.

*LOCKE, M.
  Developmental Biology Center, Case Western Reserve University, 2127 Cornell Road,
  Cleveland, Ohio, 44106, U.S.A.

*LOFTS, B.
  Department of Zoology, University of Hong Kong, Hong Kong.

LUNTZ, A. JENNIFER.
  Anti-Locust Research Centre, College House, Wright's Lane, London, W.8, England.

*MADERSON, P. F. A.
  Department of Biology, Brooklyn College, City University, New York,
  N.Y., 11210, U.S.A.

*MAETZ, J.
  Department of Biology, Zoological Station, 06 Villefranche-sur-Mer, France.

*MARTINI, L.
  Department of Pharmacology, University of Milan, Milan, Italy.

*MILNER, R. D. G.
  Medical Research Council, Tropical Metabolism Research Unit, University of the
  West Indies, Mona, Kingston 7, Jamaica, W.I.

*MORDUE, W.
  Department of Zoology and Applied Entomology, Imperial College,
  London, S.W. 7, England.

MOTTA, MARCELLA.
  Department of Pharmacology, University of Milan, Milan, Italy.

MURTON, R. K.
  Ministry of Agriculture, Fisheries and Food, Infestation Control Laboratory,
  Field Research Station, Tangley Place, Worplesdon, Guildford, Surrey, England.

MYERS, K.
Department of Physiology, Howard Florey Laboratories, University of Melbourne, Parkville, Victoria, 3052, Australia.

NELSON, J. F.
Department of Physiology, Howard Florey Laboratories, University of Melbourne, Parkville, Victoria, 3052, Australia.

*OLIVEREAU, MADELEINE.
Institute of Oceanography, 195 Rue St Jacques, Paris, France.

ORCHARD, ELSPETH.
Department of Physiology, Howard Florey Laboratories, University of Melbourne, Parkville, Victoria, 3052, Australia.

PHILLIPS, J. G.
Department of Zoology, The University, Hull, England.

PIVA, F.
Department of Pharmacology, University of Milan, Milan, Italy.

*QUAY, W. B.
Department of Zoology, University of California, Berkeley, California, 94720, U.S.A.

*REINBOTH, R.
Institute for General Zoology, Johannes Gutenberg University, Mainz, Germany.

SANDOR, T.
Hôpital Notre Dame, 1560 Est, Rue Sherbrooke, Montreal 24, Canada.

*SCOGGINS, B. A.
Department of Physiology, Howard Florey Laboratories, University of Melbourne, Parkville, Victoria, 3052, Australia.

STAINER, ISABEL M.
Department of Zoology, The University, Hull, England.

*STRAUSS, J. S.
Department of Dermatology, Boston University Medical Center, 80 East Concord Street, Boston, Massachusetts, 02118, U.S.A.

*TINDAL, J. S.
Department of Physiology, National Institute for Research in Dairying, Shinfield, Reading, England.

TIMA, L.
Department of Pharmacology, University of Milan, Milan, Italy.

*DE WILDE, J.
Entomology Laboratory, Agricultural Institute, Wageningen, Netherlands.

WRIGHT, R. D.
Department of Physiology, University of Melbourne, Howard Florey Laboratories, Parkville, Victoria, 3052, Australia.

ZANISI, MARIAROSA.
Department of Pharmacology, University of Milan, Milan, Italy.

ADDITIONAL CONTRIBUTORS TO DISCUSSIONS

BRIGGS, M. H.
Schering Chemicals Ltd., Pharmaceutical Division, Victoria Way, Burgess Hill, Sussex, England.

BUTLER, D. G.
Department of Zoology, University of Toronto, Toronto, Ontario, Canada.

CHAN, S. T. H.
Department of Zoology, University of Hong Kong, Hong Kong.

CLEGG, E. J.
Department of Human Biology and Anatomy, The University, Sheffield, England.

EVANS, C.
Department of Biology, Nottingham Regional College of Technology, Burton Street, Nottingham, England.

FENWICK, J. C.
Department of Zoology, The University, Sheffield, England.

FERGUSON, D. R.
  Department of Pharmacology, The Medical School, University Walk, Bristol,
  England.
FORSTER, M. A.
  Department of Zoology, The University, Sheffield, England.
HERBERT, J.
  Department of Anatomy, The Medical School, The University, Birmingham,
  England.
HOAR, W. S.
  Department of Zoology, University of British Columbia, Vancouver, 8, Canada.
HUTCHINSON, J. S. M.
  Department of Chemical Pathology, St Thomas's Hospital Medical School,
  London, S.E. 1, England.
HYDER, M.
  Department of Zoology, University of Nairobi, Nairobi, Kenya.
IDLER, D.R.
  Fisheries Research Board of Canada, Halifax Laboratory, P.O. Box 429, Halifax,
  Nova Scotia, Canada.
JOHNSON, ELIZABETH.
  Department of Zoology, The University, Reading, England.
MACFARLANE, N. A. A.
  Department of Biology, The University, Stirling, Scotland.
MATTY, A. J.
  Department of Biological Sciences, University of Aston in Birmingham, Gosta Green,
  Birmingham, England.
PETER, R. E.
  Department of Zoology, University of Washington, Seattle, Washington, 98105,
  U.S.A.
QURESHI, M. A.
  Department of Biological Sciences, University of Aston in Birmingham, Gosta Green,
  Birmingham, England.
SMITH, M. W.
  Agricultural Research Council, Institute of Animal Physiology, Babraham,
  Cambridge, England.
WEIR, BARBARA J.
  Wellcome Institute of Comparative Physiology, Zoological Society of London,
  Regents Park, London, N.W. 1, England.
WRIGHT, A.
  Department of Zoology, University of Hong Kong, Hong Kong.

SESSION I
THE AQUATIC ENVIRONMENT

MECHANISMS OF SALT AND WATER TRANSFER ACROSS MEMBRANES IN TELEOSTS IN RELATION TO THE AQUATIC ENVIRONMENT

By J. MAETZ

In relation to an aquatic environment ranging from fresh water to sea water and even to hypersaline media, the teleostean fish is characterized by the maintenance of an internal osmotic pressure which remains in the range of 250–350 m-osmoles. The burden of compensating for movements of water or ions due to the osmotic pressure and ionic concentration differences is borne by specialized epithelia separating internal and external media and situated in the 'effector-organs' of osmoregulation: the gills, the gut and the kidney.

The classical model drawn by Smith (1932) and completed by Krogh (1939) summarizes our knowledge of the contrasting ways by which these effector-organs come into play in hypo- and hypertonic media. In fresh water, water enters through the external boundaries of the fish and the kidney has the task of removing excess water. Urine is abundant and dilute. The renal Na and Cl loss and the passive loss of these electrolytes along the concentration gradient across the external boundaries is compensated by an active uptake of ions by the gills. As the freshwater fish drinks very little or not at all, and as the fish is able to maintain its salt-and-water balance in the prolonged absence of food intake, the gut is believed to play a negligible role in osmoregulation. In sea water, water is lost across the boundaries of the fish. The gut is definitely the site of the compensatory mechanism restoring water balance. The fish swallows the external medium and water is absorbed by the gut along with salt. This electrolyte entry is balanced by extrarenal excretion of monovalent ions, probably by the gills. The kidney function is characterized by reduction of the free-water clearance and excretion of the bivalent ions absorbed by the gut.

The purpose of this review is to emphasize for some of the teleostean effector-organs of osmoregulation, the functional differences observed when teleosts living in hypo- and hypertonic media are compared, and the functional changes, which these organs undergo when a euryhaline fish is transferred to media of various salinities. The mechanisms underlying these changes in the gut and the gills at cellular and molecular levels will be discussed, as well as the mode of action of hormones.

[3]

A. INTESTINAL SALT AND WATER ABSORPTION MECHANISM

1. *Drinking rate and external salinity*

The classical statement that freshwater fishes do not drink was first contradicted by Allee & Frank (1948), when they demonstrated the accumulation of carbon particles or colloid substances by the gut of the goldfish, when these substances are added to the external medium. The first quantitative estimate of the rate of drinking in a freshwater fish concerns *Fundulus heteroclitus* (Potts & Evans, 1967). New techniques have been developed since, permitting a rapid and accurate assessment of the amount of water swallowed in undisturbed fish. Most of them make use of radioactive labelled 'water-markers'. Table 1 presents some of the most significant results obtained and techniques employed. The following comments may be made.

Drinking rates are extremely variable, as will appear from the large standard errors reported. No explanation for this variability is available. Shock-effects following handling of the fish are certainly responsible for increased drinking rates, as suggested by our observations on the goldfish. Differences in size and experimental temperature are also of importance. The most important factor inducing changes in the drinking rate is the external salinity, as may be seen from the comparison of freshwater and marine fishes (see Table 5) or from the rates recorded in euryhaline fishes as a function of increased osmotic stress. (Table 1). Increased passive water losses are to be

Table 1. *Drinking rate and external salinity in teleosts*

Species	Medium	t (°C)	Technique	Body weight (g)	Drinking rate*	Reference
Anguilla anguilla	200% SW	20	[^{131}I]phenol red	80–225	802±182	Maetz & Skadhauge, 1968
	100% SW	—	—	—	325±34	—
	100% FW	—	—	—	135±28	—
	100% SW	15	^{198}Au	50–110	167±21	Maetz, unpublished
	35% SW	—	—	—	61±30	
Carassius auratus	FW	20	^{198}Au	95	51±28	Motais *et al.* 1969
	FW	18–22	[^{14}C]inulin	15–80	80±30	Lahlou *et al.* 1969
	35% SW	—	—	—	883±80	—
Fundulus heteroclitus	100% SW	20	[^{14}C]inulin	2–5	2300±270	Potts & Evans, 1967
	—	—	[^{35}S]sulphate	—	1540±210	—
	40% SW	—	[^{14}C]inulin	—	2000±430	—
	FW	—	[^{14}C]inulin	—	830±390	—
	FW	—	[^{35}S]sulphate	—	140±50	—
Tilapia mossambica	200% SW	21–24	[^{35}S]sulphate	0·5–3	1590±206	Potts *et al.* 1967
	100% SW	—	—	—	1110±170	—
	100% SW	—	^{125}I[PVP]	—	975±196	Evans, 1968
	40% SW	—	—	—	440±95	Potts *et al.* 1967
	FW	—	—	—	260±40	—
Xiphister atropurpureus	100% SW	13	[^{14}C]inulin	7–40	34±0·6	Evans, 1967
	10% SW	—	—	—	7±0·2	—

SW, Sea water; FW, fresh water.   See also results given in Table 5.   * μl h^{-1} (100 g)$^{-1}$.

expected which have to be compensated by increased drinking rates. In the stenohaline *Carassius auratus* also, an important increase of the drinking rate was observed during adaptation to a slightly hypersaline medium (Lahlou, Henderson & Sawyer, 1969).

Nothing is known about the internal mechanism inducing the fish to swallow increased amounts of water. Sharratt, Bellamy & Chester Jones (1964) suggest that it is the *osmolality* rather than the *salinity* of the external medium which is important. Eels are unable to adapt to hypertonic sucrose solutions. Increased drinking is however observed, because the gut is found to be distended by external fluid. Similar observations were reported by Motais (1967) for flounders kept in hypertonic mannitol solutions. Such experiments demonstrate that water absorption against an osmotic gradient can only occur in the presence of NaCl. The next section will deal with the mechanism of water absorption by the intestinal wall.

2. *Intestinal water and salt absorption capacity and external salinity*

Until recently, very few studies had appeared concerning the mechanism of water absorption in the intestine of fishes. Sharratt *et al.* (1964), House & Green (1965) and Smith (1964) concluded from their observations *in vitro* that the intestine of fishes is very similar to that of the mammals with respect to water and salt transfer. The absorption of water in the absence of an osmotic gradient is only possible in the presence of NaCl, and the relative importance of both ions is discussed by House & Green (1965). As in mammals, glucose and Na ions appear to be absorbed synergistically by the gut (Smith, 1964). Comparison of the data obtained by House and Green on the isolated intestine of the marine *Cottus scorpius* and by Smith on the gut of the freshwater goldfish show that the rate of water and salt absorption is much higher in the seawater-adapted fish and Smith comments that the difference may well be accounted for by the fact that marine teleosts have to absorb large amounts of salt to obtain water.

In the last few years, two independent groups of workers discovered that the gut of the euryhaline eel displays adaptive functional features in relation to external salinity changes. The Japanese group (Utida, Isono & Hirano, 1967; Oide, 1967), working on isolated eel-gut preparations reported first that water absorption is augmented during salt water adaptation of the fish. This increased effectiveness is observed in the absence of an osmotic gradient, and also when the lumen is filled with hypertonic saline and water first enters the gut passively; in these conditions the ability of the gut to reverse the flow of water and to start water absorption is observed earlier. In parallel, seawater adaptation is also accompanied in the gut by an increased ability to absorb salt as the net Na absorption rate is augmented by a factor of 2. As the water absorption rate is simultaneously increased by a factor of 2·5–3 it appears that the molar ratio H_2O/Na is augmented during salt-water adaptation, an increase which is of physiological significance. Table 2 gives some of the relevant data obtained by the Japanese group and compares them with the observations of the European group (Skadhauge & Maetz, 1967; Skadhauge, 1969).

The latter group studied the mechanism of water absorption not only in relation to the adaptation of the eel in waters of various salinities, but also in better-controlled thermodynamic conditions. The absorption of salt and water was studied *in vivo* by intraluminal perfusion of the intestine of European eels adapted to fresh water, sea water and double-strength sea water. When dilute sea water is used as perfusion fluid, the net absorption rate of Na and Cl ions was found to be more or less independent of the NaCl concentration, over the range 90–300 mM. When eels were transferred from fresh water to sea water, this rate doubled upon adaptation. A further augmentation was observed after transfer to double-strength sea water. The ratio influx *versus* outflux remains more-or-less constant irrespective of the adaptation medium. The salt-pumping efficiency of the gut is thus augmented in relation to increased osmotic stress, a result which confirms the above-mentioned *in vitro* observations. As may be seen in Table 2, however, the rate of absorption is about 4–5 times faster *in vivo*, although the experimental temperature was identical in both types of experiments. The rate of water absorption was measured by the European group with the help of a 'water-marker', phenol red. The intestinal epithelium is practically impermeable to this dye. The reversal of the flow is illustrated by a decrease followed by an increase in the concentration of phenol red. The 'turning-point' of the dye-concentration curve shows clearly that dilution of the intestinal content is necessary before absorption of water is possible, unless the initial osmolality of the perfusion fluid is low. The important point which emerges from the observations of the European team is that the osmolality of the perfusion fluid, which corresponds to zero net water flow (the 'turning-point osmolality') is higher than plasma osmolality by 73 m-osmoles in the freshwater animals, 126 m-osmoles in the seawater fishes and 244 m-osmoles in the hypersaline-adapted fishes. The osmotic gradient against which water absorption becomes possible increases with adaptation to higher external salinities, an observation of obvious physiological importance. A remarkably constant ratio between NaCl pumping rate and the osmotic gradient corresponding to zero net flow of water is observed, suggesting that both phenomena are closely correlated.

Perfusion experiments with impermeant solutes demonstrate that in the absence of NaCl water movement is essentially a passive process. These

Table 2. *Intestinal absorption mechanisms of the eel in relation to external salinity*

| Medium | Ionic fluxes | | | | | P_{os} | Water transport | | | |
| | Na | | Cl | | | | Solute linked water flow | | Mole ratio H₂O/Na | |
	F_{net}	F_{in}/F_{out}	F_{net}	F_{in}/F_{out}	T.P.		At TP*	At iso	At TP*	At iso
			In vivo studies (Skadhauge, 1960)							
FW	16·6±1·7	2·60±0·36	20·5±2·4	2·74±0·20	73±3	0·37±0·05	26	65	39	99
SW	36·3±3·3	2·70±0·16	42·3±3·7	2·95±0·24	126±5	0·72±0·10	90·5	162	64	114
DSW	64·0±11·0	2·95±0·23	67·6±14·9	—	244±32	—	—	—	—	—
			In vitro studies (Oide, 1967)							
FW	3·6±0·49	—	—	—	—	—	17·8±2·9	—	270	—
SW	7·0±0·52	—	—	—	—	—	41·6±3·3	—	331	—

Fluxes in μ-equiv h⁻¹ cm⁻² (at 20 °C). F_{in}, F_{out}, F_{net}: influx (mucosa serosa), outflux and net flux.

Osmotic permeability (P_{os}) in μl. h⁻¹ m-osmoles⁻¹ cm⁻².

Mole ratio in moles of water/mole⁻¹ Na absorbed.

Water flow in μl. h⁻¹ cm⁻² at T.P. or at isotonicity (iso).

T.P.: turning-point osmotic gradient (in m-osmoles) corresponding to zero-net water flow.

FW, Fresh water; SW, salt water; DSW, double salt water.

* (according to the double flow hypothesis of Skadhauge).

experiments permitted the evaluation of the passive gut osmotic permeability. The permeability is higher in the seawater-adapted eels than in the freshwater animals. 'Rectification' of flow is suggested by the fact that the serosa-to-mucosa permeability is less than that in the opposite direction (Skadhauge, 1969).

In conclusion, these observations support the view that 'solute-linked water flow' exists in the gut of fishes, as in other epithelia such as the gall bladder, secondary to the salt movement resulting from the osmotic pressure produced by local accumulation of salt within the membrane, possibly between the epithelial cells. The model proposed by Curran (1960), Whitlock & Wheeler (1964) and Diamond (1964) not only seems adequate to explain water transport in the absence of osmotic gradients or against such gradients, but also clarifies the mechanism of functional adaptation displayed by the gut of the eel. If solute-linked water flow varies proportionally with salt transport rate and osmotic permeability, the increased molar ratio (H_2O/Na) of the transported fluid and the absorption against a higher osmotic gradient may be explained on a quantitative basis.

3. *Cellular or molecular mechanisms underlying the adaptive features*

It is well known that the gut is the site of intensive cellular renewal, as shown by its high sensitivity to radiation damage, a phenomenon demonstrated in the gut of fishes (Conte, 1965). It would be of interest to compare the rate of cellular renewal in the intestine of fishes adapted to various salinities by measuring the DNA turnover with [³H]thymidine. Structural modifications may also be revealed by fine histological or electron-microscopic examination. Hirano (1967) observed that the intestinal wall seems thinner and the surface area smaller in the seawater-adapted eels. Hence, it does not seem possible that increased rate of water and salt absorption would be the result of an augmentation of the available surface.

Differences in the enzymic content of the intestinal epithelium in relation to adaptation to various external salinities have been recorded. Two enzymes have been found to have higher activities in the gut of the sea-water-adapted eel: Na–K-dependent ATPase (Oide, 1967; Jampol & Epstein, 1968), an enzyme which has been considered to be a rate-limiting factor linked with Na transport, and alkaline phosphatase, an enzyme whose role in gut function is less clear. It is of interest that NaCl activation is much more important for the latter enzyme extracted from the mucosa taken from seawater-adapted fishes than from the freshwater living species (Utida, Oide & Oide, 1968). Such studies may help to elucidate the functional significance of these enzymes in relation to water and salt transfer.

4. *Endocrine control of intestinal adaptation*

Hirano, Kamiya, Saishu & Utida (1967) and Hirano (1967) reported that after removal of the hypophysis the enhancement of water absorption induced by saltwater adaptation in the eel was no longer observed. Hirano & Utida (1968) have demonstrated that the hypophysial–interrenal axis is responsible for the adaptive changes. Among all the hormones tested for replacement therapy, only ACTH and cortisol proved capable of restoring normal water movement when administered at low doses. In parallel, cortisol has also been shown to be effective in augmenting the Na absorption rate. It may be of interest to investigate if this steroid is also acting on the passive water permeability of the intestinal epithelium. Furthermore it may be suggested that cortisol is somehow responsible for the increased enzymic activities that have been discussed above. The possible mode of action of this corticoid in relation to cellular renewal or molecular renewal will be discussed more fully in the section concerning the gill, which also appears to be a target organ for this hormone.

B. BRANCHIAL MECHANISMS OF OSMOREGULATION

The gill is the most important organ of osmoregulation. We shall first consider its role in the handling of electrolytes and secondly its regulatory function with respect to water movements. Finally, cellular and molecular processes underlying the functional adaptability of the gill in relation to external salinity changes will be discussed.

1. *Ionic transfer mechanisms and external salinity*

Recent research has shown that the functional reversal of polarity of the branchial epithelium permitting a net Na and Cl uptake in fresh water and excretion in sea water involves different types of ionic pumps.

(a) *The ionic uptake mechanism in fresh water and its functional variations*

Figs. 1 and 2 summarize the essential features of Na transport in freshwater and seawater teleosts. In fresh water the Na turnover resulting from the over-all ionic exchange represents a very small fraction of the internal Na. The net Na uptake results from a branchial influx which is about 20–50% higher than the branchial outflux and this compensates for the renal loss of salt. As seen in Fig. 3, the Na influx increases as a function of external Na concentration but the curve shows that two components of the influx have to be taken into account: a passive component which varies linearly and an active component displaying saturation kinetics. In any

case the ratio F_{in}/Na_{ext} is high, meaning that the affinity of the Na transporting mechanism is high. In contrast the ratio F_{out}/Na_{int} is very low,
emphasizing the low passive permeability of the gill in fresh water. Electrophysiological studies of the gill epithelium have shown that in fresh water

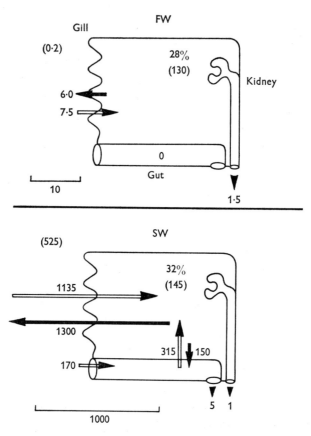

Fig. 1. Comparative Na balance in the freshwater (FW) and seawater (SW) eel. Note
different scales for the fluxes (in μ-equiv h^{-1} (100 g)$^{-1}$). External and internal Na concentrations in brackets (in m-equiv l.$^{-1}$). Size of the extracellular spaces in percentage body
weight. (According to Maetz et al. 1968).

the potential—about 20 mV negative inside—is insufficient to compensate
for the huge chemical gradient as the ratio Na or Cl_{in}/Na or Cl_{out} may be
as high as 1,000 (House, 1963; Maetz & Campanini, 1966; Evans, 1969 a).
　　Fig. 4 illustrates the model proposed by Maetz & Garcia Romeu (1964)
for the mechanism of Na$^+$ and Cl$^-$ uptake. This model confirms earlier
suggestions by Krogh (1939) and is identical to that proposed by Shaw
(1964) for the freshwater crayfish. Both ions may be taken up independently

(Garcia Romeu & Maetz, 1964) and Na^+ is exchanged against NH_4^+, the chief end-product of nitrogen metabolism, while Cl^- is exchanged against HCO_3^-, which results from the hydration of respiratory CO_2 catalysed by carbonic anhydrase (Maetz, 1956; Dejours, 1969).

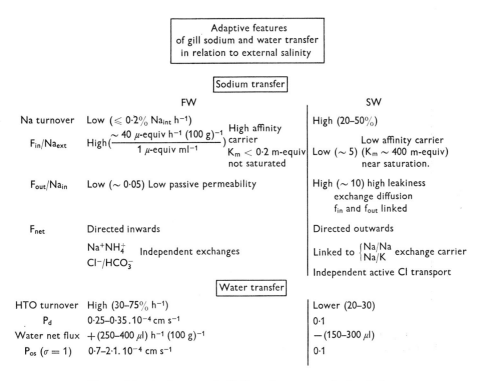

Fig. 2. Adaptive features of gill Na and water transfer in relation
to external salinity.

An important point needs to be stressed concerning the affinity of the hypothetical ionic carrier in the case of the cations for example. While this carrier has a very high affinity for Na, it has some affinity for NH_4^+, especially on the inside of the cell, thus permitting the ionic exchange. For K^+, however, affinity is not at all observed. For instance, a sudden increase of external K concentration to a level 25–50 times higher than that of Na^+ has no effect on the net uptake or influx of Na in goldfish or the eel. In seawater fish, on the contrary, a Na/K exchange mechanism is operative which will be discussed below.

The ion uptake mechanism displays functional variations in relation to the regulatory needs of the fish. One type of variation is related to the maintenance of the acid-base balance by branchial elimination of H^+ or

NH_4^+ ions or HCO_3^- ions against Na^+ or Cl^- as suggested by the model given above. A second type of variation is related to osmoregulatory needs of the fish. The gill responds to internal salt depletion by increasing the pumping rate permitting the readjustment of the internal sodium chloride level. Such a response is observed after keeping the fish in de-ionized water or after injection of hypotonic saline into the intraperitoneal cavity (Krogh, 1939; Maetz, 1964; Bourguet, Lahlou & Maetz, 1964). Conversely, the

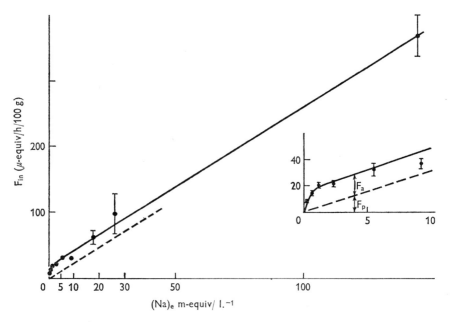

Fig. 3. Na influx of the flounder in fresh water as a function of external sodium concentrations. Ordinates: influx (F_{in} in μ-equiv/h/100 g). Abscissae: $(Na)_e$, external sodium concentration in m-equiv $l.^{-1}$. Temperature of the flux measurements: 16 °C. Inset fig.: expanded scale for the lower values; F_a, active component; F_p, passive component represented by a dashed line. The standard errors of the flux values are given. (Maetz & Zwingelstein, unpublished.)

response to a salt load, whether by keeping the fish in a saline solution or by injection of hypersaline Ringer solution, is an inhibition of the Na uptake and a simultaneous increase of the branchial outflow. When certain euryhaline fishes such as *Fundulus heteroclitus* adapted to sea water are transferred to fresh water the influx is instantaneously reduced to a level comparable to that observed in long-term adapted freshwater fishes, while in other species such as the eel it is reduced to a level significantly lower than that found for long-term adapted freshwater fish (see Maetz, Mayer & Chartier-Barduc, 1967, and Maetz, Sawyer, Pickford & Mayer, 1967).

The outflux readjustment pattern which occurs simultaneously during seawater to freshwater transfer will be discussed in the next section.

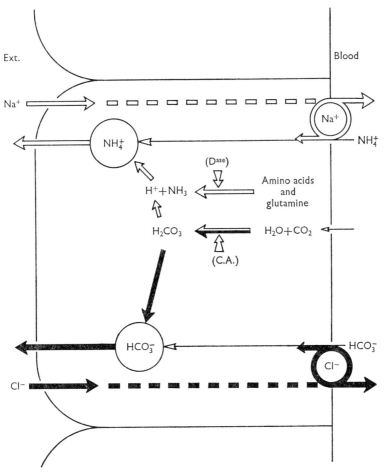

Fig. 4. Schematic representation of ionic exchanges in the branchial cell of *Carassius*, according to Maetz & Garcia Romeu, 1964, slightly modified. (D^{ase}) stands for deamidation and deamination enzymes contributing to the cellular pool of NH_3. C.A. stands for carbonic anhydrase contributing to the cellular pool of HCO_3^- ions. NH_4^+ and HCO_3^- may also result from plasma clearance. The location of the exchange pumps on the external or internal face of the epithelial cell is not known.

(b) The ion excretion pump in sea water and its functional variations

Figs. 1 and 2 indicate that the Na (or Cl) exchange across the gill is very fast. The high turnover rate of the internal NaCl can only be observed with the help of radioactive tracers (Mullins, 1950; Motais, 1967; Maetz, 1968; Potts, 1968). This fast exchange involves 10–30 times more salt than can be accounted for by drinking, gut absorption and gill net excretion rates. Table 3 presents the values found for the Na turnover rate compared with those calculated from drinking-rate studies and net excretion rates. In some fishes both these parameters have been compared in waters of various salinities and it may be seen that the turnover rate is correlated with the net pumping rate (see also Maetz & Skadhauge, 1968).

Table 3. *Comparison of the unidirectional (exchange) sodium fluxes and the net excretion flux*

Species	Medium	Na turnover rate ($\%$ h^{-1})	Exchange flux (μM h^{-1} (100 g)$^{-1}$)	Net flux (μm h^{-1} (100 g)$^{-1}$)	Reference
Anguilla anguilla	200 % SW	61·2±7·9	3550	− 670	Maetz & Skadhauge, 1968
	100 % SW	27·2±3·5	1320	− 85 to 170	Maetz *et al.* 1967
	50 % SW	6·6±0·4	260	—	Maetz, unpublished
	35 % SW	—	—	− 10	—
	FW (0·1 mM)	≃0·1	5	+ 1·5	—
Fundulus heteroclitus	100 % SW	46·2±2·4	3640	− 835	Potts & Evans, 1967
	40 % SW	13·5±2·7	820	− 340	—
	FW (1 mM)	1·7±0·3	80	—	—
Tilapia mossambica	200 % SW	116±22	7000	− 1340	Potts *et al.* 1967
	100 % SW	55·5±3·95	3330	− 470	—
	40 % SW	9·9±2·3	500	− 75	—
	FW (0·25 mM)	4·9±0·6	250	—	—
	FW (0·1 mM)	≃0·3	15	—	—
Opsanus tau	100 % SW	15·6±1·9	805	− 30	Lahlou & Sawyer,1969a, b
	50 % SW	4·9	255	—	—
	FW	<0·3	5–15	− 10	—

SW, Sea water; FW, fresh water.

Fig. 5 illustrates the model suggested recently by Maetz (unpublished) to account for this correlation. According to this model, the central mechanism permitting Na extrusion is a Na–K exchange pump. Most of the experimental evidence has been obtained on the flounder (see Maetz, 1969 b). Rapid transfer experiments were performed on the fish kept in artificial sea water alternately with and without K$^+$. A small but statistically significant reduction of the outflux is seen in K-free seawater, resulting in suppression of the net Na excretion. If the fish is kept in K-free sea water, the internal Na level increases as well as the internal Na space, suggesting a failure of the Na pump. Upon return to normal sea water, the internal Na level returns to normal as if the extrusion mechanism has been 'turned on' again as soon as K is made available. In the absence of external K, the Na turnover rate remains high but this Na–Na exchange is not the essential mechanism for the maintenance of salt balance. Additional experiments suggest that the K ions thus taken up in exchange for Na ions compensate for the K$^+$ loss through the gills as both K fluxes amount to about 120 μM h^{-1} (100 g)$^{-1}$ and the renal K loss and K uptake by the gut is negligible. As may be seen from Fig. 5, the electrical gradient across the gill epithelium (+ 18 mV inside according to Maetz & Campanini, 1966) more or less compensates for the chemical gradient for both the Na and K ions, and the flux ratio being 1. The Na and K exchanges may therefore be considered as passive in contrast to the Cl exchanges, which are certainly active. More work is necessary in order to investigate the Cl extrusion mechanism. In any case, the correlation between the Na–Na exchange and Na–K exchange is considered to result from competition between both ionic species for a common carrier.

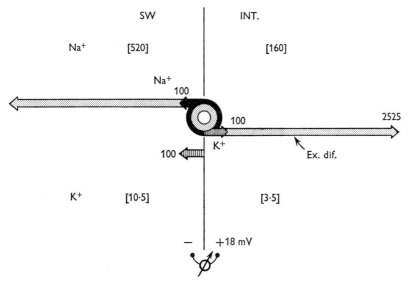

Fig. 5. A model for the ionic exchanges across the seawater teleostean gill. In the centre, the Na$^+$–K$^+$ exchange corresponding to about 100 μ-equiv fluxes (h^{-1} (100 g)$^{-1}$). The K$^+$ influx is compensated by an equivalent K$^+$ loss. The exchange diffusion component (Ex.dif.) corresponding to a one-by-one Na$^+$ exchange is also depicted, representing about a 2,500 μ-equiv flux. External and internal Na$^+$ and K$^+$ concentrations (in m-equiv l.$^{-1}$) are given in parentheses. The electrical potential difference is also represented (see Maetz & Campanini, 1966).

Independent evidence suggesting the existence of a Na carrier has been obtained by Motais, Garcia Romeu & Maetz, (1966) in the flounder. This euryhaline fish was submitted to rapid-transfer experiments from the adaptation medium to media of various salinities. Both Na influx and out-flux readjustments were followed after transfer. For all transfers a dual pattern of readjustments was observed for both the influx and the outflux: an *instantaneous* increase or decrease followed by a *delayed progressive* change. If the influxes and outfluxes resulting from the instantaneous readjustments are plotted against the external Na concentrations of the transfer media, a hyperbolic function would be obtained similar to that expressing enzyme activity as a function of substrate concentration. Fig. 6

illustrates such curves for flounders adapted to various salinities and gives
the corresponding equations of the Michaelis–Menten type. Motais and
his colleagues suggest that the Na influx and most of the Na outflux
are linked through a common carrier whose apparent affinity for Na is low
(K_m about 400 mM). Recently Maetz (1969 b) showed in the same fish, by

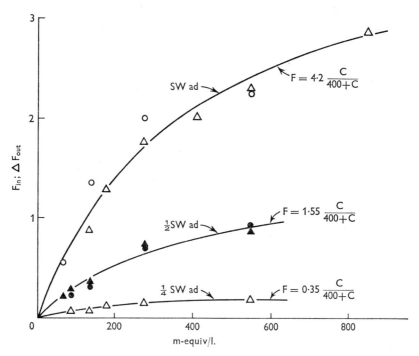

Fig. 6. Influx and exchange diffusion component of the outflux (F_{in} and $\triangle$ F_{out}) as
function of external Na concentrations (in m-equiv/l.) according to Motais et al. (1966).
$\bigcirc$, Influxes; $\triangle$, outfluxes, in m-equiv/h/100 g. The curves have been calculated from
the Michaelis–Menten equations to fit empirically the experimental data. Observe that
the equations corresponding to the different adaptation media differ in their F_{max} values
but not in the K_m values. Ad, Adapted; SW, sea water.

rapid transfer experiments from sea water to KCl solutions of various
concentrations, that at low concentrations K is much more effective than
Na to 'drive Na out' of the fish. These experiments confirm that Na and
K compete for a common carrier, but its apparent affinity for K ions is
very high (K_m less than 5 mM).

The observations of Motais et al. (1966) also concern fishes adapted to
various salinities. As illustrated in Fig. 6, the flux variations observed for
half- and quarter-seawater-adapted fishes also obey saturation kinetics but
the maximal flux observed is less when the carrier is saturated, the apparent

affinity of the carrier remaining unchanged. These curves suggest that adaptation to lower salinities is accompanied by a reduction of either the number of carrier sites or molecules, or the synthesis of a 'non-competitive' inhibitor of the transport carrier in the gill.

The hypothetical linkage between unidirectional fluxes through a common carrier was first proposed by Ussing (1947) to account for the high Na exchange observed for the isolated frog sartorius muscle and was called 'exchange-diffusion'. This phenomenon is observed very clearly in the flounder and the eel and accounts for the huge outflux reduction observed immediately upon transfer to fresh water.

Other fishes such as the stenohaline *Serranus* or the euryhaline *Fundulus*, however, do not seem to possess an exchange diffusion carrier although they exhibit very rapid Na turnovers in sea water. In these fishes the influx also varies as a function of the external Na concentration according to a Michaelis–Menten curve. But it appears to be dissociated from the outflux. These fishes, when adapted to media of lower salinity, show a secondary delayed regulation and a progressive reduction in the rate of Na exchange. When transferred to freshwater, however, this mechanism invariably fails in stenohaline fishes but still operates in the euryhaline *Fundulus heteroclitus* (Motais *et al.* 1966; Motais, 1967). The 'linkage' between the unidirectional fluxes may also be broken down in the eel, as observed recently by Rankin (1969) in isolated perfused gills and by Mayer & Nibelle (in preparation) *in vivo*. Experimental changes of the Na concentration of the perfusion fluid have no effect on the Na influx. Intravenous injections of hyper- or hyposaline solutions in the eel induce respectively an important increase or decrease of the Na outflux, while the Na influx remains unaffected. Incidently this experiment demonstrates functional variations of the branchial ionic exchange mechanism permitting a rapid restoration of the initial plasma Na level. Further investigation is necessary to get a better understanding of the ion exchange phenomena across the gill epithelium.

2. *Branchial water permeability in relation to external salinity*

In the preceding section we discussed the great flexibility of the gill in the handling of the monovalent ions as shown by modification of the passive permeability properties and also by the alternative use of widely different pumping mechanisms. Water transfer across the gills is most probably an entirely passive phenomenon. Nevertheless, we shall see that this passive process is under control in relation to the osmotic stress encountered by the fish.

Two types of water permeabilities (see Maetz, 1968) may be measured across biological membranes. The *osmotic permeability*, measured by following the net flow of water across the membrane, corresponds to a transfer by 'bulk flow', the driving force being the difference of osmotic pressure of the fluids separated by the membrane, provided the membrane is impermeable to the solutes responsible for the osmotic pressure. Bulk flow is a type of flow in which water moves in an organized fashion in clusters of water molecules. The *diffusional water permeability* corresponds to a transfer by random movement of individual water molecules. It is measured by the use of labelled water, tritiated water (HTO) being the most useful. The concentration gradient to be taken into account when diffusional unidirectional water flux is measured is the molar concentration of water (55 moles l.$^{-1}$).

Table 4. *Diffusional water flux deduced from the HTO turnover experiments*

External medium	Species	t (°C)	λ (h^{-1}%)	Unidirectional flux	
				ml h^{-1} (100 g)$^{-1}$	μl. h^{-1} cm^2
FW	*Carassius*	20	74·0±7·0	48·6±4·7	121
FW	*Anguilla*	20	42·4±4·7	27·3±3·0	91
FW	*Platichthys*	16	31·1±2·0	19·5±1·3	97
SW	*Platichthys*	16	19·8±1·6**	12·3±1·0**	61
SW	*Anguilla*	20	29·3±2·4*	19·1±1·5*	64
SW	*Serranus*	20	20·7±1·6	13·3±1·0	33

The means±s.e. are given. λ is the sum of the external and internal fraction renewed per hour. The outflux is given relative to the body weight or the gill surface area.
Statistical analysis of the comparison FW and SW for the euryhaline fishes: ** $P <$ 0·01, * $P <$ 0·05.
External medium: FW, fresh water; SW, sea water.
See Motais *et al.* (1969), for details.

The diffusional water permeability of the boundaries of the fish has been measured *in vivo* after injection of the tracer into the intraperitoneal cavity and the analysis of the cumulative appearance curve of the isotope in the outside medium. The turnover rate corresponds to the fraction of the internal water compartment (about 70% of the body weight) renewed per hour (see Evans, 1969a, b; Motais, Isaia, Rankin & Maetz, 1969). Table 4 presents the values obtained by Motais *et al.* (1969). It may be seen that the freshwater species show a higher diffusional permeability than the seawater species, especially when the euryhaline fishes are compared in both media. Evans (1969a, b) arrives at a similar conclusion, although in some cases identical permeabilities are found. Additional experiments on the eel, by Motais *et al.* (1969), show definitely that the gill is the major route of water exchange, the HTO clearance by the gill accounting for all the water turnover observed. Incidently these clearance experiments involve the determination of the cardiac output and demonstrate that the blood flow across the gill is not a limiting factor in diffusional water exchange.

The measurement of the osmotic water flow across the gill *in vivo* can only be made indirectly. On the one hand, the drinking rate minus the urine flow is taken as the rate of water flow across the gill of the seawater fish, the assumption being made that all the water swallowed is absorbed. Assessment of rectal water loss is difficult as anal catheterization (see Hickman, 1968) prevents final reabsorption. On the other hand, the urine flow minus the drinking rate corresponds to the osmotic inflow of water in the freshwater fish. It may be noted that urinary catheterization also introduces systematic errors as the bladder has been shown to play a role in water reabsorption (Lahlou, 1967). Table 5 indicates the values of the net water fluxes thus measured in the species used for the HTO turnover experiments. It may be seen that despite a greater osmotic gradient across the gill, the net water loss is less than the net water gain when seawater fishes are compared with freshwater fishes.

Table 5. *Osmotic water flux as deduced from*
drinking rate and urine flow

External medium	Species	t (°C)	Drinking rate	Urine flow	Net flux μl. h^{-1} (100 g)$^{-1}$	μl. h^{-1} cm^{-2}	Osmotic gradient
FW	*Carassius*	20	51 ± 28	1445 ± 132	$+1394$	$+3\cdot5$	$+250$
FW	*Anguilla*	20	135 ± 28	538 ± 32	$+403$	$+1\cdot35$	$+255$
FW	*Platichthys*	16	37 ± 11	287 ± 27	$+250$	$+1\cdot25$	$+265$
SW	*Platichthys*	16	192 ± 35	47 ± 9	-145	$-0\cdot72$	-825
SW	*Anguilla*	20	325 ± 34	31 ± 9	-294	$-0\cdot98$	-815
SW	*Serranus*	20	277 ± 35	70 (2)	-207	$-0\cdot52$	-820

The means ± S.E. are given.
Drinking rate and urine flow in μl. h^{-1} (100 g)$^{-1}$. Osmotic gradient in m-osmoles. kg^{-1}
External medium: 10 and 1150 m-osmoles.
See Motais *et al.* (1969) for technical details.

Moreover, for any biological membranes, comparison of both permeability coefficients expressed in the same unit (in cm s^{-1} usually) is of interest because it throws some light on the mechanism of water transfer. Unfortunately, for the calculation of the osmotic permeability coefficient, P_{os}, it is necessary to know the degree of semi-permeability of the membrane. If the gill is assumed to be semi-permeable in both freshwater and seawater fishes, it turns out that in freshwater fishes P_{os} is higher than P_d, the diffusional permeability coefficient, the ratio being 2·5–6. For the gills of seawater fishes, however, this ratio is 1. This result interpreted in terms of thermodynamic considerations means that water crosses the gill of these fishes only by diffusion and never by bulk flow. More work is necessary to ascertain if there is no interaction of water flow and ion flow in the gill of seawater fish.

In conclusion, our observations on the water permeabilities of the gill suggest that these parameters are modified in relation to the osmotic gradient encountered by the fish. A decreased permeability is observed in response to adaptation to increased osmotic gradient. The physiological significance of this defence mechanism is evident. The high diffusional permeabilities observed in marine crustacea (Rudy, 1967) or elasmobranchs (Payan & Maetz, unpublished), animals whose internal medium is more or less isotonic to sea water, confirm our view.

3. Cellular and molecular processes underlying the adaptive changes of the gill

For the gills as for the gut the functional modifications observed in relation to the osmoregulatory needs of the fish may be the result of either cellular renewal or molecular renewal of the existing active cells. It is probable that in the gill both processes coexist. Fig. 7 illustrates the possible events which may occur during saltwater adaptation.

According to Conte & Lin (1967) cell division is required for the initiation of the new specialized function of gill epithelium in salt water. Autoradiographic observations with the help of [³H]thymidine serving as a marker of cell division show in the gill of *Oncorhynchus* an increased cellular activity induced by salt water. Conte and Lin suggest that the newly formed and differentiated cells are the 'chloride-cells' or mitochondriarich cells whose role in osmoregulation has been the object of controversy (see Parry, 1966). Recently Motais & Masoni (in preparation) extended the observations of Conte and Lin to the European eel. Furthermore, an elevated DNA turnover in the gill of seawater-adapted *Oncorhynchus* is observed in comparison with freshwater-adapted fishes. Conte & Lin (1967) suggest that epithelial cells 'wear out' and are replaced faster in sea water. Earlier observations of Conte (1965) on the same fish also suggest a more rapid cellular turnover in sea water as shown by a much higher radiation sensitivity in this medium. Histopathological changes of the gill epithelium and osmotic imbalance both strongly suggest failure of the gill salt-excreting cells, while in fresh water the osmoregulatory unbalance produced by X-rays is very slight.

Parallel experiments have been recently made on the eel in our laboratory with the use of several antibiotic inhibitors of cell division and of protein synthesis (Maetz, Nibelle, Bornancin & Motais, 1969). In sea water, actinomycin D produces a complete failure of the salt excretion mechanism associated with a progressive reduction of the Na turnover rate. Death follows within a week. In fresh water, on the contrary, the active component of the Na influx remains unchanged while the outflux is increased. The osmoregulatory unbalance is, however, slight and the fish survives much longer. Furthermore, actinomycin D retards or blocks the slow outflux readjustment which intervenes when the fish is transferred from fresh water to sea water or vice versa. Actinomycin D is known to inhibit mRNA polymerase, thus blocking the transcription mechanism in the

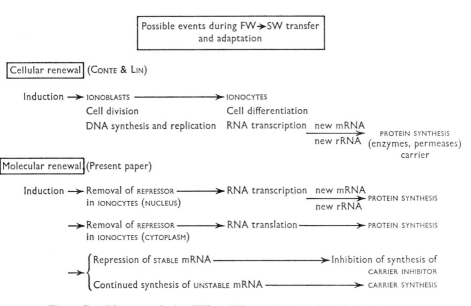

Fig. 7. Possible events during FW to SW transfer and adaptation. Schematic representation of the two major hypotheses. See Maetz *et al.* (1969).

nucleus, which intervenes only indirectly in the ribosomal translation process necessary for protein synthesis. Maetz and his colleagues suggested that in fresh water the active cells are characterized by a slow molecular renewal rate into the seawater-adapted cells. However, actinomycin is also known to block cellular division indirectly and the observed effects on osmoregulation may still reflect differences in cellular renewal rates in accordance with the suggestion of Conte and Lin. Injections of mitomycin in the seawater eel remain, however, without effect on the Na turnover rate of fish in contrast to the powerful action of actinomycin. However, it remains to be seen if during salt-water adaptation when a 'wave' of cellular division is observed in the gill, mitomycin may not retard the outflux readjustment pattern.

Complementary observations by Maetz *et al.* (1969) concerning the effects of two inhibitors of protein synthesis—puromycin and cyclo-heximide—both of which affect *directly* the ribosomal translation mechanism suggest moreover that the molecular processes are even more complicated than suspected. These chemicals produce an *acceleration* of the Na transport mechanism in sea water or the outflux readjustment pattern during saltwater adaptation. A model is proposed suggesting two mRNAs with different half-lives acting as templates for the synthesis of two proteins associated with salt transport: the carrier of exchange diffusion and Na transport and an inhibitor of this carrier. The half-life of the inhibitor would be much shorter than that of the carrier. Thus the flexibility of the gill transport mechanisms in relation to external salinity changes would be realized by two processes acting in opposite directions.

In any case, new molecular complexes, a transport carrier or enzyme directly or indirectly linked with salt transport, appear to be produced as a result of external salinity changes whether by the means of cell division and differentiation or by molecular renewal in pre-existing cells. Recent observations by Conte & Morita (1968) reveal the existence of such 'salt-induced' proteins as shown by the antigenic reactions of gill filament extracts. Among the possible candidates for such 'salt-induced' proteins, rate-limiting factors for salt transport such as enzymes have been the subject of intensive investigations. The most interesting discovery in this field concerns the level of the Na–K activated ATPase, which appears to be much higher in the gill extracts of seawater-adapted teleosts (Kamiya & Utida, 1968; Epstein, Katz & Pickford, 1967; Jampol & Epstein, 1968; Motais, 1969). Other enzymes such as carbonic anhydrases which are also involved in ion exchanges have also been found in higher activity levels in seawater fishes (Maetz, 1956).

Recent results obtained by Motais (1969) are particularly interesting in relation to the effects of actinomycin D produced in freshwater and seawater eels. Injection of actinomycin remains without effect on the gill Na–K ATPase activity level in the freshwater eel, while it produces a significant reduction of this level in sea water. A comparison with the earlier reported absence of effect on the Na pump in fresh water and huge impairment of the salt pump in sea water indicates that the enzyme plays a major role in the branchial ion excretion mechanism in sea water and that the half-life of this enzyme molecule or of the corresponding mRNA necessary for its synthesis is relatively short in sea water. The implication of Na–K-dependent ATPase fits particularly well with the model we proposed for the Na pump in sea water, which turns out to be probably a Na–K exchange pump associated with a Na–Na exchange pump comparable to the one observed for the red cells.

Several points, however, remain obscure concerning the role of this enzyme. The time-scale of the S-shaped curve describing the progressive increase of the enzyme level in the Japanese eel when transferred from fresh water to sea water (Kamiya & Utida, 1968) indicates a very slow process, while Maetz et al. (1969) observe that the ouflux readjustment is completed in 48 h in the European eel, although these latter experiments were completed at a lower temperature.

4. Endocrine control of the branchial mechanism of osmoregulation

Our purpose is to limit ourselves to a few points as this subject has been covered by us before (Maetz, 1968, 1969a; Maetz, Motais & Mayer, 1968; Maetz & Rankin, 1969).

Neurohypophysial and urophysial hormones have been implicated in the control of branchial ion uptake (see review by Fridberg & Bern, 1968; Maetz, 1968). These endocrine principles, which have most probably a short half-life, might also produce effects of short duration. The long-protracted effects observed earlier with neurohypophysial hormones, notably by Maetz (1963), were possibly the result of adenohypophysial stimulation by the large doses injected. Recent observations by Chester Jones, Chan & Rankin (1969) and by Maetz & Rankin (1969) show that these hormones also produce haemodynamic effects affecting the blood flow inside the gills. It may well be that the alterations in the ion uptake mechanisms observed in freshwater and seawater fishes (Maetz, Bourguet & Lahlou, 1964; Maetz, Bourguet, Lahlou & Hourdry, 1964; Motais & Maetz, 1964) may reflect these haemodynamic changes since the presumably active cells involved in ion transport are located at the base of the lamellae on the sides of the by-pass channels discovered by Steen & Kruysse (1964). These channels are preferentially irrigated when neuro-hypophysial hormones are added to the perfusion medium (Maetz & Rankin, 1969).

Other endocrine principles have been implicated in long-term actions on the gill ionic pumps or water permeability. Of particular interest are the corticosteroids secreted by the interrenals under ACTH stimulation and the prolactin-like factor secreted by the adenohypophysis. Both hormones have been implicated in mammalian target-organs such as the liver (synthesis of gluconeogenetic enzymes by cortisol) or the mammary gland (synthesis of milk proteins). They probably act on the nuclear level by inducing new mRNA transcription followed by new ribosomal synthesis of proteins (Tomkins, Thompson, Hayashi, Gelehrter, Granner & Peterkofsky, 1966; Turkington, 1968a, b; Bern & Nicoll, 1968). Both hormones have been shown to have long-lasting effects on the gill ionic turnover in seawater fishes, either by increasing for several days the Na pump activity, an effect produced typically by cortisol or ACTH (Maetz, 1969 a) or decreasing the Na pump and Na exchange, an effect produced by mammalian prolactin (Maetz et al. 1968). We suggest as a working hypothesis that cortisol controls the synthesis of the transport carrier or enzymes associated with Na transport such as Na–K ATPase, while prolactin controls the synthesis of the inhibitor of the transport carrier. Concordant with this hypothesis are the results of Epstein et al. (1967) showing that hypophysectomy produces in the seawater killifish a significant decrease in the Na–K ATPase activity. This parallels our observation that hypophysectomy results in a partial inactivation of the branchial salt pumping activity, a defect which is repaired by ACTH or cortisol injection and not by prolactin treatment (Maetz et al. 1968).

Both these hormones have also been recently implicated in the control of the branchial water permeability. Lahlou & Giordan (in preparation) have recently observed in my laboratory that hypophysectomy of the goldfish produces an important reduction of the water permeability, as seen by a diminution of the HTO turnover. Both prolactin and cortisol are able to repair this defect and to restore normal HTO turnover. Long-term action has also been observed in this case.

In conclusion, we have indulged in a great deal of speculation concerning the mechanisms of osmoregulation as well as the cellular or molecular processes underlying the adaptive flexibility displayed by the gut and the gill in response to the osmoregulatory needs of the fish. The object of this review has been precisely to propose new models in order to induce more investigators to become interested in fish physiology and endocrinology. More work with the help of new tools will no doubt result in better models which will replace those presented herein.

SUMMARY

In relation to the salinity of the aquatic environment, several 'effector-organs' of osmoregulation exhibit widely different functional features. This observation stems from comparison of stenohaline freshwater and seawater species and from the adaptive alterations observed when euryhaline teleosts are kept in waters of various salinities.

1. Differences in the drinking rate are observed which are accompanied by modifications of the intestinal water-pumping efficiency. Parallel changes in the ion-absorption capacity and in the passive osmotic water permeability are responsible for the increased solute-linked water flow, increased gradient against which absorption is accomplished and the augmented molar ratio of water v. salt absorbed, all these changes characterizing adaptation to higher salinities.

2. In the gills, transfer from hypo- to hypertonic media is accompanied by an increase of the Na and Cl exchange rates. The salt absorption pump is replaced by a salt excretion pump, the two ionic pumps having completely different mechanisms. Decreased passive water permeability is observed during salt-water adaptation.

All the alterations induced by external salinity changes involve cellular processes such as cell division and differentiation, and molecular processes such as protein synthesis, particularly of enzyme systems associated with ion pumps and carriers. It is evident that a better understanding of the mechanisms thus revealed will further our insight into the mode of action of hormones in so far as endocrine control is important in osmoregulation.

REFERENCES

ALLEE, W. C. & FRANK, P. (1948). Ingestion of colloidal material and water by goldfish. *Physiol. Zool.* **21**, 381–390.

BERN, H. A. & NICOLL, C. S. (1968). The comparative endocrinology of prolactin. *Recent Prog. Horm. Res.* **24**, 681–720.

BOURGUET, J., LAHLOUH, B. & MAETZ, J. (1964). Modifications expérimentales de l'équilibre hydrominéral et osmorégulation chez *Carassius auratus*. *Gen. comp. Endocrin.* **4**, 563–76.

CHESTER JONES, I., CHAN, O. K. O. & RANKIN, J. C. (1969). Renal function in the European eel (*Anguilla anguilla* L.): effects of the caudal neurosecretory system, corpuscles of Stannius, neurohypophysial peptides and vasoactive substances. *J. Endocr.* **43**, 21–31.

CONTE, F. P. (1965). Effects of ionizing radiation on osmoregulation in fish *Oncorhynchus kisutch*. *Comp. Biochem. Physiol.* **15**, 293–302.

CONTE, F. P. & LIN, D. H. Y. (1967). Kinetics of cellular morphogenesis in gill epithelium during sea water adaptation of *Oncorhynchus* (Walbaum). *Comp. Biochem. Physiol.* **23**, 945–57.

CONTE, F. P. & MORITA, T. N. (1968). Immunological study of cell differentiation in gill epithelium of euryhaline *Oncorhynchus* (Walbaum). *Comp. Biochem. Physiol.* **24**, 445–54.

CURRAN, P. F. (1960). Na, Cl and water transport by rat ileum *in vitro*. *J. Gen. Physiol.* **43**, 1137–48.

DEJOURS, P. (1969). Variations of CO_2 output of a fresh-water teleost upon change of the ionic composition of water. *J. Physiol., Lond.,* **202**, 113–14P (abstr.).

DIAMOND, J. M. (1964). The mechanism of isotonic water transport. *J. gen. Physiol.* **48**, 15–42.

EPSTEIN, F. H., KATZ, A. I. & PICKFORD, G. E. (1967). Sodium- and potassium-activated adenosine triphosphatase of gills: Role in adaptation of teleosts to salt water. *Science, N.Y.* **156**, 1245–7.

EVANS, D. H. (1967). Sodium, chloride and water balance of the intertidal teleost, *Xiphister atropurpureus*. II. The role of the kidney and the gut. *J. exp. Biol.* **47**, 519–23.

EVANS, D. H. (1968). Measurement of drinking rates in fish. *Comp. Biochem. Physiol.* **25**, 751–3.

EVANS, D. H. (1969a). Sodium, chloride and water balance of the intertidal teleost, *Pholis gunnellus*. *J. exp. Biol.* **50**, 179–90.

EVANS, D. H. (1969b). Studies on the permeability to water of selected marine, freshwater and euryhaline teleosts. *J. exp. Biol.* **50**, 689–703.

FRIDBERG, G. & BERN, H. A. (1968). The urophysis and the caudal neurosecretory system of fishes. *Biol. Rev.* **43**, 175–99.

GARCIA ROMEU, F. & MAETZ, J. (1964). The mechanism of sodium and chloride uptake by the gills of a freshwater fish, *Carassius auratus*. I. Evidence for an independent uptake of sodium and chloride ions. *J. gen. Physiol.* **47**, 1195–207.

HICKMAN, C. P. (1968). Ingestion, intestinal absorption, and elimination of sea-water and salts in the southern flounder, *Paralichthys lethostigma. Can. J. Zool.* **46**, 457–67.

HIRANO, T. (1967). Effect of hypophysectomy on water transport in isolated intestine of the eel, *Anguilla japonica. Proc. Japan Acad.* **43**, 793–6.

HIRANO, T., KAMIYA, M., SAISHU, S. & UTIDA, S. (1967). Effects of hypophysectomy and urophysectomy on water and sodium transport in isolated intestine and gills of japanese eels (*Anguilla japonica*). *Endocr. jap.* **14**, 182–6.

HIRANO, T. & UTIDA, S. (1968). Effects of ACTH and cortisol on water movement in isolated intestine of the eel, *Anguilla japonica. Gen. comp. Endocrin.* **11**, 373–80.

HOUSE, C. R. (1963). Osmotic regulation in the brackish water teleost, *Blennius pholis. J. exp. Biol.*, **40**, 87–104.

HOUSE, C. R. & GREEN, K. (1965). Ion and water transport in isolated intestine of the marine teleost, *Cottus scorpius. J. exp. Biol.* **42**, 177–89.

JAMPOL, L. M. & EPSTEIN, F. H. (1968). Role of Na–K-ATPase in osmotic regulation by marine vertebrates. *Bull. Mt Desert Isl. biol. Lab.* **8**, 32–4.

KAMIYA, M. & UTIDA, S. (1968). Changes in activity of sodium–potassium-activated adenosinetriphosphatase in gills during adaptation of the Japanese eel to sea water. *Comp. Biochem. Physiol.* **26**, 675–85.

KROGH, A. (1939). *Osmotic Regulation in Aquatic Animals.* Cambridge University Press.

LAHLOU, B. (1967). Excrétion rénale chez un poisson euryhalin le flet (*Platichthys flesus*). Caractéristiques de l'urine normale en eau douce et en eau de mer et effet des changements de milieu. *Comp. Biochem. Physiol.* **20**, 925–38.

LAHLOU, B., HENDERSON, I. W. & SAWYER, W. H. (1969). Sodium exchanges in goldfish (*Carassius auratus* L.) adapted to a hypertonic saline solution. *Comp. Biochem. Physiol.* **28**, 1427–33.

LAHLOU, B. & SAWYER, W. H. (1969a). Adaptation des échanges d'eau en fonction de la salinité externe chez *Opsanus tau*, téléostéen algomérulaire euryhalin. *J. Physiol.*, *Paris.* **61**, 143 (abstr.).

LAHLOU, B. & SAWYER, W. H. (1969b). Sodium exchanges in the toadfish, *Opsanus tau*, a euryhaline aglomerular teleost. *Am. J. Physiol.* **216**, 1273–78.

MAETZ, J. (1956). Le rôle biologique de l'anhydrase carbonique chez quelques téléostéens. *Bull. biol. Fr. Belg.* (*Suppl.*) **40**, 1–129.

MAETZ, J. (1963). Physiological aspects of neurohypophysial function in fishes with some reference to the Amphibia. *Symp. Zool. Soc. Lond.* **9**, 107–40.

MAETZ, J. (1964). Recherches sur la perméabilité aux électrolytes de la branchie des poissons et sa régulation endocrinienne. *Bull. Inf. Scient. Techn. Saclay* **86**, 11–70.

MAETZ, J. (1968). Salt and water metabolism. In *Perspectives in Endocrinology: Hormones in the Lives of Lower Vertebrates*, pp. 47–162 (eds. E. J. W. Barrington and C. B. Jørgensen). London and New York: Academic Press.

MAETZ, J. (1969a). Observations on the role of the pituitary–interrenal axis in the ion regulation of the eel and other teleosts. (Symposium of Comparative Endocrinology, New Delhi.) *Gen. Comp. Endocr. Suppl.* **2**, 299–316.

MAETZ, J. (1969b). Sea water teleosts: evidence for a Na–K exchange in the branchial sodium excreting pump. *Science, N.Y.* **166**, 613–615.

MAETZ, J., BOURGUET, J. & LAHLOU, B. (1964). Urophyse et osmorégulation chez *Carassius auratus*. *Gen. comp. Endocr.* **4**, 401–14.

MAETZ, J., BOURGUET, J., LAHLOU, B. & HOURDRY, J. (1964). Peptides neurohypophysaires et osmorégulation chez *Carassius auratus*. *Gen. comp. Endocr.* **4**, 508–522.

MAETZ, J. & CAMPANINI, G. (1966). Potentiels transépithéliaux de la branchie d'Anguille *in vivo* en eau douce et en eau de mer. *J. Physiol., Paris.* **58**, 248 (abstract).

MAETZ, J. & GARCIA ROMEU, F. (1964). The mechanism of sodium and chloride uptake by the gills of a fresh-water fish, *Carassius auratus*. II Evidence for NH_4^+/Na^+ and HCO_3^-/Cl^- exchanges. *J. gen. Physiol.* **47**, 1209–27.

MAETZ, J., MAYER, N. & CHARTIER-BARADUC, M. M. (1967). La balance minérale du sodium chez *Anguilla anguilla* en eau de mer, en eau douce et au cours de transfert d'un milieu à l'autre: Effets de l'hypophysectomie et de la prolactine. *Gen. comp. Endocr.* **8**, 177–88.

MAETZ, J., MOTAIS, R. & MAYER, N. (1968). Isotopic kinetic studies on the endocrine control of teleostean ionoregulation. (International Congress of Endocrinology, Mexico.) *Excerpta medica* **184**, 225–232.

MAETZ, J., NIBELLE, J. M., BORNANCIN, M. & MOTAIS, H. (1969). Action sur l'osmorégulation de l'anguille de divers antibiotiques inhibiteurs de la synthèse des protéines ou du renouvellement cellulaire. *Comp. Biochem. Physiol.* **30**, 1125–51.

MAETZ, J. & RANKIN, J. C. (1969). Quelques aspects du rôle biologique des hormones neurohypophysaires chez les poissons. *Colloques du C.N.R.S.* **177**, 45–55.

MAETZ, J., SAWYER, W. H., PICKFORD, G. E. & MAYER, N. (1967). Evolution de la balance minérale du sodium chez *Fundulus heteroclitus* au cours du transfert d'eau de mer en eau douce: Effets de l'hypophysectomie et de la prolactine. *Gen. comp. Endocr.* **8**, 163–76.

MAETZ, J. & SKADHAUGE, E. (1968). Drinking rates and gill ionic turnover in relation to external salinities in the eel. *Nature, Lond.* **217**, 371–3.

MOTAIS, R. (1967). Les mécanismes d'échanges ioniques branchiaux chez les Téléostéens. *Annls. Inst Océanog. Monaco* **45**, 1–83.

MOTAIS, R. (1969). Na–K activated adenosine triphosphatase of gills: evidence for two forms of this enzyme in sea water adapted teleost. *Comp. Biochem. Physiol.*, in press.

MOTAIS, R., GARCIA ROMEU, F. & MAETZ, J. (1966). Exchange diffusion effect and euryhalinity in teleosts. *J. gen. Physiol.* **50**, 391–422.

MOTAIS, R., ISAIA, J., RANKIN, J. C. & MAETZ, J. (1969). Adaptive changes of the water permeability of the teleostean gill epithelium in relation to external salinity. *J. exp. Biol.* **51**, 529–546.

MOTAIS, R. & MAETZ, J. (1964). Action des hormones neurohypophysaires sur les échanges de sodium (mesurés à l'aide du radiosodium ^{24}Na) chez un téléostéen euryhalin, *Platichthys flesus* L. *Gen. comp. Endocr.* **4**, 210–24.

MULLINS, L. J. (1950). Osmotic regulation in fish as studied with radioisotopes. *Acta physiol. scand.*, **21**, 303–14.

OIDE, M. (1967). Effects of inhibitors on transport of water and ion in isolated intestine and $Na^+–K^+$ ATPase in intestinal mucosa of the eel. *Ann. Zool. Japon.* **40**, 130–5.

PARRY, G. H. (1966). Osmotic adaptation in fishes. *Biol. Rev.* **41**, 392–444.

POTTS, W. T. W. (1968). Osmotic and ionic regulation. *Ann. Rev. Physiol.* **30**, 73–104.

POTTS, W. T. W. & EVANS, D. H. (1967). Sodium and chloride balance in the killifish *Fundulus heteroclitus*. *Biol. Bull mar. biol. Lab.*, *Woods Hole* 133, 411–25.

POTTS, W. T. W., FOSTER, M. A., RUDY, P. P. & PARRY, G. H. (1967). Sodium and water balance in the cichlid teleost, *Tilapia mossambica*. *J. exp. Biol.* 47, 461–70.

RANKIN, J. C. (1969). Sodium exchange across a perfused teleostean gill preparation. Effect of ouabain and ethacrynic acid. In preparation.

RUDY, P. P. (1967). Water permeability in selected decapod crustacea. *Comp. Biochem. Physiol.* 22, 581–9.

SHARRATT, B. M., BELLAMY, D. & CHESTER JONES, I. (1964). Adaptation of the silver eel (*Anguilla anguilla* L.) to sea water and to artificial media together with observations on the role of the gut. *Comp. Biochem. Physiol.* 11, 19–30.

SHAW, J. (1964). The control of salt balance in the crustacea. *Symp. Soc. exp. Biol.* 18, 237–56.

SKADHAUGE, E. (1969). The mechanism of salt and water absorption in the intestine of the eel (*Anguilla anguilla*) adapted to waters of various salinities. *J. Physiol.*, *Lond.* 204, 135–158.

SKADHAUGE, E. & MAETZ, J. (1967). Etude *in vivo* de l'absorption intestinale d'eau et d'électrolytes chez *Anguilla anguilla* adapté à des milieux de salinités diverses. *C. r. hebd. Séanc. Acad. Sci.*, *Paris* 265, 347–50, 923.

SMITH, H. W. (1932). Water regulation and its evolution in fishes. *Q. Rev. Biol.* 7, 1–26.

SMITH, M. W. (1964). The *in vitro* absorption of water and solutes from the intestine of goldfish, *Carassius auratus*. *J. Physiol.*, *Lond.* 175, 38–49.

STEEN, J. B. & KRUYSSE, E. D. (1964). The respiratory function of the teleostean gills. *Comp. Biochem. Physiol.* 12, 127–42.

TOMKINS, G. M., THOMPSON, S., HAYASHI, S., GELEHRTER, T., GRANNER, D. & PETERKOFSKY, B. (1966). Tyrosine transaminase induction in mammalian cells in tissue culture. In *The Genetic Code. Symp. Quantitative Biol.* 31, 349–60.

TURKINGTON, R. W. (1968a). Hormone-induced synthesis of DNA by mammary gland *in vitro*. *Endocrinology*, 82, 540–6.

TURKINGTON, R. W. (1968b). Induction of milk protein synthesis by placental lactogen and prolactin *in vitro*. *Endocrinology*, 82, 575–83.

USSING, H. H. (1947). Interpretation of the exchange of radiosodium in isolated muscle. *Nature*, *Lond.* 160, 262–3.

UTIDA, S., ISONO, N. & HIRANO, T. (1967). Water movement in isolated intestine of the eel adapted to fresh water or sea water. *Zool. Mag.* 76, 203–4.

UTIDA, S., OIDE, M. & OIDE, H. (1968). Ionic effects on alkaline phosphatase activity in intestinal mucosa with special reference to sea-water adaptation of the japanese eel, *Anguilla japonica*. *Comp. Biochem. Physiol.* 27, 239–49.

WHITLOCK, R. T. & WHEELER, H. O. (1964). Coupled transport of solute and water across rabbit gallbladder epithelium. *J. clin. Invest.* 43, 2249–65.

Groupe de Biologie Marine du Département de Biologie du Commissariat à l'Energie Atomique Station Zoologique,
06-Villefranche-sur-Mer,
France.

DISCUSSION

Quay: I would appreciate your opinion, Dr Maetz, on the possible localization of the electrolyte carrier protein that you discussed. To what extent might cellular membrane systems reveal some deficiencies after your treatment with actinomycin?

Maetz: The question is rather complicated because we still don't know if we have to deal with molecular renewal in pre-existing cells or cellular renewal, new cells being formed for seawater-life adaptation. This latter hypothesis has been forwarded by Conte and Lin. Now we don't even know which of the cells do the transfer. I am personally very much in favour of the chloride cells, which are mitochondria-rich in both freshwater and seawater gills, but there is still a lot to be done to confirm this.

With Motais, we have only just started studies by tagging the cells with tritium-labelled thymidine, but we have not gone very far in that direction. If the carrier is Na–K-dependent ATPase and if it is ouabain-sensitive then we could perhaps use radioactive ouabain and pin it down by autoradiography. However, sodium exchange is not blocked by ouabain, even at very high doses. It will kill the fish whether we inject the inhibitor or add it outside.

THE ADRENAL CORTEX AND
OSMOREGULATION IN TELEOSTS

By I. W. HENDERSON, D. K. O. CHAN,
T. SANDOR and I. CHESTER JONES

INTRODUCTION

Teleostei occupy a wide variety of environmental niches. Whilst some genera remain in constant environments with little attendant stress on these grounds, others migrate short or long distances with consequent demands on their osmoregulatory capacities. The broad sweep of the general physiological and morphological bases for the selective movement of water and ions has been established. Maetz, in this Symposium, has reviewed some of the mechanisms involved—cellular, enzymic and physiological—particularly as regards the function of extra-renal tissues in both fresh water and marine teleosts. Maetz has clearly shown both the extent of and the lacunae in our knowledge. Amongst the wide gamut of factors which necessarily must impinge on the osmoregulatory processes, hormones play a significant part. In this account only one set of hormones will be primarily considered, namely those arising from adrenocortical tissue. Here a comprehensive literature survey is not intended, this will be found in Chester Jones, Chan, Henderson & Ball (1969); rather a drawing together of some of the relevant data will be attempted.

The general pattern to be considered comprises, in freshwater fish, the osmotic influx of water across permeable surfaces bathed in environmental waters hypotonic to the body fluids. Associated with this is a relatively high output of urine related to the glomerular filtration rate. Although there is intense renal tubular reabsorption of electrolytes, some must be obligatorily lost, offset by the active uptake of ions by the gill epithelium from the environmental water. In marine teleosts the situation is, in many ways, reversed. The tendency of the seawater fish to lose water by osmotic outflux across the permeable surfaces, due to the environmental water being hypertonic to the body fluids is lessened by reduced membrane permeability. Accompanying this is a much reduced glomerular filtration rate and urine flow. The fishes drink the sea water and obligatorily gain electrolytes in the process of water absorption from the gastrointestinal tract; these electrolytes are actively extruded by the gills, thereby donating osmotically free water to the animal. Whilst the kidney shows reabsorption of the monovalent ions, one major function would appear

[31]

to be the excretion of divalent cations, absorbed by the gut, and which are not extruded by the branchial epithelium (Maetz, 1970; this symposium, page 3).

Basic physiological information was first given in the early studies of Smith (1930, 1932), Keys (1933) and Krogh (1939). Recent literature is replete with data describing these processes in many different species with the employment of more sophisticated methods of analysis. Experiments indicating an endocrine control of these processes are, however, not numerous. Indeed there are but few studies which can really be judged to demonstrate a physiological action of hormones in this regard. By analogy with mammals, the adrenal steroid hormones would seem to be among those involved in a major way with the regulation of water and electrolyte homeostasis. Direct extrapolation of activities in mammals to those in fish is often dangerous as the target organs may differ, the steroid spectrum be dissimilar, and also fishes possess additional refinements, relevant to their specialized niches, of their endocrine systems. Bern (1967) emphasizes this when he says: 'In their endocrine system as in all other aspects of their anatomy and physiology, the fishes reveal a broader range of variation and longer history of adaptation than do tetrapod vertebrates.'

In this paper we shall first assess the present knowledge of the adrenocortical secretory products and their possible biosynthetic pathways in teleost fish and then go on to consider some aspects of the involvement of these steroids in osmoregulatory processes.

ADRENOCORTICOSTEROIDS IN TELEOSTEI

There is little doubt at present that quantitatively cortisol is the major corticosteroid secreted by the teleost adrenal cortex. This steroid has been demonstrated in the peripheral plasma of a variety of bony fish. (Bondy, Upton & Pickford, 1957; Phillips & Chester Jones, 1957; Phillips, 1959; Hane & Robertson, 1959; Chester Jones, Phillips & Holmes, 1959; Idler, Ronald & Schmidt, 1959; Schmidt & Idler, 1962; Idler, Freeman & Truscott, 1964; Leloup-Hatey, 1964a; Donaldson & McBride, 1967; Bradshaw & Fontaine-Bertrand, 1968). In the rainbow trout (*Salmo gairdneri*) the peripheral plasma cortisol concentration is about 8 μg/100 ml, whilst in the European eel (*Anguilla anguilla*) in both sea water and fresh water the concentration is about 2–3 μg/100 ml (M. E. Forster, unpublished data). It should be pointed out that apart from preliminary data, secretion rates of this hormone have yet to be firmly established. The amount of cortisol in the peripheral circulation represents the product of secretion and metabolism and does not quantitatively reflect biosynthesis.

Another approach to the study of corticosteroidogenesis in fish has been the use of *in vitro* incubation techniques. Adrenocortical tissue has been incubated with or without added precursors, and these studies have largely confirmed the *in vivo* observations. Extraction of adrenocortical tissue of *Salmo salar* revealed the presence of cortisol (Fontaine & Leloup-Hatey, 1959). Incubation of adrenal tissue without exogenous precursors again resulted in cortisol formation in *Salmo gairdneri*, *Tilapia mossambica* and *Bodianus bilunulatus* (Nandi & Bern, 1960, 1965). Incubation of European eel (*Anguilla anguilla*) adrenal cortical tissue with a variety of steroidal substrates yielded cortisol as major metabolite (Butler, 1965; Leloup-Hatey, 1966, 1968; Sandor *et al.* 1966; Sandor, Henderson, Chester Jones & Lanthier, 1967*a*; Sandor, Lanthier, Henderson & Chester Jones, 1967*b*). Similarly, cortisol was found to be the major metabolite of exogenous steroid precursors with adrenal tissue of the marine eel (*Conger conger*, Butler, 1965), the rainbow trout (*Salmo gairdnerii*, Arai, Tajima & Tamaoki, 1969), a killifish (*Fundulus kansae*) and the bowfin (*Amia calva*; Holostei) (Henderson & Sandor, unpublished data). The presence of cortisone was demonstrated in almost all instances, whenever cortisol was isolated. In regard to the presence of 17-deoxycorticosteroids and 18-oxygenated corticosteroids, this is a debated question and we will return to it later.

The biosynthesis of cortisol in the interrenals of the European eel has been investigated in some detail. Cholesterol, pregnenolone, progesterone, 17α-hydroxyprogesterone all gave rise to cortisol under *in vitro* conditions. In addition, when working with adrenal tissue homogenates in the presence of NADPH, 11-deoxy-17-hydroxycorticosterone was found to be a quantitatively important transformation product from all the above-mentioned substrates, in addition to cortisol and cortisone. It would seem that in the eel, as in other vertebrates, including man, cortisol synthesis is accomplished by at least two biosynthetic routes; the first one consists of a pregnenolone – progesterone – 17α-hydroxyprogesterone – 11-deoxy-17-hydroxycorticosterone–cortisol pathway, while the second, and probably more important, one follows the pregnenolone–17α-hydroxypregnenolone–17α-hydroxyprogesterone – 11-deoxy-17-hydroxy-corticosterone – cortisol route. (Sandor, *et al.* 1966, 1967*a*, *b*; Sandor, 1969).

Eel adrenal tissue preparations failed to incorporate 1C-fragments from acetate into cholesterol and corticosteroids. While this transformation is very effectively performed by mammalian (Dorfman & Ungar, 1965) and avian (Sandor, Lamoureux & Lanthier, 1965) adrenal tissue *in vitro*, recent reports by Mehdi, Carballeira & Brown (1969) have shown that, like the eel, reptilian and amphibian adrenal cortices also failed to transform exogenous acetate to corticosteroids *in vitro*. Thus, in the light of present knowledge, it is conceivable that the source of corticosteroid precursor in the eel is the plasma cholesterol.

A detailed study of the biosynthesis of cortisol by rainbow-trout adrenal tissue *in vitro* has shown that here also, cortisol biosynthesis proceeded through the routes established for the eel (Arai *et al.* 1969).

According to data obtained by various groups in the course of the last ten years and presented in this review, teleost fish adrenal tissue possesses very active 17α-hydroxylase systems. However, right from the beginning of the investigation of fish adrenocortical secretion the question of the eventual biosynthesis of aldosterone by these adrenals was raised. Physiologically the sodium-retaining capacity of aldosterone would have contributed a logical element to the electrolyte regulating mechanisms of the fish. However, until quite recently, aldosterone biosynthesis by teleost adrenals was a very much debated question, for the following reasons: in mammals, birds, reptiles and amphibians, which all biosynthesize aldosterone as a normal and sometimes major component of their adrenal steroid secretion pattern, the biosynthesis of aldosterone can be demonstrated, with considerable ease, by *in vitro* experiments using cholesterol, pregnenolone, progesterone or 21-hydroxy-17-deoxycorticosteroids, such as 11-deoxycorticosterone or corticosterone as exogenous substrate. In the teleost fish, in spite of numerous attempts by several research groups, the above substrates failed to yield aldosterone. Not only was aldosterone absent, but its constant and apparently obligatory companion, 18-hydroxycorticosterone, secreted by the adrenals of all aldosterone producing species, was equally non-detectable (Butler, 1965; Leloup-Hatey, 1966, 1968; Sandor, *et al.* 1966, 1967*b*). According to present views, corticosterone can be regarded as the obligatory precursor of 18-hydroxycorticosterone and aldosterone (Sandor, 1969). Incubation of corticosterone with eel adrenal tissue minces yielded cortisol (Sandor *et al.* 1966), a reaction not described for any other vertebrate adrenal (Dorfman & Ungar, 1965). While there were isolated reports concerning the presence of corticosterone in teleost plasma (Phillips, 1959; Idler *et al.* 1959; Leloup-Hatey, 1964*a*) and in teleost adrenal tissue incubates (Nandi & Bern, 1965; Leloup-Hatey, 1966), the appearance of this compound was far from consistent and the consensus is that under normal conditions the very active 17-hydroxylase system will eliminate the accumulation of 17-deoxy compounds (Bern, 1967). The reported presence of aldosterone in the plasma of the salmon *Oncorhynchus nerka* by Phillips, Holmes & Bondy in 1959 and its biosynthesis from progesterone by the adrenal tissue of a killifish *Fundulus heteroclitus* by Phillips & Mulrow (1959) were not immediately confirmed (Idler *et al.* 1959).

In the last few years, however, important advances have been achieved on these problems. Arai (1967) showed the transformation of progesterone to 11-deoxycorticosterone and corticosterone by rainbow-trout adrenal tissue *in vitro*, and Arai & Tamaoki (1967) reported the 18-hydroxylation of 11-deoxycorticosterone by the adrenal tissue of the same fish. These results were confirmed and expanded subsequently (Arai *et al.* 1969). In 1968, Truscott and Idler demonstrated the transformation *in vitro* of exogenous corticosterone to 18-hydroxycorticosterone and aldosterone by the adrenal tissue of the herring (*Clupea harengus harengus*). This report was followed by isolation and identification of aldosterone from the peripheral blood of this species (Truscott & Idler, 1969). Under ordinary circumstances these reports could have been taken as positive proof that teleost fish do indeed biosynthesize 17-deoxycorticosteroids, and in addition these 21-hydroxy-17-deoxy steroid substances can undergo 18-oxygenation. However, the picture is complicated by the capability of teleost adrenals to 17α-hydroxylate corticosterone. This reaction, mentioned above and observed by us in the eel (Sandor *et al.* 1966) was confirmed by Truscott & Idler (1968) working with herring adrenal tissue. Thus the question arises, if aldosterone biosynthesis is to take place in the teleost adrenal, at what level of hydroxylation will the potential aldosterone precursor(s) 'escape' 17-hydroxylation? In addition, what are the conditions which will inhibit 17-hydroxylation of a given precursor? In the mammalian adrenal, which elaborates both 18-hydroxy and 17-deoxycorticosteroids, it is believed that a functional zonation of the gland would separate the two biosynthetic chains (Stachenko & Giroud, 1959). However, in the teleost adrenal tissue, which does not even form a distinct encapsulated organ, such histological differences are not seen. Clearly, if it is admitted that teleosts will biosynthesize both cortisol and aldosterone as normal components of their adrenocortical secretion, one should look for a biochemical regulating factor which will change the sequence of hydroxylation from 17–21–11 to 21–11–18. Another question to be resolved is the possibility of species differences in the adrenal steroid biosynthesis pathway of bony fish. Of the thousands of living species, only a small handful have been critically examined.

It is believed that partial answers to these questions could be obtained by the systematic study of steroid hydroxylating enzyme systems in teleost adrenal tissue. Such an investigation has been initiated by us and preliminary findings have shown that, in the eel, a low-activity 18-hydroxylating enzyme is present in the mitochondrial fraction of the adrenal tissue. This system is NADPH-dependent and will transform exogenous corticosterone to 18-hydroxycorticosterone. However, no evidence of aldo-

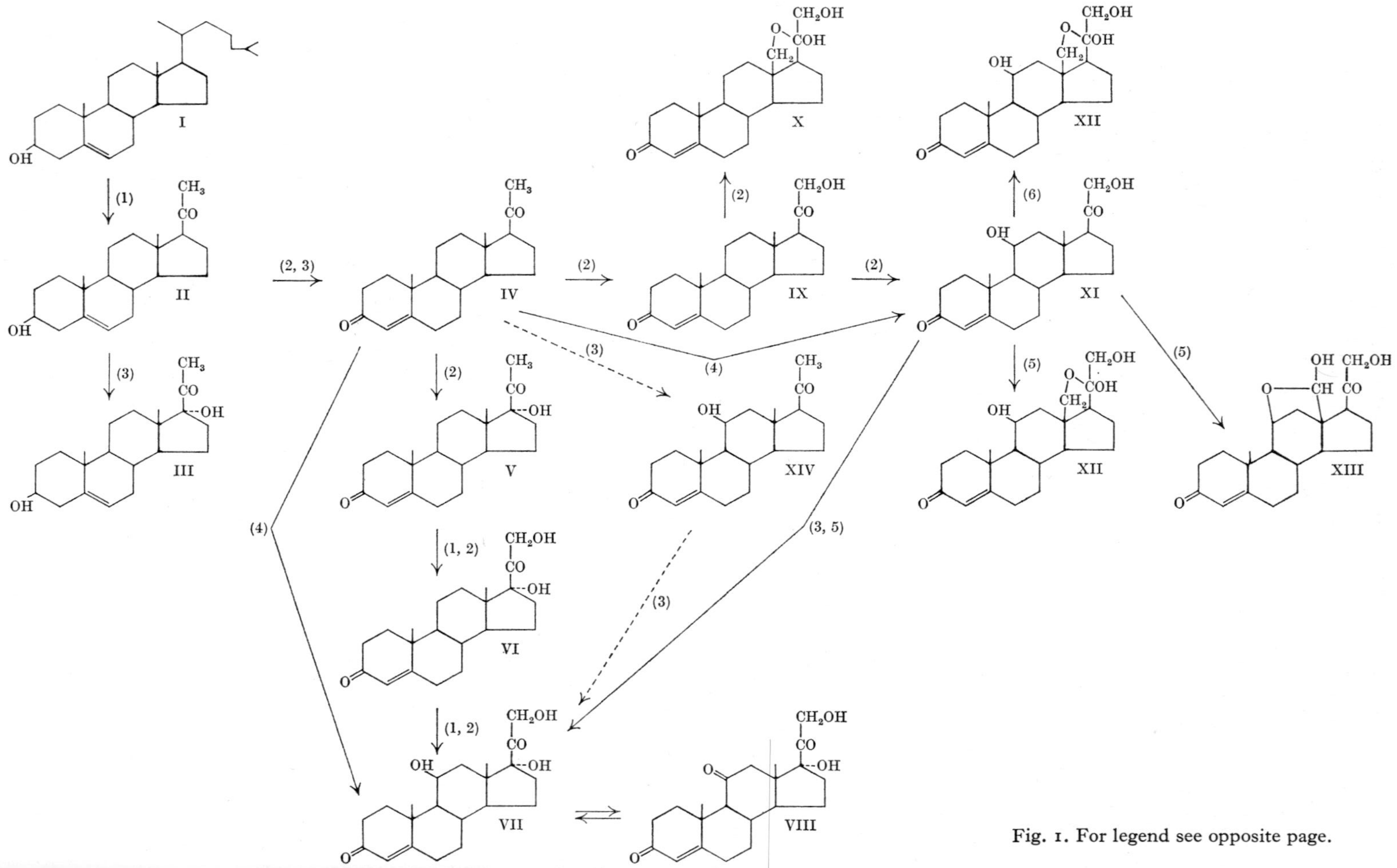

Fig. 1. For legend see opposite page.

sterone biosynthesis was obtained (Sandor, Henderson & Chester Jones, 1969).

It is quite clear that the question of corticosteroid biosynthesis in the teleost adrenal is far from resolved and a large amount of systematic work has yet to be done.

To summarize our present knowledge on the steroid biogenesis in teleost adrenals, the validated steroid synthetic pathways are shown in Fig. 1.

Although this account of corticosteroidogenesis is, for the most part, qualitative and incomplete, it suggests which corticoids may be used to approximate physiological requirements in experimental conditions.

COMPOSITION AND DISTRIBUTION OF BODY FLUIDS

Injection of adrenal steroids or ACTH have almost always been found to alter body composition. Two reports differ from this generalization (Lockley, 1957; Edelman, Young & Harris, 1960). These authors investigated the survival of various stenohaline and euryhaline species in different salinities. However, the negative observations arose from the employment of osmotic stress to stenohaline fish, transferred from either sea water to fresh water or vice versa, an inappropriately severe challenge to investigate the physiological role of steroids.

The direction of changes of electrolyte composition in plasma and tissues after injection of corticosteroids is not always predictable. Much depends on the doses used. In relatively large amounts, cortisol, corticosterone, aldosterone, deoxycorticosterone and cortisone—all, in general, cause a reduction in the concentration of plasma sodium of fresh water fishes (Chester Jones, 1957; Holmes & Butler, 1963; Holmes, Phillips & Chester Jones, 1963; Chester Jones, Henderson & Butler, 1965). The concentration of plasma potassium either remains unchanged or falls. In muscle the concomitant changes in the relationships of intracellular–extracellular water influence, to some degree, the alterations in muscle electrolyte concentrations, rendering the results difficult to assess (Chan,

---

Fig. 1. Schematic representation of validated corticosteroid biosynthetic reactions in teleost fishes. Most of the data were obtained on the European eel (*Anguilla anguilla*), the rainbow trout (*Salmo gairdneri*) and the herring (*Clupea harengus*). Compounds VII, VIII and XIII were also identified in teleost peripheral blood. I, cholesterol; II, pregnenolone; III, 17α-hydroxypregnenolone; IV, progesterone; V, 17α-hydroxy-progesterone; VI, 11-deoxy-17-hydroxycorticosterone; VII, cortisol; VIII, cortisone; IX, 11-deoxycorticosterone; X, 18-hydroxy-11-deoxycorticosterone; XI, corticosterone; XII, 18-hydroxycorticosterone; XIII, aldosterone; XIV, 11β-hydroxyprogesterone. References: (1) Sandor *et al.* 1967a; (2) Arai *et al.* 1969; (3) Sandor *et al.* 1966; (4) Leloup-Hatey, 1966, 1968; (5) Truscott & Idler, 1968; (6) Sandor *et al.* 1969.

Chester Jones, Henderson & Rankin, 1967). Thus, for example, a fall in electrolyte concentration does not necessarily indicate a net loss from the tissue. Another variable to take into consideration is the length of time after injections that the analyses are made (Holmes & Butler, 1963).

Mammalian adrenocorticotrophin (ACTH) administered to freshwater teleosts is followed by an elevation of the sodium concentration of plasma (Chester Jones, 1956; Fleming, Stanley & Meier, 1964; Stanley & Fleming, 1967). Either, then, ACTH stimulates one of the known corticosteroids in appropriate amounts to give the effects, or another, undefined corticoid produces the elevation of plasma sodium concentration. It would be profitable, therefore, to attempt correlation of adrenal secretory activity with plasma sodium changes utilizing all steroids produced *in vivo* after ACTH administration. Ideally the ACTH to be used should be of teleostean origin.

Other plasma electrolytes are also affected by corticosteroids. Injection of cortisol into hypophysectomized eels has a slight hypercalcaemic action (Chan, Chester Jones & Mosley, 1968). ACTH has a hypocalcaemic one in *Fundulus kansae* maintained in deionized water during the winter months (Fleming, Stanley & Meier, 1964). The experimental conditions were too different to resolve this apparent dichotomy of action as between ACTH and exogenous cortisol.

Data obtained from the eel, after removal of the known adrenocortical tissue to give partial or total adrenalectomy, possibly provide the most direct evidence for the role of corticosteroids in the regulation of body fluids. The concentrations of electrolytes, the total electrolyte content and the volumes of distributions are then disrupted (Chester Jones *et al.* 1965; Chester Jones, Henderson, Chan & Rankin, 1967; Chan *et al.* 1967). Although adrenalectomy of the goldfish has been reported (Etoh & Egami, 1963), *Anguilla* sp. is the only teleostean genus so far extensively investigated.

Following adrenalectomy of the freshwater eel, there is an increase in intracellular water and a fall in the concentration of plasma sodium. The concentration of plasma potassium remains within the normal range—in part a reflexion of the accumulation of body water. This too is shown by the increase in body weight dependent on the decline in urine volume with continued influx of water, perhaps because of enhanced branchial permeability (Chan *et al.* 1967). In one experimental series, Butler & Langford (1967) did not observe similar changes in the adrenalectomized freshwater North American eel (*Anguilla rostrata*). The operation of adrenalectomy presents, however, certain technical difficulties not always readily resolved.

In the European eel, adapted to seawater, adrenalectomy results in a decrease in body weight suggesting water loss perhaps aided by a change in permeability of the gill epithelium. There is a diminuition of extra- and intra-cellular fluid volumes and, in turn, an increased concentration of intra- and extracellular sodium. In terms of absolute amounts, there is an actual increase in the intracellular content of sodium.

RENAL FUNCTION

Data concerning actions of adrenocorticosteroids on renal function in teleosts are scanty and information at present available is far from definitive. The problem of obtaining sufficient urine from marine teleosts has resulted in the major research efforts being directed towards investigations of kidney function in freshwater forms. In earlier work, Chester Jones *et al.* (1959) and Holmes (1959) described experiments which indicated that deoxycorticosterone and cortisol can cause renal sodium conservation in freshwater trout loaded with sodium chloride solution. Subsequently, Holmes & McBean (1963), by complex analyses of rates of inulin excretion into the environment of trout maintained in fresh water, concluded that the glomerular filtration rate was depressed by corticosterone but were unable to detect an effect of aldosterone. In a recent study, Chan, Rankin & Chester Jones (1969) tested a number of adrenocorticosteroids—cortisol, 11-deoxycortisol, aldosterone and 11-deoxycorticosterone—on adrenalectomized freshwater eels in which renal function is impaired. It was found that, of the steroids tested and in the doses used, cortisol most nearly restored urinary electrolyte composition and urine flow to normal. Stanley & Fleming (1964, 1967) used mammalian ACTH in *Fundulus kansae* with the rationale of thereby stimulating cortisol secretion; in the event there were no effects on urine flow in the intact or hypophysectomized fish.

The possible effects of adrenal steroids on renal function in marine and freshwater teleosts are difficult to assess critically. It is clear that lack of adrenocorticosteroids in the freshwater eel reduces the ability of the kidney to excrete water (Chan *et al.* 1969), with impaired sodium re-absorption by the renal tubule. The actions of these steroids on the kidney of marine teleosts remain largely unexamined. Clearly the phenomenon of glomerular intermittency in mesonephric kidneys (Hickman, 1965; Lahlou, 1966) could play a significant role in the renal adaptation to different environmental tonicities. Here, too, the vasoactive hormones would be expected to play a crucial role (Chester Jones, Chan & Rankin, 1969 *a*, *b*). We have only begun the examination of the many interplaying factors influencing renal function. It may be, originally, one factor—the adrenocorticosteroids primarily influenced the tubular handling of electrolytes, that is, ionic regulation.

A further point worth noting regarding renal excretory processes is that teleosts possess a urinary bladder in which urine may be stored for long periods (Smith, 1953). It seems that the bladder of at least certain species is capable of modifying formed urine (Lahlou, 1967; Lahlou, Henderson & Sawyer, 1969). This would then provide an additional site of action of steroids in the control of electrolytes in teleosts. Although the teleostean bladder is mesodermal, unlike that of the tetrapod, which is endodermal, a function similar to that of the amphibian bladder may possibly be ascribed to it (Bentley, 1966).

BRANCHIAL FUNCTION

The rates of movement of sodium and chloride ions through the gills of freshwater and marine teleosts are affected by administration of adrenocortical steroids, though the results may often reflect pharmacological rather than physiological actions (see review by Chester Jones et al. 1969).

Sexton (1955) first demonstrated that sodium uptake by the gills of goldfish in freshwater was decreased by deoxycorticosterone acetate (DOCA); cortisone was without effect. Later Chester Jones et al. (1959) and Holmes (1959) showed that freshwater trout excreted a load of sodium chloride solution more rapidly via extrarenal routes when pretreated with DOCA. It was suggested that the steroids employed inhibited, to some degree, the uptake mechanism or increased the permeability to passive efflux of water. On the other hand, lack of adrenal hormones, as in the adrenalectomized freshwater eel, results in impaired net uptake of sodium by the gills. Aldosterone, not so far shown to occur in the eel, was capable of restoring the rate of uptake to normal or higher than normal values. In larger doses aldosterone given to intact eels reduced sodium uptake (Chester Jones, Phillips & Bellamy, 1962). Motais (1967) demonstrated that aldosterone reduced the efflux of sodium in flounders when transferred from fresh water to sea water and that this steroid (or an endogenous one that mimicked its action) was of importance during the transfer from sea water to fresh water and not the reverse; that is, it was sodium-conserving.

Cortisol, the predominant teleostean steroid, also affects extrarenal function, its action depending on the amounts used. Thus with mg quantities/100 g body weight, intact freshwater eels go into profound negative extrarenal sodium balance (Henderson & Chester Jones, 1967). Smaller doses, of the order of 20 μg/100 g body weight, increase sodium uptake in intact fish and rectify the impaired uptake of adrenalectomized fish (Chan *et al.* 1969). Cortisol is active in seawater eels. The adrenalectomized seawater eel has a much reduced total sodium turnover rate, with a depressed sodium extrusion mechanism in the gills. These activities are restored to normal values by use of cortisol in the smaller dose range (Mayer, Maetz, Chan, Forster & Chester Jones, 1967; Mayer & Maetz, 1967).

The branchial response to adrenocorticosteroids have thus been found to be variable. We have noted the importance of the doses used to provide inhibition or stimulation of sodium uptake. Another factor would be undoubtedly the physiological state of the animal at the time of experimentation (Chester Jones *et al.* 1962; Maetz, 1964; Henderson & Chester Jones, 1967; Chan *et al.* 1967). It would now appear to be clear that in seawater fish adrenocortical steroids aid in the maintenance of normal sodium efflux and sodium turnover rates. Data obtained from hypophysectomized eels, with lower than normal amounts of circulating cortisol (M. E. Forster, unpublished data), from adrenalectomized eels and from replacement injections of cortisol, all indicate that cortisol is certainly one of the major hormones concerned in sodium metabolism (Mayer *et al.* 1967; Maetz *et al.* 1967).

GASTROINTESTINAL FUNCTION

The nature of adrenocortical steroidal control of drinking and the rate of absorption from the gut are incompletely known. A few data are available. Skadhauge & Maetz (1967) and Hickman (1965, 1968) have clearly defined some of the parameters which may or may not be acted upon by humoral mechanisms. Thus in marine teleosts between 90% and 100% of the imbibed sodium, potassium and chloride are absorbed, but much smaller percentages of the sulphates and phosphates of magnesium. A significant quantity of calcium is removed by the gastrointestinal tract. The movement of water appears to be secondary to these electrolyte movements. Such a system for fluid and electrolyte passage is very likely to be acted upon by corticoids.

The specific drinking of the environmental water in freshwater teleosts has recently been demonstrated (Maetz & Skadhauge, 1968; Gaitskell, 1969). This is to be distinguished from intake of water incidental to feeding. Hypophysectomy appreciably reduces the rate of drinking in freshwater eels, though the operation is without significant effect in seawater fish (Gaitskell, 1969). The significance of drinking in freshwater forms is difficult to understand as it would appear to aggravate certain problems in osmoregulation. Indeed the problem has been dismissed in the past on the erroneous supposition that freshwater teleosts ingest only slight amounts of water.

Gastrointestinal permeability may be altered following hypophysectomy or adrenalectomy. The intestine of the hypophysectomized seawater eel, examined *in vitro*, is less permeable to water compared with that of the intact animal (Hirano, 1967; Hirano & Utida, 1968). In the same studies, it was found that cortisol or ACTH returned the permeability to water towards normal values, though aldosterone was without effect. Oide & Utida (1967) suggested that ACTH (via cortisol) acts to stimulate sodium movement across the mucosa whilst a pars distalis hormone (prolactin?) changes permeability to water. Gaitskell (1969) has recently shown that everted sacs of the intestine of freshwater eels were less permeable to water following adrenalectomy and restoration to normal could be achieved by cortisol administration. In addition, cortisol was found to stimulate water movement in the gut sacs of intact freshwater eels.

THE VASCULAR SYSTEM

The normal activities of fishes in their environment depend on the integrity of the cardio-vascular system and its appropriate changes to demand. Systemic blood pressure is lower in seawater eels than in freshwater fish with concomitant influence on glomerular filtration rates. Adrenalectomy of freshwater eels is followed by lowering of systemic blood pressure and, as we have noted earlier, urine volume declines. The possibility exists that some of the vascular changes might be the result of the surgery required for removal of the adrenocortical tissue. However in the first place some of the effects of suppression of adrenal function by blocking agents mimick those of the surgical operation (Chan *et al.* 1967). Secondly, cortisol is capable of rectifying all changes, so far examined, in the adrenalectomized animal.

Blood-flow distributions within the branchial tissue is of profound functional importance (Steen & Kruysse, 1964; Richards & Fromm, 1969). For example, intravenous injection of neurohypophysial and urophysial extracts which given an immediate increase in the blood pressure in the ventral aorta (Chan & Chester Jones 1967, 1969; Chester Jones *et al.* 1969*b*; Chan, Chester Jones & Ponniah, 1969) cause blood to be forced into the central compartments of the gill lamellae; here the cells, active in sodium movement, lie in the interlamellar epithelium (Chan, unpublished data). Both neurohypophysial peptides and urophysial extracts increase the rate of sodium influx through the gills of freshwater goldfish (Maetz, Bourguet & Lahlou, 1964; Maetz, Bourguet, Lahlou & Hourdry, 1964).

In contrast, adrenaline forces blood into the respiratory lamellae, by-passing the central channels in the filament (Chan, unpublished data). It will be seen, therefore, that hormones can act not only on the cells and their system but also in a secondary fashion on the blood flow in unit time and thus the quantities of ions presented to any cellular system.

CHANGES IN ADRENOCORTICAL FUNCTION

A pituitary–adrenal cortex axis has been demonstrated by histological means and by direct determination of circulating cortisol levels. Following hypophysectomy, cortical tissue atrophies (Van Overbeeke & Ahsan, 1966; Basu, Nandi & Bern, 1965; Hanke & Chester Jones, 1966; Roy, 1964; Handin, Nandi & Bern, 1964; Sundararaj & Goswami, 1968), and the changes rectified by mammalian ACTH or by fish pituitaries (Van Overbeeke & Ahsan, 1966; Ball & Olivereau, 1966; Hanke, Bergerhoff & Chan, 1969). See Olivereau & Ball (this symposium, p. 57) for further details. The level of plasma cortisol is lowered following hypophysectomy (Donaldson & McBride, 1967; Bradshaw & Fontaine-Bertrand, 1968; Butler, Donaldson & Clarke, 1969). Either fluorimetry or the competitive pro-tein binding method is used for the determination of cortisol and the proviso must be mentioned that the estimations are not necessarily always rigorous. Confirmatory evidence comes from the use of metyrapone, an adrenocortical blocking agent, when plasma cortisol levels are reduced (Fagerlund, McBride & Donaldson, 1968).

A variety of stresses such as forced swimming, reduced water levels in holding tanks, tactile agitation, noxious ions have been reported to increase the concentration of cortisol in the plasma (Donaldson & McBride, 1967; Hill & Fromm, 1968; Wedemeyer, 1969). This largely confirms earlier work (Leloup-Hatey, 1958, 1960; 1964c; Fontaine & Leloup-Hatey, 1960). The increased levels of plasma cortisol induced by stress may be suppressed by hypophysectomy (Donaldson & McBride, 1967). All the evidence therefore suggests an adenohypophysial–adrenocortical inter-relationship and we would anticipate a piscine ACTH not unlike that of mammals coupled with some form of hypothalamic control.

In nature, if we accept the traditional theory, the silver eel swims some 3,000 miles from Europe to the Sargasso Sea, where spawning takes place. The swimming of this long distance by the silver eel which does not feed must constitute a form of stress. In this laboratory, silver eels have been forced to swim the equivalent of 2,500 miles. This results in a marked in-crease in the level of plasma cortisol and of liver glycogen (J. C. Fenwick & M. E. Forster, unpublished data).

CORPUSCLES OF STANNIUS

Changes occurring after the removal of the corpuscles of Stannius from freshwater eels show similarities to and important differences from those following adrenalectomy. These eels show loss of sodium and a shift of water into the cells, but blood potassium concentration rises and there is no significant increase in body weight. This means that there is no over-all water retention and this is reflected in the clear diminution of extracellular fluid volume, contributing to the increased blood potassium concentration, the actual amount remaining within the control range. The increase in calcium both in blood and muscle, after removal of the corpuscles, was an absolute one, not seen in the adrenalectomized eel, and this was coupled with increased renal retention (Fontaine, 1964; 1967; Chester Jones et al. 1965; Chan et al. 1967; Chan, 1969). Removal of the corpuscles from sea-water eels brings about some changes similar to those following adrenalectomy but they are less pronounced. There is some decline in body weight and the extra- and intracellular fluid volumes are smaller. Intracellular sodium increases both in amount and concentration. Extracellular sodium also increases in concentration but the intracellular amount is sometimes less. Increased concentration of intracellular potassium is due to the shift of water out of the cells. As regards calcium both adrenalectomy and removal of the corpuscles result in an increase in amount and concentration of calcium in blood and muscle. In both freshwater and seawater eels the alterations in calcium after removal of the corpuscles return to normal and this appears due to the activity of the ultimobranchial body with presumptive increased calcitonin secretion (Chan, 1969, and unpublished data).

The status of the corpuscles of Stannius as steroid-producing organs in normal physiological conditions is dubious (Nandi, 1967). Some of the changes in sodium metabolism following their removal are rectified by aldosterone administration (Leloup-Hatey, 1964b), and in addition they contain a pressor material with certain characteristics of renin (Chester Jones et al. 1966). Renin-like material is also present in the kidneys of a variety of teleost fishes (Kaley & Donshik, 1965; Chester Jones et al. 1966; Mizogami, Oguri, Sokabe, & Nishimura, 1968). A renin angiotensin–adrenocortical interrelationship is possible but not yet shown, though there is some histological evidence for an adrenocortical–corpuscular axis (Hanke & Chester Jones, 1966; Olivereau & Olivereau, 1968; Hanke et al. 1969). Variations in renal renin content in marine and freshwater fish have not given consistent patterns and interpretation must await definitive experiments (Kaley, Robison & Lubben, 1963; Sokabe, 1968; Sokabe, Mizogami & Sato, 1968).

SEXUAL MATURITY

Studies have shown that concomitant with the reproductive migration of salmonids, there are increasing levels of blood 17-hydroxycorticosteroids and cortisol (Fontaine & Hatey, 1954; Idler *et al.* 1959; Hane & Robertson, 1959; Robertson *et al.* 1961 *a*, *b*; Schmidt & Idler, 1962). Adrenocortical function thus may be altered during the various stages of the onset of spawning. The gland becomes less responsive to stress and administration of ACTH (Hane, Robertson, Wexler & Krupp, 1966). However, this increased secretion is not always the case in salmonids (Fagerlund, 1967). Differences would be expected between those species which die after spawning and those that do not. In addition to variations in the secretion rate of adrenal hormones, demonstrated or suspected, the rate of clearance of hormones from the circulation, the biological half-life, the degree of protein binding, as well as the volume of distribution, are all modified (Idler & Truscott, 1963; Seal & Doe, 1966; Freeman & Idler, 1966; Donaldson & Fagerlund, 1968 *a*, *b*). Again both androgenic (Fagerlund & Donaldson, 1969) and oestrogenic (Donaldson & Fagerlund, 1969) steroids alter the metabolism of cortisol in salmonids. Alterations of steroid metabolism can be expected irrespective of the sexual state of the animal. Thus Idler & Freeman (1965) showed an impaired hormone clearance in moribund cod (*Gadus morhua*) which were neither sexually mature nor approaching maturity. The authors make the valid observation that impaired steroidal metabolism in some species at sexual maturity may merely be coincidental rather than a directly related phenomenon. Little is known of other species. Though in this laboratory the eel can be caused to swim for the equivalent of long distances, the definitive experiments of associated sexual maturity have not yet been achieved.

RHYTHMS

Fish reveal both diurnal (Boehlke, Church, Tiemeier & Elftheriou, 1966) and seasonal changes though little is known of the endocrinology involved. Siddiqi (1966) observed a seasonal variation in the cholesterol content of various tissues of *Ophiocephalus* and, as cholesterol is often a steroid hormone precursor (see p. 34), this may be reflected in changed adrenal hormonal output. The problems of the influence of breeding seasons, light, variations in temperature, salinity, availability of food and population densities on adrenal activity clearly require to be investigated and little has been done.

CONCLUSIONS

Much of mammalian endocrinology had its inception in clinical observations and the laboratory use of a few species. Just as there the vista is widening to take in wild species in relation to their environment, so investigation of the few teleost species must be extended to more of the some 23,000 species known. Here is a rich and fascinating area of research. Even in those species so far examined, the exact nature of the spectrum of adrenocorticosteroids remains to be established. The tantalizing problems of the presence or absence of aldosterone, its apparent sporadic occurrence and its physiological role remain. Indeed, the general role of adrenal steroids in water and electrolyte movements has yet to be clearly delineated and their actions at the cellular level explored. Chester Jones (1956) in the First International Symposium of Comparative Endocrinology, July 1954, posed the question 'Does a galaxy of steroids, similar to that of the Eutheria, act upon those tissues to preserve the *milieu interieur* so that we can suppose electrocortin (aldosterone) to support sodium conservation in the freshwater fish by its catalysis of transport systems in the gill?' We are but a little way along the road to the answer.

REFERENCES

ARAI, R. (1967). *In vitro* corticosteroidogenesis by the head kidney of the rainbow trout. *Annotnes zool. japon.* **40**, 136–45.

ARAI, R., TAJIMA, H. & TAMAOKI, B. (1969). *In vitro* transformation of steroids by the head kidney, the body kidney and the corpuscles of Stannius of the rainbow trout (*Salmo gairdnerii*). *Gen. comp. Endocr.* **12**, 99–109.

ARAI, R. & TAMAOKI, B. (1967). Biosynthesis *in vitro* of 18-hydroxy-11-deoxy-corticosterone by the interrenal tissue of the rainbow trout. *J. Endocr.* **39**, 453–4.

BALL, J. N. & OLIVEREAU, M. (1966). Identification of ACTH cells in the pituitary of two teleosts, *Poecilia latipinna* and *Anguilla anguilla*: correlated changes in the interrenal and the pars distalis resulting from administration of metopirone (SU 4885). *Gen. comp. Endocr.* **6**, 5–18.

BASU, J., NANDI, J. & BERN, H. A. (1965). The homolog of the pituitary–adrenocortical axis in the teleost fish *Tilapia mossambica*. *J. exp. Zool.* **159**, 347–56.

BENTLEY, P. J. (1966). The physiology of the urinary bladder of amphibia. *Biol. Rev.* **41**, 275–316.

BERN, H. A. (1967). Hormones and endocrine glands of fishes. *Science, N.Y.* **158**, 455–62.

BOEHLKE, K. W., CHURCH, R. L., TIEMEIER, O. W. & ELEFTHERIOU, B. E. (1966). Diurnal rhythm in plasma glucocorticoid levels in Channel catfish, *Ictalurus punctatus*. *Gen. comp. Endocr.* **7**, 18–21.

BONDY, P. K., UPTON, G. V. & PICKFORD, G. E. (1957). Demonstration of cortisol in fish blood. *Nature, Lond.* **179**, 1354.

BRADSHAW, S. D. & FONTAINE-BERTRAND, E. (1968). Le cortisol dans le plasma de l'Anguille, dosé par fluorimetrie et par inhibition compétitive de la liaison spécifique cortisol-transcortine. *C. r. hebd. Séanc. Acad. Sci., Paris*, **267**, 894–7.

BUTLER, D. G. (1965). Adrenocortical steroid production by the interrenal tissue of the freshwater European silver eel (*Anguilla anguilla*) and the marine eel (*Conger conger*). *Comp. Biochem. Physiol.* **16**, 583–88.

BUTLER, D. G., DONALDSON, E. M. & CLARKE, W. C. (1969). Physiological evidence for a pituitary–adrenocortical feedback mechanism in the eel, *Anguilla rostrata. Gen. comp. Endocr.* **12**, 173–5.

BUTLER, D. G. & LANGFORD, R. W. (1967). Tissue electrolyte composition of the freshwater eel, *Anguilla rostrata*, following partial surgical adrenalectomy. *Comp. Biochem. Physiol.* **22**, 309–312.

CHAN, D. K. O. (1969). Endocrine regulation of calcium and inorganic phosphate balance in the freshwater adapted teleost fish, *Anguilla anguilla* and *A. japonica*. *Proc. 3rd Int. Congr. Endocr.*, Mexico City 1968, pp. 709–16. Excerpta Med. Found., Int. Congr. Series. No. 184.

CHAN, D. K. O. & CHESTER JONES, I. (1967). The regulation of blood pressure in the European eel, *Anguilla anguilla* L. *Gen. comp. Endocr.* **9**, 439.

CHAN, D. K. O. & CHESTER JONES, I. (1969). Pressor effects of neurohypophysial peptides in the eel, *Anguilla anguilla* L., with some reference to their interaction with adrenergic and cholinergic receptors. *J. Endocr.* **45**, 161–74.

CHAN, D. K. O., CHESTER JONES, I., HENDERSON, I. W. & RANKIN, J. C. (1967). Studies on the experimental alteration of water and electrolyte composition of the eel, *Anguilla anguilla* L. *J. Endocr.* **37**, 297–317.

CHAN, D. K. O., CHESTER JONES, I. & MOSLEY, W. (1968). Pituitary and adrenocortical factors in the control of water and electrolyte composition of the freshwater European eel, *Anguilla anguilla* L. *J. Endocr.* **42**, 91–8.

CHAN, D. K. O., CHESTER JONES, I. & PONNIAH, S. (1969). Studies on the pressor substances of the caudal neurosectetory system of teleost fish: bioassay and fractionation. *J. Endocr.* **45**, 151–9.

CHAN, D. K. O., RANKIN, J. C. & CHESTER JONES, I. (1969). Influences of the adrenal cortex and the corpuscles of Stannius on osmoregulation in the European eel, *Anguilla anguilla* L., adapted to fresh water. *Gen. comp. Endocr.* Suppl. 2, pp. 342–53.

CHESTER JONES, I. (1956). The role of the adrenal cortex in the control of water and salt-electrolyte metabolism in vertebrates. *Mem. Soc. Endocr.* **5**, 102–24.

CHESTER JONES, I. (1957). *The Adrenal Cortex.* Cambridge University Press.

CHESTER JONES, I., CHAN, D. K. O., HENDERSON, I. W. & BALL, J. N. (1969). The adrenal cortex, adrenocorticotropin and the corpuscles of Stannius. In *Physiology of Fishes* (eds. W. S. Hoar and D. J. Randall). New York: Academic Press. Vol. II. pp. 321–376.

CHESTER JONES, I. CHAN, D. K. O. & RANKIN, J. C. (1969a). Renal function in the European eel (*Anguilla anguilla* L): changes in blood pressure and renal function of the freshwater eel transferred to sea water. *J. Endocr.* **43**, 9–19.

CHESTER JONES, I., CHAN, D. K. O. & RANKIN, J. C. (1969b). Renal function in the European eel (*Anguilla anguilla* L.): effects of the caudal neurosecretory system, corpuscles of Stannius, neurohypophysial peptides and vasoactive substances. *J. Endocr.* **43**, 21–31.

CHESTER JONES, I., HENDERSON, I. W. & BUTLER, D. G. (1965). Osmoregulation in teleost fish with special reference to the European eel, *Anguilla anguilla* L. *Archs Anat. microsc. Morph. exp.* **54**, 453–69.

CHESTER JONES, I., HENDERSON, I. W., CHAN, D. K. O. & RANKIN, J. C. (1967). Steroids and pressor substances in bony fish with special reference to the adrenal cortex and corpuscles of Stannius of the eel (*Anguilla anguilla* L.). *Proc. IInd Int. Congr. Hormonal Steroids*, Milan 1966, pp. 136–45. Excerpta Med. Found., Int. Ser. no. 132.

CHESTER JONES, I., HENDERSON, I. W., CHAN, D. K. O., RANKIN, J. C., MOSLEY, W., BROWN, J. J., LEVER, A. F., ROBERTSON, J. I. S. & TREE, M. (1966). Pressor activity in extracts of the corpuscles of Stannius from the European eel. (*Anguilla anguilla* L.) *J. Endocr.* **34**, 393–408.

CHESTER JONES, I., PHILLIPS, J. G. & BELLAMY, D. (1962). Studies on water and electrolytes in cyclostomes and teleosts with special reference to *Myxine glutinosa* (the hagfish) and *Anguilla anguilla* L. *Gen. comp. Endocr.* Suppl. 1, pp. 36–47.

CHESTER JONES, I., PHILLIPS, J. G. & HOLMES, W. N. (1959). Comparative physiology of the adrenal cortex. In *Comparative Endocrinology*, pp. 582–612. (ed. A. Gorbman), New York: J. Wiley & Son.

DONALDSON, E. M. & FAGERLUND, U. H. M. (1968a). Changes in the cortisol dynamics of sockeye salmon (*Oncorhynchus nerka*) resulting from sexual maturation. *Gen. comp. Endocr.* **11**, 552–61.

DONALDSON, E. M. & FAGERLUND, U. H. M. (1968b). Changes resulting from sexual maturation in the half life ($T\frac{1}{2}$) and volume of distribution (V) of radioactivity after injection of C^{14}-cortisol (F) into salmon. Abstracts, *IIIrd Int. Congr. Endocr.* Mexico City 1968. Excerpta Medica, Int. Congr. Ser., no. 157. Abstract, 348.

DONALDSON, E. M. & FAGERLUND, U. H. M. (1969). Cortisol secretion rate in gonadectomized female sockeye salmon (*Oncorhynchus nerka*): effects of estrogen and cortisol treatment. *J. Fish. Res. Bd Can.* **26**, 1789–99.

DONALDSON, E. M. & McBRIDE, J. R. (1967). The effect of hypophysectomy in the rainbow trout, *Salmo gairdnerii* (Rich), with special reference to the pituitary–interrenal axis. *Gen. comp. Endocr.* **9**, 93–101.

DORFMAN, R. I. & UNGAR, F. (1965). *Metabolism of Steroid Hormones*. New York: Academic Press.

EDELMAN, I. S., YOUNG, H. L. & HARRIS, J. B. (1960). Effects of corticosteroids on electrolyte metabolism during osmoregulation in teleosts. *Am. J. Physiol.* **199**, 666–70.

ETOH, H. & EGAMI, N. (1963). Effect of adrenalectomy and hypophysectomy on the length of survival time after X-irradiation in the goldfish, *Carassius auratus*. *Proc. Japan. Acad.* **39**, 503–6.

FAGERLUND, U. H. M. (1967). Plasma cortisol concentration in relation to stress in adult sockeye salmon during the freshwater stage of their life cycle. *Gen. comp. Endocr.* **8**, 197–207.

FAGERLUND, U. H. M. & DONALDSON, E. M. (1969). The effect of androgens on the distribution and secretion of cortisol in gonadectomised male sockeye salmon (*Oncorhynchus nerka*). *Gen. comp. Endocr.* **12**, 438–48.

FAGERLUND, U. H. M., McBRIDE, J. R. & DONALDSON, E. M. (1968). Effect of metopirone on pituitary–interrenal function in two teleosts: sockeye salmon, *Oncorhynchus nerka*, and rainbow trout *Salmo gairdnerii*. *J. Fish. Res. Bd Can.* **25**, 1465–74.

FLEMING, W. R., STANLEY, J. G. & MEIER, A. H. (1964). Seasonal effects of external calcium, estradiol and ACTH on serum calcium and sodium levels of *Fundulus kansae*. *Gen. comp. Endocr.* **4**, 61–7.

FONTAINE, M. (1964). Corpuscules de Stannius et régulation ionique (Ca, K et Na) du milieu intérieur de l'anguille (*Anguilla anguilla* L.). *C. r. hebd. Séanc. Acad. Sci., Paris* **259**, 875–8.

FONTAINE, M. (1967). Intervention des corpuscules de Stannius dans l'équilibre phosphocalcique du milieu intérieur d'un poisson Téléostéen, l'Anguille. *C. r. hebd. Séanc. Acad. Sci., Paris*, **264**, 736–7.

FONTAINE, M. & HATEY, J. (1954). Sur la teneur en 17-hydroxy corticostéroïdes du plasma du saumon (*Salmo salar* L.) *C. r. hebd. Séanc. Acad. Sci., Paris* **239**, 319–21.

FONTAINE, M. & LELOUP-HATEY, J. (1959). Mise en évidence de corticostéroïdes dans l'intérrenal d'un Téléostéen: le saumon (*Salmo salar* L.). *J. Physiol., Paris*. **51**, 464–9.

FONTAINE, M. & LELOUP-HATEY, J. (1960). Influence de l'activité motrice (nagé a contre-courant) sur la 17-hydroxycorticostéroïdémie de la Truite arc-en-ciel (*Salmo gairdneri* Rich). Intervention probable de ce facteur dans l'activation de l'intérrenal anterieur du jeune saumon (*Salmo salar*) pendant sa migration d'avalaison. *C. r. hebd. Séanc. Acad. Sci., Paris*, **250**, 3089–94.

FREEMAN, H. C. & IDLER, D. R. (1966). Transcortin binding of cortisol in atlantic salmon, *Salmo salar*, plasma. *Gen. comp. Endocr.* **7**, 37–43.

GAITSKELL, R. E. (1969). Osmotic and ionic regulation in the European eel, *Anguilla anguilla* L., with special reference to the role of the gut. Ph.D. thesis, University of Sheffield, England.

HANDIN, R. I., NANDI, J. & BERN, H. A. (1964). Effect of hypophysectomy on survival and on thyroid and interrenal histology of the cichlid teleost, *Tilapia mossambica*. *J. exp. Zool.* **157**, 339–44.

HANE, S. & ROBERTSON, O. H. (1959). Changes in plasma 17-hydroxycorticosteroids accompanying sexual maturation and spawning of the pacific salmon (*Oncorhychus tshawytscha*) and rainbow trout (*Salmo gairdnerii*). *Proc. natn. Acad. Sci. U.S.A.* **45**, 886–93.

HANE, S. ROBERTSON, O. H., WEXLER, B. C. & KRUPP, M. A. (1966). Adrenocortical response to stress and ACTH in Pacific salmon, *Oncorhynchus tschawytscha* and steel head trout, *Salmo gairdnerii* at successive stages in the sexual cycle. *Endocrinology*, **78**, 791–800.

HANKE, W., BERGERHOFF, K. & CHAN, D. K. O. (1969). Histological and histochemical changes of organs associated with adjustment to the environment. *Gen. comp. Endocr.* Suppl. 2, pp 331–41.

HANKE, W. & CHESTER JONES, I. (1966). Histological and histochemical studies on the adrenal cortex and the corpuscles of Stannius of the European eel (*Anguilla anguilla* L.). *Gen. comp. Endocr.* **7**, 166–78.

HENDERSON, I. W. & CHESTER JONES, I. (1967). Endocrine influences on the net extrarenal fluxes of sodium and potassium in the European eel (*Anguilla anguilla* L.). *J. Endocr.* **37**, 319–25.

HICKMAN, C. P. (1965). Studies on renal function in freshwater teleost fish. *Trans. R. Soc. Can.* **3**, ser. 4, pp. 213–36.

HICKMAN, C. P. (1968). Ingestion, intestinal absorption and elimination of sea water and salts in the southern flounder. *Paralichthys lethostigma. Can. J. Zool.* **46**, 457–66.

HILL, C. W. & FROMM, P. O. (1968). Response of the interrenal gland of rainbow trout (*Salmo gairdnerii*) to stress. *Gen. comp. Endocr.* **11**, 69–77.

HIRANO, T. (1967). Effect of hypophysectomy on water transport in isolated intestine of the eel, *Anguilla japonica. Proc. Japan. Acad.* **43**, 793–6.

HIRANO, T. & UTIDA, S. (1968). Effects of ACTH and cortisol on water transport in isolated intestine of the eel, *Anguilla japonica. Gen. comp. Endocr.* **11**, 373–80.

HOLMES, W. N. (1959). Studies on the hormonal control of sodium metabolism in the rainbow trout (*Salmo gairdneri*). *Acta endocr., Copenh.* **31**, 587–602.

HOLMES, W. N. & BUTLER, D. G. (1963). The effect of adrenocortical steroids on the tissue electrolyte composition of fresh water rainbow trout (*Salmo gairdnerii*). *J. Endocr.* **25**, 457–64.

HOLMES, W. N. & MCBEAN, R. L. (1963). Studies on the glomerular filtration rate of rainbow trout (*Salmo gairdnerii*). *J. exp. Biol.* **40**, 335–41.

HOLMES, W. N., PHILLIPS, J. G. & CHESTER JONES, I. (1963). Adrenocortical factors associated with adaptation of vertebrates to marine environments. *Recent Progr. Hormone. Res.* **19**, 619–72.

IDLER, D. R. & FREEMAN, H. C. (1965). A demonstration of an impaired hormone metabolism in moribund Atlantic cod (*Gadus morhua* L.). *J. Fish. Res. Bd Can.* **23**, 1249–55.

IDLER, D. R., FREEMAN, H. C. & TRUSCOTT, B. (1964). Steroid hormones in the plasma of the spawned Atlantic salmon (*Salmo salar*) and a comparison of their determination by chemical and biological assay methods. *Can. J. Biochem.* **42**, 211–18.

IDLER, D. R., RONALD, A. P. & SCHMIDT, P. J. (1959). Biochemical studies on the sockeye salmon during spawning migration. VII. Steroid hormones in plasma. *Can. J. Biochem. Physiol.* **37**, 1227–38.

IDLER, D. R. & TRUSCOTT, B. (1963). *In vivo* metabolism of steroid hormones by sockeye salmon. (A) Impaired hormone clearance in mature and spawned pacific salmon (*O. nerka*). (B) Precursors of 11-keto-testosterone. *Can. J. Biochem. Physiol.* **41**, 875–87.

KALEY, G. & DONSHIK, P. C. (1965). Specificity and quantitative aspects of the 'renin-angiotensin' system in lower vertebrates. *Biol. Bull. mar. biol. Lab., Woods Hole*, **129**, 411.

KALEY, G., ROBISON, G. A. & LUBBEN, B. (1963). Comparative aspects of the 'renin-angiotensin' system. *Biol. Bull. mar. biol. Lab., Woods Hole*, **125**, 381.

KEYS, A. B. (1933). The mechanism of adaptation to varying salinities in the common eel and the general problem of osmotic regulation in fishes. *Proc. Roy Soc. Lond.* B **112**, 184–99.

KROGH, A. (1939). *Osmotic Regulation in Aquatic Animals.* Cambridge University Press.

LAHLOU, B. (1966). Mise en évidence d'un recrutement glomerulaire dans le rein des Téléostéens d'après la mesure du T_m glucose. *C. r. hebd. Séanc. Acad. Sci., Paris*, **272**, 1356–8.

LAHLOU, B. (1967). Excretion rénale chez un poisson euryhalin, le flet (*Platychthys flesus* L): caractéristiques de l'urine normale en eau douce et en eau de mer et effets des changements de milieu. *Comp. Biochem. Physiol.* **20**, 925–38.

LAHLOU, B., HENDERSON, I. W. & SAWYER, W. H. (1969). Renal adaptations by *Opsanus tau*, a euryhaline aglomerular teleost, to dilute media. *Am. J. Physiol.* **216**, 1266–72.

LELOUP-HATEY, J. (1958). Influence de l'agitation motrice sur la teneur du plasma en 17-hydroxycorticostéroïdes d'un Téléostéen: la carpe (*Cyprinus carpio* L). *C. r. hebd. Séanc. Acad. Sic., Paris*, **246**, 1088–91.

LELOUP-HATEY, J. (1960). Influence de l'agitation motrice sur la teneur en corticostéroïdes du plasma d'un Téléostéen: la carpe (*Cyprinus carpio* L). *J. Physiol., Paris*, **52**, 145–6.

LELOUP-HATEY, J. (1964*a*). Fonctionnement de l'interrénal anterieur de deux Téléostéens: le saumon atlantique et l'Anguille européenne. *Annls Inst. Océanogr., Paris*, **42**, 221–37.

LELOUP-HATEY, J. (1964*b*). Influence d'un apport sodique et de l'aldostérone sur le déséquilibre minéral consécutif à l'ablation des corpuscules de Stannius chez l'anguille *Anguilla anguilla* L. *C. r. Séanc. Soc. Biol., Paris*, **158**, 991–4.

LELOUP-HATEY, J. (1964*c*). Etude du déterminisme de l'activation de l'interrénal anterieur observée chez quelques Téléostéens soumis à un accroissement de la salinité du milieu extérieur. *Archs Sci. physiol.* **18**, 293–324.

LELOUP-HATEY, J. (1966). Etude *in vitro* de la corticostéroïdogénèse dans l'interrenal de l'Anguille européenne (*Anguilla anguilla* L.). *Comp. Biochem. Physiol.* **19**, 63–74.

LELOUP-HATEY, J. (1968). Contrôle corticotrope de la corticostéroïdogénèse intér-renaliénne chez les vertébrés inférieurs (reptiles, Téléostéens). *Comp. Biochem. Physiol.* **26**, 997–1013.

LOCKLEY, A. S. (1957). Adrenal cortical hormones and osmotic stress in three species of fishes. *Copeia* 241.

MAETZ, J. (1964). Recherches sur la perméabilité aux électrolytes de la branchie des poissons et sa régulation endocrinienne. Extrait du *Bull. Infs scient. tech. Commt Energ. atom.* **86**, 11–70.

MAETZ, J. (1970). Mechanisms of salt and water transfer across membranes in relation to the aquatic environment in teleosts. *Mem. Soc. Endocr.* **18**, 3–28.

MAETZ, J., BOURGUET, J. & LAHLOU, B. (1964). Urophyse et osmorégulation chez *Carassius auratus*. *Gen. comp. Endocr.* **4**, 401–14.

MAETZ, J., BOURGUET, J., LAHLOU, B. & HOURDRY, J. (1964). Peptides neurohypo-physaires et osmorégulation chez *Carassius auratus*. *Gen. comp. Endocr.* **4**, 508–22.

MAETZ, J., MAYER, N., FORSTER, M. E. & CHAN, D. K. O. (1967). Axe hypophyso-interrénalien et osmorégulation chez les poissons d'eau de mer. *Gen. comp. Endocr.* **9**, 471.

MAETZ, J. & SKADHAUGE, E. (1968). Drinking rates and gill ionic turnover in rela-tion to external salinities in the eel. *Nature, Lond.* **217**, 371–3.

MAYER, N. & MAETZ, J. (1967). Axe hypophyso-interrénalien et osmorégulation chez l'Anguille en eau de mer. Etude de l'animal intact avec notes sur les perturbations de l'équilibre minéral produites par l'effet de 'choc'. *C. r. hebd. Séanc. Acad. Sci., Paris*, **264**, 1632–5.

MAYER, N., MAETZ, J., CHAN, D. K. O., FORSTER, M. & CHESTER JONES, I. (1967). Cortisol, a sodium excreting factor in the eel (*Anguilla anguilla* L.) adapted to sea water. *Nature, Lond.* **214**, 1118–20.

MEHDI, A., CARBALLEIRA, A. & BROWN, J. S. L. (1969). Steroid biosynthesis by frog and turtle interrenals. *Gen. comp. Endocr.* **13**, 497.

MIZOGAMI, S., OGURI, M., SOKABE, H. & NISHIMURA, H. (1968). Presence of renin in glomerular and aglomerular kidney of marine teleosts. *Am. J. Physiol.* **215**, 991–4.

MOTAIS, R. (1967). Les mécanismes d'échanges ionique branchiaux chez les Téléostéens. *Annls. Inst. Océanogr., Paris*, **45**, 1–83.

NANDI, J. (1967). Comparative endocrinology of steroid hormones in vertebrates. *Am. Zoologist* **7**, 115–33.

NANDI, J. & BERN, H. A. (1960). Corticosteroid production by interrenal tissue of teleost fishes. *Endocrinology*, **66**, 295–303.

NANDI, J. & BERN, H. A. (1965). Chromatography of corticosteroids from teleost fishes. *Gen. comp. Endocr.* **5**, 1–15.

OIDE, M. & UTIDA, S. (1967). Changes in water and ion transport in isolated intestines of the eel during salt adaptation and migration. *Marine Biol.* **1**, 102–6.

OLIVEREAU, M. & BALL, J. N. (1970). Pituitary influences on osmoregulation in teleosts. *Mem. Soc. Endocr.* **18**, 57–82.

OLIVEREAU, M. & OLIVEREAU, J. (1968). Effets de l'interrénalectomie sur la structure histologique de l'hypophyse et de quelques tissus de l'Anguille. *Z. Zellforsch. mikrosk. Anat.* **83**, 44–58.

PHILLIPS, J. G. (1959). Adrenocorticosteroids in fish. *J. Endocr.* **18**, xxxvii–xxxix.

PHILLIPS, J. G. & CHESTER JONES, I. (1957). The identity of adrenocortical steroids in lower vertebrates. *J. Endocr.* **16**, iii.

PHILLIPS, J. G., HOLMES, W. N. & BONDY, P. K. (1959). Adrenocorticosteroids in salmon plasma (*Oncorhynchus nerka*). *Endocrinology* **65**, 811–18.

PHILLIPS, J. G. & MULROW, P. J. (1959). Corticosteroid production *in vitro* by the interrenal tissue of the killifish, *Fundulus heteroclitus* (Linn). *Proc. Soc. exp. Biol. Med.* **101**, 262–4.

RICHARDS, B. D. & FROMM, P. O. (1969). Patterns of blood flow through filaments and lamellae of isolated-perfused rainbow trout (*Salmo gairdnerii*) gills. *Comp. Biochem. Physiol.* **29**, 1063–70.

ROBERTSON, O. H., KRUPP, M. A., FAVOUR, C. B., HANE, S. & THOMAS, S. F. (1961*a*). Physiological changes occurring in the blood of the pacific salmon, *Oncorhynchus tschawytscha*, accompanying sexual maturation and spawning. *Endocrinology* **68**, 733–46.

ROBERTSON, O. H., KRUPP, M. A., THOMAS, S. F., FAVOUR, C. B., HANE, S. & WEXLER, B. C. (1961*b*). Hyperadrenocorticism in spawning migratory and non-migratory rainbow trout (*Salmo gairdnerii*); comparison with Pacific salmon (*genus Oncorhynchus*). *Gen. comp. Endocr.* **1**, 473–84.

ROY, B. B. (1964). Production of corticosteroids *in vitro* in some Indian fishes with experimental, histological and biochemical studies of the adrenal cortex with general observations on gonads after hypophysectomy in *O. punctatus*. *Calcutta Med. J.* **61**, 223–44.

SANDOR, T. (1969). A comparative survey of steroids and steroidogenic pathways throughout the vertebrates. *Gen. comp. Endocr.* Suppl. 2, pp. 284–98.

SANDOR, T., HENDERSON, I. W. & CHESTER JONES, I. (1969). Biosynthesis *in vitro* of 18-hydroxycorticosterone from corticosterone by eel (*Anguilla anguilla* L.) adrenal mitochondria. *Gen. comp. Endocr.* **13**, 529.

SANDOR, T., HENDERSON, I. W., CHESTER JONES, I. & LANTHIER, A. (1967*a*). Biogenesis *in vitro* of corticosteroids by eel (*Anguilla anguilla* L.) adrenocortical preparations. *Gen. comp. Endocr.* **9**, 490.

SANDOR, T., LAMOUREUX, J. & LANTHIER, A. (1965). Adrenal cortical function in birds. The *in vitro* transformation of sodium acetate-1-^{14}C and cholesterol-4-^{14}C by adrenal gland preparations of the domestic duck (*Anas platyrhynchos*) and the goose (*Anser anser*). *Steroids* **6**, 143–57.

SANDOR, T., LANTHIER, A., HENDERSON, I. W. & CHESTER JONES, I. (1967*b*). Steroidogenesis *in vitro* by homogenates of adrenocortical tissue of the European eel (*Anguilla anguilla* L.). *Endocrinology* **81**, 904–12.

SANDOR, T., VINSON, G. P., CHESTER JONES, I., HENDERSON, I. W. & WHITEHOUSE, B. J. (1966). Biogenesis of corticosteroids in the European eel, *Anguilla anguilla* L. *J. Endocr.* **34**, 105–15.

SCHMIDT, P. J. & IDLER, D. R. (1962). Steroid hormones in the plasma of salmon at various stages of maturation. *Gen. comp. Endocr.* **2**, 204–14.

SEAL, U. S. & DOE, R. P. (1966). Vertebrate distribution of corticosteroid-binding globulin and some endocrine effects on concentration. *Steroids* **5**, 827–41.

SEXTON, A. W. (1955). Factors influencing the uptake of sodium against a diffusion gradient in the goldfish gill. *Diss. Abstr.* **15**, 2270–1.

SIDDIQI, M. A. (1966). Seasonal variation in total cholesterol content in different tissues of *Ophiocephalus punctatus*. *Indian J. exp. Biol.* **4**, 122–3.

SKADHAUGE, E. & MAETZ, J. (1967). Etude *in vitro* de l'absorption intestinale d'eau et d'électrolytes chez *Anguilla anguilla* adapté à des milieux diverses. *C. r. hebd. Séanc. Acad. Sci., Paris* **265**, 347–50.

SMITH, H. W. (1930). The absorption and excretion of water and salts by marine teleosts. *Am. J. Physiol.* **93**, 480–505.

SMITH, H. W. (1932). Evolution of fish kidneys. *Q. Rev. Biol.* **7**, 1–26.

SMITH, H. W. (1953). *From Fish to Philosopher.* Boston: Little Brown and Co.

SOKABE, H. (1968). Comparative physiology of the renin-angiotensin system. *J. Japan med. Ass.* **59**, 502–12.

SOKABE, H., MIZOGAMI, S. & SATO, A. (1968). Role of renin in adaptation to sea water in euryhaline fishes. *Japan. J. Pharmac.* **18**, 332–43.

STACHENKO, J. & GIROUD, C. J. P. (1959). Functional zonation of the adrenal cortex. I. Pathways of corticosteroid biogenesis. *Endocrinology* **64**, 730–42.

STANLEY, J. G. & FLEMING, W. R. (1964). The effect of ACTH and adrenal steroids on the sodium metabolism of *Fundulus kansae. Am. Zool.* **4**, 236.

STANLEY, J. G. & FLEMING, W. R. (1967). Effect of prolactin and ACTH on the serum and urine sodium levels of *Fundulus kansae. Comp. Biochem. Physiol.* **20**, 199–208.

STEEN, J. B. & KRUYSSE, A. (1964). The respiratory function of teleostean gills. *Comp. Biochem. Physiol.* **12**, 127–42.

SUNDARARAJ, B. I. & GOSWAMI, S. V. (1968). Effect of long term hypophysectomy on the ovary and interrenal of catfish *Heteropneustes fossilis* (Bloch). *J. exp. Zool.* **168**, 85–104.

TRUSCOTT, B. & IDLER, D. R. (1968). Biosynthesis of aldosterone and 18-hydroxy-corticosterone from corticosterone by interrenal tissue of a teleost fish. (*Clupea harengus harengus*). *J. Fish. Res. Bd Can.* **25**, 431–5.

TRUSTCOTT, B. & IDLER, D. R. (1969). Identification and quantification of aldo-sterone in blood of herring (*Clupea harengus harengus* L.). *Gen. comp. Endocr.* **13**, 535.

VAN OVERBEEKE, A. P. & AHSAN, S. N. (1966). ACTH effect of pituitary glands of Pacific salmon demonstrated in the hypophysectomized *Couesius plumbeus. Can. J. Zool.* **44**, 969–79.

WEDEMEYER, G. (1969). Stress-induced ascorbic acid depletion and cortisol pro-duction in two salmonid fishes. *Comp. Biochem. Physiol.* **29**, 1247–51.

DISCUSSION

QUAY: I was wondering whether you would care to comment on the possibility that the cortisol effect on electrolytes in the adrenalectomized eel might be, in part, due to a metabolic reconstitution, since it took a week, apparently, for the effect of cortisol to be shown.

CHAN: We have very little data to give us an answer to your question, but in some experiments where we have the adrenalectomized animal and a cannulated bladder for collection of urine there is a gradual decline of urine output from these animals. After say ten days when the urine volume is low you can start injecting cortisol and on the first day there is in fact no effect, but on the second day and third day the gradual rise in urine output brings it back to normal. Now on the third day if you stop injections the urine volume will stay up for another two days and then it will drop abruptly. Perhaps that answers your question.

BERN: When you mentioned that Forster found that the level of cortisol before and after transfer is the same, I wondered whether the real issue is what happens during the time of transfer.

FORSTER: We find levels of 2–3 μg% in both freshwater and seawater-adapted eels, using Murphy's competitive protein binding technique. This is after three weeks in either fresh water or sea water. There is indeed an elevation of plasma cortisol levels from 3 to 9 μg% two to three days after transfer from fresh water to sea water. This returns to normal by the eighth day.

MAETZ: I would like to know if there is a difference of half-life of cortisol in the eel adapted to sea water or fresh water.

FORSTER: Using a single isotope injection technique, we find a biological half-life for cortisol of 6·0 hr in fresh water and 5·3 hr in sea water. There is considerable individual variation and the difference between the two groups is not significant.

VAN OORDT: Dr Chan, you said that in the eel 17-hydroxylation of steroids is more important than 18-hydroxylation. It seems that in adult amphibians the reverse situation occurs. Do you know what situation exists in larval amphibia? Do they resemble the adults or rather freshwater teleosts with respect to corticosteroid formation?

CHAN: I did not work on this myself, but perhaps somebody in the audience might be able to answer your question.

IDLER: With regard to Professor van Oordt's question, the absence of significant 17-hydroxylation in amphibians may account for an apparently consistent occurrence of aldosterone in these animals in contrast to the irregular occurrence in some fishes.

Large blood samples collected from herring at different times have given inconsistent results for the occurrence of aldosterone. When aldosterone was present the levels were generally in the range reported for humans, but frequently no aldosterone was detected (i.e. less than 0·5 ng/100 ml). Limited samples of plasma from salmon, alewives and cod (full salinity and 50% salinity) have thus far failed to contain measurable aldosterone.

When we incubated corticosterone with herring head kidney tissue, aldosterone was readily isolated, but such was not the case when pregnenolone or progesterone were incubated. In other experiments the latter steroids were precursors of Reichsteins S and deoxycorticosterone was a precursor of corticosterone. Therefore corticosterone synthesis and its 17α-hydroxylation become important factors in deciding the degree of aldosterone synthesis in fish. Thus, because of 17-hydroxylation of corticosterone, some fish have a mechanism for regulating aldosterone synthesis which appears not to be significant in amphibians, adult rodents or many elasmobranchs.

SANDOR: In reply to Dr Idler, it is quite possible that the shunt operating in teleost adrenals between 17-deoxy- and 17-hydroxy-corticosteroids might operate earlier in the pathway than corticosterone. Thus, under most circumstances, pregnenolone and progesterone are already 17-hydroxylated, thus corticosterone is not formed at all and aldosterone formation is prevented. On the other hand, the 17-hydroxylation of corticosterone to cortisol does occur in teleosts.

IDLER: Yes this is correct. Certainly the earlier pathway, whether you go through Reichstein's S or 21-deoxycortisol, is going to affect the amount of corticosterone formed and therefore as Dr Sandor says this is going to influence the possibility for aldosterone synthesis. However, 17-hydroxylation of corticosterone remains as an apparently unique control on aldosterone synthesis in some fish.

IDLER: I wonder if I might just ask Mr Forster about the cortisol levels in the eel in fresh water and in salt water. Are these preliminary results or did you carry out a controlled experiment in which similar transfers were made, say, between fresh water and fresh water, in view of the studies of Leloup-Hatey in which some of these transfers did, in fact, bring about quite significant effects.

FORSTER: There are short-term effects, described by Leloup-Hatey as 'le choc osmotique'; an effect within a few hours, later falling. Ours were controlled experiments over two weeks. Control animals were transferred from fresh water to fresh water and there was a complementary experiment transferring fish from sea water to fresh water. The only change occurs on transfer from fresh water to sea water and there is a peak of cortisol levels at two to three days. This is associated with the time of maximum osmotic stress for the animal and occurs at the time of greatest change in several parameters, including body weight and plasma sodium. It coincides with the osmotic crisis of the animal transferred to sea water.

PITUITARY INFLUENCES ON OSMOREGULATION IN TELEOSTS

By MADELEINE OLIVEREAU and J. N. BALL

INTRODUCTION

The possibility of pituitary involvement in teleostean osmoregulation was first considered by Fontaine, Callamand & Olivereau (1949), working on the eel (*Anguilla anguilla*). At that time, as Callamand, Fontaine, Olivereau & Raffy (1951) emphasised: 'Le problème de l'intervention des glandes endocrines dans l'osmorégulation chez les Poissons est à peine posé.' Numerous papers published since then suggest that teleosts may be placed in two categories: one in which the pituitary appears to play no *essential* role in osmotic adjustments, the other in which the fish cannot live in fresh water without the gland. Virtually no information is available for non-teleostean fishes. One line of research, initiated by Dr G. E. Pickford's laboratory (see Pickford & Phillips, 1959), has demonstrated the role of some pituitary factor, a 'freshwater survival factor', which has been identified as a prolactin-like hormone. It is this factor that is essential to the second type of teleost for osmoregulation in freshwater (Ball & Olivereau, 1964).

In this review we shall consider changes in the pituitary when teleosts are transferred between sea water (SW), fresh water (FW) and distilled water (DW). The results of hypophysectomy in different media will be reviewed, and the effects of hormonal treatments. We shall in particular attempt to correlate the histological changes described in the hypothalamo-hypophysial system with concurrent changes in osmotic parameters.

EFFECTS OF SALINITY CHANGES ON THE PITUITARY

Fresh water to sea water

Effects on the prolactin cells

The presence of prolactin cells, located in the rostral pars distalis (pro-adenohypophysis) of the eel pituitary, was first suspected on a tinctorial basis by Olivereau & Herlant (1960). These cells contain numerous erythrosinophilic granules, similar to those in the mammalian prolactin cell. Experimental evidence in support of this functional identity of these eta cells has since been obtained, mainly in the eel, *Fundulus heteroclitus*, *Poecilia latipinna* and *P. formosa* (reviews, Olivereau, 1969a; Ball & Baker,

1969). In summary, the prolactin cells in FW fish are well granulated, with an abundant cytoplasmic RNA and all the classical signs of high activity. In those teleosts (e.g. salmonids, eel) with a follicular organization of the rostral pars distalis, the prolactin cells in FW form a tall columnar epithelium, with prominent nuclei. Ultrastructural studies on these cells in *Tilapia* have also demonstrated marked activity in FW fish (Dharmamba & Nishioka, 1968).

When fish are transferred from FW to SW, the prolactin cells become less active. In species with a follicular arrangement, the cell height is reduced and the follicular lumen often enlarges. In both European and Japanese eels, after 3–6 days in SW the cells are degranulated, nuclei and nucleoli smaller, and the nucleus lies more basally as the endoplasmic reticulum is reduced. The rostral pars distalis shrinks in volume and its blood supply diminishes (Olivereau, 1966*a*, and unpublished). In *Poecilia latipinna*, by 72 h after transfer from FW to ⅓ SW the prolactin cells and their nuclei are significantly smaller than in FW, with less pronounced nucleoli and Golgi images. These differences persist in fish adapted to the two media, and the volume of the rostral pars distalis is less in ⅓ SW than in FW. During the 72 h adaptation period in ⅓ SW, plasma sodium shows a transitory increase at 24 h, then drops to levels typical of fish adapted to ⅓ SW (higher than in FW). The prolactin cells appear even less active in full SW (Ball, unpublished). The electron microscope shows that in *Tilapia* the secretory granules are smaller and more irregularly shaped after 21 days in SW, and both Golgi system and endoplasmic reticulum are reduced (Dharmamba & Nishioka, 1968). A similar reduction in granule size was found in eels caught at sea (Knowles & Vollrath, 1966).

Thus transfer to SW induces hypoactivity of the prolactin cells, though not necessarily complete inactivity. After 21 months in SW, the prolactin cells of steelhead trout are poorly granulated, but still exhibit some mitotic activity, and atrophy of this follicular region in SW is less pronounced in trout than in the eel (Olivereau, unpublished).

Effects on the corticotrophic (ACTH) cells

In *Salmo gairdneri* kept in SW, the ACTH cells are generally larger and better granulated than in FW. This correlates with the transitory stimulation of the interrenal, demonstrated histologically (Olivereau, 1962) and biochemically (Leloup-Hatey, 1964) in *Salmo salar* smolts.

In the silver eel, Hanke, Bergerhoff & Chan (1967) reported that ACTH cell nuclei increased in size after 4 weeks in SW, but, surprisingly, the interrenal nuclei concurrently *decreased* in size. Similar trends, though not statistically significant, were found in yellow eels. It is difficult to understand this apparent inhibition of the interrenal in the presence of increased activity of the ACTH cells. No change in the granulation of the ACTH cells was observed. It is perhaps more informative to study such changes as a function of time after transfer to SW. In such a study on male European silver eels, the ACTH cells were initially stimulated, with partial de-granulation; a slow adaptation occurred later. The interrenal was also transitorily stimulated, showing larger nuclei and cells (Olivereau, un-published). However, after 3 days in SW, these reactions were no longer discernible (Olivereau, 1966b). These findings are to be correlated with those of Hirano & Utida (1968), who found a transitory increase in plasma cortisol levels during the first 24 h after Japanese eels were transferred to SW, and with the unpublished results of M. E. Forster (1969), showing elevated plasma cortisol levels during the first 2–4 days after the European silver or yellow eel enters SW. However, Leloup-Hatey (1964) found no increase in cortisol levels in transferred European female silver eels.

Effects on the thyrotrophic (TSH) cells

In aquarium eels, the TSH cells often stain more intensely (i.e. are better-granulated) in SW than in FW, a feature also noted by Knowles & Vollrath (1966) in eels caught in the sea. However, SW induces no clear histological changes in the eel thyroid. Possibly the high iodine content of SW (about 60 μg/l), which reduces iodine uptake and alters iodine meta-bolism in various teleosts, without clear effects on thyroid histology (Oli-vereau, 1955; Hickman, 1959; Leloup and Fontaine, 1960), may act on the pituitary and slightly reduce the rate of TSH secretion.

The activation of the TSH–thyroid axis during smoltification in *Salmo salar* (Fontaine, Leloup & Olivereau, 1952) may be involved in the associ-ated increase in osmoregulatory capacity. The TSH cells become activated and degranulated just before the migration to SW, and regranulate if the fish are kept in SW for several days (Olivereau, 1954).

Effects on the growth hormone (GH) cells

Eels placed in SW display a slight reduction in the activity of the GH cells, an effect more pronounced in cultured Japanese eels than in the European eel, and more marked after 48 days than at 16 days (Olivereau, unpublished). It is difficult to relate this response to osmoregulation. Chartier (1959) demonstrated some osmoregulatory effects of GH in intact trout, in particular a retention of muscle potassium. Possibly in SW, rich in potassium, such a potassium-retaining action of GH would be less essential than in FW.

Migratory salmonids have a phase of rapid growth in the sea. Studies on wild Pacific salmon (*Oncorhynchus tschawytscha*) showed that the GH cells are very active during this marine growth phase (Olivereau & Ridgway, 1962); in contrast, the silver eel migrates to the sea having probably completed its growth, but GH secretion can be stimulated in male silver eels by prolonged starvation (Olivereau, unpublished). These observations most probably relate to the metabolic actions of GH, and an osmoregulatory role for this hormone in SW has not been established.

Effects on the gonadotrophic cells

Gonadotrophins have not been shown definitely to influence osmoregulation. It is true that silver eels, with their better-developed gonads, are more euryhaline than yellow eels; but this increased osmoregulatory capacity of silver eels probably results from simultaneous changes in other endocrine organs and in the gills and perhaps the kidney, rather than from gonad development. The gonadotrophs are scarcely more differentiated or active in the silver eel than in the yellow, in contrast to the marked activation of these cells in eels artificially brought to sexual maturity (Olivereau, 1967 a).

Similarly, adjustment to SW is much more efficient in the salmon smolt (migratory form) than in the parr (sedentary FW form); but about 60% of the male parr become paedogenetically sexually mature in December, and then the following March–April they become smoltified and osmotically tolerant. At this time the testes and gonadotrophs are usually completely regressed (Olivereau, 1954), although a few smolts may have mature testes (M. Fontaine, personal communication).

Effects on the pars intermedia

The teleostean pars intermedia contains two cell-types. One stains with lead-haematoxylin (PbH), and in the eel this may secrete intermedin (MSH) (Olivereau, 1969 b, c); its nuclear diameter is not altered by salinity changes. The other cell-type, generally less abundant, is periodic acid–Schiff positive (PAS +ve), and smaller and less numerous in SW eels than in FW animals (Olivereau & Olivereau, 1968). The function of this PAS +ve cell is uncertain, but recent observations on eels and the medaka (*Oryzias latipes*) suggest some osmoregulatory role.

When European silver eels are transferred to SW, the PAS +ve cells progressively show nuclear shrinkage, slow degranulation and atrophy (Olivereau, 1969 c). Preliminary results of Kawashima & Hirano (1968) indicated a similar response in cultured Japanese eels; subsequently, a more detailed study of these Japanese fish has confirmed that during periods of 16 or 48 days in SW, the PAS +ve cells display nuclear shrinkage, decrease in cell size, and degranulation (Plate 1, c, d), so that the cells become difficult to distinguish (Olivereau, 1969c). In *Oryzias* transferred to SW, Kawashima & Hirano (1968) noted a progressive decrease in the number and PAS-stainability of these cells, but the nuclear diameter increased, in contrast to the eel.

According to Knowles & Vollrath (1966), neurohypophysial fibres do not actually contact the pars intermedia cells in the eel, but terminate on extravascular channels not visible with the light microscope and lying in the connective tissue bordering the pars intermedia. It is perhaps significant that the granules in both cell-types are always located close to this connective tissue border between pars intermedia and neurohypophysis, and it is possible that the granules are discharged into the extravascular space when the cells are stimulated. Furthermore, when the PAS +ve cells are inactive, as in SW, their contact with the connective tissue border is lost (Plate 1, d). The physiological significance of this reaction remains uncertain (Olivereau, 1969b, c).

A suggestive comparison has been made between Japanese and European eels kept in FW in Tokyo and Paris respectively. Amongst several differences in the pituitary, one of the most striking is that the pars intermedia PAS +ve cells are more numerous and much larger in the Japanese animals, with larger nuclei (Plate 1, a, c). The electrolyte composition of FW is approximately identical in Tokyo and in Paris, except that calcium content is about 5 times higher in Paris. This raises the possibility that the PAS +ve cells may be concerned in calcium regulation, a hypothesis that merits investigation (Olivereau, 1969c), and which is supported by observations on European eels kept in de-ionized water (Olivereau, 1967b).

Effects on the hypothalamo-neurohypophysial system

Many workers have studied this system in relation to external salinity. The main references are Arvy, Fontaine & Gabe (1959) and Schiebler & Hartmann (1963), both on the eel. In *Ammodytes lanceolatus* and *Callionymus lyra* placed in hypertonic SW (150 parts per thousand), the nucleus preopticus (NPO) is completely depleted of neurosecretory material within 30 min (Arvy & Gabe, 1954), while in *Oryzias* after a progressive transfer to $\frac{1}{2}$ SW the neurohypophysis shows a transitory decrease of neurosecretory material (Kawashimo & Hirano, 1968).

The literature has recently been reviewed (Ball & Baker, 1969) and discussed in the light of new observations on eels (Olivereau, 1969c). During adaptation to SW, there is a rapid depletion of neurosecretory material in the eel NPO and neurohypophysis; repletion of the material occurs rapidly in the NPO but more slowly in the neurohypophysis

(Olivereau, 1969 c). This proximo-distal migration of hypothalamic neuro-secretory matter was not, however, detected by *in toto* studies with the Braak technique in eels taken after 1, 4 and 24 h in SW (Leatherland, 1967). Leray (1963) found some uptake of radio-cysteine in the NPO of (probably) brackish-water *Mugil*, but no evidence of migration of the active material to the neurohypophysis. In FW eels injected with radio-cysteine or serine, the NPO showed maximal radioactivity between 5 h and 5 days; no activity was present at 6 days, but it reappeared at 8 and 12 days. Neurohypophysial labelling appeared on the 2nd day, was maximal on the 3rd and disappeared by the 5th day. Eels adapted for 2–4 weeks to SW showed labelling only irregularly in the NPO (maximum at 1 day), and neurohypophysial activity appeared in only 1 animal, at 6 days. Eels transferred to SW, and injected with radio-cysteine at transfer, showed NPO labelling between 5 h and 5 days (maximum, 17 h); neurohypophysial labelling appeared after 5 h and was maximal at 4 days. Extreme variability makes difficult the evaluation of the physiological meaning of these findings (Leatherland, 1967).

Neurohypophysial bioassay data is available for trout during transfer to SW. Transfer of FW *Salmo gairdneri* to 60% SW was accompanied by a significant decline in pituitary antidiuretic activity (possibly arginine vasotocin) during the first 3 h, with return to normal levels during the next 3 h. The level of pituitary oxytocic activity simultaneously increased and then returned to control levels. Total hypothalamic oxytocic activity did not change significantly during this period, but it is possible that hypothalamic levels of a chromatographically distinct oxytocic component increased during the first few hours (Carlson & Holmes, 1962). After 2 h in full SW, the arginine vasotocin content of *S. irideus* pituitary fell by 50%, but the ichthyotocin (isotocin) content did not change appreciably. After 4–8 h in SW, the pituitary content of both neurohypophysial hormones was similar to that in FW controls. After 2 h in SW there was depletion of osmiophilic material in the elementary granules in some neurohypophysial fibre terminals, associated with increased abundance of elementary granules in the NPO. After 4–8 h, the ultrastructure of the neurohypophysis and NPO resembled that of FW controls (Lederis, 1963, 1964).

When goldfish were transferred to saline (0·85 %) the pituitary natriferic activity decreased on the first day and remained below FW levels for a week. Conversely, hypothalamic activity increased after 1 day and then declined below FW levels after 8 days (Favre, 1960).

These bioassay data agree in general with the histological findings, although it would seem that the return of hormone content to normal levels is more rapid in salmonids than the repletion of neurosecretory material in the eel. However, the usual 'neurosecretion' stains probably do not stain the actual neurohypophysial hormones but rather the carrier proteins (neurophysins).

Sea water to fresh water

Effects on the prolactin cells

Activation of these cells on entering FW was first reported by von Hagen (1936), who studied the eel in the elver stage, when it enters estuaries and ascends rivers. Vollrath (1966) confirmed these changes in the elver, and similar activation changes occur in the yellow eel transferred from SW to FW (Olivereau & Olivereau, 1968).

Changes in the salmon (*Salmo salar*) prolactin cells during migration from estuaries into rivers have been described (Fontaine & Olivereau, 1949; Olivereau, 1954). In the estuary during April the cells are active and proliferating; a few weeks later, in the rivers, the cells are less stimulated, and they become relatively inactive during the spawning period (December).

In *Fundulus heteroclitus* the prolactin cells are more numerous and active in FW than in SW (Ball & Pickford, 1964; Emmart, Pickford & Wilhelmi, 1966), and the same is true of *Poecilia latipinna* (Olivereau & Ball, 1964). The activation of the prolactin cells during transfer of *P. latipinna* from $\frac{1}{3}$ SW to FW can be correlated with changes in plasma sodium levels. After intact fish enter FW, plasma sodium falls to its lowest level at 18 h, and then rises so that by 72 h it is in the normal FW range (Ball, 1969a); this restoration of plasma sodium depends on secretion of prolactin by the pituitary (Ball & Ensor, 1967). The prolactin cells show signs of progressive activation during this period (Ball, 1969a). At 18 h the cells are degranulated, suggesting that the output of hormone from the cells has increased relatively more than synthesis, but by 72 h the granulation is returning, indicating that synthesis has caught up with output, as it were, allowing some of the product once again to be stored as visible granules; by this time, the pituitary prolactin content is double its $\frac{1}{3}$ SW level (Ensor & Ball, 1968a).

In mullets (*Mugil auratus*) collected in different coastal environments, the volume of the prolactin cell zone increases as external salinity decreases, with prolactin cell nuclear diameters increasing in parallel. When mullets are gradually transferred to FW from SW over a period of 7 days, the nuclear size progressively increases, the synthesis of cytoplasmic RNA increases, and it seems that granule discharge is accelerated, no storage being detectable when the fish have been in FW for 24 h (Olivereau, 1968).

These, and other data (reviews, Olivereau, 1969a; Ball & Baker, 1969) show that in fish transferred to FW the prolactin cells are rapidly activated, with cellular, nuclear and nucleolar hypertrophy, increase in cytoplasmic RNA, hyperaemia and sometimes increased mitotic activity. Studies with the electron microscope in *Tilapia mossambica* have confirmed these findings at the fine structural level (Dharmamba & Nishioka, 1968).

Effects on the ACTH cells

In *Poecilia* these cells are not obviously altered on transfer to FW (Ball, unpublished), but in *Mugil auratus* the endoplasmic reticulum enlarges, with an increase of RNA in the basal part of the cell; nuclei and nucleoli are also larger in fish collected in coastal lagoons with lowered salinity (Olivereau, 1968).

Effects on the pars intermedia

The PAS +ve cells are stimulated in *Mugil* adapted to FW or brackish waters, with increased RNA in the apical zone (Olivereau, 1968).

Effects on the hypothalamo-neurohypophysial system

Recent reviews (Olivereau, 1967*b*; Ball & Baker, 1969) have dealt with the relevant data, which are often conflicting. For example, Korn (1960) found a reduction of stainable material in the neurohypophysis of *Mugil* trans-ferred from SW to brackish water, with degenerative changes in this lobe, whereas Olivereau (1968) observed no such changes during the progressive transfer of *Mugil* to FW.

Gobius displayed more neurosecretory material in the neurohypophysis when kept in dilute SW than in full SW, together with degeneration of some hypothalamic neurones (Korn, 1960). In *Ammodytes lanceolatus* and *Callionymus lyra*, neurosecretory material accumulated in both NPO and neurohypophysis $2\frac{1}{2}$ h after transfer to $\frac{1}{2}$ SW or $\frac{3}{4}$ SW from full SW (Arvy & Gabe, 1954). One hour after *Chromis chromis* is transferred from SW to dilute SW, neurosecretory material disappears from both neurohypophysis and hyperactive NPO; it reappears 4 h later, and after 8 h is more abundant than in SW (Srebo, 1963). In contrast, the total amount of neurosecretory material in the NPO-neurohypophysial system of the eel was unchanged 1, 4 and 24 h after a SW–FW transfer. The maximal uptake of [³⁵S]cysteine in the NPO was at 3 h, much earlier than in SW, but no labelling appeared in the neurohypophysis (Leatherland, 1967). Deminatti (1969) found no labelling in the neurohypophysis of FW *Carassius* and *Poecilia sphenops* after radio-methionine injection.

It is difficult to generalize about this field. Probably some of the diver-gent findings reflect species differences, perhaps correlated with degree of steno- or euryhalinity, while some of the discrepancies may stem from the different experimental procedures employed: a progressive transfer will more closely mimic natural changes than a direct transfer, and will probably induce a more prolonged, but perhaps less intense, stimulation of the hypothalamo-neurohypophysial system.

There seems to be no information about changes in biological activities in the system during SW–FW transfer. Details of the electrolyte regulative processes during transfers in both directions have been recently reviewed by Maetz (1968, 1970; Motais, 1967).

Fresh water to de-ionized water

Adenohypophysial reactions to this transfer have so far been studied only in the European eel.

Effects on the prolactin cells

Surprisingly, these cells are regressed and chromophobic after long-term (35–46 days) adaptation to de-ionized or distilled water (DW), looking similar to the regressed cells in SW eels (Olivereau, 1967b).

Effects on the TSH cells

Despite the low iodine content of DW, no typical and constant response appears in these cells, although the thyroid may appear quite active (Olivereau, 1967b and unpublished).

Effects on the ACTH cells

In male silver eels, the ACTH cells are slightly stimulated and partially degranulated in DW, displaying a transitory mitotic activity; these changes correlate with cellular hypertrophy in the interrenal, though this is not accompanied by nuclear enlargement (Olivereau, 1966c). In yellow eels adapted to DW for 4 weeks, the ACTH cells display a slight decrease in nuclear diameter, but remain well granulated with larger nucleoli than FW controls (Hanke et al. 1967); in this experiment the animals received ten injections of 0·5 ml 0·9% saline, which may have interfered with the processes of 'demineralization' in DW.

Effects on the GH cells

The GH cells are strongly stimulated in DW, well granulated with cellular, nuclear and nucleolar hypertrophy and marked development of the endoplasmic reticulum (Olivereau, 1967b). This reaction may be related to the potassium retention induced in trout by GH (Chartier, 1959).

5

Effects on the pars intermedia (Plate 1, a, b)

The PAS +ve cells rapidly enlarge and multiply, and their cytoplasm becomes degranulated. After a month or more in DW this cell-type predominates in the pars intermedia, which is much larger than in FW controls (Olivereau, 1967b).

Effects on the hypothalamo-neurohypophysial system

In the eel kept in DW, the neurohypophysis becomes smaller, with narrower ramifications into the adenohypophysis. Neurosecretory material increases initially but then decreases. The NPO does not appear stimulated (Olivereau, 1967b). Similarly, stainable material increased after short-term exposure to DW ($\leqslant$ 48 h) in the NPO and neurohypophysis of the stickleback, minnow (Tuurala, 1957) and *Misgurnus fossilis* (Szabó & Molnár, 1965). However, after 6 h exposure the nuclei in the parvocellular area of the minnow NPO become smaller, with decreased cytoplasmic granulation (Szabó, 1965).

Goldfish kept in DW display increased pituitary natriferic activity by the 11th day, but the activity in the hypothalamus remains normal (Favre, 1960).

Direct data on the effects of DW on electrolyte metabolism come mainly from the eel. Intact or hypophysectomized eels in DW display reduced plasma levels of sodium and calcium (Chester Jones, Henderson & Butler, 1965) and elevated plasma potassium (Olivereau, 1966c), this rise in potassium appearing more rapidly in hypophysectomized animals (Olivereau & Chartier, 1966). There is a fall in muscle sodium and potassium (Chester Jones & Henderson, 1965), and an increased diuresis (Chester Jones *et al.* 1965). At the gills, sodium influx and potassium outflux are both increased when the animals are tested on return to tap water (Henderson & Chester Jones, 1967), and sodium influx is similarly stimulated in *Carassius* (references in Olivereau, 1967b).

The correlations between the electrolyte data and pituitary histophysiological changes in response to DW have been discussed elsewhere (Olivereau, 1967b).

EFFECTS OF HYPOPHYSECTOMY

In sea water

In the eel and *Fundulus heteroclitus*, hypophysectomy reduces the rate of sodium turnover in SW; this deficiency is not corrected by prolactin injections, but ACTH augments the turnover rate in both cases (Maetz, Sawyer, Pickford & Mayer, 1967c; Maetz, Mayer & Chartier-Baraduc,

1967*a*; Maetz, Mayer, Forster & Chan, 1967*b*). *Poecilia latipinna* behaves rather differently: hypophysectomy in SW produces a rise in plasma sodium after 7 days (Ball & Ensor, 1969), but even at 10 days there is no detectable change in the sodium turnover rate (Ball, unpublished).

Sodium–potassium-activated ATPase is implicated in the sodium-extrusion mechanism of the gill in SW (see Maetz, 1968, 1970); hypophysectomy reduces the level of this enzyme in the gill of SW *F. heteroclitus* (Epstein, Katz & Pickford, 1967), though not of SW Japanese eels (Utida, Hirano, Oide, Kamiya, Saisyu & Oide, 1966).

When Japanese eels are transferred from FW to SW there is an increase in the rate of water and sodium transport across the wall of the intestine, and this increase is abolished by hypophysectomy (Hirano & Utida, 1968).

In dilute sea water

In a medium adjusted to have approximately the same sodium concentration as the fish plasma, hypophysectomy elevates the sodium turnover rate, although plasma sodium levels are not altered (Ball & Ensor, 1969).

In fresh water

The effects of hypophysectomy in various species on sodium fluxes, blood and urine electrolytes and osmolality, and on FW survival have been reviewed recently (Olivereau & Lemoine, 1969; Ball, 1969*a*, *b*; Ball & Ensor, 1969). Hypophysectomy may lead to a rapid fall in plasma electrolytes and an increase in sodium outflux, resulting in death within 24–48 h in *Poecilia latipinna* (Ball & Ensor, 1969), or within a few days in other atheriniform fish (Schreibman & Kallman, 1966, 1968). Other teleosts (e.g. *Carassius*) tolerate FW for much longer periods after hypophysectomy— for example 18 months at least in the case of *Anguilla anguilla* (Olivereau & Fontaine, 1965)—in which the fall in plasma electrolytes is very slow (Fontaine, 1956; Olivereau & Chartier, 1966), although hypophysectomy of the eel in FW increases sodium outflux (Maetz *et al.* 1967*a*) as in *Fundulus heteroclitus* (Maetz *et al.* 1967*c*) and *P. latipinna* (Ensor & Ball, 1968*b*).

Recently, Lahlou & Sawyer (1969*a*) demonstrated that hypophysectomized goldfish (*Carassius*) also show a sharp drop in plasma sodium and chloride levels (33% or more) and of total exchangeable sodium (20%). Extrarenal sodium outflux is greatly increased, and since water permeability is actually slightly decreased by hypophysectomy (Lahlou & Sawyer, 1969*b*), demineralization of the operated fish is to be attributed to the increased rate of sodium loss (influx being unchanged) rather than to hydration. Donaldson, Yamazaki & Clarke (1968) also showed that hypophysectomized goldfish suffer an exaggerated fall in plasma osmolarity when transferred from dilute saline to FW. Thus the survival of hypophysectomized goldfish in FW for long periods must depend on a particular resistance of the tissues to dilution of the internal medium.

In de-ionized water

Hypophysectomized European eels can survive in DW for at least 40 days, although asthenia may appear over longer periods (Olivereau, 1967b). As in intact eels, there is a slow decline in plasma sodium and calcium, and an increase in plasma potassium (Olivereau & Chartier, 1966, and unpublished). In the same species, hypophysectomy after 2 weeks in DW results in reduced urine output and increased renal loss of potassium (Chester Jones *et al.* 1965); sodium uptake by the gills is reduced, and the net extrarenal loss of potassium is either unchanged (Henderson & Chester Jones, 1967) or increased (Chester Jones *et al.* 1965).

EFFECTS OF HORMONES

Pituitary hormones, being peptidic in nature, display to a marked degree the phenomenon of phylogenetic specificity (Fontaine, 1969). For this reason, we shall distinguish between experiments in which exogenous (mammalian) hormones have been given to fishes, and those in which the selective release of an endogenous (fish) hormone has been induced in relation to osmotic adjustment.

Exogenous hormones

Prolactin

(a) *Effects on freshwater survival and electrolytes.* Purified ovine prolactin is able to promote FW survival in those hypophysectomized cyprinodonts which otherwise cannot tolerate this medium. This was first demonstrated for *Fundulus heteroclitus* (Pickford & Phillips, 1959), and is a specific effect which cannot be duplicated by any other pituitary hormone (Pickford, Robertson & Sawyer, 1965), apart from primate growth hormone, which is known to possess intrinsic prolactin activity. A similar specific effect of ovine prolactin on FW survival and sodium retention has been established for *Poecilia latipinna*: a single injection of ovine prolactin prevents the fall in plasma sodium that occurs when fish are hypophysectomized in FW (Ball & Ensor, 1965) or when hypophysectomized fish are transferred from $\frac{1}{3}$ SW to FW (Ensor & Ball, 1968a). In both cases, the maintained level of plasma sodium is proportional to the dose of prolactin. In the transfer experiments, the associated drop in plasma chloride was also prevented by prolactin (Ball & Ensor, 1969).

A single injection of prolactin reduces sodium outflux to normal in hypophysectomized *P. latipinna* transferred to FW, but cortisol is not effective (Ball & Ensor, 1969). In isotonic $\frac{1}{3}$ SW, chronic treatment of hypophysectomized fish with prolactin maintains the normal low sodium turnover rate (Ball & Ensor, 1969). Prolactin had previously been shown to reduce the excessive sodium outflux in hypophysectomized *F. heteroclitus* and *Anguilla* (Potts & Evans, 1966; Maetz *et al.* 1967*a, c*); in all these cases, prolactin had no action on sodium influx.

Ovine prolactin attenuates the slow decrease in plasma sodium in the hypophysectomized FW eel (Chartier & Olivereau, 1965). This protective action of prolactin is more strikingly demonstrated if the eels are kept in DW, in which medium the fall in plasma sodium and calcium is accelerated. However, prolactin is effective only when injections are started immediately after hypophysectomy ('maintenance' treatment) and it has no effect if given several weeks after hypophysectomy ('repair' treatment). Maintenance treatment for 11 (DW) or 15 (FW) days prevents the fall in sodium and calcium, but enhances the increase in plasma potassium (Olivereau & Chartier, 1966).

Hypophysectomized *Tilapia mossambica* show a marked decrease in plasma osmotic pressure when transferred from SW to FW, and this is minimized by a single injection of prolactin given 24 h before transfer (Dharmamba, 1968).

Pertinent reviews are available dealing with the numerous papers published over the past few years reporting effects of mammalian prolactin on sodium fluxes, plasma and urine electrolytes and osmolality, survival, etc., in various hypophysectomized fish in FW (Olivereau & Lemoine, 1969; Ball, 1969*a, b*). Recently, Lahlou & Sawyer (1969*a*) showed that repeated injections of ovine prolactin, starting immediately after the operation, partially prevented the decline in plasma electrolyte concentrations in hypophysectomized FW goldfish, and Donaldson *et al.* (1968) found that pretreatment with ovine prolactin on a salmon pituitary 'prolactin' fraction largely prevented the fall in plasma osmolality that occurs when hypophysectomized goldfish are transferred from dilute saline to FW.

An interesting case is the intact stickleback, *Gasterosteus aculeatus* form *trachurus*, which is stenohaline during the winter but euryhaline during the summer. If this marine form is placed in FW during the autumn or early winter, it does not survive (Smith, 1962); the mortality of 50% within 3–5 days is attributed to failure of the pituitary to secrete fish prolactin, the animals being considered 'physiologically hypophysectomized' although inactivation of the prolactin cells in winter fish has not yet been verified histologically. Ovine prolactin markedly improved FW tolerance in winter fish, mortality being less than 50% even after 9 days (Lam & Leatherland, 1969a). When intact early-winter marine sticklebacks are transferred from SW to FW, plasma sodium and chloride levels rapidly fall and the urine is only slightly hypo-osmotic. A single injection of ovine prolactin 24 h before transfer reduces this fall in plasma electrolytes, though having no effect on plasma potassium (Lam, 1968). Similar data were obtained by measuring plasma osmolality; again, prolactin reduced the drop of plasma osmolality in transferred fish and enabled the fish to excrete a highly hypo-osmotic urine (Lam & Hoar, 1967). Late-spring fish, control-injected with the prolactin solvent, were transferred to FW and showed changes in plasma and urine osmolalities essentially the same as those found in the prolactin-injected winter fish, and prolactin had no significant action on transferred late-spring fish. The data suggest a seasonal increase in fish prolactin secretion, associated with the spring migration from SW to FW (Lam & Hoar, 1967). More recently it was found that winter sticklebacks can live for only 12 h in DW, and that this survival can be prolonged by very low doses of ovine prolactin. High doses ($\geqslant$ 30 μg/g) seem less effective than intermediate doses (Leatherland & Lam, 1969).

(b) *Effects in sea water.* Stanley & Fleming (1967) found that prolactin promotes sodium retention in SW-adapted *Fundulus kansae*, whether intact or hypophysectomized. Thus prolactin seems to be 'toxic' to *F. kansae* in SW (Fleming, 1968), though not to SW *F. heteroclitus* (Pickford *et al.* 1965). Unlike ACTH, prolactin does not increase the reduced sodium turnover rate in hypophysectomized *Anguilla* and *F. heteroclitus* in SW (Maetz, *et al.* 1967a, b, c); in fact prolactin has recently been found to *reduce* the high rate of sodium pumping in the gill and of water transport in the intestine of the SW eel, in opposition to ACTH (see Maetz, 1969; Utida, Hirano & Kamiya, 1969).

(c) *Effects on the gills.* The reduction by prolactin of the excessive sodium outflux in hypophysectomized FW fish minimizes the negative sodium balance, so conserving sodium and permitting survival. For *Anguilla* and *F. heteroclitus* this effect is largely extra-renal, presumably on the gills (Potts & Evans, 1966; Maetz *et al.* 1967a, c). In *Gasterosteus* in FW, preliminary studies showed that prolactin reduces extra-renal sodium loss (Lam, 1968), an effect which may be related to the stimulatory action of prolactin on the gill mucous cells (Leatherland & Lam, 1969).

(d) *Effects on mucous cells.* Egami & Ishii (1962) were the first to demonstrate a stimulatory action of prolactin on epidermal mucous cells, in intact *Symphysodon.* Similar results have been obtained in various intact teleosts, and also a thickening of the skin in cichlids (Blüm & Fiedler, 1965; Blüm, 1966: Fiedler, Osewold & Blüm, cited by Blüm, 1968; Mattheij, Sprangers & van Oordt, 1969). In contrast, no such effect could be demonstrated in *Pterophyllum* (Blüm & Fiedler, 1965) and *Tilapia* (Bern, 1967), nor, for gill mucous cells, in *Poecilia latipinna* (Ball, unpublished).

In correlation, hypophysectomy causes a reduction in number of skin mucous cells in some teleosts (Schreibman & Kallman, 1965; Donaldson & McBride, 1967), with a simultaneous reduction in cell size in the goldfish (Ogawa & Johansen, 1967). There is indirect evidence for such an effect in the eel (Olivereau & Lemoine, unpublished). Again, this could not be demonstrated in *Tilapia* (Bern, 1967; Bern & Nicoll, 1968) and the gills of *P. latipinna* (Ball, unpublished).

(*e*) *Effects on water metabolism.* While prolactin does not enhance water transfer across the isolated eel intestine (Hirano & Utida, 1968), it increases diuresis in *Fundulus kansae* (Stanley & Fleming, 1966), the glomerular filtration rate being reduced after hypophysectomy (Stanley & Fleming, 1967), as in the eel (Chester Jones *et al.* 1965). Prolactin also increases branchial water permeability in the goldfish (see Maetz, 1970), and bovine prolactin corrects the hyperhydration of the muscles in the hypophysectomized FW eel (Chan, Chester Jones & Mosley, 1968).

(*f*) *Effects on the kidney.* In the intact eel, ovine prolactin stimulates the renal tubule, inducing significant increases in cell height and nuclear diameter, in width of the tubules and of the lumen, and in mitotic activity in various segments of the tubule; the greatest changes occur in the initial collecting segment. Numerous 'buds' form close to the initial collecting segment and differentiate to form new active tubules (Olivereau & Lemoine, 1968).

Kidney histology is not much altered in the eel several weeks after hypophysectomy (Olivereau, 1965); however, the tubule nuclei are significantly reduced. Ovine prolactin given to hypophysectomized eels increases cell height and tubule diameter, but does not restore nuclear size. The responsiveness of the initial collecting segment appears to be reduced by hypophysectomy, but prolactin nevertheless does induce 'bud' formation as in intact fish. Probably some pituitary factor(s) other than prolactin must be present to maintain full responsiveness to prolactin (Olivereau, & Lemoine, 1969).

In the light of these observations, it is interesting that Blüm (1968) found that injected ovine prolactin is still present in the kidney of *Tilapia mariae* up to 10 h after it has disappeared from the blood.

In the late-spring marine *Gasterosteus*, transfer from SW to FW induced a slight increase in size of Bowman's capsule and glomerular tuft; a lesser response was observed in late-autumn fish. This was interpreted as indicating a higher output of prolactin in the former group (Ogawa, 1968), although it is not clear why the autumn fish should respond at all, on the hypothesis that they are 'physiologically hypophysectomized' as regards prolactin (Lam & Hoar, 1967).

Subsequent experiments with late-autumn and early-winter sticklebacks showed that prolactin injections in SW *reduce* the size of Bowman's capsule and the glomerular tuft. Transfer of control fish to FW also reduces the sizes of these structures, in contrast to Ogawa's finding (Lam & Leatherland, 1969 b). Prolactin-treated fish were transferred from SW to FW. Expressing the mean glomerular size in prolactin- and solvent-treated fish as percentages of the initial values in SW, it appears that prolactin enlarges the glomeruli in FW progressively up to 7 days, whereas in the solvent-controls the glomeruli at first shrank and were restored to their original diameter only by day 4. Prolactin also reduces the intracapsular space, and apparently increases the glomerular filtration rate (Lam & Leatherland, 1969 b), and may reduce renal electrolyte loss (Lam & Hoar, 1967).

ACTH

FW survival of hypophysectomized *Gambusia* can be promoted with ACTH as well as prolactin (Chambolle, 1966, 1967). In *Anguilla japonica* ACTH, like cortisol, increases intestinal water transport, which is reduced after hypophysectomy (Hirano & Utida, 1968). However, ACTH has no effect on urine and plasma sodium levels in hypophysectomized *Fundulus kansae* (Stanley & Fleming, 1967), or plasma levels in *Poecilia latipinna* (Ball & Ensor, 1967), nor does it reduce the elevated sodium turnover rate in hypophysectomized *P. latipinna* in $\frac{1}{3}$ SW, although it clearly stimulates the interrenal (Ball & Ensor, 1969). On the other hand, in the eel, cortisol augments the sodium turnover rate in SW, which is reduced by inter-renalectomy or hypophysectomy, and ACTH also augments sodium exchanges in hypophysectomized eels and *Fundulus heteroclitus* in SW (Maetz *et al.* 1967 a, b, c; Mayer, Maetz, Chan, Forster & Chester Jones, 1967).

TSH, growth hormone, gonadotrophins and alpha-MSH

These adenohypophysial hormones were unable to promote FW survival in hypophysectomized *Fundulus heteroclitus* (Pickford *et al.* 1965) and *Xiphophorus maculatus* (Schreibman & Kallman, 1966), nor did they promote sodium retention in hypophysectomized *Poecilia latipinna* in FW (Ball & Ensor, 1967).

Neurohypophysial hormones

After hypophysectomy, teleostean or mammalian neurohypophysial peptides did not prolong FW survival in *Fundulus heteroclitus* or *Xiphophorus maculatus*, and did not promote sodium retention in *Poecilia latipinna* (Pickford, *et al.* 1965; Schreibman & Kallman, 1966; Ball & Ensor, 1967). On the other hand, these peptides have been shown to exert osmoregulatory effects in various intact teleosts. Thus, in intact goldfish, arginine vasotocin (AVT), oxytocin and fish hypophysial extracts promote diuresis and increase renal electrolyte loss. Isotocin may have reverse effects. All three peptides and the hypophysial extract may increase sodium influx at the gill in intact FW teleosts; in the goldfish, AVT, oxytocin and the extract simultaneously increased sodium outflux. In SW teleosts there is some evidence that neurohypophysial peptides may promote sodium outflux (Julien, 1960; Maetz, Bourguet, Lahlou & Hourdry, 1964; Motais & Maetz, 1967). These effects may possibly result from alteration in branchial haemodynamics induced by the peptides (see Maetz, 1970).

Recent work on the FW eel also suggests that teleostean neurohypophysial peptides may play an important role in controlling renal function, stimulating glomerular filtration rate and sodium reabsorbtion (Chester Jones, Chan & Rankin, 1969).

Endogenous hormones

Fish prolactin

Surgical removal of most of the prolactin cell zone (rostral pars distalis), like complete hypophysectomy, prevents FW survival in *Poecilia latipinna* (Ball, 1965*a*). Ectopic autotransplants of the rostral pars distalis in this species, composed mainly of prolactin cells, maintained the fish in FW, but transplants of the proximal pars distalis plus neurointermediate lobe, containing few prolactin cells, were not effective (Ball, 1965*b*). Thus the native fish prolactin is the essential FW survival factor. Single ectopic pituitary transplants in both *P. formosa* and *P. latipinna* maintained the fish alive in FW for long periods (up to 15 months), confirming the continued secretion of fish prolactin by the transplants (Ball, Olivereau, Slicher & Kallman, 1965; Ball & Olivereau, 1965; Ball, unpublished). The transplanted pituitary in *P. latipinna* reduces sodium outflux in FW and sodium turnover rate in $\frac{1}{3}$ SW, indicating that fish prolactin has similar biological actions to the exogenous mammalian hormone (Ball & Ensor, unpublished).

The prolactin cells in *P. latipinna* pituitary transplants display signs of activation following transfer from $\frac{1}{3}$ SW to FW which are similar to the changes in these cells in the *in situ* pituitary following the transfer;

thus hypothalamic connections are not essential for the increase in the secretory rate of these cells in FW (Ball & Olivereau, 1965; Ball, 1969a). The prolactin cells remain very active in transplants in *P. latipinna* after 8 days in FW (Ball & Olivereau, 1965), and retain their normal appearance in transplants in *P. formosa* and *P. latipinna* living in ⅓ SW (Olivereau & Ball, 1966; Ball, unpublished).

The various data on teleostean prolactin cells cultured *in vitro* have recently been reviewed (Olivereau, 1967c, 1969a; Ball & Baker, 1969). Dilution of the culture medium appears to stimulate the cells (Sage, 1968), indicating that they may normally be activated by a transient fall in plasma sodium such as occurs when *P. latipinna* first enters FW (Ball, 1969a).

In the FW eel, 2 months after autotransplantation of the pituitary the prolactin cells are always well preserved (Olivereau & Dimovska, 1968), and appear more active than in the *in situ* gland, judging by their larger nuclei and intense degranulation (Olivereau & Dimovska, 1969). This suggests an inhibitory hypothalamic control of prolactin, similar to that demonstrated in mammals (see Olivereau, 1969d). The effects of the eel autotransplants on blood electrolytes are still under investigation.

In goldfish, the number of epidermal mucous cells declines after hypophysectomy and is maintained at normal levels by pituitary autotransplants, perhaps reflecting prolactin secretion by the grafts (Ogawa & Johansen, 1967; Johansen, 1967).

DISCUSSION AND CONCLUSIONS

It will be seen that three components of the hypothalamo-hypophysial system undergo marked changes in relation to alterations in external salinity: the prolactin cells, the NPO-neurohypophysial tract, and the PAS +ve cells of the pars intermedia. Other pituitary factors, such as growth hormone, may well participate in osmoregulatory control, and a physiological role for ACTH in SW sodium regulation appears likely.

In various cyprinodonts, fish prolactin is apparently indispensible for electrolyte regulation in FW, in particular for sodium conservation. This action is specific to prolactin, is dose-dependent, and is produced by fish, frog and mammalian prolactins (Ball, 1969a, b; Ball & Ensor, 1969).

However, in other species (eel, goldfish) prolactin is not essential for prolonged survival in FW, although it has effects on electrolyte regulation qualitatively similar to its action in cyprinodonts. In the FW eel, the prolactin-dependent mechanism seems to be of only minor importance in the total electrolyte economy, while in the FW goldfish prolactin seems to be a major factor in ionic regulation, but the tissues are able to tolerate large disturbances in internal electrolyte composition.

Prolactin cell activity is strongly correlated with blood sodium levels in *Poecilia* and other cyprinodonts. In the eel, the correlation holds good in FW and SW, but not in DW, where despite hyponatraemia the prolactin cells are inactive. However, the simultaneous hypocalcaemia and raised plasma potassium levels in DW may interfere with the normal regulation of prolactin secretion, since prolactin itself increases potassium levels in hypophysectomized eels in both FW and DW (Olivereau & Chartier, 1966).

The role of neurohypophysial hormones is more difficult to assess, because hypothalamic neurosecretory material continues to accumulate in the pituitary stalk after hypophysectomy (Stutinsky, 1953), and neurohypophysial peptides, released from this material, may still be available for osmoregulatory function despite the absence of the neurohypophysis. In addition, in the hypophysectomized goldfish it has been shown that the sectioned pituitary stalk regenerates to form a kind of isolated vascularized neurohypophysis (Sathyanesan & Gorbman, 1965).

The exact role of the PAS +ve cells of the pars intermedia in *Anguilla* and *Oryzias* is not yet established, but the striking changes in these cells during adaptation to DW and to SW indicates some osmoregulatory function for their secretory product. Comparison of these cells in Japanese and European eels suggests as a tentative working hypothesis some role in calcium regulation.

This review has indicated that prolactin may act on the kidney and on extra-renal sites, presumably the gills; any action on the skin (mucous cells) is perhaps only minor since consideration of relative surface areas shows that the skin must be insignificant as a site for osmoregulatory exchanges compared with the gills, and skin permeability probably varies widely in different teleosts (Reid & Townsley, 1961; Maetz, 1968). The physiological mechanisms modulated by prolactin are unknown, but at the gill the hormone tends to limit sodium loss in FW, and at the kidney it may increase urine flow and possibly reduce renal electrolyte loss (see reviews, Olivereau & Lemoine, 1969; Ball, 1969a, b; cf. Maetz, 1970).

It is important to guard against the assumption that osmoregulation in any teleost is influenced only by the pituitary, since other endocrine organs are undoubtedly involved in mineral metabolism. Furthermore, different species differ in the details of their osmoregulatory equipment (Maetz, 1968, 1970), and it is unrealistic to expect that all teleosts should have absolutely identical patterns of hormonal control of osmoregulation. It is more likely that different species will be found to emphasize different aspects of endocrine control, depending on the osmotic challenges offered by their natural environments. A recent illustration of the variability that may occur even within the one genus is the finding (Griffith, 1969) that the FW species *Fundulus rathbuni* and *F. notatus* can survive in FW after hypophysectomy (unlike the brackish-water *F. heteroclitus* and the marine *F. majalis*), but are less tolerant of hypertonic media than intact fish, indicating a pituitary role in acclimation to hypertonic media.

REFERENCES

ARVY, L., FONTAINE, M. & GABE, M. (1959). La voie neurosécrétrice hypothalamo-hypophysaire des Téléostéens. *J. Physiol.*, *Paris* **51**, 1031–85.

ARVY, L. & GABE, M. (1954). Modifications du système hypothalamo-hypophysaire chez *Callionymus lyra* L. et *Ammodytes laceolatus*. Les au cours des variations de l'équilibre osmotique. *C. r. Ass. Anat.* **41**, 843–9.

BALL, J. N. (1965a). Partial hypophysectomy in the teleost *Poecilia*: separate identities of teleostean growth hormone and teleostean prolactin-like hormone. *Gen. comp. Endocr.* **5**, 654–61.

BALL, J. N. (1965b). Effects of autotransplantation of different regions of the pituitary gland on freshwater survival in the teleost *Poecilia latipinna*. *J. Endocr.* **33**, v–vi.

BALL, J. N. (1969a). Prolactin and osmoregulation in teleost fishes: a review. *Gen. comp. Endocr. Suppl.* 2, pp. 10–25.

BALL, J. N. (1969b). Prolactin (fish prolactin or paralactin) and growth hormone. In *Fish Physiology*, vol. 2, pp. 207–40 (eds. W. S. Hoar and D. J. Randall). New York: Academic Press.

BALL, J. N. & BAKER, B. I. (1969). The pituitary gland: anatomy and physiology. In *Fish Physiology*, vol. 2, pp. 1–110 (eds. W. S. Hoar and D. J. Randall). New York: Academic Press.

BALL, J. N. & ENSOR, D. M. (1965). Effects of prolactin on plasma sodium in the teleost *Poecilia latipinna*. *J. Endocr.* **32**, 269–70.

BALL, J. N. & ENSOR, D. M. (1967). Specific action of prolactin on plasma sodium levels in hypophysectomized *Poecilia latipinna* (Teleostei). *Gen. comp. Endocr.* **8**, 432–40.

BALL, J. N. & ENSOR, D. M. (1969). Aspects of the action of prolactin on sodium metabolism in cyprinodont fishes. In *La spécificité zoologique des hormones hypophysaires et de leurs activités* (ed. M. Fontaine). Paris. Colloque International du C.N.R.S. no. 177. 215–224.

BALL, J. N. & OLIVEREAU, M. (1964). Rôle de la prolactine dans la survie en eau douce de *Poecilia latipinna* hypophysectomisé et arguments en faveur de sa synthèse par les cellules érythrosinophiles *eta* de l'hypophyse des Téléostéens. *C. r. hebd. Séanc., Acad. Sci., Paris* **259** 1443–6.

BALL, J. N. & OLIVEREAU, M. (1965). Pituitary autotransplants and freshwater adaptation in the teleost *Poecilia latipinna*. *Am. Zool.* **5**, 232–3.

BALL, J. N., OLIVEREAU, M., SLICHER, A. M. & KALLMAN, K. D. (1965). Functional capacity of ectopic pituitary transplants in the teleost *Poecilia formosa*, with a comparative discussion on the transplanted pituitary. *Phil. Trans. Roy. Soc. Lond.* B **249**, 69–99.

BALL, J. N. & PICKFORD, G. E. (1964). Pituitary cytology and freshwater adaptation in *Fundulus heteroclitus*. *Anat. Rec.* **148**, 358.

BERN, H. A. (1967). Hormones and endocrine glands of fishes. *Science, N.Y.* **158**, 455–62.

BERN, H. A. & NICOLL, C. S. (1968). The comparative endocrinology of prolactin. *Rec. Prog. Horm. Res.* **24**, 681–720.

BLÜM, V. (1966). Zur hormonalen Steuerung der Brutpflege einiger Cichliden. *Zool. Jb. Physiol.* **72**, 264–94.

BLÜM, V. (1968). Immunological determination of injected mammalian prolactin in cichlid fishes. *Gen. comp. Endocr.* **11**, 595–602.

BLÜM, V. & FIEDLER, K. (1965). Hormonal control of reproductive behaviour in some cichlid fish. *Gen. comp. Endocr.* **5**, 186–96.

CALLAMAND, O., FONTAINE, M., OLIVEREAU, M. & RAFFY, A. (1951). Hypophyse et osmorégulation chez les Poissons. *Bull. Inst. Océanogr, Monaco* no. 984, pp. 1–8.

CARLSON, I. H. & HOLMES, W. N. (1962). Changes in the hormone content of the hypothalamo-hypophysial system of the rainbow trout (*Salmo gairdnerii*). *J. Endocr.* **24**, 23–32.

CHAMBOLLE, P. (1966). Recherches sur l'allongement de la durée de survie après hypophysectomie chez *Gambusia* sp. *C. r. hebd. Séanc. Acad. Sci., Paris*, D **262**, 1750–3.

CHAMBOLLE, P. (1967). Influence de l'injection d'A.C.T.H. sur la survie de *Gambusia* sp. (Poisson Téléostéen) privé d'hypophyse. *C. r. hebd. Séanc. Acad. Sci., Paris*, D **264**, 1464–6.

CHAN, D. K. O., CHESTER JONES, I. & MOSLEY, W. (1968). Pituitary and adrenocortical factors in the control of the water and electrolyte composition of the freshwater European eel (*Anguilla anguilla* L.). *J. Endocr.* **42**, 91–8.

CHARTIER, M. M. (1959). Influence de l'hormone somatotrope sur les teneurs en eau et en électrolytes du plasma et du muscle de la Truite arc-en-ciel (*Salmo gairdnerii*). *C. r. Séanc. Soc. Biol.* **153**, 1757–61.

CHARTIER, M. M. & OLIVEREAU, M. (1965). Prolactin and osmoregulation in eel. *Am. Zool.* **5**, 724.

CHESTER JONES, I., CHAN, D. K. O. & RANKIN, J. C. (1969). Renal function in the European eel (*Anguilla anguilla* L): effects of the caudal neurosecretory system, corpuscles of Stannius, neurohypophysial peptides and vasoactive substances. *J. Endocr.* **43**, 21–31.

CHESTER JONES, I. & HENDERSON, I. W. (1965). Electrolyte changes in the European eel. *J. Endocr.* **32**, iii–iv.

CHESTER JONES, I. HENDERSON, I. W. & BUTLER, D. G. (1965). Water and electrolyte flux in the European eel (*Anguilla anguilla* L.). *Archs Anat. microsc. Morph. exp.* **54**, 453–70.

DEMINATTI, M. M. (1969). Recherches sur la présence de groupements sulfates et leur participation métabolique dans les cellules préhypophysaires. Thesis, Doct. Sci. Nat. University of Strasbourg. Pp. 105.

DHARMAMBA, M. (1968). Effects of prolactin on hypophysectomized *Tilapia mossambica*. *Am. Zool.* **8**, 760.

DHARMAMBA, M. & NISHIOKA, R. S. (1968). Response of the 'prolactin-secreting' cells of *Tilapia mossambica* to environmental salinity. *Gen. comp. Endocr.* **10**, 409–20.

DONALDSON, E. M. & McBRIDE, J. R. (1967). The effects of hypophysectomy in the rainbow trout *Salmo gairdnerii* (Rich.) with special reference to the pituitary–interrenal axis. *Gen comp. Endocr.* **9**, 93–101.

DONALDSON, E. M., YAMAZAKI, F. & CLARKE, W. C. (1968). Effect of hypophysectomy on plasma osmolarity in goldfish and its reversal by ovine prolactin and a preparation of salmon pituitary 'prolactin'. *J. Fish. Res. Bd Can.* **25**, 1497–1500.

EGAMI, N. & ISHII, S. (1962). Hypophyseal control of reproductive function in teleost fishes. *Gen comp. Endocr.* Suppl. 1, pp. 248–53.

EMMART, E. W., PICKFORD, G. E. & WILHELMI, A. E. (1966). Localization of prolactin within the pituitary of a cyprinodont fish, *Fundulus heteroclitus* (Linnaeus), by specific fluorescent antiovine prolactin globulin. *Gen. comp. Endocr.* **7**, 571–83.

ENSOR, D. M. & BALL, J. N. (1968a). A bioassay for fish prolactin (paralactin). *Gen. comp. Endocr.* **11**, 104–10.

ENSOR, D. M. & BALL, J. N. (1968b). Prolactin and freshwater sodium fluxes in *Poecilia latipinna* (Teleostei). *J. Endocr.* **41**, xvi.

EPSTEIN, F. H., KATZ, A. I. & PICKFORD, G. E. (1967). Sodium- and potassium-activated adenosine triphosphatase of gills: role in adaptation of teleosts to salt water. *Science, N.Y.* **156**, 1245–7.

FAVRE, L. C. (1960). Recherches sur l'adaptation de cyprins dorés à des milieux de salinité diverse et sur son déterminisme endocrinien. *Dipl. Et. Sup. Univ. Paris.* Pp. 54.

FLEMING, W. R. (1968). Involvement of pituitary–interrenal axis in fish osmoregulation. In *U.S.–Japan Seminar on Endocrine Glands and Osmoregulation in Fishes*, Tokyo. Pp. 16–22.

FONTAINE, M. (1956). The hormonal control of water and salt-electrolyte metabolism in fish. *Mem. Soc. Endocr.* **5**, 69–82.

FONTAINE, M. (ed.) (1969). *La spécificité zoologique des hormones hypophysaires et de leurs activités.* Paris: Colloque International du C.N.R.S. no. 177, p. 412.

FONTAINE, M., CALLAMAND, O. & OLIVEREAU, M. (1949). Hypophyse et euryhalinité chez l'anguille. *C. r. hebd. Séanc. Acad. Sci., Paris* **228**, 513–14.

FONTAINE, M., LELOUP, J. & OLIVEREAU, M. (1952). La fonction thyroïdienne du jeune Saumon, *Salmo salar* L. (parr et smolt) et son intervention possible dans la migration d'avalaison. *Archs Sci. physiol.* **6**, 83–104.

FONTAINE, M. & OLIVEREAU, M. (1949). L'hypophyse du Saumon (*Salmo salar* L.) à diverses étapes de sa migration. *C. r. hebd. Séanc. Acad. Sci., Paris* **228**, 772–4.

GRIFFITH, R. W. (1969). Hypophysectomy and salinity tolerance in some killifishes of the genus *Fundulus. Am. Zool.* **9**, 574.

HAGEN, F. VON (1936). Die wichtigsten Endokrinen des Flussaals. Thyroidea, Thymus, und Hypophyse in Lebenszyklus des Flussaals (*Anguilla vulgaris*) nebst einigen Untersuchungen über das chromophile und chromophobe Kolloid der Thyroidea. *Zool. Jb. (Abt. Anat.)*, **61**, 467–538.

HANKE, W., BERGERHOFF, K. & CHAN, D. K. O. (1967). Histological observations on pituitary ACTH-cells, adrenal cortex, and the corpuscles of Stannius of the European eel (*Anguilla anguilla* L.). *Gen. comp. Endocr.* **9**, 64–75.

HENDERSON, I. W. & CHESTER JONES, I. (1967). Endocrine influences on the net extrarenal fluxes of sodium and potassium in the European eel (*Anguilla anguilla* L.). *J. Endocr.* **37**, 319–25.

HICKMAN, C. P. JR. (1959). The osmoregulatory role of the thyroid gland and radioiodine metabolism of yearling steelhead trout, *Salmo gairdneri. Can. J. Zool.* **37**, 997–1060.

HIRANO, T. & UTIDA, S. (1968). Effects of ACTH and cortisol on water movement in isolated intestine of the eel, *Anguilla japonica. Gen. comp. Endocr.* **11**, 373–80.

JOHANSEN, P. H. (1967). The role of the pituitary in the resistance of the goldfish (*Carassius auratus* L.) to a high temperature. *Can. J. Zool.* **45**, 329–45.

JULIEN, M. (1960). Action des hormones neurohypophysaires sur les échanges de sodium de *Carassius auratus* L. *Dipl. Et. Sup. Univ. Paris.* Pp. 41.

KAWASHIMA, S. & HIRANO, T. (1968). A preliminary report on the effect of increased salinity on cytology of the pars intermedia of the medaka and the eel. *U.S.–Japan Seminar on Endocrine Glands and Osmoregulation in Fishes*, Tokyo.

KNOWLES, F. & VOLLRATH, L. (1966). Neurosecretory innervation of the pituitary of the eels *Anguilla* and *Conger*. I. The structure and ultrastructure of the neuro-intermediate lobe under normal and experimental conditions. *Phil. Trans. Roy. Soc. Lond.* B **250**, 311–27.

KORN, H. (1960). Über die Einwirkung hypotonischer Medien auf das Zwischen-hirn-Hypophysensystem von *Mugil* und *Gobius*. *Z. Zellforsch. mikrosk. Anat.* **52**, 45–59.

LAHLOU, B. & SAWYER, W. H. (1969a). Electrolyte balance in hypophysectomized goldfish, *Carassius auratus* L. *Gen. comp. Endocr.* **12**, 370–7.

LAHLOU, B. & SAWYER, W. H. (1969b). Influence de l'hypophysectomie sur le renouvellement d'eau interne (étudié à l'aide de l'eau tritiée) chez le poisson rouge, *Carassius auratus* L. *C. r. hebd. Séanc. Acad. Sci., Paris*, D **268**, 725–8.

LAM, T. J. (1968). Effect of prolactin on plasma electrolytes of the early-winter marine threespine stickleback, *Gasterosteus aculeatus*, form *trachurus*, following transfer from sea- to fresh water. *Can. J. Zool.* **46**, 1095–8.

LAM, T. J. & HOAR, W. S. (1967). Seasonal effects of prolactin on freshwater osmoregulation of the marine form (*trachurus*) of the stickleback *Gasterosteus aculeatus*. *Can. J. Zool.* **45**, 509–16.

LAM, T. J. & LEATHERLAND, J. F. (1969a). Effects of prolactin on freshwater survival of the marine form (*trachurus*) of the three spine stickleback, *Gasterosteus aculeatus*, in the early winter. *Gen. comp. Endocr.* **12**, 385–7.

LAM, T. J. & LEATHERLAND, J. F. (1969b). Effect of prolactin on the glomerulus of the marine threespine stickleback *Gasterosteus aculeatus* L., form *trachurus*, after transfer from seawater to freshwater, during the late autumn and early winter. *Can. J. Zool.* **47**, 245–50.

LEATHERLAND, J. F. (1967). Structure and function in the hypothalamo-neuro-hypophysial complex and associated ependymal structures in the freshwater eel *Anguilla anguilla* L. Ph.D. Thesis, University of Leeds. Pp. 224.

LEATHERLAND, J. F. & LAM, T. J. (1969). Personal communication of results in press. *Can. J. Zool.*

LEDERIS, K. (1963). Effects of salinity on hormone content and on ultrastructure of trout neurohypophysis. *J. Endocr.* **26**, xxi–xxii.

LEDERIS, K. (1964). Fine structure and hormone content of the hypothalamo-neurohypophysial system of the rainbow trout (*Salmo irideus*) exposed to sea water. *Gen. comp. Endocr.* **4**, 638–61.

LELOUP, J. & FONTAINE, M. (1960). Iodine metabolism in lower vertebrates. *Ann. N.Y. Acad. Sci.* **86**, 316–53.

LELOUP-HATEY, J. (1964). Etude du déterminisme de l'activation de l'interrénal antérieur observée chez quelques Téléostéens soumis à un accroissement de la salinité du milieu extérieur. *Archs Sci. physiol.* **18**, 293–324.

LERAY, C. (1963). Etude de l'incorporation de cystéine marquée au soufre 35 dans le système hypothalamo-hypophysaire et plus spécialement dans l'adéno-hypophyse chez un Téléostéen: *Mugil cephalus* L. *C. r. hebd. Séanc. Acad. Sci., Paris* **256**, 795–8.

MAETZ, J. (1968). Salt and water metabolism. In *Perspectives in Endocrinology*, pp. 47–162, (eds. E. J. W. Barrington and C. B. Jørgensen). London and New York: Academic Press.

MAETZ, J. (1970). Mechanisms of salt and water transfer across membranes in relation to the aquatic environment in teleosts. *Mem. Soc. Endocr.* **18**, 3–28.

MAETZ, J., BOURGUET, J., LAHLOU, B. & HOURDRY, J. (1964). Peptides neurohypo-physaires et osmorégulation chez *Carassius*. *Gen. comp. Endocr.* **4**, 508–22.

MAETZ, J., MAYER, N. & CHARTIER-BARADUC, M. M. (1967a). La balance minérale du sodium chez *Anguilla anguilla* en eau de mer, en eau douce et au cours du transfert d'un milieu à l'autre: effets de l'hypophysectomie et de la prolactine. *Gen. comp. Endocr.* **8**, 177–88.

MAETZ, J., MAYER, N., FORSTER, M. E. & CHAN, D. K. O. (1967*b*). Axe hypophyso-interrenalien et osmorégulation chez les poissons d'eau de mer. *Gen. comp. Endocr.* **9**, 471.

MAETZ, J., SAWYER, W. H., PICKFORD, G. E. & MAYER, N. (1967*c*). Evolution de la balance minérale du sodium chez *Fundulus heteroclitus* au cours du transfert d'eau mer en eau douce: effets de l'hypophysectomie et de la prolactine. *Gen. comp. Endocr.* **8**, 163–76.

MATTHEIJ, J. A. M., SPRANGERS, J. S. P. & VAN OORDT, P. G. W. J. (1969). The site of prolactin synthesis in the pituitary gland of *Anoptichthys jordani* and the influence of this hormone on mucous cells. *Gen. comp. Endocr.* **13**, 519.

MAYER, N., MAETZ, J., CHAN, D. K. O., FORSTER, M. E. & CHESTER JONES, I. (1967). Cortisol, a sodium excreting factor in the eel (*Anguilla anguilla* L.) adapted to sea water. *Nature, Lond.* **214**, 118–20.

MOTAIS, R. (1967). Les mécanismes d'échanges ioniques branchiaux chez les Téléostéens. *Annls Inst. océanogr., Monaco* (N.S.) **45**, 1–84.

MOTAIS, R. & MAETZ, J. (1967). Arginine vasotocine et évolution de la perméabilité branchiale au sodium au cours du passage d'eau douce en eau de mer chez le Flet. *J. Physiol., Paris* **59**, 271.

OGAWA, M. (1968). Seasonal differences of glomerular change of the marine form of the stickleback, *Gasterosteus aculeatus* L., after transfer into freshwater. *Sci. Rep. Saitama Univ.* B **5**, 117–23.

OGAWA, M. & JOHANSEN, P. H. (1967). A note on the effect of hypophysectomy on the mucous cells of the goldfish, *Carassius auratus* L. *Can. J. Zool.* **45**, 885–6.

OLIVEREAU, M. (1954). Hypophyse et glande thyroïde chez les Poissons. Étude histophysiologique de quelques corrélations endocriniennes, en particulier chez *Salmo salar* L. *Annls Inst. océanogr., Monaco* (N.S.) **29**, 95–296.

OLIVEREAU, M. (1955). Hormone thyréotrope et température chez la Truite arc-en-ciel. (*Salmo gairdnerii* L.) et l'Anguille (*Anguilla anguilla* L.) *Archs Anat. microsc. Morph. exp.* **44**, 236–64.

OLIVEREAU, M. (1962). Modifications de l'interrénal du smolt (*Salmo salar* L.) au cours du passage d'eau douce en eau de mer. *Gen. comp. Endocr.* **2**, 565–73.

OLIVEREAU, M. (1965). Action de la métopirone chez l'Anguille normale et hypophysectomisée, en particulier sur le système hypophyso-corticosurrénalien. *Gen. comp. Endocr.* **5**, 109–28.

OLIVEREAU, M. (1966*a*). Modifications of the 'prolactin cells' in sea water eels. *Am. Zool.* **6**, 598.

OLIVEREAU, M. (1966*b*). Problèmes posés par l'étude histophysiologique quantitative de quelques glandes endocrines chez les Téléostéens. *Helgoländer wiss. Meeresunters.* **14**, 422–38.

OLIVEREAU, M. (1966*c*). Influence d'un séjour en eau demineralisée sur le système hypophyso-surrénalien de l'Anguille. *Ann. Endocr., Paris* **27**, 665–78.

OLIVEREAU, M. (1967*a*). Observations sur l'hypophyse de l'Anguille femelle, en particulier lors de la maturation sexuelle. *Z. Zellforsch. mikrosk Anat.* **80**, 286–306.

OLIVEREAU, M. (1967*b*). Réactions observées chez l'Anguille maintenue dans un milieu privé d'électrolytes, en particulier au niveau du système hypothalamo-hypophysaire. *Z. Zellforsch. mikrosk. Anat.* **80**, 264–85.

OLIVEREAU, M. (1967*c*). Notions actuelles sur le contrôle hypothalamique des fonctions hypophysaires chez les Poissons. *Rev. Européenne Endocr.* **4**, 175–96.

OLIVEREAU, M. (1968). Étude cytologique de l'hypophyse du Muge, en particulier en relation avec la salinité extérieure. *Z. Zellforsch. mikrosk. Anat.* **87**, 545–61.

OLIVEREAU, M. (1969*a*). Functional cytology of prolactin secreting cells. *Gen. comp. Endocr.* Suppl. 2, 32–41.

OLIVEREAU, M. (1969*b*). Activité de la pars intermedia de l'hypophyse autotransplantée chez l'Anguille. *Z. Zellforsch. mikrosk. Anat.* **98**, 74–87.

OLIVEREAU, M. (1969*c*). Complexe neuro-intermédiaire et osmorégulation: comparaison chez l'Anguille européenne et chez l'Anguille japonaise d'élevage au cours du transfert en eau de mer. *Z. Zellforsch. mikrosk. Anat.* **99**, 389–410.

OLIVEREAU, M. (1969*d*). Cytologie de l'hypophyse autotransplantée chez l'Anguille. Comparaison avec celle de *Poecilia*. In *Colloque National du C.N.R.S. sur la Neuroendocrinologie*, ed. J. Benoit. Paris: Editions du C.N.R.S.

OLIVEREAU, M. & BALL, J. N. (1964). Contribution à l'histophysiologie de l'hypophyse des Téléostéens, en particulier de celle de *Poecilia* species. *Gen. comp. Endocr.* **4**, 523–32.

OLIVEREAU, M. & BALL, J. N. (1966). Histological study of functional ectopic pituitary transplants in a teleost fish (*Poecilia formosa*). *Proc. Roy. Soc. Lond.* B **164**, 106–24.

OLIVEREAU, M. & CHARTIER-BARADUC, M. M. (1966). Action de la prolactine chez l'Anguille intacte et hypophysectomisée. II. Effets sur les électrolytes plasmatiques (sodium, potassium et calcium). *Gen. comp. Endocr.* **7**, 27–36.

OLIVEREAU, M. & DIMOVSKA, A. (1968). Histological study of autotransplanted pituitary in eels. *Proc. Symp. Comp. Endocr. Varanasi*, pp. 26–7.

OLIVEREAU, M. & DIMOVSKA, A. (1969). Prolactin-secreting cells in the autotransplanted pituitary of the eel. *Gen. comp. Endocr.* (in the Press.)

OLIVEREAU, M. & FONTAINE, M. (1965). Effet de l'hypophysectomie sur les corpuscles de Stannius de l'Anguille. *C. r. hebd. Séanc. Acad. Sci., Paris*, **261**, 2003–9.

OLIVEREAU, M. & HERLANT, M. (1960). Etude de l'hypophyse de l'Anguille mâle au cours de la reproduction. *C. r. Séanc. Soc. Biol.* **154**, 706–9.

OLIVEREAU, M. & LEMOINE, A. M. (1968). Action de la prolactine chez l'Anguille intacte. III. Effet sur la structure histologique du rein. *Z. Zellforsch. mikrosk. Anat.* **88**, 576–90.

OLIVEREAU, M. & LEMOINE, A. M. (1969). Action de la prolactine chez l'Anguille. V. Effet sur la structure histologique du rein après hypophysectomie. *Z. Zellforsch. mikrosk. Anat.* **95**, 361–76.

OLIVEREAU, M. & OLIVEREAU, J. (1968). Effets de l'interrénalectomie sur la structure histologique de l'hypophyse et de quelques tissus de l'Anguille. *Z. Zellforsch. mikrosk. Anat.* **84**, 44–58.

OLIVEREAU, M. & RIDGWAY, G. J. (1962). Cytologie hypophysaire et antigène sérique en relation avec la maturation sexuelle chez *Oncorhynchus* species. *C. r. hebd. Séanc. Acad. Sci., Paris* **254** 753–5.

PICKFORD, G. E. & PHILLIPS, J. G. (1959). Prolactin, a factor in promoting survival of hypophysectomized killifish in fresh water. *Science, N.Y.* **130**, 454–5.

PICKFORD, G. E., ROBERTSON, E. E. & SAWYER, W. H. (1965). Hypophysectomy, replacement therapy, and the tolerance of the euryhaline killifish, *Fundulus heteroclitus*, to hypotonic media. *Gen. comp. Endocr.* **5**, 160–80.

POTTS, W. T. W. & EVANS, D. H. (1966). The effects of hypophysectomy and bovine prolactin on salt fluxes in fresh-water-adapted *Fundulus heteroclitus*. *Biol. Bull. mar. biol. Lab., Woods Hole*, **131**, 362–9.

REID, D. F. & TOWNSLEY, S. J. (1961). Discussion of fish osmoregulation. *Proc. Xth Pacif. Sci. Congr.* pp. 172–3.

SAGE, M. (1968). Responses to osmotic stimuli of *Xiphophorus* prolactin cells in organ culture. *Gen. comp. Endocr.* **10**, 70–4.

SATHAYNESAN, A. G. & GORBMAN, A. (1965). Typical and atypical regeneration and overgrowth of hypothalamo-hypophysial neurosecretory tract after partial or complete hypophysectomy in the goldfish. *Gen. comp. Endocr.* **5**, 456–63.

SCHIEBLER, T. H. & HARTMANN, J. (1963). Histologische und histochemische Untersuchungen am neurosekretorischen Zwischenhirn-hypophysen-system von Teleostiern unter normalen und experimentellen Bedingungen. *Z. Zellforsch. mikrosk. Anat.* **60**, 89–146.

SCHREIBMAN, M. P. & KALLMAN, K. D. (1965). The effect of hypophysectomy on freshwater survival and scale mucous cells in two species of freshwater fishes. *Am. Zool.* **5**, 728.

SCHREIBMAN, M. P. & KALLMAN, K. D. (1966). Endocrine control of freshwater tolerance in teleosts. *Gen. comp. Endocr.* **6**, 144–55.

SCHREIBMAN, M. P. & KALLMAN, K. D. (1968). Pituitary control of freshwater survival in five species of Atheriniformes. *Am. Zool.* **8**, 760.

SMITH, J. (1962). Mortality of anadromous stickleback (*Gasterosteus aculeatus*) in fresh water. B.A. Thesis, Department of Zoology, University of British Columbia.

SREBO, Z. (1963). The hypothalamo-hypophysial neurosecretory system of *Chromis chromis* in laboratory and hypotonic conditions. *Folia Biol., Krakow,* **11**, 451–64.

STANLEY, J. G. & FLEMING, W. R. (1966). Effect of hypophysectomy on the function of the kidney of the euryhaline teleost, *Fundulus kansae. Biol. Bull. mar. biol. Lab., Woods Hole* **130**, 430–41.

STANLEY, J. G. & FLEMING, W. R. (1967). Effect of prolactin and ACTH on the serum and urine sodium levels of *Fundulus kansae. Comp. Biochem. Physiol.* **20**, 199–208.

STUTINSKY, F. (1953). La neurosécrétion chez l'Anguille normale et hypophysectomisée. *Z. Zellforsch. mikrosk. Anat.* **39**, 276–97.

SZABÓ, ZS. (1965). L'activité neuro-sécrétoire du noyau préoptique chez le Vairon (*Phoxinus phoxinus* L.) dans les conditions normales et expérimentales. *Rev. Roumaine Embryol. Cytol.* **2**, 159–62.

SZABÓ, ZS. & MOLNÁR, B. (1965). Experimental investigations on neurosecretion in mudfish (*Misgurnus fossilis* L.). *Acta biol. acad. sci. hung.* **15**, 383–92.

TUURALA, O. (1957). Über den Einfluss des osmotischen Belastung auf die neurosekretion den Kleinfische *Gasterosteus aculeatus* L. und *Phoxinus laevis* Agass. aus dem Brackwasser des finnischen Meerbusens. *Ann. Acad. Sci. Fenn.* A IV (Biol.) **36**, 1–9.

UTIDA, S., HIRANO, T. & KAMIYA, M. (1969). Seasonal variations in the adjustive responses to seawater in the intestine and gills of the Japanese cultured eel, *Anguilla japonica. Proc. Japan. Acad.* **45**, 293–7.

UTIDA, S., HIRANO, T., OIDE, M., KAMIYA, M., SAISYU, S. & OIDE, H. (1966). $Na^+–K^+$-activated adenosinetriphosphatase in gills and Cl^--activated alkaline phosphatase in intestinal mucosa with special reference to salt adaptation of eels. *Proc. XIth Pacif. Sci. Congr.* **7**, 5.

VOLLRATH, L. (1966). The ultrastructure of the eel pituitary at the elver stage with special reference to its neurosecretory innervation. *Z. Zellforsch. mikrosk. Anat.* **73**, 107–31.

DISCUSSION

MAETZ: Although I have to congratulate you on the excellence of your presentation, I must say that I regret that you add one more possible endocrine tissue which may be concerned with control of osmoregulation: the PAS-positive cells of the intermediate lobe. We have already too many glands involved in osmoregulation and things get more and more confusing.

Now I have a few comments to add concerning work done in my group. First of all, Lahlou and Giordan have shown that in the goldfish, passive water permeability of the gill is under endocrine control. Hypophysectomy brings about a reduction of the permeability and both ACTH-cortisol *and* prolactin are able to repair this defect. On the other hand hypophysectomy is followed by a reduction of the GFR and urine flow and an elevation of the urine sodium concentration. Only prolactin is able to repair this defect, not cortisol. In relation to the physiological effect of prolactin, I was very interested in your finding that prolactin produces morphogenetic effects in the kidney of the eel. I believe that also on the gill there must be morphogenetic effects of prolactin as well as cortisol.

Secondly, I would like to mention that Miss Dharmamba, who has been working with us for three months on *Tilapia*, has shown that after hypophysectomy of this fish in fresh water there is, as in other teleosts, a reduction of the sodium influx and increase of the sodium efflux. Now prolactin is able not only to bring back the sodium efflux to normal as in *Poecilia* and *Fundulus heteroclitus*, but also to augment the sodium influx which is restored to normal. I remember that Dr Ball has observed a similar dual effect of prolactin in *Fundulus kansae*.

Thirdly, I would like to mention recent work of McFarlane on hypophysectomy of the flounder. It is quite possible that this fish survives in fresh water much longer than *Fundulus heteroclitus* or *Tilapia*. It is possible that the reason for this may be the fact that the sodium exchange mechanisms are so different in the eel and flounder compared to *Fundulus*. Perhaps Dr McFarlane would like to comment on this.

OLIVEREAU: Thank you very much for your excellent comments.

MACFARLANE: Concerning Dr Maetz's comment on the hypophysectomized flounder in fresh water, at the moment I would not like to say whether the hypophysectomized flounder can survive for very long after transfer to fresh water, since after only 48 h, operated flounders show a significantly greater decrease in serum osmolality and plasma sodium levels than do sham-operated fish. The effects of longer periods in fresh water are being investigated.

BUTLER: Will you comment on your finding that the injection of mammalian prolactin into hypophysectomized eels must be started soon after the operation in order to lower the rate of decline in plasma sodium? Why weren't plasma sodium levels re-established if prolactin injections were started several days after hypophysectomy?

OLIVEREAU: What we found is that ovine prolactin prevents plasma sodium decline in hypophysectomized silver European eels if injected immediately after the operation. When the treatment begins a month after hypophysectomy, we have not been able to increase blood sodium to normal values (Olivereau & Chartier, 1966). It is possible that some target organs are affected several weeks after hypophysectomy as we showed for the adrenals or the thyroid, and perhaps the gills too. This might prevent a normal response to prolactin in the eel. A similar observation was noted by Lahlou in the goldfish. However, according to Dr Ball's findings, in *Poecilia*, it seems that prolactin may still correct alterations in sodium metabolism a few weeks after hypophysectomy.

CHAN, D. K. O: Perhaps I could add a few words from our own findings. In *Anguilla anguilla* maintained in fresh water, injection of prolactin into hypophysectomized animals could not prevent the drop in plasma sodium concentration or muscle sodium content, but it could abolish the rise in muscle water content. On the other hand, injection of ACTH or cortisol prevented the decline in muscle sodium content but was without effect on muscle water. However, injection of cortisol and prolactin together could prevent changes in electrolyte and water content of both plasma and muscle.

BALL: With regard to Dr Chan's point, in his experiments he used bovine prolactin. In other experiments in which prolactin influenced sodium, Dr Olivereau's included, it was ovine prolactin which was used. We know that dose for dose (that is, international unit for international unit) bovine prolactin is less effective in terms of freshwater survival than the ovine hormone (Pickford *et al.* 1965). This may be a dose effect in other words. Perhaps if you used a higher dose of bovine prolactin you might have got an effect on sodium.

PETER: In the pars intermedia of the goldfish, *Carassius auratus*, PAS-positive cells border the stainable neurosecretion of the pars nervosa. These same cells also stain strongly with ponceau xylidine. After lesioning the preoptic nucleus, or the preoptic nucleus tracts, all stainable neurosecretion disappears from the neurointermediate lobe and the PAS positive cells and ponceau xylidine stained cells also disappear. If the lesion of the preoptic nucleus is incomplete so that between 200–500 cells remain in the nucleus, there is only a small amount of stainable neurosecretion present in

PLATE I

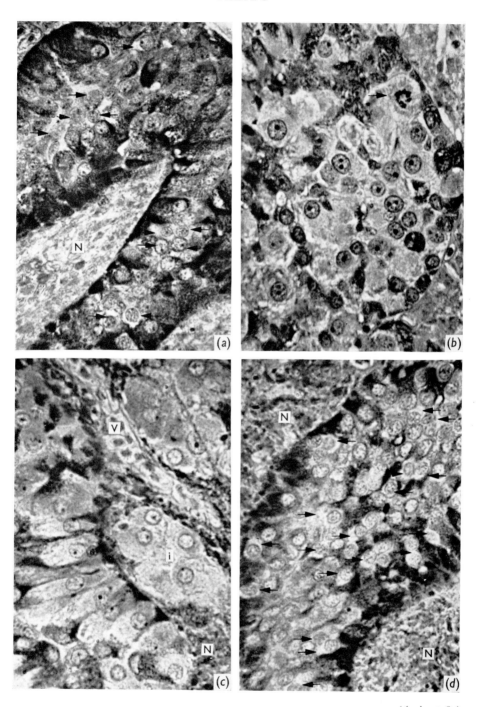

For explanation see p. 85

the neurointermediate lobe. The PAS-positive cells, and ponceau xylidine-stained cells, border these small amounts of stainable neurosecretion. These observations suggest that the stainable neurosecretion from the preoptic nucleus regulates these cells in the goldfish. Do the PAS-positive cells that you have observed in the pars intermedia of the eel correlate with the levels of stainable neurosecretion in the neurointermediate lobe?

OLIVEREAU: In the eel kept in fresh water, sea water or de-ionized water, there is no clear correlation between the amount of neurosecretory material and the activity of the PAS-positive cells. However, I did autotransplantation of the pituitary gland into the dorsal muscle in eels. I found that the pigmentation was normal or subnormal in the majority of the transplanted animals and that the lead-haematoxylin positive cells, quite numerous mainly in the darker eels, were rather well preserved. The PAS-positive cells regress or may completely disappear in the transplant after 2 months and no neurosecretory material is detected. Then it seems that the correlation between the presence of the PAS-positive cells and the amount of stainable neurosecretion in the neurohypophysis is valid in the eel as in the goldfish.

DESCRIPTION OF PLATE

Pars intermedia of the eel. Green filter, PAS-PbH staining. ×925.

(a) European eel (*Anguilla anguilla*), weight 50 g, in FW in Paris. Among the dark lead haematoxylin (PbH) positive cells are some PAS +ve cells (arrowed). N, Neurohypophysis.

(b) European eel kept in de-ionized water for 37 days. The PAS +ve cells are enlarged, with large nuclei, one or two large nucleoli, and show a mitosis (arrow); their apical endoplasmic reticulum is usually well developed. The PbH +ve cells are more-or-less compressed by the enlarged PAS +ve cells.

(c) Cultured Japanese eel (*Anguilla japonica*), weight 150 g, in FW in Tokyo. The PAS +ve cells are more numerous and much larger than in the European eel (fig. a). They make contact with the limiting membrane, and appear highly active. A small islet of PAS +ve cells (i) protrudes into the neurohypophysis (N) without an apparent limiting membrane. V, Blood vessel.

(d) Cultured Japanese eel kept in SW for 48 days. The PbH +ve cells are not clearly modified, but the PAS +ve cells (arrows) are much smaller than in FW (fig. c), with atrophied nuclei and insignificant nucleoli. Contact with the limiting membrane has been lost in many cases.

THE CONTROL OF SOME ENDOCRINE MECHANISMS ASSOCIATED WITH SALT REGULATION IN AQUATIC BIRDS*

By W. N. HOLMES, MO-YIN CHAN, J. S. BRADLEY AND I. M. STAINER

INTRODUCTION

The successful adaptation of freshwater tetrapod species to a marine environment is largely dependent upon functional changes which occur in the excretory mechanisms of the organism. The nephron of the lower vertebrate kidney, however, possesses poorly developed urine concentrating properties. For example, experiments in these laboratories have shown that although the duck maintained on hypertonic saline invariably produced urine which was hypertonic with respect to plasma, more than 50% of the available osmotic space was occupied by NH_4^+ (Holmes, Fletcher & Stewart, 1968). Consequently, the kidney of this species is severely restricted in its ability to gain osmotically free water from ingested solutions of electrolytes such as sea water.

Certain avian and reptilian species, however, have evolved efficient accessory excretory organs which have the ability to actively transport electrolytes out of the extracellular compartment. Thus, these extra-renal excretory pathways, working in concert with the renal pathway, effectively maintain homeostasis and permit the organisms to make either short or prolonged excursions into arid or marine environments.

In the bird, the extra-renal excretory organs are known as the nasal glands. These organs are situated along the dorsal margins of orbits and their excretory fluid is, in most species, discharged via ducts into the anterior nasal cavity. In the Pelecaniformes, where the external nares are extremely small or absent, the ducts continue to run along the inner surface of the upper beak and the fluid is discharged at the tip of the beak.

The functional significance of the nasal gland in the marine bird was first described by Schmidt-Nielsen, Jorgensen & Osaki in 1958. These workers demonstrated that the glands secreted an extremely concentrated solution of Na^+, K^+ and Cl^- after the oral or parenteral administration of an electrolyte load. Furthermore, it was shown that the secretion of nasal gland fluid was dependent upon an intact parasympathetic pathway asso-

* This work was supported by funds from the National Science Foundation (grant no. GB 7945) and the Committee on Research, University of California.

ciated with the VIIth cranial nerve. Later investigations on the duck
suggested that an intact pituitary–adrenal axis was also necessary for the
functional integrity of the nasal glands (Holmes, Phillips & Butler, 1961;
Phillips, Holmes & Butler, 1961; Phillips & Bellamy, 1962; Wright,
Phillips & Huang, 1966).

The exact role of the adrenal steroids in the control of nasal gland
function is not clear. In the acutely saline-loaded duck which had been
treated with ACTH, a 30% increase in the blood glucose concentration
accompanied the increased secretion of nasal gland fluid (Peaker, Wright

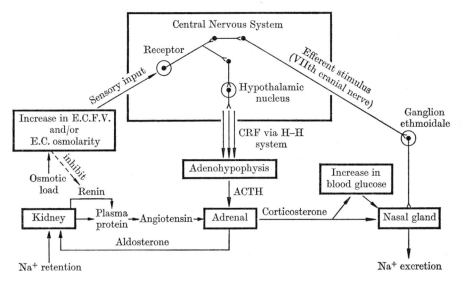

Fig. 1. A schematic representation of the possible pathways involved in the regulation of
excretion in marine birds.

& Phillips, unpublished). Moreover, these investigators observed that the
extra-renal excretory capacity was enhanced in the intact bird when the
hypertonic saline load was preceded by the intravenous administration of
glucose. Conversely, insulin-induced hypoglycaemia was accompanied by a
reduced extra-renal response. These data have been reviewed by Holmes,
Phillips & Wright (1969). The changes which occur in the structure
(Ernst & Ellis, 1969), the protein and nucleic acid composition (Holmes &
Stewart, 1968) and the Na^+ transporting properties (Ernst, Goertemiller
& Ellis, 1967; Fletcher, Stainer & Holmes, 1967) of the nasal gland tissue
in the duck following exposure to hypertonic saline, suggest that extremely
complex mechanisms are initiated to control the development of the gland.
Again, the adrenal steroids may be involved in the control of these develop-

mental mechanisms. Certainly the intracellular accumulation of cortisol-4-[^{14}C] or corticosterone-1,2-[^{3}H] in the cells of the actively secreting nasal gland suggest this possibility (Bellamy & Phillips, 1966; Phillips & Bellamy, 1967).

The interrelationships which might exist between the central nervous system, the pituitary–adrenal axis and the excretory organs of the duck are summarized in Fig. 1. We have previously reviewed the possible role of the adrenal steroids in maintaining homeostasis in birds exposed to large intakes of electrolytes, and the structural and functional changes which occur in the nasal glands at this time (Holmes, Phillips & Chester Jones, 1963; Holmes, 1965; Holmes *et al.* 1969; Holmes & Wright, 1969). In the present discussion, however, we would like to confine our considerations to the factors associated with the control of pituitary and adrenal function in birds.

HYPOTHALAMIC CONTROL OF PITUITARY FUNCTION

Evidence that the central nervous system is closely associated with the control of adenohypophysial function in birds has been well documented. Indeed, the photo-sexual reflex may well be an essential component in the control of gonadotrophic activity in birds. Changes in the distribution of neurosecretory material in the hypothalamus have been correlated with both the photoperiodic exposure of the bird and its concomitant gonadal function (Legait, 1959; Farner, 1962; Benoit & Assenmacher, 1955; Benoit, Assenmacher & Brard, 1956). However, the paradoxical effects of removal, stalk section and ectopic transplantation of the adenohypophysis have led to the suggestion that the central control of corticotrophic function in birds may differ from that in mammals (Elton, Zarrow & Zarrow, 1959; Newcomer, 1959; Resko, Norton & Nalbandov, 1964; Boissin, Baylé & Assenmacher, 1966; Frankel, Cook, Graber & Nalbandov, 1967).

We have recently attempted to measure the effects *in vitro* of simple acetic acid extracts of central nervous tissue from the duck on the release of ACTH from duck pituitaries during a short-term incubation period (Stainer & Holmes, 1969). The presence of hypothalamic extract in the incubation media of pituitary-halves significantly enhanced the release of ACTH when compared to the release from the contralateral pituitary-halves incubated in buffer alone (Fig. 2). Although the duck adrenal tissue responded to added ACTH, there was no evidence to suggest that the extracts of the hypothalamic tissue contained any ACTH (Fig. 2). In other experimental groups, however, similar amounts of cerebral and spinal cord tissue extracts contained equally potent corticotrophin releasing material

(Stainer & Holmes, 1969). In a later series of experiments the small portion of the hypothalamus corresponding to the median eminence was extracted and bioassayed for corticotrophin releasing activity. When the amount of ACTH released from the contralateral pituitary-halves incubated in buffer was subtracted from the amount released from the pituitary-halves simultaneously incubated in the presence of median eminence extract, then the

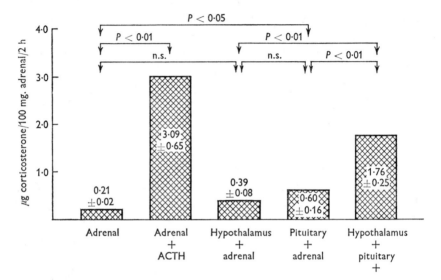

Fig. 2. A comparison of the stimulatory action of duck hypothalamic extract (five replicate experiments) on ACTH released from duck pituitary-halves *in vitro* with the ACTH releasing activity of their contralateral pituitary halves incubated in buffer alone (five replicated experiments). ACTH activities of the pituitary media were assayed by *in vitro* incubation with duck adrenal slices and the corticosterone concentrations of these incubation media were measured by a fluorometric method. These rates of corticosterone release were compared to control adrenals incubated in the presence of buffer alone or in the presence of buffer and hypothalamic extract. One international unit ACTH (mammalian) was added to a sample of adrenal slices incubated alone in each experimental run. All values were expressed as means ± S.E. and mean values were compared according to the two-tailed *t* test (Stainer & Holmes, 1969).

release of ACTH was significantly greater than zero (Fig. 3). This was true for both the high and low dose levels of median eminence. Once more, cerebral tissue contained a significant amount of corticotrophin releasing material (Stainer & Holmes, 1969).

More recently, we have used an *in vitro* pituitary-culture technique (Nicoll & Meites, 1962) to re-examine the corticotrophin releasing activity of central nervous tissue from the duck. The details of this experimental plan are outlined in Fig. 4. During the first 24 h of incubation the rates of ACTH released from the two sets of pituitary-halves were the same (Fig. 5). For the next 2 days one set of pituitary-halves was incubated in the

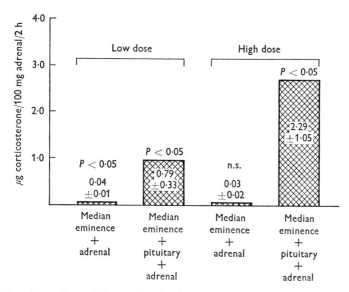

Fig. 3. The effect of two different dose levels of duck median eminence extract on the release *in vitro* of ACTH from duck pituitary-halves. The high dose of median eminence extract contained the equivalent of 4·2 median eminences per four pituitary-halves incubated (seven replicate experiments) and the low dose of extract similarly corresponded to 2·1 median eminences (seven replicate experiments). The rate of ACTH release from the pituitary-halves incubated with median eminence extract was compared to the rate of release from the contralateral halves incubated in buffer alone. The ACTH activities of the pituitary media were assayed by incubation *in vitro* with duck adrenal slices and subsequent analysis for corticosterone. The corresponding pituitary blank value was subtracted from each individual rate of ACTH release obtained from the pituitary-halves incubated in the presence of median eminence extract. Similarly, the corresponding adrenal blank values were subtracted from each value obtained from adrenal slices incubated in the presence of median eminence extract. All values were expressed as means ±S.E. and mean values were compared according to the single-tailed *t* test (Stainer & Holmes, 1969).

Experimental Plan

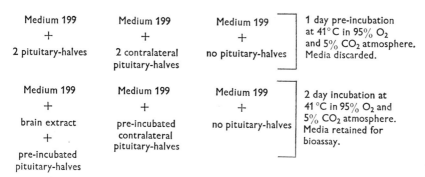

Fig. 4. Plan of experiments using organ-culture technique (Nicoll & Meites, 1962) to determine the C.R.F. activity of central nervous tissues from the duck. ACTH activities of incubation media were bioassayed according to (1) the stimulation *in vitro* of corticosterone synthesis by duck adrenal slices (Stainer & Holmes, 1969), or (2) the *in vivo* depletion of adrenal ascorbic acid in the hypophysectomized rat (Sayers, Sayers & Woodbury, 1948).

presence of median eminence extract. These pituitaries released significantly more ACTH than the contralateral pituitary-halves incubated alone
(Fig. 5). Furthermore, when estimated by the *in vivo* adrenal ascorbic acid
depletion assay in hypophysectomized rats, the rate of ACTH release
from the pituitary-halves incubated in the presence of median eminence
extract was significantly enhanced (Fig. 6). Cerebral extracts, however,
contained no corticotrophin-releasing activity when assayed in this system.

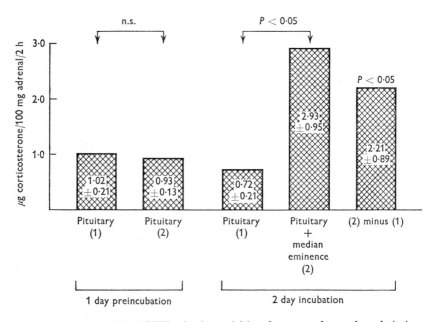

Fig. 5. A comparison of the ACTH-releasing activities of two sets of contralateral pituitary
from the duck halves (six replicate experiments) incubated for 1 day in Medium 199
alone. During the following 2 days, Medium 199 containing duck median eminence
extract was added to one set of pituitary-halves (six replicate experiments) and the other
set were incubated in Medium 199 alone (six replicate experiments). The ACTH activities
of the incubation media were assayed by *in vitro* incubation with duck adrenal slices and
the media were subsequently analysed for corticosterone.

These observations are in agreement with those of Péczely & Zboray
(1967) and Péczely (1969) on the pigeon, *Columba livia*. These workers
have concluded that the pigeon adrenal is under the control of the hypothalamo-hypophysial system. They have also presented evidence for the
presence of an ACTH-like material in the median eminence of the pigeon
(Péczely & Zboray, 1967). We have been unable to detect any ACTH-
like material in either the hypothalamus or the median eminence of the
duck.

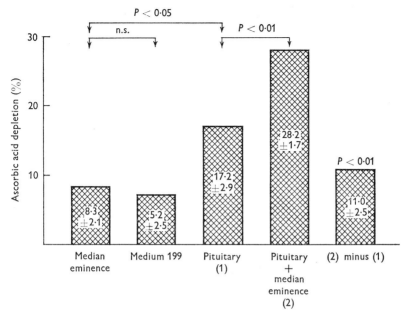

Fig. 6. The effect of duck median eminence extract on the release of ACTH from duck pituitary-halves cultured *in vitro*. The ACTH released in the presence of the extract was compared to the ACTH released from the contralateral pituitary-halves cultured in Medium 199 alone (six replicated experiments). The ACTH activities of the pituitary media were measured by the adrenal ascorbic acid depletion assay in hypophysectomized rats.

PITUITARY CONTROL OF ADRENAL FUNCTION

Hypophysectomy causes some structural degeneration of the bird adrenal (Miller & Riddle, 1942; Nalbandov & Card, 1943; Assenmacher, 1958; Miller, 1961) and the decline in plasma corticoid concentration suggests that a partial biosynthetic dysfunction also occurs. The relatively high level of circulating corticoids which remain after hypophysectomy (Resko *et al.* 1964; Frankel *et al.* 1967) and particularly after ectopic transplantation of the pituitary (Boissin *et al.* 1966) has been interpreted as evidence for a high degree of 'adrenal autonomy' in birds. The implication is that not only does the steroidogenic pathway in the bird adrenal appear to be largely independent of pituitary ACTH but also that the central control over the release of pituitary ACTH may be less than that suggested by studies on mammals.

Earlier studies have suggested that the volume in which corticosterone is distributed in the duck may vary considerably according to the physiological state of the bird (Donaldson & Holmes, 1965). Indeed, the apparent volume of distribution (A.V.D.) may reflect a significant increase or decrease in the total corticoid in the bird but little or no change may occur in the peripheral plasma concentration of the steroids. The lability of the A.V.D. in birds may be a general phenomenon. For example, the A.V.D. of corti-

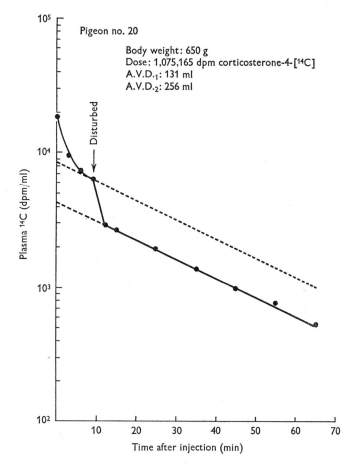

Fig. 7. The disappearance of corticosterone-4-[^{14}C] from the plasma of an unanaesthetized pigeon. The bird was accidentally disturbed at the point indicated when a heavy door in the room was slammed. The change in the apparent volume of distribution (A.V.D.) was estimated on the assumption that if the disturbance had not occurred, the disappearance curve (dotted line) would have been parallel to new one established following the abrupt decline in plasma radioactivity.

costerone in the pigeon also changes rapidly in response to stress or disturbance (Fig. 7).

We have therefore attempted to examine the distribution and metabolism of corticosteroids following hypophysectomy of the duck. The doses of corticosterone-4-[^{14}C] administered to the intact and sham-operated ducks became rapidly distributed into a complex compartment which probably consisted, in part, of the intravascular, the interstitial and the intracellular spaces of the body. Indeed, the initial distribution within the first intravascular compartment of the two-compartment model system (Tait, Tait, Little & Lumas, 1961) was so rapid it could not be detected by the techniques used (Fig. 8). In the hypophysectomized duck, however, the distribution and disappearance of the labelled corticosterone was much slower and its distribution within two compartments could be identified (Fig. 8). Even so, the initial phases of distribution in the first compartment could not be accurately determined. All analyses of these data, therefore, were conducted according to a single compartment model system (Tait *et al.* 1961).

The disappearance curves of corticosterone for the three birds illustrated in Fig. 8 are typical of the responses we have observed following hypophysectomy in the duck. The establishment of an extremely long biological half-life, an increase in the A.V.D. and a marked decline in the metabolic clearance rate (M.C.R.) are typical responses of birds examined 14 days after hypophysectomy.

Although Fig. 8 represents the disappearance of total plasma ^{14}C from each bird, all of the methylene chloride extractable radioactivity in the plasma at the end of the experimental period was in the form of corticosterone-4-[^{14}C]. As in mammals, most of the circulating corticosterone in these birds was probably bound to a transcortin-like plasma protein. The decline in plasma radioactivity, therefore, would reflect the steady dissociation of the protein-bound moiety to compensate for the metabolism and removal of free corticosterone from the plasma. The present calculations of the MCR for corticosterone in the duck were based upon the total ^{14}C activities in plasma rather than plasma radioactivities due to free corticosterone-4-[^{14}C] or methylene chloride extractable ^{14}C. We must emphasize, therefore, that the MCR values reported for the intact, sham-operated and hypohysectomized ducks may have only relative significance. Nevertheless, we find that, while the estimated MCR of corticosterone in the intact and sham-operated birds were 4·84 and 7·00 ml/kg/min respectively, that of the hypophysectomized bird was 0·661 ml/kg/min. This rate represented declines in the MCR of corticosterone to 13·7% and 9·4% of the rates in the intact and sham-operated birds.

A similar series of experiments has measured the rate of disappearance of total ^{14}C activity in plasma following the injection of aldosterone-4-[^{14}C] into intact, sham-operated and hypophysectomized ducks (Fig. 9). Only a single compartment of distribution for aldosterone was evident and in all birds the A.V.D. of aldosterone was approximately twice that of

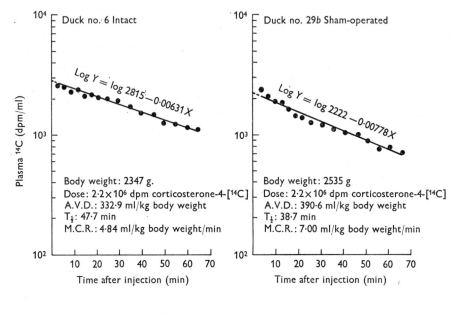

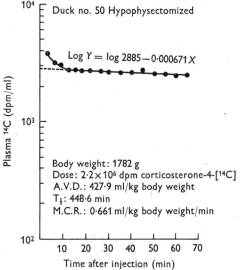

Fig. 8. The typical patterns of disappearance of corticosterone-4-[^{14}C] from the plasma of intact, sham-operated and hypophysectomized ducks 14 days after surgery. A.V.D. = apparent volume of distribution; $T_{\frac{1}{2}}$ = biological half-life; M.C.R. = metabolic clearance rate.

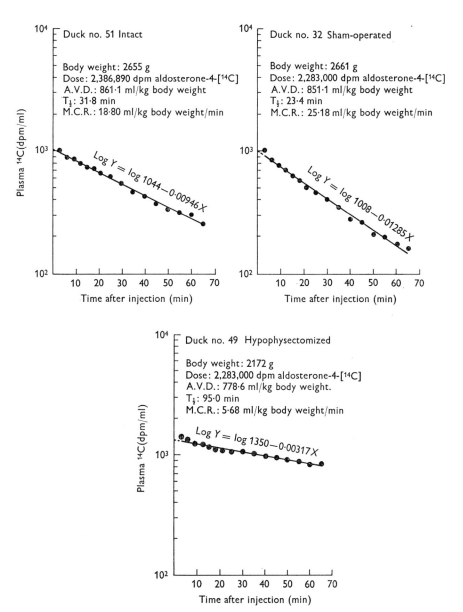

Fig. 9. The typical patterns of disappearance of aldosterone-4-[14C] from the plasma of intact, sham-operated and hypophysectomized ducks 14 days after surgery. A.V.D. = apparent volume of distribution; $T_{\frac{1}{2}}$ = biological half-life; M.C.R. = metabolic clearance rate.

corticosterone. But, while the A.V.D. of corticosterone increased following hypophysectomy, that of aldosterone decreased. The biological half-life of aldosterone was prolonged in the hypophysectomized birds but the increase was not as great as that of corticosterone (cf. Figs. 8, 9).

In each case, the M.C.R. of aldosterone was high when compared to the corresponding M.C.R. for corticosterone. The decline in the M.C.R. of aldosterone following hypophysectomy, however, was of the same order as that observed for corticosterone (Figs. 8, 9).

Clearly therefore the adenohypophysis profoundly influences the metabolism of both corticosterone and aldosterone in the duck.

THE RENIN–ANGIOTENSIN SYSTEM

The renin–angiotensin system seems to influence, either directly or indirectly, the secretion of aldosterone in several mammalian species. Data on the renin–angiotensin system in non-mammalian species, however, are scattered and incomplete. In the birds, Schaffenburg, Haas & Goldblatt (1960) have shown the presence of a pressor substance in the chicken kidney. Preliminary studies in these laboratories have shown that pigeon and duck plasma substrates were hydrolysed in the presence of hog renin to produce an angiotensin-like material. But, under the same conditions, renin extracts of pigeon and duck kidney tissue would not hydrolyse hog plasma substrate. We have found, however, that renin extracts of duck and pigeon kidney tissue do yield pressor substances when incubated with nephrectomized rat plasma (Table 1). Similarly, significant concentrations of renin and low concentrations of angiotensin have been identified in plasma from both the duck and the pigeon (Table 1). The high value for the plasma renin activity in the duck (Table 1) was derived from mixed blood collected after decapitation and complete exsanguination of the birds. Bleeding profoundly increased the release of renin in the pigeon (Fig. 10) and a similar phenomenon in the duck almost certainly accounted for the high value for the plasma renin activity recorded in the duck (Table 1). The intravenous administration of valine–angiotensin into the anaesthetized pigeon caused a rise in arterial blood pressure and the response pattern was similar to that of angiotensin in the rat (Fig. 11).

Since hypophysectomy is more easily performed on the pigeon than on the duck we have confined our experimental studies of the renin-angiotensin system to the pigeon. During a 10-day period following hypophysectomy the plasma renin activity in the pigeon showed a fourfold increase whereas no significant change occurred in the sham-operated birds during this period (Fig. 12). The daily administration of ACTH for 1 week, starting

10 days after hypophysectomy, restored the plasma renin activities to the levels observed in the intact or sham-operated birds. We have not detected any decrease in the blood or plasma volume of the hypophysectomized pigeon. Furthermore, in the hypophysectomized pigeon the pattern of renin release in response to haemorrhage was quite different from that in the intact bird. Thus, there appeared to be an increase in the absolute amount of circulating renin in the pigeon following hypophysectomy.

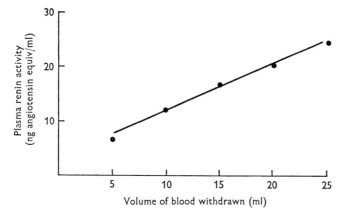

Fig. 10. The effect of haemorrhage on the plasma renin activities in the intact pigeon. Each point represents the mean plasma renin activity of a pooled blood sample from birds.

Table 1. *The renin activities in kidney tissue, and the renin and angiotensin activities in blood from the duck and the pigeon*

The plasma renin activities were determined from plasma samples (10 ml) incubated for 3 h at 41 °C in the presence of EDTA (0·15 ml 3·8 %) and Dowex resin (3 ml). The angiotensin concentrations in plasma were determined from non-incubated plasma samples. The activities of renin extracts from kidney tissue were determined indirectly as angiotensin equivalents after incubation with nephrectomized rat plasma for 3 h at 41 °C. Following purification (Boucher *et al.* 1964), all angiotensin activities were bioassayed in nephrectomized, pentolinium-treated rats. Numerals in parentheses indicate the number of birds used for each determination and all mean values are reported ±S.E.

	Plasma concentrations		Kidney concentrations
	Angiotensin (ng/ml)	Renin (ng angiotensin equiv./ml)	Renin (μg angiotensin equiv./g wet tissue)
Duck (mixed blood)	7·9±0·9 (9)	40·5±4·2 (10)	1·21±0·28 (4)
Pigeon (arterial blood)	—	5·8±1·3 (5)	1·08±0·53 (5)

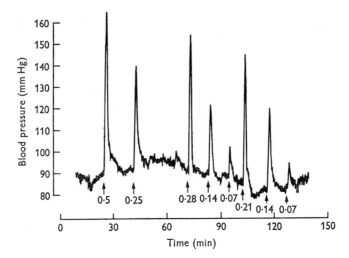

Fig. 11. The effect of valine-angiotensin (CIBA) on the arterial blood pressure of the anaesthetized pigeon (0·4 mg Nembutal i.v./100 g body weight) maintained on a respirator at 30 inhalations/min. Each arrow represents the intravenous administration of the indicated dose of angiotensin in nanogrammes.

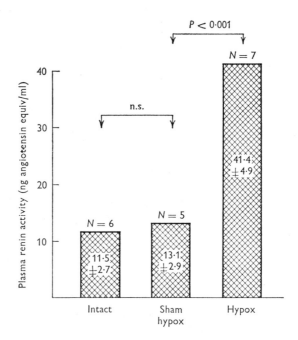

Fig. 12. Renin activities of venous blood from control and sham-operated and hypophysectomized pigeons 10 days after surgery. Plasma samples (10 ml) were incubated for 3 h in the presence of EDTA (0·5 ml 3·8 %) and Dowex resin (3 ml) at 41 °C. Following purification (Boucher et al. 1964), angiotensin activities were bioassayed in nephrectomized, pentolinium-treated rats. Mean values are reported ±S.E.

These studies suggest that a reciprocal relationship exists between the circulating levels of ACTH or corticoids and the activity of a renin-like enzyme in the plasma. In the rat, adrenalectomy has been shown to be followed by an increase in the plasma renin content (Gross, 1964) and hypophysectomy caused an increase in the kidney renin content (Bruinvels, van Houten & van Noorwijk, 1964). Hypophysectomy of the frog, how-ever, has been reported to cause depletion in juxta-glomerular cell granula-tion (Yoshimura, Hoshiko & Suzuki, 1951). Clearly, the exact significance of these data must await further investigation.

EFFECT OF ACTH ON THE ADRENAL

The adrenal cortex shows at least two distinct patterns of response to ACTH stimulation. The prolonged effects of ACTH involve the differen-tiation of new adrenocortical tissue to sustain the higher rate of steroido-genesis. During the first 4 days of treatment with ACTH a period of adrenal hypertrophy in the rat is characterized by a progressive increase in the protein and RNA content of the adrenals (Farese, 1964). During the next few days of ACTH treatment an increase in the adrenal DNA con-tent reflects a hyperplasia of the adrenocortical tissue (Farese, 1964).

The short-term effect of ACTH is recognizable both *in vitro* and *in vivo* and the response is characterized by the synthesis and release of adrenal steroids within 1 h after exposure of the tissue to ACTH. Although pro-tein synthesis is considered to be a prerequisite of all ACTH-induced corticosteroid synthesis, the effects of ACTH on adrenal protein synthesis during the first few hours are still uncertain (see review, Farese, 1968).

It was recently reported that the steroidogenic pathways of adrenals and ovaries from rats could be altered by culturing the glands in the presence of RNA extracted from other steroid-producing endocrine glands. The pattern of steroid hormones synthesized in the cultural tissue reflected the origin of the RNA (Villee, 1967). In the following experiments, there-fore, answers were sought to two questions. First, would the culture of duck adrenals in the presence of adrenal gland RNA extracted from ACTH-treated ducks influence the acute response of this tissue to ACTH? Secondly, could the glucocorticoid synthesis in the duck adrenal be modified to produce cortisol by culturing the glands in the presence of adrenal RNA from a species known to produce cortisol as the principal glucocorticoid? We emphasize, however, that these data are preliminary and the interpre-tation is tentative.

Using the dodecyl sulphate and phenol method of Hiatt (1962), RNA was extracted from the adrenals of hamsters and ducks which had received

University of Pittsburgh
Bradford Campus Library

2·0 i.u. ACTH/100 g body weight (Armour, Achtar gel) twice daily for 2 days. Explants of adrenal tissue from untreated ducks were then cultured for 24 h in the presence of ACTH (1·0 i.u.) and/or the adrenal RNA (300 μg). The explants were subsequently homogenized and incubated for 1 h in phosphate buffer containing 0·1 μc pregnenolone-4-[^{14}C] (s.a. 52·3

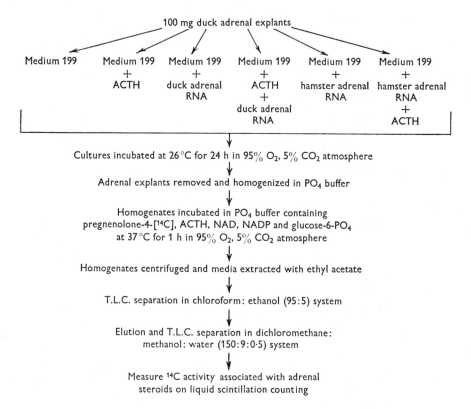

Fig. 13. The experimental plan used to examine the effects *in vitro* of ACTH and adrenal RNA on the duck adrenal maintained in sterile organ culture (Holmes & Rice, unpublished).

mc/mM), 1·0 I.U., 2·7 μM NAD, 2·7 μM NADP and 2·4 μM glucose-6-phosphate (Villee, 1966). The radioactivity associated with corticosterone was then measured. The experimental plan is summarized in Fig. 13.

The adrenal explants which had been cultured for 24 h in Medium 199 alone still showed some response to the ACTH added to the phosphate buffer medium of the tissue homogenate. When the explants were cultured for 24 h in the presence of ACTH, a three-fold increase in the yield of ^{14}C-labelled material associated with corticosterone occurred. The presence of RNA from ACTH-treated duck adrenals, however, was equally effective in enhancing the conversion from pregnenolone-4-[^{14}C], but no additive effect was detected when explants were cultured in the presence of ACTH and duck adrenal RNA (Fig. 14). Ribonucleic acid which had been extracted from ACTH-treated hamster adrenals had no effect on the subsequent response of the duck adrenal explants to ACTH. Furthermore, no increase

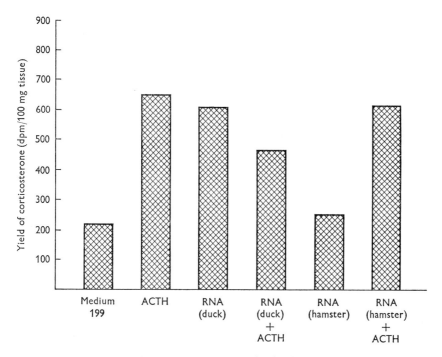

Fig. 14. The effect of ACTH on the conversion *in vitro* by homogenates of duck adrenal explants of pregnenolone-4-[^{14}C] to corticosterone-4-[^{14}C]. The duck adrenal explants were previously cultured simultaneously for 24 h in Medium 199 containing the indicated additives. Radioactive steroid material having the same chromatographic mobility as corticosterone was subsequently acetylated with acetic anhydride and the resultant ^{14}C-labelled steroid acetate was isopolar with authentic corticosterone acetate (Holmes & Rice, unpublished).

in the small amount of radioactivity always found in association with the cortisol spot was observed. The explants, however, were still viable since those which had been cultured with hamster adrenal RNA and ACTH still responded to the ACTH added to the phosphate buffer in which the tissue was homogenized (Fig. 14). A replicate series of cultured duck adrenal explants showed precisely similar effects (Holmes & Rice, unpublished).

The significance of these data is not yet clear. The adrenal RNA extracts were crude and some contamination of these extracts with DNA or protein cannot be discounted. Nevertheless, following the chronic exposure of the adrenal explants to duck adrenal RNA, or to ACTH, the acute response of the homogenized explants to ACTH was significantly increased. Therefore, the RNA from the adrenals of ducks treated *in vivo* with ACTH contained information similar to that elicited by the ACTH *in vitro* during the culture of the adrenal explants. Furthermore, the enhanced response to ACTH of the RNA-treated explants suggested that the acute response to ACTH may involve the expression of the information contained in the RNA. More than one locus for the action of ACTH on adrenal function may therefore exist. Hamster adrenal RNA elicited no response in the duck adrenal explants and had no influence on their pattern of glucocorticoid synthesis.

REFERENCES

ASSENMACHER, I. (1958). Recherches sur le contrôle hypothalamique de la fonction gonadotrope préhypophysaire chez le canard. *Archs Anat. microsc. Morph. exp.* **47**, 447–572.

BELLAMY, D. & PHILLIPS, J. G. (1966). Effect of the administration of sodium chloride solutions on the concentration of radioactivity in the nasal glands of ducks (*Anas platyrhynchos*) injected with ^{3}H-corticosterone. *J. Endocr.* **36**, 97–8.

BOUCHER, R., VEYRAT, R., DE CHAMPLAIN, J. & GENEST, J. (1964). New procedures for measurement of human angiotensin and renin activity levels. *Can. med. Ass. J.* **90**, 194–201.

BENOIT, J. & ASSENMACHER, I. (1955). Le contrôle hypothalamique de l'activité prehypophysaire gonadotrope. *J. Physiol., Paris* **47**, 427–567.

BENOIT, J., ASSENMACHER, I. & BRARD E. (1956). Etude de l'évolution testiculaire du canard domestique soumis très jeune à un éclairement artificiel permanent pendant deux ans. *C. r. hebd. Séanc. Acad. Sci., Paris* **242**, 3113–15.

BOISSIN, J., BAYLÉ, J. D. & ASSENMACHER, I. (1966). Le fonctionement cortico-surrenalien du canard mâle après préhypophysectomie ou autogreffe hypophysaire ectopique. *C. r. hebd. Séanc. Acad. Sci., Paris* **263**, 1127–9.

BRUINVELS, J., VAN HOUTEN, J. C. & VAN NOORWIJK, J. (1964). Influence of pinealectomy and hypophysectomy on the renin content of rat kidneys. *Q. Jl exp. Physiol.* **49**, 95.

COGHLAN, J. P. & SCOGGINS, B. A. (1967). Measurement of aldosterone in peripheral blood of man and sheep. *J. Clin. Endocrinol.* **27**, 1470–86.

DONALDSON, E. M. & HOLMES, W. N. (1965). Corticosteroidogenesis in the freshwater and saline-maintained duck (*Anas platyrhynchos*). *J. Endocr.* **32**, 329–36.

ELTON, E. L., ZARROW, I. G. & ZARROW, M. X. (1959). Depletion of adrenal ascorbic acid and cholesterol. A comparative study. *Endocrinology* **65**, 152–60.

ERNST, S. A., GOERTEMILLER, C. C. & ELLIS, R. A. (1967). The effect of salt regimens on the development of (Na$^+$ + K$^+$)-dependent ATP-ase activity during growth of salt glands of ducklings. *Biochim. biophys. Acta* **135**, 682.

ERNST, S. A. & ELLIS, R. A. (1969). The development of surface specialization in the secretory epithelium of the avian salt gland in response to osmotic stress. *J. Cell Biol.* **40**, 305–21.

FARESE, R. V. (1964). Changes in ^{14}C-glycine incorporating activities of rat adrenal microsomes and soluble cell fraction during prolonged adrenocorticotrophin administration. *Biochim. biophys. Acta* **91**, 515–21.

FARESE, R. V. (1968). Regulation of adrenal growth and steroidogenesis by ACTH. In *Functions of the Adrenal Cortex*, vol. 1, pp. 539–81. (ed. K. W. McKerna). New York: Appleton-Century-Crofts.

FARNER, D. (1962). Hypothalamic neurosecretion and phosphatase activity in relation to the photoperiodic control of the testicular cycle of *Zonotrichia leucophrys gambelii*. In *Progress in Comparative Endocrinology, Gen. comp. Endocr.* Suppl. 1, pp. 160–7.

FLETCHER, G. L., STAINER, I. M. & HOLMES, W. N. (1967). Sequential changes in the adenosinetriphosphatase activity and the electrolyte excretory capacity of the nasal glands of the duck (*Anas platyrhynchos*) during the period of adaptation to hypertonic saline. *J. exp. Biol.* **47**, 375–92.

FRANKEL, A. I., COOK, B., GRABER, J. W. & NALBANDOV, A. V. (1967). Determination of corticosterone in plasma by fluorometric techniques. *Endocrinology* **80**, 181–94.

GROSS, F. (1964). Differentiation of effects mediated by aldosterone and renin angiotensin in rats with experimental hypertension. In *Aldosterone* (eds. E. E. Baulieu and P. Robel), pp. 307–20. Oxford: Blackwell.

HIATT, H. H. (1962). A rapidly labeled RNA in rat liver nuclei. *J. molec. Biol.* **5**, 217–29.

HOLMES, W. N. (1965). Some aspects of osmoregulation in reptiles and birds. In *IVth Symposium d'Endocrinologie Comparée, Paris. Archs Anat. microsc. Morph. exp.* **54**, 491–514.

HOLMES, W. N., FLETCHER, G. L. & STEWART, D. J. (1968). The patterns of renal electrolyte excretion in the duck (*Anas platyrhynchos*) maintained on freshwater and on hypertonic saline. *J. exp. Biol.* **48**, 487–508.

HOLMES, W. N., PHILLIPS, J. G. & BUTLER, D. G. (1961). The effect of adrenocortical steroids on the renal and extra-renal responses of the domestic duck (*Anas platyrhynchos*) after hypertonic saline loading. *Endocrinology* **69**, 483–95.

HOLMES, W. N., PHILLIPS, J. G. & CHESTER JONES, I. (1963). Adrenocortical factors associated with adaptation of vertebrates to marine environments. *Recent Progr. Hormone Res.* **19**, 619–72.

HOLMES, W. N., PHILLIPS, J. G. & WRIGHT, A. (1969). The control of extra-renal excretion in the duck (*Anas platyrhynchos*) with special reference to the pituitary–adrenal axis. *Gen. comp. Endocrinol.* Suppl. 2, pp. 358–73.

HOLMES, W. N. & STEWART, D. J. (1968). Changes in the nucleic acid and protein composition of the nasal glands from the duck (*Anas platyrhynchos*) during the period of adaptation to hypertonic saline. *J. exp. Biol.* **48**, 509–20.

HOLMES, W. N. & WRIGHT, A. (1969). Some aspects of the control of osmoregulation and homeostasis in birds. Proc. *IIIrd Int. Congr. Endoc. Excerpta Medica, Int. Congr. Series*, no. 184, 199–210.

LEGAIT, H. (1959). Contribution à l'étude morphologique du système hypothalamo-neurohypophysaire de la Poule Rhode Island. Thesis, University of Louvain-Nancy.

MILLER, R. A. (1961). Hypertrophic adrenals and their response to stress after lesions in the median eminence of totally hypophysectomized pigeons. *Acta Endocr.* **37**, *Copenh.* 565–76.

MILLER, R. A. & RIDDLE, O. (1942). The cytology of the adrenal cortex of normal pigeons and experimentally induced atrophy and hypertrophy. *Am. J. Anat.* **71**, 311–35.

NALBANDOV, A. V. & CARD, L. E. (1943). Effects of hypophysectomy on growing chicks. *J. exp. Zool.* **94**, 387–409.

NEWCOMER, W. S. (1959). Effects of hypophysectomy on some functional aspects of the domestic pigeon. *Endocrinology* **65**, 133–41.

NICOLL, C. S. & MEITES, J. (1962). Estrogen stimulation of prolactin production by rat adenohypophysis *in vitro*. *Endocrinology* **70**, 272–7.

PHILLIPS, J. G. & BELLAMY, D. (1962). Aspects of the hormonal control of nasal gland secretion in birds. *J. Endocrin.* **24**, vi–vii.

PHILLIPS, J. G. & BELLAMY, D. (1967). The control of nasal glands function, with special reference to the role of adrenocorticosteroids. *Proc. 2nd Int. Symp. Hormonal Steroids. Excerpta Medica Int. Congr. Series*, no. 132, pp. 877–81.

PHILLIPS, J. G., HOLMES, W. N. & BUTLER, D. G. (1961). The effect of total and subtotal adrenalectomy on the renal and extra-renal response of the domestic duck (*Anas platyrhynchos*). *Endocrinology* **69**, 958–69.

PÉCZELEY, P. (1969). Effect of the median eminence of the pigeon (*Columba livia domestica* L.) on the regulation of adenohypophysial corticotropin secretion. *Acta physiol. hung.* **35** (1), 47–57.

PÉCZELY, P. & ZBORAY, G. (1967). CRF and ACTH activity in the median eminence of the pigeon. *Acta physiol. hung.* **32** (3), 229–39.

RESKO, J. A., NORTON, H. W. & NALBANDOV, A. V. (1964). Endocrine control of the adrenal in chickens. *Endocrinology* **75**, 192–200.

VAN ROSSUM, G. D. V. (1966). Movements of Na^+ and K^+ in slices of herring-gull salt gland. *Biochem. biophys. Acta* **126**, 338–49.

SAYERS, M. A., SAYERS, G. & WOODBURY, J. A. (1948). The assay of adreno-corticotropic hormone by the adrenal ascorbic acid depletion method. *Endocrinology* **42**, 379–93.

SCHAFFENBURG, C. A., HAAS, E. & GOLDBLATT, H. (1960). Concentration of renin in kidneys and angiotensinogen in serum of various species. *Am. J. Physiol.* **199**, 788–92.

SCHMIDT-NIELSEN, K., JORGENSEN, C. B. & OSAKI, H. (1958). Extrarenal salt excretion in birds. *Am. J. Physiol.* **193**, 101–7.

STAINER, I. M. & HOLMES, W. N. (1969). Some evidence for the presence of a corticotrophin releasing factor (CRF) in the duck (*Anas platyrhyncos*). *Gen. comp. Endocrin.* **12** (2), 350–9.

TAIT, J. F., TAIT, S. A. S., LITTLE, B. & LUMAS, K. B. (1961). The disappearance of 7-H^3-d-aldosterone in the plasma of normal subjects. *J. Clin. Invest.* **40**, 72–80.

VILLEE, D. B. (1966). Effects of progesterone on enzyme activity of adrenals in organ culture. *Adv. Enzyme Reg.* **4**, 269–80.

VILLEE, D. B. (1967). Ribonucleic acid: Control of steroid synthesis in endocrine tissue. *Science, N.Y.* **58**, 652–3.

WRIGHT, A., PHILLIPS, J. G. & HUANG, D. P. (1966). The effect of adenohypo-physectomy on the extra-renal and renal excretion of the saline-loaded duck (*Anas platyrhynchos*). *J. Endocrin.* **36**, 249–56.

YOSHIMURA, F., HOSHIKO, N. & SUZUKI, Y., (1951). Endocrinological study of renin. I. The relationships between anterior pituitary and renin of frogs. *Nisshin Igaku* **26**, 631.

DISCUSSION

HENDERSON: In your experiments showing an increase in plasma renin activity after hypophysectomy, and a return to normal following ACTH, was there much of a change in the animals electrolyte balance.

HOLMES: There were no changes in the plasma concentrations of Na^+ and K^+ or in the $Na^+:K^+$ ratios in the plasma which could be correlated with the concomitant renin activities of the plasma.

We have not yet developed a satisfactory metabolism cage for the pigeon and so we have no data at this time on the intake and output of electrolytes by these birds.

PHILLIPS: Dr Holmes, I wonder whether you would like to comment on two things. First, what are your ideas about the role of the nervous system in the control of nasal gland secretion. And secondly, if I could refer to your earlier slide with your scheme of control, you probably know that Peaker has recently shown an action of prolactin on nasal gland activity and David Ensor in Hull has recently shown in the duck that prolactin levels in the pituitary increase by about 50% on day 2 following the feeding of saline (0·3 M-NaCl) thereafter decreasing to day five when they are subnormal.

Interestingly further studies revealed a difference between the duck and the gull. Gulls maintained on 0·3 M NaCl showed no change in pituitary prolactin levels, but transfer to 0·7 M NaCl resulted in a similar depression of prolactin levels on day 5. It may be, therefore, that the gull is better adapted to saline conditions than is the duck and this may reflect its capacity to cope with a natural marine environment.

HOLMES: The effects of the parasympathetic nerve pathways are certainly not restricted to vasomotor effects; following a saline load the blood flow to the nasal gland does indeed increase. Van Rossum (1966), however, has demonstrated that incubation of nasal gland tissues with acetylcholine does change the Na^+ flux. So, there does appear to be an effect which is fundamentally related to the active transport of Na^+ and distinct from the control of blood flow through the gland.

With respect to the release of prolactin, this may well be true. We have not worked on this aspect of the problem, but my reaction would be to enquire whether the prolactin did contain any ACTH-like activity.

PHILLIPS: Peaker did show, in fact, that he could not duplicate the effect of prolactin using mammalian ACTH; the pattern was different and so he thinks that prolactin is not mimicking ACTH.

SANDOR: In the experiments using duck adrenal explants with hamster adrenal RNA, did you find any evidence of cortisol formation?

HOLMES: We did find a very small amount of ^{14}C associated with material in the region of the cortisol spots on the chromatograms of all duck adrenal incubates irrespective of their treatment. We reclaimed this material from the scintillation fluid by the method of Coghlan & Scoggins (1967) and when acetylated it had the same R_F as authentic cortisol acetate. If you consider that this is sufficient evidence to establish that cortisol is a natural product of incubated duck adrenals, then a very small amount was produced under the conditions of these experiments.

BERN: As I think Dr Holmes knows, Dr Nicol, Mr William Shirley and I have been interested in the same kinds of experiments that were reported towards the end of his talk with regard to the crop-sac response to prolactin and the use of RNA's. I think there is one thing only that worries me about the material that he showed us and that is, I'd be happier if you had used a homologous RNA from liver or something like that.

HOLMES: We did, and we found that duck liver RNA had no effect on cultured duck adrenal explants.

BERN: It has no effect, O.K., bless you!

HOLMES: An intriguing aspect of these and other similar experiments is that, presumably, the RNA gets into the cell from the culture medium and yet little is ever said concerning the nature of this transport mechanism.

BERN: I think as much as ten years ago or so there were experiments of this kind done, with regard to embryonic development, by Brachet where again RNA apparently does get into cells in that kind of a system.

HOLMES: Yes, but how?

MAETZ: It might be entering by a 'corkscrew' effect, as it is a thin and elongated coiled molecule and you don't need very wide pores to get the molecule into the cell. It certainly goes through the nuclear membrane.

HOLMES: But the nuclear membrane is fenestrated.

MAETZ: Not necessarily. There is a potential across the nuclear membrane and the electron-microscope picture does not give a definitive answer to the question of pore size. RNA may well pass through also by this 'corkscrew' effect, although I am not sure whether any work has been done on its permability through the cellular membrane.

SESSION II
THE TERRESTRIAL ENVIRONMENT

ENVIRONMENTAL EFFECTS UPON ENDOCRINE-MEDIATED PROCESSES IN LOCUSTS

By W. MORDUE, K. C. HIGHNAM, L. HILL and A. J. LUNTZ

LOCUST GROWTH AND REPRODUCTION

It is axiomatic that all organisms respond to changes in their environment. Where these responses involve alterations in developmental processes, the controlling endocrine mechanism is very likely to act as a mediator between the environmental change and the individual response. Analysis of development in locusts provides much useful information about the ways in which environmental fluctuations exert profound influences upon these terrestrial organisms.

Adult female locusts attain a 'basic weight', during their somatic growth period, before egg development begins (Phipps, 1950; Norris, 1954; Hill, Luntz & Steele, 1968). In reaching this 'basic weight' the female increases in dry weight by some 112%, due particularly to the growth of the cuticle, muscles and to the increased concentration of blood metabolites. The second growth period of the adult female locust is concerned with reproductive development: during this period the reproductive tissues, i.e. ovaries and accessory glands, show marked increases in dry weight (Hill et al. 1968). About 70% of the yolk deposited during egg development is protein, synthesized mainly outside the ovary (Telfer, 1965). The site of synthesis is now known to be the fat body (Coles, 1965; Hill, 1965; Minks, 1967). The fat body constituents in *Schistocerca* show cyclical variations which are related to ovarian growth, but amounts present in the fat body are of course not representative of the large amount of metabolite turnover by the fat body during reproductive growth (Hill et al. 1968). The two growth phases are exclusive, probably because of their similar high demands for metabolites: reproductive development does not normally begin until somatic growth is finished.

The amount of protein laid down in the tissues during both somatic and reproductive growth is approximately 175 mg in each period (Hill et al. 1968). However, the requirements for carbohydrate are very different in the two periods and 12 times as much carbohydrate is deposited during somatic growth than during reproductive growth (Luntz, 1968). During

[111]

the somatic growth period locusts have an intense feeding period (Hill, Mordue & Highnam, 1966; Strong, 1967a). During this feeding peak, in *Schistocerca* at least, bran is selected in preference to lettuce (Hill *et al.* 1968). The protein requirements for both somatic and reproductive growth are easily met by the normal food intake, and it seems likely that the preference for bran lies in the provision of easily available carbohydrate without the problem of water loading (Luntz, 1968). The utilization of

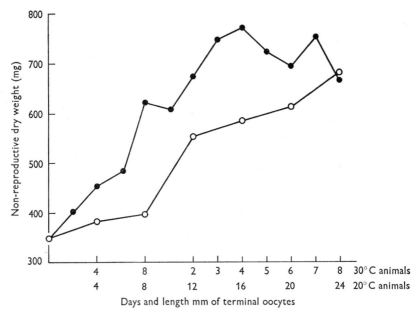

Fig. 1. Changes in weight during somatic growth and the 1st gonotrophic cycle in *Schisto-cerca* females reared at 30 °C (●) and during somatic growth in females reared at 20 °C (○). Oocyte growth begins at 10–12 days in the 30 °C animals and at about 24 days in the 20 °C animals.

food and the efficiency of conversion of food to body substance differ in the two growth phases. Food conversion is particularly affected by factors which influence the amount of energy devoted to physiological functions or to the support of activity. Food is very efficiently converted to body substance during ovarian growth and during somatic growth more of the digested food is channelled into energy production (Luntz, 1968).

The relationships between the somatic and reproductive growth periods can be altered by environmental conditions. Females reared at 20 °C show slower growth, take longer to reach their basic weight and the onset of reproductive development is delayed compared with females reared at 30 °C (Fig. 1) (Hill *et al.* 1968). Females reared without mature males show patterns of growth and development similar to those of low-temperature females, but females reared with mature males show accelerated growth phases (Fig. 2) (Norris, 1954). Oocyte growth itself is remarkably sensitive to environmental events. In *Schistocerca* the growth of oocytes is more

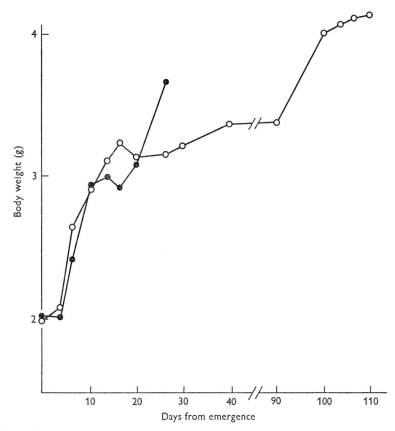

Fig. 2. Changes in weight from emergence to just before oviposition in two female *Schistocerca*, one reared with an immature male (○), the other with a mature male (●) (After Norris, 1954).

rapid in crowded compared with isolated individuals, and flying the females on a turntable accelerates oocyte growth in both crowded and isolated forms (Highnam & Haskell, 1964). Accelerated reproductive development occurs in female desert locusts reared with mature males (Norris, 1954; Highnam & Lusis, 1962; Highnam & Haskell, 1964) (Fig. 2). These females grow more rapidly than females reared without males. In *Locusta* the reverse is true and environmental factors such as crowding retard maturation (Norris, 1964b; Highnam & Haskell, 1964). However, flying accelerates oocyte growth in crowded *Locusta* (Highnam & Haskell, 1964).

Both feeding and diet are of the utmost importance in regulating development. In *Schistocerca* starvation, or semi-starvation, retards growth and the onset of reproductive development (Hill *et al.* 1966; Highnam, Hill & Mordue, 1966). A diet of lettuce alone reduces somatic and reproductive growth in *Schistocerca* and the onset of maturation is delayed compared with locusts fed lettuce but in addition given whole-meal bran (Hill *et al.* 1968; Hill & Steele, unpublished observations). This availability of the correct kind of food is of obvious importance to the animal when it is realized how the utilization of food may vary throughout the adult instar (Luntz, 1968).

ENDOCRINES, GROWTH AND REPRODUCTION IN LOCUSTS

The ultimate regulation of reproductive development is by hormones released from the cerebral neurosecretory system and the corpora allata (Highnam, 1962 *a*, *b*; Highnam, Lusis & Hill, 1963 *a*, *b*), but the role of endocrines in regulating somatic growth is not yet elucidated fully. The processes of somatic and ovarian growth do not overlap; and ovarian growth only occurs when materials become available after somatic growth has finished. It is possible that the neuroendocrine system operates a switch mechanism which channels metabolism from somatic to ovarian growth. It is supposed that slight fluctuations, or longer-term changes in the environment, which affect oocyte growth and reproductive development do so by altering the hormone balance. What evidence is there for this view?

The measurement of changes in hormone balance is a problem common to all endocrinologists, but is particularly acute for the insect endocrinologist. Not only must he attempt to estimate changes in the level of hormones, the nature and structure of some of which are still unknown, but he has to contend with very small amounts of hormone-producing tissues and even smaller amounts of hormones. A number of methods have been devised to measure activity levels and hormone titres, and as far as the neurosecretory system is concerned these methods are still necessarily indirect. However, it is now possible to estimate directly the levels of corpus allatum hormone in the blood (de Wilde, Staal, de Kort & Baard, 1968).

HISTOLOGY OF LOCUST NEUROSECRETORY SYSTEMS

Variations in the activity of the neurosecretory systems of locusts have been followed in a number of ways, with differing degrees of success. Histological methods alone have only a limited use in the analysis of the dynamics of neurosecretory activity and more sophisticated techniques employing radioactively labelled sulphur amino acids are necessary for full and detailed studies on the interrelationships between synthesis, transport and release of neurosecretory material (see the section following). The limitations of histological methods in the analysis and separation of these three parameters is discussed in detail by Highnam (1965). These limitations are made obvious by analyses of neurosecretory activity in insects such as *Tenebrio*, where it is possible to obtain histological pictures which convey nothing of the secretory activity of the neurosecretory cell (Mordue, 1967*a*). However, in some instances (and within carefully defined limits) the histology of the neurosecretory system can provide useful information about the effect of the environment upon the neurosecretory system and subsequently upon growth and reproduction.

We have seen that if *Schistocerca* are reared without males or in low-density conditions oocyte growth is retarded: these slowly developing females have large amounts of stainable material in their neurosecretory systems. Both crowding and mating accelerate oocyte development in *Schistocerca* and these rapidly maturing females possess neurosecretory systems which are histologically almost devoid of stainable material (Highnam, 1962*a*; Highnam & Lusis, 1962; Highnam & Haskell, 1964). The proportion of 'full' to 'empty' neurosecretory cells in the pars intercerebralis is related fairly closely to oocyte development (Fig. 3) (Highnam, 1966). However, in *Locusta*, as mentioned previously, crowding retards oocyte growth and this restriction is associated with an accumulation of stainable material in the neurosecretory system (Highnam & Haskell, 1964). Obviously, environmental conditions markedly affect the histological appearance of the neurosecretory system. Further histological changes in the neurosecretory cells caused by environmental factors can be seen if slowly developing *Locusta* and *Schistocerca* are forced to fly. In both species, oocyte development is accelerated after flying and the flight activity produces a change from a full to an empty neurosecretory system (Highnam & Haskell, 1964).

As outlined earlier, feeding stimuli and diet exert profound effects both upon growth and reproductive development. Feeding stimuli are important in that they are potent regulators of neurosecretory activity. When females of *Schistocerca* are starved neurosecretory material accumulates within the perikarya of the neurosecretory cells, along the cell axons and within the corpora cardiaca. This accumulation is followed by a decrease in the volume of the nuclei of the neurosecretory cells (Highnam *et al.* 1966; Highnam, 1966). When *Locusta* are starved, material accumulates similarly within the neurosecretory cell axons and the corpora cardiaca, and the nuclei of the neurosecretory cells decrease in size. However, there is no marked accumulation within the perikarya of the neurosecretory cells

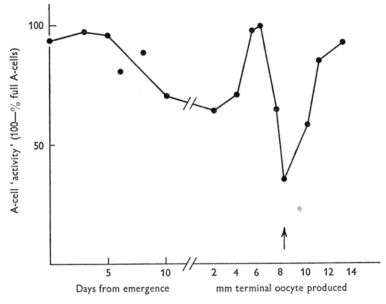

Fig. 3. Changes in the number of empty 'A'-cells (i.e. A-cell activity) in the pars inter-cerebralis of *Schistocerca* during the somatic growth period and the first and second gonotropic cycles. ↑, Oviposition.

because of the much increased volume of the cell axons in *Locusta* compared with *Schistocerca* (West & Highnam, unpublished observations). When starved animals of both species are allowed to feed there is a rapid deple-tion of stainable material from all parts of the neurosecretory system (Highnam *et al.* 1966). This depletion occurs without an immediate increase in the size of the cell nuclei. However, after continuous feeding the nuclei increase in size and remain large (Highnam, 1966).

If these histological changes in the neurosecretory systems reflect differences in activity, clearly environmental factors such as feeding would have profound effects upon neurosecretory activity. However, similar effects upon neurosecretory activity in *Schistocerca* and *Locusta* can be masked by such histological analysis: the neurosecretory cells of starved *Schistocerca* and *Locusta* look quite different but they are both inactive. These histological methods allow the different contents of stainable material in the neurosecretory systems to be distinguished in particular species and perhaps allow distinctions to be made between systems exhibiting different activities. The interrelationships between synthesis (as judged by nuclear volume) and release of neurosecretion under certain specialized conditions of feeding and starvation can also be realised from such histological methods. However, little can be deduced, by these methods, of the amounts of neurosecretion synthesized and released into the blood. The rate of synthesis of neurosecretion can vary independently of the rate of release of material (see also Highnam, 1967), although over the long term the two processes may well be linked. Consequently, estimates of activity based upon isolated observations of the histological appearance of the neurosecretory system are at the best uncertain and frequently most inaccurate.

RADIOISOTOPES AND MEASUREMENT OF NEUROSECRETORY ACTIVITY

Autoradiographic and liquid scintillation counting methods which follow the incorporation of ^{14}C- and ^{35}S-labelled cysteine or methionine into neurosecretory protein allow more certain estimates of neurosecretory activity to be made. These methods are of particular importance in following long-term changes in neurosecretory activity, but they also give more precise determinations of short-term changes. For example, the relative variations in activity of the neurosecretory system during somatic growth and ovarian growth can be accurately followed by autoradiography. During rapid oocyte development the neurosecretory cells contain only small amounts of stainable material (Highnam, 1962a, b) compared with the large amounts of material present in slowly developing females. In both groups of females the incorporation of [^{35}S]cysteine into the neurosecretory cells is much greater than the incorporation into the association neurones of the mushroom bodies.

During rapid oocyte development there is a rapid and high rate of incorporation of the label into the neurosecretory cells which is greater than that in females with slowly developing oocytes (Fig. 4a, b). In females with rapidly developing oocytes, ^{35}S incorporation into the centre of the corpus cardiacum follows the pattern of incorporation into the neurosecretory cells, whereas in slowly developing females there is progressive loss of ^{35}S-labelled material from the centre of the corpora cardiaca. These differences in incorporation rate are probably due to the rapid transport of

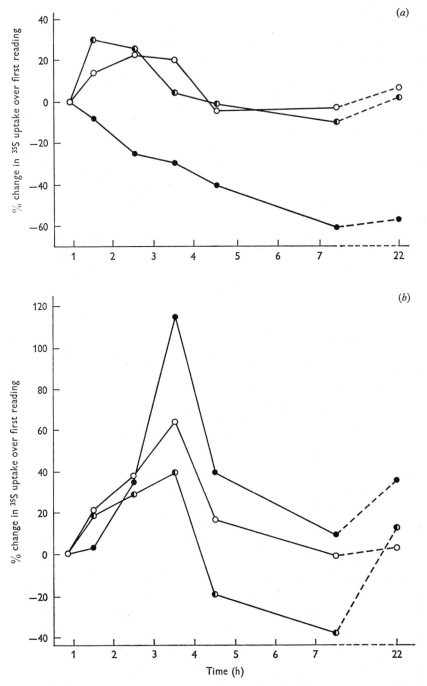

Fig. 4. The incorporation of [³⁵S]cysteine into the cerebral neurosecretory cells (○), centre of the corpus cardiacum (●) and periphery of the corpus cardiacum (◑) in females of *Schistocerca* with slowly developing oocytes (*a*) and rapidly developing oocytes (*b*).

newly synthesized labelled neurosecretion into the corpora cardiaca of females with rapid oocyte growth, but in slowly developing females the reduced synthesis of neurosecretion and its transport into the corpus cardiacum is unable to keep pace with the loss of neurosecretion from the glands.

The increase in label at the periphery of the corpora cardiaca of rapidly developing females is never as great as the increase which occurs in the centre of the gland. By contrast, the increase in radioactive label in the periphery of the corpora cardiaca of slowly developing females is always greater than in the centre of the glands. These observations suggest that the release of neurosecretion from corpora cardiaca into the blood is greater in females undergoing rapid oocyte maturation than in slowly developing females. Thus the application of autoradiographic methods amplifies and quantifies the histological analysis of long-term changes in activity.

More information can be gleaned from autoradiographic analysis of short-term changes in activity, such as those resulting from feeding and starvation. There is some synthesis of material in starved animals, but the incorporation rate of [^{35}S]cysteine is quite low. The corpora cardiaca in starved animals are initially labelled due to adsorption and interchange of the label with the neurosecretion already present in the gland. This initial labelling is quickly lost as a result of the combination of the effects of equilibration with the falling levels of label in the blood (Luntz, 1968) and the slow release of neurosecretion into the blood (Highnam, 1966; Highnam & Luntz, 1969). Labelled, freshly synthesized neurosecretion only appears in the corpora cardiaca some 7–9 h after injection of the label. Thus the inference from histological observations that starved animals have inactive neurosecretory systems receives considerable support. Moreover, estimates of the amounts of neurosecretion synthesized and released can be made. When starved animals are allowed to feed, the rapid change which occurs in the activity of the neurosecretory system can be more fully understood in autoradiographic analyses. Starved animals which are allowed to feed at the time of injection of the label show apparent low incorporation rates, but the incorporation continues for some 7 h. The apparently low incorporation rate results from the immediate transport from the neurosecretory cells of the newly labelled material. If the starved animals are not allowed to feed until 2 h after injection of the label, then the measured incorporation of the label into the neurosecretory cells is very high.

Analysis of the corpora cardiaca in these starved animals, which have subsequently been allowed to feed, reveals that large amounts of neurosecretion are released from the corpora cardiaca in response to the feeding stimuli. Moreover, there is a rapid movement into the corpora cardiaca of labelled newly synthesized neurosecretion. Thus feeding stimuli are potent in effecting release of neurosecretion and there follows a lapse of time before the stimuli are effective in initiating synthesis of neurosecretion. From these autoradiographic studies (Highnam & Luntz, 1969) and from histological observations (Highnam et al. 1966) it can be seen that when starved animals, with full neurosecretory systems, are allowed to feed, the whole neurosecretory system becomes devoid of stainable material within some 3 h. Thus from the relative sizes of the different parts of the neurosecretory system (Highnam, 1966; Highnam & Luntz, 1969), it is estimated that at least one corpora cardiacaful of material has been released in this time.

These observations on the variations in neurosecretory activity together with the different content of material within the neurosecretory system which result from environmental fluctuations are amply supported by direct measurements of the amount of label incorporated into the neurosecretory system by liquid scintillation counting methods (Luntz, 1968). This method has proved particularly suitable for analysing the dynamics of transport into, storage within and release from the corpora cardiaca of neurosecretion. Furthermore, it is possible to make use of the initial labelling of the corpora cardiaca, which occurs after injection of the label, to measure the different amounts of neurosecretion stored within the corpora cardiaca. Liquid scintillation methods give quantitative estimates of the amount of neurosecretion present within the corpora cardiaca at any one time and these direct measurements are obviously less tedious and considerably more reliable than subjective estimates of the degree of staining of the gland.

CORPUS ALLATUM ACTIVITY

Juvenile hormone from the corpora allata is necessary for normal oocyte growth in locusts and many other insects (Highnam, 1964); removal of the glands prevents yolk deposition in growing oocytes and causes the resorption of vitellogenic oocytes (Highnam et al. 1963a, b). Juvenile hormone may also be important in locusts in regulating the production of proteins necessary for yolk formation (Minks, 1967). It is difficult to estimate variations in corpus allatum activity which may result from changing environmental conditions. Corpus allatum activity, until recently, could only be judged by measuring the rate of change in a process, such as yolk deposition, which had been shown to be dependent upon continued secretion of the hormone. Estimates of juvenile hormone titres and corpus allatum activity in larval and adult forms are often spurious since the only indications of gland activity used have been gross changes in the animal's morphology. Gland size has often been used as an indicator of corpus allatum activity in locusts and other insects.

It is now clear that it is often invalid to equate the secretory activity of the corpus allatum with its size (Johannson, 1958; Staal, 1961; Mordue 1965a–c; 1967a; Lea & Thomsen, 1969). In *Locusta*, Staal (1961) has demonstrated that glands of varying size show either similar or quite different secretory activities. Changes in environmental factors such as crowding or flying produce marked changes in corpus allatum activity, and these changes are not always associated with a change in the volume of the gland (Highnam & Haskell, 1964). Reductions in feeding which affect yolk deposition and corpus allatum activity do not always result in a reduction in the size of the gland (Highnam *et al.* 1966). The variations in environmental stimuli which either reduce or accelerate the activity of the neuro-secretory system have similar effects upon the secretory activity of the corpus allatum. This similarity in effect is perhaps not surprising when it is remembered that the corpora allata may be primed and in some instances their activity maintained and regulated by neurosecretion (see Highnam, 1964; Mordue, 1965a–c, 1967a; de Wilde & de Boer, 1969).

Girardie (1963, 1965) provides evidence that in *Locusta* the corpora allata are directly stimulated and inhibited by specific trophic hormones from different cerebral neurosecretory cells. Highnam (1967) has reviewed the importance of other factors such as nervous stimuli or levels of haemolymph metabolites in regulating the production and release of corpus allatum hormone (see also Mordue, 1967a). A full understanding of variations in activity of the corpora allata in response to environmental stimuli is frequently obscured by the variety of factors which are said to regulate the activity of the glands in different insects. It is likely that the contradictory results of similar experiments in different insects, designed to elucidate control over the activity of the corpora allata, may be explained by the action of sequential regulators (Highnam, 1967; Mordue, 1967a). The relative importance of these sequentially acting regulators may differ in some insects, but it is probable that the corpora allata have to be primed, in the same way as the prothoracic glands, by a trophic factor from the neurosecretory system (Scharrer, 1964; Mordue, 1965a–c, 1967a; Strong, 1965; Odhiambo, 1966; Highnam, 1967; Girardie, 1963, 1965).

With the introduction of bioassay techniques, such as the *Galleria* test, which have been successfully applied to other insects (de Wilde *et al.* 1968), it is now possible to measure directly hormone output and circulatory levels of juvenile hormone in locusts (Johnson, unpublished observations).

These measurements should clarify the problems of the regulation of corpus allatum activity and the effects of environmental factors upon this activity.

THE ENVIRONMENT AND DIRECT OR NERVOUS CONTROL OF ENDOCRINE ACTIVITY

The effects of the environment upon the neuroendocrine system may be direct, or mediated via nervous pathways. Temperature can obviously have a direct effect upon the rates of developmental and physiological reactions, but there is evidence from some insects that this environmental regulator acts directly upon the neurosecretory cells (Williams, 1952; Church, 1955; Schneiderman & Horwitz, 1958). Photoperiod is known to be important in regulating the onset and rate of reproductive development: short day-length produces a reproductive diapause in *Nomadacris* (Norris, 1962a), *Schistocerca* (Norris, 1957) and *Anacridium* (Norris, 1964a). In *Nomadacris* reproductive diapause is thought to result from cessation of activity of the corpora allata (Strong, 1967b), although this was judged only from the small size of the glands.

An inactive neurosecretory system could well have been the prime factor in initiating the onset of diapause. In *Anacridium* (Geldiay, 1967) reproductive diapause has been shown to be a result of cessation of neurosecretory activity. (The relationships between the environment and diapause are discussed by de Wilde, this symposium, p. 487). It is of interest that in some insects light can act directly upon the brain, almost certainly upon neurones close to the cerebral neurosecretory cells, and consequently affect development (Adkisson, 1964; Williams & Adkisson, 1964; Williams Adkisson & Walcott, 1965; Lees, 1964; de Wilde, Duintjer & Mook, 1959; de Wilde & de Boer, 1969). However, the directness or otherwise of these environmental factors upon the neuroendocrine system in locusts is not clear. This problem of a direct or indirect regulation of endocrine activity by environmental factors is complicated (Highnam, 1967). It is well established, particularly in relation to certain developmental events such as the intervention of a period of diapause, that environmental effects can disrupt endocrine activity a long time subsequent to their initial perception by the insect. During this lapse of time the endocrine system functions normally.

The stimulus of crowding upon growth and development in locusts could exert its effect in a number of different ways. Crowding obviously influences hormone balance in the larval stages as well as in adults since crowded hoppers of *Schistocerca* and *Nomadacris* always have one instar less than isolated hoppers (Mossop, 1933; Burnett, 1951; Albrecht, 1955) and in *Anacridium* this phenomenon is often though not always seen (Volkonsky, 1937). Moreover, the rates of larval growth are more rapid in isolated hoppers than in crowded ones (Staal, 1961; Hunter Jones, in Uvarov, 1966). These effects of crowding upon development in larvae and adults may operate in a number of different ways: directly, as a result of increased individual activity (Ellis, 1951, 1964 a, b), via pheromones (Norris, 1954, 1964 b; Loher, 1960; Haskell, 1962) which may be either stimulatory or inhibitory (see Barth, this Symposium, p. 373, for an account of pheromone regulated processes) or via feeding (Norris, 1961, has shown that crowding stimulates food intake). The effects of crowding upon general activity in locusts are also rather obscure, but it has been suggested that humoral factors may be of importance (Ellis & Hoyle, 1954). The over-all activity of crowded *Schistocerca* is greater than that of isolated individuals, particularly prior to maturation (Norris, 1962 b).

Crowded hoppers of the gregarious phase also show increased marching activity compared with solitary hoppers (Ellis, 1951, 1964 a, b). Moreover, this activity is increased if the hoppers are starved. It is possible that activity is in some way related to hormone balance since during the phases of highest activity, during starvation, and before maturation, the levels of neurosecretory factors are low (Highnam, 1962 a; Highnam & Luntz, 1969; Mordue, 1966, 1969 a). The corpus allatum hormone has been implicated in the regulation of activity of adult locusts (Odhiambo, 1966) but the exact role of this hormone seems uncertain (Strong, 1968).

Feeding stimuli and the correct diet accelerate, whereas starvation or feeding sugar water reduce, endocrine activity in a number of insects (see Highnam, 1964; Mordue, 1967 a), but it is not known exactly how these factors operate. In *Locusta*, feeding stimuli are thought to be relayed via the frontal ganglia to the brain (and hence the neurosecretory cells). Removal of the frontal ganglion prevents impulses from stretch receptors passing to the brain and results in inactivity of the neurosecretory system, the locusts consequently failing to grow or undergo sexual maturation (Clarke & Langley, 1961, 1963 a–c; Gillot, 1964; Clarke & Gillot, 1965). However, in *Schistocerca* removal of the frontal ganglion prevents normal feeding (feeding is said to be normal following this operation in *Locusta* (Clarke & Gillot, 1965)) and prevents the fore-gut from emptying. The effects of frontal ganglion removal upon endocrine activity are therefore more than likely the result of semi-starvation; this is supported by the histology of the neurosecretory system following frontal ganglion removal, which suggests that the activity of the system lies between that of starved animals and that of normal animals (Highnam *et al.* 1966; Hill *et al.* 1966).

Thus feeding stimuli affect endocrine activity not simply by the frontal ganglion acting as a relay centre for the passage of nervous impulses to the brain (though undoubtedly this could be one of its functions), but by the intervention of other factors. The role of feeding as a regulator of endocrine activity would also seem to depend upon the inherent feeding pattern which exists both in the larval and adult forms of many insects. The prime importance of the single blood meal in each instar of *Rhodnius* in regulating development (Wigglesworth, 1936) is now axiomatic. Feeding patterns exist in both larval and adult locusts (Hill *et al.* 1966; Strong, 1967*a*; Hill & Goldsworthy, 1968) and in adult *Schistocerca* this pattern is not affected by experimental alteration of the endocrine balance (Hill *et al.* 1966), though the actual amount of food eaten is dependent upon factors such as crowding which *do* influence endocrine activity (Norris, 1961). Many insect larvae will delay moulting or metamorphosis if starved in the early part of the instar, but these events are accelerated if the larvae are starved after a certain critical time (Bounhiol, 1938; Fukuda, 1944; Friden, 1958; Johannson, 1958). This is probably due to lack of activation of the prothoracic glands in the early part of the instar and perhaps lack of activation of the corpora allata in the later part of the instar (Highnam, 1967).

Further evidence for the influence of feeding and diet upon hormone balance is gained from the effects of these factors upon larval growth and development in locusts (see Uvarov, 1966, for review). It is of particular interest that in *Melanoplus* individuals fed on oats have an extra instar in their developmental cycle. Thus as well as starvation, nutritionally deficient food (*Melanoplus* fed on oats have a high mortality rate) may upset the endocrine balance. Similarly, in many adult insects starvation, or lack of suitable food may alter the hormone balance and affect oocyte development (Johannson, 1958; Thomsen, 1965; Mordue, 1967*a*). In some insects such as *Calliphora* (Lea & Thomsen, 1962), *Rhodnius* (Wigglesworth, 1936), *Tenebrio* (Mordue 1965*a–c*, 1967*a*) and perhaps in *Schistocerca* (Mordue, unpublished observations) the endocrine system becomes active for a short time just after the imaginal moult. This activity is perhaps initiated by the moulting itself (Mordue 1967*a*). It is this inherent activity which is acted upon and modified by the variety of environmental factors which sustain hormone secretion and result in oocyte growth (Mordue, 1967*a*).

WATER BALANCE IN LOCUSTS

To satisfy their energy requirements and the demands of growth and reproduction locusts eat approximately their own weight of food each day. On a laboratory diet of fresh grass or lettuce the daily intake of water is between 1,500–2,000 μl (Hill *et al.* 1968). Osmotic problems posed by this water intake are alleviated by the action of a diuretic hormone produced by the neurosecretory system (Highnam, Hill & Gingell, 1965; Mordue, 1966, 1969 a) which is present in the storage lobe of the corpora cardiaca (Mordue & Goldsworthy, 1969). The diuretic hormone is probably a peptide (Mordue & Goldsworthy, 1969) and stimulates excretion through the Malpighian tubes. It also restricts the reabsorption of water from the gut lumen through the rectal wall back into the blood. The net result is that water loss through the excretory system is increased.

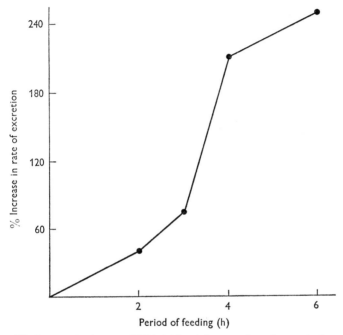

Fig. 5. The increase in the rate of Malpighian-tube excretion when maturing male *Schistocerca*, which had been starved for 6 days, were allowed to feed for different times (five animals/time interval).

Locusts with active neurosecretory systems, produced by the variety of environmental stimuli described above, have higher rates of excretion than animals with less active systems (Mordue, 1966, 1969 a). Since the diuretic hormone is associated with neurosecretion from the cerebral neurosecretory system it is possible to analyse neurosecretory activity using the physiological parameter of excretory rate. The validity of the principle is well shown by the effects of feeding and starvation upon the release of diuretic factors from the cerebral neurosecretory system (Mordue, 1966, 1969a). Malpighian tube excretory rate is low in starved animals and when such animals are allowed to feed, the rate of excretion increases markedly (Fig. 5).

From the rate of excretion at different time intervals after feeding commences together with the dose response curves of rates of excretion against concentration of diuretic hormone (Mordue & Goldsworthy, 1969) it is possible to quantify this hormone release. Within 3 h of feeding the equivalent volume, of one at least, corpora cardiaca of neurosecretion have been released into the haemolymph. This estimate of neurosecretory activity based upon physiological criteria is in very close agreement with the estimate based upon histological and autoradiographic methods (see above). Moreover, the differences in neurosecretory activity throughout the periods of growth and reproductive development can be estimated from measurements of excretory rate coupled with determinations of the rate at which the hormone is removed from the blood (Mordue, 1966, 1969 a). Females with developing oocytes release the equivalent, by volume, of one corpora cardiaca of neurosecretion every 4–6 h compared with a release rate, by volume, of 1 gland every 10–11 h in immature females (Mordue, 1966, 1969 a). Feeding stimuli are known to be important in causing the release of diuretic hormones in *Rhodnius* (Maddrell, 1964) and *Dysdercus* (Berridge, 1966). In *Schistocerca* measurements of the water content of faeces at different times after feeding show that the water content is highest during and immediately following feeding and falls during non-feeding periods (Norris, 1961). This is in close agreement with the observations that feeding releases the diuretic hormone in *Schistocerca*. Since environmental stimuli such as crowding and mating increase the levels of food intake and the output of the diuretic hormone, it seems possible that since feeding is such a potent factor in bringing about the rapid release of the diuretic factor (Mordue 1966, 1969 a) and other neurosecretions (Highnam *et al.* 1966) that increased feeding may act as a link between the environment and the endocrine system in some instances.

During periods of water conservation an anti-diuretic hormone is probably released from the glandular lobe of the corpus cardiacum (Mordue, 1969 b) (Fig. 6). This hormone has no effect upon Malpighian-tube function (Mordue & Goldsworthy, 1969) but produces a marked increase in reabsorption through the rectal wall. Thus water balance can be maintained or water content increased by the action of this anti-diuretic hormone, without restricting Malpighian tube functioning. The advantages of this system are obvious. In *Locusta* anti-diuretic and diuretic hormones have also been shown to control water balance and excretion (Cazal & Girardie, 1968). However, in *Locusta* a factor has been demonstrated which restricts Malpighian-tube functioning in *in vitro* systems. In intact locusts of both species storage lobes of the corpus cardiacum increase excretion and glandular lobes are without effect (Mordue, 1967 b;

Mordue & Goldsworthy, 1969). It is of interest that as far as the rectum is concerned different effects can be obtained with whole corpora cardiaca of *Schistocerca* and *Locusta* upon reabsorption from the gut lumen (Mordue, 1969*b*). Whole glands from *Schistocerca* depress rectal reabsorption in both species, whereas whole glands from *Locusta* stimulate reabsorption in both species.

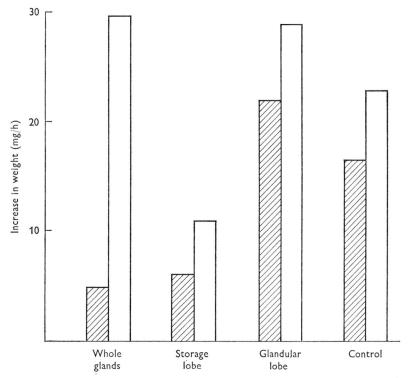

Fig. 6. The effect of whole corpora cardiaca, storage and glandular lobes upon reabsorption in isolated inverted recta of *Schistocerca* (▨) and *Locusta* (□). At least four preparations/experiment.

Obviously the effects of whole gland extracts upon physiological systems in locusts, and in other insects, should be analysed with caution. The exogenous or endogenous stimuli which elicit the release of the anti-diuretic hormone in locusts are not yet known and neither are the temporal variations in the content of this hormone within the glandular lobe of the corpora cardiaca. It seems probable that the anti-diuretic hormone will be released during non-feeding periods and at other times such as during the early part of the adult instar when the blood volume increases markedly (Mordue, 1966, 1969*a*; Strong, 1967*a*; Hill *et al.* 1968). During this time the locusts have relatively inactive neurosecretory systems (Luntz, 1968) and little diuretic hormone is released (Mordue, 1969a). The elucidation of the stimuli which cause the release of anti-diuretic hormone will be of considerable interest, since the hormonal regulation of the excretory system seems to be so versatile. Feeding—that is, water intake or factors which increase food intake—cause release of the hormone responsible for restoring water balance.

This review has been concerned particularly with endocrine mechanisms controlling development and diuresis. But neurosecretion in insects regulates a multiplicity of other developmental and physiological processes. (see Highnam, 1965). Since the amount and rate of release of the hormone(s) into the blood is regulated by environmental factors, these factors are able to exert their very pronounced effect over a wide range of events in the insect. Whether or not these diverse events are controlled by many or few neurosecretory factors is not clear. It may be that the same factor is variously modified by the different target tissues. However, in locusts only two biologically active fractions can be isolated from the corpora cardiaca and only one of these is associated with neurosecretion (Mordue & Goldsworthy, 1969). These observations are very similar to the extensive researches of Gersch and his co-workers who have also extracted two neurohormones from the nervous systems of a number of insects (Gersch, Unger & Fischer, 1956; Gersch, Fischer, Unger & Koch, 1960).

The number and role of the hormonal factors is further confused by the fact that whole corpora cardiaca are often used as a source of neurosecretion and it is now well established that these glands produce their own intrinsic secretions and whole glands may exert similar or different effects compared with neurosecretion when tested upon the same tissue (Mordue & Goldsworthy, 1969; Mordue, 1969b). Neurosecretion or corpus cardiacum factors regulate specific functions or can exert a general effect over certain processes. Specific functions such as Malpighian tube excretion may be regulated by neurosecretion from the brain in *Schistocerca* (Mordue, 1966, 1969a; Mordue & Goldsworthy, 1969), *Locusta* (Cazal & Girardie, 1968; Mordue, 1967b, 1969b) or in insects other than locusts such as *Anisotarsus* (Nunez, 1956), *Dysdercus* (Berridge, 1966), and *Carausius* (Unger, 1965; Vietinghoff, 1967). More general effects of the hormones are upon the stimulation of heart beat, gut movement and Malpighian tube movement in *Periplaneta* and other insects (see Davey, 1964).

In locusts Malpighian-tube movement (Mordue, unpublished observations) and heart beat are also stimulated by corpus cardiacum factors (Mordue & Goldsworthy, 1969). However, these general effects upon motility and heart beat are judged from *in vitro* studies and it perhaps should not be assumed that they play this role in the intact insect (Brown, 1965; Mordue & Goldsworthy, 1969). The corpora cardiaca exert a hyperglycaemic effect upon blood sugars in a number of insects (Steele, 1961, 1963; Bowers & Friedman, 1963; Wiens & Gilbert, 1967; Natalizi & Frontali, 1966; Friedman, 1967; Goldsworthy, 1968; Mordue & Goldsworthy, 1969) and these factors are restricted to the glandular lobe in locusts (Goldsworthy, 1968; Mordue & Goldsworthy, 1969). Part of this specific function of the corpus cardiacum factor is the stimulation of the production of active phosphorylase in the fat body tissue (Steele, 1963;

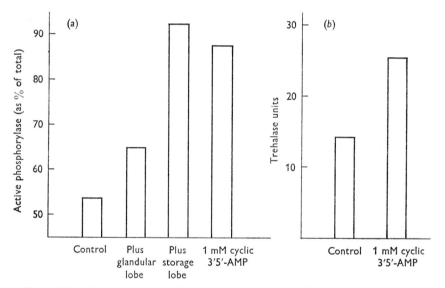

Fig. 7. The effects of corpus cardiacum extracts and cyclic 3′,5′-AMP upon enzyme levels in Malpighian tubes of *Schistocerca*. (*a*) Active phosphorylase levels, (*b*) trehalase levels.

Goldsworthy, 1968; Mordue & Goldsworthy, 1969). It is of interest that neurosecretion can elevate active phosphorylase levels and trehalase levels in locust Malpighian tubes (Mordue, 1969*c*).

It seems very likely that the action of the diuretic hormone upon the Malpighian tubes, in increasing enzyme levels is mediated via cyclic 3′,5′-AMP (Fig. 7*a, b*) (Mordue, 1969*c*). Neurosecretion may also exert a general effect upon metabolic events, and Hill (1962, 1965) has shown that protein synthesis by the fat body of *Schistocerca* is controlled by neurosecretion. For the secretion and production of particular proteins, such as yolk protein, the action of the corpus allatum hormone is needed in addition to the neurosecretion (Minks, 1967). The possibility that different target tissues may respond to different titres of hormone is not supported by the observations that protein synthesis, excretion, carbohydrate regulation and other events in locusts all require the same amounts of hormones to elicit responses in the different target tissues; they have thresholds of about 0·2 pairs of corpora cardiaca/300 μl haemolymph (Mordue & Goldsworthy, 1969).

REFERENCES

ADKISSON, P. L. (1964). Action of photoperiod in controlling insect diapause. *Am. Nat.* **98**, 357–74.

ALBRECHT, F. O. (1955). La densité des populations et la croissance chez *Schistocerca gregaria* (Forsk) et *Nomadacris septemfasciata* (Serv); la mue d'ajustement. *J. Agric. trop. Bot. appl.* **2**, 109–92.

BERRIDGE, M. J. (1966). The physiology of excretion in the cotton stainer, *Dysdercus fasciatus* Signoret. IV. Hormonal control of excretion. *J. exp. Biol.* **44**, 533–66.

BOUNHIOL, J. J. (1938). Recherches expérimentales sur le déterminisme de la métamorphose chez les Lépidoptères. *Bull. biol. Fr. Belg.* (Suppl.) **24**, 1–199.

BOWERS, W. S. & FRIEDMAN, S. (1963). Mobilization of fat body glycogen by an extract of corpus cardiacum. *Nature, Lond.* **198**, 685.

BROWN, B. E. (1965). Pharmacologically active constituents of the cockroach corpus cardiacum: resolution and some characteristics. *Gen. comp. Endocr.* **5**, 387–401.

BURNETT, G. F. (1951). Observations on the life history of the red locust, *Nomadacris septemfasciata* in the solitary phase. *Bull. ent. Res.* **42**, 473–90.

CAZAL, M. & GIRARDIE, A. (1968). Contrôle humoral de l'équilibre hydrique chez *Locusta migratoria migratorioides*. *J. Insect Physiol.* **14**, 655–68.

CHURCH, N. S. (1955). Hormones and the termination and reinduction of diapause in *Cephus cinctus* Vort. (Hymenoptera: Cephidae). *Can. J. Zool.* **33**, 339–69.

CLARKE, K. U. & GILLOT, C. (1965). Relationship between the removal of the frontal ganglion and protein starvation in *Locusta migratoria* L. *Nature, Lond.* **208**, 808–9.

CLARKE, K. U. & LANGLEY, P. (1961). Effect of removal of the frontal ganglion on the development of the gonads in *Locusta migratoria* L. *Nature, Lond.* **198**, 811.

CLARKE, K. U. & LANGLEY, P. (1963a). Studies on the initiation of growth and moulting in *Locusta migratoria migratorioides* R. & F. II. The role of the stomatogastric nervous system. *J. Insect Physiol.* **9**, 363–73.

CLARKE, K. U. & LANGLEY, P. (1963b). III. The role of the frontal ganglion. *J. Insect Physiol.* **9**, 411–21.

CLARKE, K. U. & LANGLEY, P. (1963c). IV. The relationship between the stomatogastric nervous system and neurosecretion. *J. Insect Physiol.* **9**, 423–30.

COLES, G. C. (1965). Haemolymph proteins and yolk formation in *Rhodnius prolixus* Stal. *J. exp. Biol.* **43**, 425–31.

DAVEY, K. G. (1964). The control of visceral muscles in insects. *Adv. Insect Physiol.* **2**, 219–45.

ELLIS, P. E. (1951). The marching behaviour of hoppers of the African migratory locust (*Locusta migratoria migratorioides* R. & F.) in the laboratory. *Anti-Locust Bull.* no. 7.

ELLIS, P. E. (1964a). Marching and colour in locust hoppers in relation to social factors. *Behaviour* **23**, 177–92.

ELLIS, P. E. (1964b). Changes in the marching of locusts with rearing conditions. *Behaviour* **23**, 193–202.

ELLIS, P. E. & HOYLE, G. (1954). A physiological interpretation of the marching of hoppers of the African migratory locust (*Locusta migratoria migratorioides* R. & F.). *J. exp. Biol.* **31**, 271–9.

FRIDEN, F. (1958). *Frass Drop Frequency in Lepidoptera.* Uppsala: Almquist and Wiksells.

FRIEDMAN, S. (1967). The control of trehalase synthesis in the blowfly *Phormia regina* Meig. *J. Insect Physiol.* **13**, 397–405.

FUKUDA, S. (1944). The hormonal mechanism of larval moulting in the silkworm. *J. Fac. Sci. Tokyo Univ.* **6**, 477–532.

GELDIAY, S. (1967). Hormonal control of adult reproductive diapause in the Egyptian grasshopper, *Anacridium aegyptium* L. *J. Endocr.* **37**, 63–71.

GERSCH, M., FISCHER, F., UNGER, H. & KOCH, H. (1960). Die Isolierung neurohormonaler Faktoren aus dem Nervensystem der Kuchenschabe *Periplaneta americana*. *Z. Naturf.* **15**b, 319–22.

GERSCH, M., UNGER, H. & FISCHER, F. (1956). Die Isolierung eines Neurohormons aus dem Nervensystem von *Periplaneta americana* und einige biol. Testverfahren. *Wiss. Z. Friedrich Schiller-Univ. Jena*, Heft 3/4, 125–9.

GILLOT, C. (1964). The role of the frontal ganglion in the control of protein metabolism in *Locusta migratoria*. *Helgoländer wiss. Meersunters.* **9**, 141–9.

GIRARDIE, A. (1963). Action de la pars intercerebralis sur le développement de *Locusta migratoria* L. *J. Insect Physiol.* **10**, 599–609.

GIRARDIE, A. (1965). Contribution à l'étude du contrôle de l'activité des corpora allata par la pars intercerebralis chez *Locusta migratoria* L. *C. r. hebd. Séanc. Acad. Sci., Paris* **261**, 4876–8.

GOLDSWORTHY, G. J. (1968). The action of hyperglycaemic factors from the corpora cardiaca of *Locusta migratoria*. Ph.D. Thesis, University of Sheffield.

HASKELL, P. T. (1962). Sensory factors influencing phase change in locusts. *Coll. int. Cent. nat. Rech. sci.* **114**, 145–63.

HIGHNAM, K. C. (1962a). Neurosecretory control of ovarian development in *Schistocerca gregaria*. *Q. Jl microsc. Sci.* **103**, 57–72.

HIGHNAM, K. C. (1962b). Neurosecretory control of ovarian development in the desert locust. *Mem. Soc. Endocr.* **12**, 379–90.

HIGHNAM, K. C. (1964). Endocrine relationships in insect reproduction. In *Insect Reproduction*, pp. 26–42. *Symp. no. 2 Royal Ent. Soc. Lond.* 1964.

HIGHNAM, K. C. (1965). Some aspects of neurosecretion in arthropods. *Zool. Jb. Physiol.* **71**, 558–82.

HIGHNAM, K. C. (1966). Estimates of neurosecretory activity during maturation in locusts. In *Insect Endocrines*, ed. V. J. A. Novak. Prague.

HIGHNAM, K. C. (1967). Insect hormones, *J. Endocr.* **39**, 123–50.

HIGHNAM, K. C. & HASKELL, P. T. (1964). The endocrine system of isolated *Locusta* and *Schistocerca* in relation to oocyte growth, and the effects of flying upon maturation. *J. Insect Physiol.* **10**, 849–64.

HIGHNAM, K. C., HILL, L. & GINGELL, D. (1965). Neurosecretion and water balance in the male desert locust (*Schistocerca gregaria*). *J. Zool.* **147**, 201–315.

HIGHNAM, K. C., HILL, L. & MORDUE, W. (1966). The endocrine system and oocyte growth in *Schistocerca* in relation to starvation and frontal ganglionectomy. *J. Insect Physiol.* **12**, 977–94.

HIGHNAM, K. C. & LUNTZ, A. J. (1969). Autoradiographic analysis of neurosecretory activity in locusts. (In preparation.)

HIGHNAM, K. C. & LUSIS, O. (1962). The effect of mature males on the neurosecretory control of ovarian development in the desert locust. *Q. Jl microsc. Sci.* **103**, 73–83.

HIGHNAM, K. C., LUSIS, O. & HILL, L. (1963a). The role of the corpora allata during oocyte growth in the desert locust *Schistocerca gregaria* Forsk. *J. Insect Physiol.* **9**, 587–96.

HIGHNAM, K. C., LUSIS, O. & HILL, L. (1963b). Factors affecting oocyte resorption in the desert locust *Schistocerca gregaria* Forsk. *J. Insect Physiol.* **9**, 827–37.

HILL, L. (1962). Neurosecretory control of haemolymph protein concentration during ovarian development in the desert locust. *J. Insect Physiol.* **8**, 609–19.

HILL, L. (1965). The incorporation of C^{14}-glycine into the proteins of the fat body of the desert locust during ovarian development. *J. Insect Physiol.* **11**, 1605–15.

HILL, L. & GOLDSWORTHY, G. J. (1968). Growth, feeding activity, and utilisation of reserves in larvae of *Locusta*. *J. Insect Physiol.* **14**, 1085–98.

HILL, L., LUNTZ, A. J. & STEELE, P. A. (1968). The relationships between somatic growth, ovarian growth, and feeding activity in the adult desert locust. *J. Insect Physiol.* **14**, 1–20.

HILL, L., MORDUE, W. & HIGHNAM, K. C. (1966). The endocrine system, frontal ganglion and feeding during maturation in the female desert locust. *J. Insect Physiol.* **12**, 1197–208.

JOHANNSON, A. S. (1958). Relation of nutrition to endocrine-reproductive function in the milkweed bug *Oncopeltus fasciatus* (Dallas) (*Heteroptera: Lygaeidae*). *Nytt. Mag. Zool.* **7**, 3–132.

LEA, A. O. & THOMSEN, E. (1962). Cycles in the synthetic activity of the medial neurosecretory cells of *Calliphora erythrocephala* and their regulation. *Mem. Soc. Endocr.* **12**, 345–7.

LEA, A. O. & THOMSEN, E. (1969). Size independent secretion by the corpus allatum of *Calliphora erythrocephala*. *J. Insect Physiol.* **15**, 477–82.

LEES, A. D. (1964). The location of the photoperiodic receptors in the aphid *Megoura viciae* Buckton. *J. exp. Biol.* **41**, 119–33.

LOHER, W. (1960). The chemical acceleration of the maturation process and its hormonal control in the male desert locust. *Proc. Roy. Soc. Lond.* B **153**, 380–97.

LUNTZ, A. J. (1968). Neurosecretory activity and growth during reproductive development in *Schistocerca gregaria* Forsk. M.Sc. thesis, University of Sheffield.

MADDRELL, S. H. P. (1964). Excretion in the blood sucking bug, *Rhodnius prolixus* Stal. III. The control of the release of the diuretic hormone. *J. exp. Biol.* **41**, 459–72.

MINKS, A. K. (1967). Biochemical aspects of juvenile hormone action in the adult *Locusta migratoria*. *Archs. néerl. Zool.* **17**, 175–258.

MORDUE, W. (1965a). Studies on oocyte production and associated histological changes in the neuroendocrine system in *Tenebrio molitor* L. *J. Insect Physiol.* **11**, 493–503.

MORDUE, W. (1965b). The neuroendocrine control of oocyte production in *Tenebrio molitor* L. *J. Insect Physiol.* **11**, 505–11.

MORDUE, W. (1965c). Neuroendocrine factors in the control of oocyte production in *Tenebrio molitor* L. *J. Insect Physiol.* **11**, 617–29.

MORDUE, W. (1966). Hormones and water balance in locusts. In *Insect Endocrines*, ed. V. J. A. Novak, Prague.

MORDUE, W. (1967a). The influence of feeding upon the activity of neuro-endocrine system during oocyte growth in *Tenebrio molitor*. *Gen. comp. Endocr.* **9**, 406–15.

MORDUE, W. (1967b). Some physiological effects of corpus cardiacum factors in locusts. *Gen. comp. Endocr.* **9**, 475.

MORDUE, W. (1969a). Hormonal control of Malpighian tube and rectal function in the desert locust, *Schistocerca gregaria*. *J. Insect Physiol.* **15**, 273–85.

MORDUE, W. (1969b). Evidence for the existence of diuretic and anti-diuretic hormones in locusts. *J. Endocr.* (in the Press).

MORDUE, W. (1969c). A possible mode of action of the diuretic factor in locusts. (In preparation).

MORDUE, W. & GOLDSWORTHY, G. J. (1969). The physiological effects of corpus cardiacum extracts in locusts. *Gen. comp. Endocr.* **12**, 360–9.

MOSSOP, M. C. (1933). Description of hopper instars of the red locust, *Nomadacris septemfasciata* Serv. phase gregaria and some changes in adult colouration. *Proc. Rhod. Sci. Ass.* **32**, 113–18.

NATALIZI, C. M. & FRONTALI, N. (1966). Purification of insect hyperglycaemic and heart accelerating hormones. *J. Insect Physiol.* **12**, 1279–87.

NORRIS, M. J. (1954). Sexual maturation in the desert locust (*Schistocerca gregaria* (Forsk.) with special reference to the effects of grouping. *Anti-Locust Bull.* no. 18.

NORRIS, M. J. (1957). Factors affecting the rate of sexual maturation of the desert locust (*Schistocerca gregaria* Forsk.) in the laboratory. *Anti-Locust Bull.* no. 28.

NORRIS, M. J. (1961). Group effect on feeding in adult males of the adult desert locust, *Schistocerca gregaria* (Forsk.), in relation to sexual maturation. *Bull. ent. Res.* **51**, 731–53.

NORRIS, M. J. (1962a). Diapause induced by photoperiod in a tropical locust, *Nomadacris septemfasciata* (Serv.). *Ann. appl. Biol.* **50**, 600–3.

NORRIS, M. J. (1962b). The effects of density and grouping on sexual maturation, feeding and activity in caged *Schistocerca gregaria*. *Coll. int. Cent. nat. Rech. sci.* **114**, 23–35.

NORRIS, M. J. (1964a). Reproduction of the grasshopper *Anacridium aegyptium* L. in the laboratory. *Proc. R. ent. Soc. Lond.* **40**, 19–29.

NORRIS, M. J. (1964b). Accelerating and inhibiting effects of crowding on sexual maturation in two species of locusts. *Nature, Lond.* **203**, 784–5.

NUNEZ, J. A. (1956). Untersuchungen über die Regelung des Wasserhaushaltes bei *Anisotarsus cupripennis* Germ. *Z. vergl. Physiol.* **38**, 341–54.

ODHIAMBO, T. R. (1966). The metabolic effects of the corpus allatum hormone in the male desert locust. II. Spontaneous locomotor activity. *J. exp. Biol.* **45**, 45–50.

PHIPPS, J. (1950). The maturation of ovaries and relation between weight and maturity in *Locusta migratoria migratorioides* (R. & F.). *Bull. ent. Res.* **40**, 539–57.

SCHARRER, B. (1964). Histophysiological studies on the corpus allatum of *Leucophaea maderae*. IV. Ultrastructure during normal activity cycle. *Z. Zellforsch. mikrosk. Anat.* **64**, 301–26.

SCHNEIDERMAN, H. A. & HORWITZ, J. (1958). The induction and termination of facultative diapause in the chalcid wasps *Mormoniella vitripennis* (Walker) *Tritheptis kluggii* (Ratzeburg). *J. exp. Biol.* **35**, 520–51.

STAAL, G. B. (1961). Studies on the physiology of phase induction in *Locusta migratoria migratorioides* R. & F. Ph.D. thesis, Wageningen, Veenman and Zonen.

STEELE, J. E. (1961). Occurrence of a hyperglycaemic factor in the corpus cardiacum of an insect. *Nature, Lond.* **192**, 680–1.

STEELE, J. E. (1963). The site of action of insect hyperglycaemic hormone. *Gen. comp. Endocr.* **3**, 46–52.

STRONG, L. (1965). The relationships between the brain, corpora allata and oocyte growth in the Central American locust, *Schistocerca* sp. II. The innervation of the corpora allata, the lateral neurosecretory system and oocyte growth. *J. Insect Physiol.* **11**, 271–80.

STRONG, L. (1967a). Feeding activity, sexual maturation, hormones, and water balance in the female African Migratory locust. *J. Insect Physiol.* **13**, 495–507.

STRONG, L. (1967b). Endocrinology of diapause in the female red locust, *Nomadacris septemfasciata* (Serv). *Nature, Lond.* **212**, 1276–7.

STRONG, L. (1968). The effect of enforced locomotor activity on lipid content in allatectomized males of *Locusta migratoria migratorioides*. *J. exp. Biol.* **48**, 625–30.

TELFER, W. H. (1965). The mechanism and control of yolk formation. *Ann. Rev. Ent.* **10**, 161–84.

THOMSEN, M. (1965). The neurosecretory system of the adult *Calliphora erythrocephala*. II. Histology of the neurosecretory cells of the brain and some related structures. *Z. Zellforsch. mikrosk. Anat.* **67**, 693–717.

UNGER, H. (1965). Der Einfluss der Neurohormone C und D auf die Farbstoffabsorptions-fähigkeit der Malpighischen Gefässe (und des Darmes) der Stabheuschrecke *Carausius morosus* (Br) *in vitro*. *Zool. Jb. Physiol.* **71**, 710–17.

UVAROV, B. (1966). *Grasshoppers and Locusts*, vol. I. Cambridge University Press.

VIETINGHOFF, U. (1967). Neurohormonal control of renal function in *Carausius morosus*. *Gen. comp. Endocr.* **9**, 503.

VOLKONSKY, M. A. (1937). Elevage et croissance larvaire du criquet égyptien (*Anacridium aegyptium*). *C. r. Séanc. Soc. Biol.* **125**, 739–42.

WIENS, A. W. & GILBERT, L. I. (1967). Regulation of carbohydrate mobilization and utilization in *Leucophaea maderae*. *J. Insect Physiol.* **13**, 779–94.

WIGGLESWORTH, V. B. (1936). The function of the corpus allatum in growth and reproduction in *Rhodnius prolixus* (Hemiptera). *Q. Jl microsc. Sci.* **79**, 91–121.

DE WILDE, J. & DE BOER, J. A. (1969). Humoral and nervous pathways in the photoperiodic induction of diapause in *Leptinotarsa decemlineata J. Insect Physiol.* **15**, 661–75.

DE WILDE, J., DUINTJER, C. S. & MOOK, L. (1959). Physiology of diapause in the adult Colorado beetle. I. The photoperiod as a controlling factor. *J. Insect Physiol.* **3**, 75–85.

DE WILDE, J., STAAL, G. B., DE KORT, C. A. D. & BAARD, G. (1968). Juvenile hormone titre in the haemolymph as a function of photoperiodic treatment in the adult Colorado beetle (*Leptinotarsa decemlineata* Say.). *Proc. K. ned. Akad. Wet.* C **71**, 321–6.

WILLIAMS, C. M. (1952). Physiology of insect diapause. IV. The brain and prothoracic glands as an endocrine system in the cecropia silkworm. *Biol. Bull. mar. biol. Lab., Woods Hole* **103**, 120–38.

WILLIAMS, C. M. & ADKISSON, P. L. (1964). Physiology of insect diapause XIV. An endocrine mechanism for the photoperiodic control of pupal diapause in the oak silkworm, *Antheraea pernyi*. *Biol. Bull. mar. biol. Lab., Woods Hole* **127**, 511–25.

WILLIAMS, C. M., ADKISSON, P. L. & WALCOTT, C. (1965). Physiology of insect diapause. XV. The transmission of photoperiod signals to the brain of the oak silkworm, *Antheraea pernyi*. *Biol. Bull. mar. biol. Lab., Woods Hole* **128**, 497–507.

DISCUSSION

BERN: Is the diuretic factor that you discussed active on the abluminal side; in other words is the insect rectum comparable to the toad bladder in this respect?

MORDUE: I think it might work through the gut lumen side but I haven't done sufficient work on that aspect; normally it works on the blood side rather than the epithelial side.

GREEP: I wonder what effect variations in some other aspects of the environment such as temperature and particularly photoperiodicity may have on the reactions you have described?

MORDUE: The animals are maintained under standard conditions of photoperiod and temperature. We haven't actually done a lot of this basic work on the effects of photoperiod and other environmental factors in Sheffield, but a considerable amount of work has been done on the effects of varying photoperiod upon rates of reproductive development in a large number of different types of locusts. The effects of environmental factors on inducing diapause in a number of different insects is now well understood, and I think that the effects of photoperiod in particular will be dealt with by Professor de Wilde in his paper.

LEATHEM: Did you have any special purpose in using [^{35}S]cysteine for your studies and have you looked at other sulphur labelled amino acids, since this incorporation may be different depending upon the amino acid that you are using, and the protein into which it is being incorporated.

MORDUE: I would like to pass that one to Dr Highnam.

HIGHNAM: We used [^{35}S]cysteine initially because it was easily available, it was cheap and had been used by Sloper (*J. Endocr.* **20**, 9, 1960) to follow the incorporation into neurosecretory protein in rats. We have used both cysteine and methionine and we now use [^{14}C]cysteine, because one can follow this isotope with liquid emulsion autoradiography which permits a much more exact location of the label within the cells, than when ^{35}S label is used. The patterns of incorporation, transport and release of neurosecretion we obtain with [^{14}C]cysteine are almost identical to those obtained with [^{35}S]cysteine.

BERN: As long as the significance of the uptake data has been raised, I wonder whether one couldn't suggest an alternative possibility for the data on the first slide. The 'rapid transport' would imply a rate of movement greater than a couple of millimetres within a couple of hours in order to account for that burst of radioactivity within the corpus cardiacum. I was interested in the fact that approximately 20 h later there was a peak which was as large as the initial peak 3 h after injection of the material.

Could there not be an initial burst of synthesis *in situ* with the peak occurring 20 h later as the result of transport to that area? The difficulty with this kind of experiment is that one operates with some preconceptions and then one tends to interpret the graph to fit the preconceptions.

HIGHNAM: My preconception when this work was started, which was derived from a whole lot of insect neurosecretory work in the literature, was that the rates of synthesis, transport and release of neurosecretion would be very slow indeed. The actual results showed a fairly rapid incorporation into the cells and the fairly rapid build-up at the ends of the neurosecretory axons and in certain experimental circumstances a rapid release from the corpus cardiacum and these rapid rates were entirely against the initial preconception.

I was rather interested two years ago in Brno to talk to Dr Niemierko. She had been studying the transport of choline esterases down axons of ordinary neurones and was getting rates of transport which were very similar to the rates of transport which occur in locust neurosecretory cells, and it seems to me that in explaining these rates of transport in ordinary neurones and those in locust neurosecretory cells the idea of axoplasmic movement really cannot apply and perhaps we have to think in terms of a microtubal arrangement of transport channels within the axons of the kind demonstrated by Wigglesworth some years ago in *Rhodnius* neurones.

Both the autoradiographic and the histological methods and what I might call Dr Mordue's experimental results suggest that in the locust there is a very rapid turnover in neurosecretion with fairly high concentrations of neurosecretory hormones within the blood. These concentrations are fairly high even when the system is inactive and are extremely high when the system is active. All this evidence points to a very rapid turnover in the neurosecretory system and this rapidly changing system is perhaps more sensitive to modulation by exogenous and endogenous events. Our results are certainly contrary to the whole concept of insect neurosecretion up to three or four years ago, when it was thought that the neurosecretory system turned over extremely slowly and the concentration of hormones in the blood was very low indeed. But certainly all our experience in the locust indicates quite the contrary, and, I repeat, is contrary to our preconceived ideas.

THE ENDOCRINE MECHANISM
OF AMPHIBIAN METAMORPHOSIS, AN
EVOLUTIONARY ACHIEVEMENT

By WILLIAM ETKIN

Some 300 million years ago a line of Crossopterygian fish gave rise to the first amphibians, commonly called the Stegocephalians. The Crossopterygians apparently lived in shallow fresh waters subject to seasonal drying. In adaptation to this environment they had evolved lungs and lobe-fins for crawling along the bottom (Szarski, 1962, Romer, 1966, Schmalhausen, 1968). The step to the amphibian was therefore a small one, involving chiefly the development of the typical tetrapod limb. The mode of life of the early stegocephalian could not have differed greatly from that of their fish ancestors. In fact the earliest known fossil form, Ichthyostega, has been described as a 'four-footed fish', for in habits and structure, except for the limbs, it closely resembled its fish ancestors. There is even a hint of the persistence of gills into the adult stage.

The importance of this concept of the origin of the amphibia for our present discussion is that it indicates that the earliest amphibians had a direct development with no specialized metamorphosis. The subsequent evolution of the two principal groups of amphibia, the urodeles (salamanders) and the anurans (frogs and toads), was characterized by the adaptation of the early or larval stages for life in specialized niches in the water environment while the adults evolved divergent adaptations for terrestrial living. This tendency for separate evolution in the larva and the adult necessitated the development of a transitional period in which the specializations of the larva are eliminated and the characteristics of the adult emerge. This is the period we call metamorphosis.* It is most highly developed in the anurans, where the tadpole larva is adapted to the life of a swimming herbivore and the adult is the jumping carnivorous frog. We shall concentrate most of our attention on the anurans because the very complexity and precision of their metamorphosis gives us numerous 'handles' for experimental analysis.

---

* The term metamorphosis in amphibians is still used in widely different senses. Here we mean the period of transformation, under thyroid control, of larva to adult. Earlier transformations of embryo to tadpole are excluded as are the minor changes during tadpole development which occur in the thyroidectomized animal. The criterion of thyroid dependence establishes a clear-cut distinction in most anurans between the premetamorphic period and early or prometamorphosis. This is often unrecognized by some experimentalists, who refer only to climax changes.

The early development of the limbs usual in vertebrate embryos has been suppressed during the larval stage of the frog to be released at metamorphosis. The ancestral carnivorous mouth and digestive apparatus has likewise been suppressed and a specialized larval mouth and gut substituted.

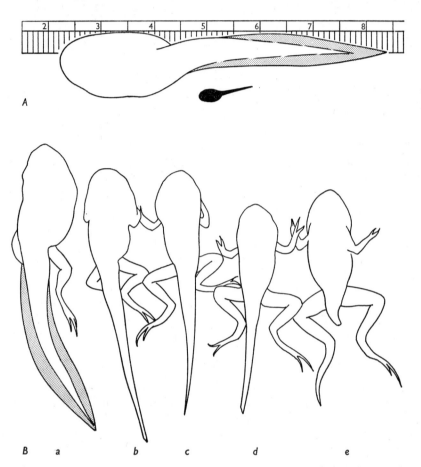

Fig. 1. Normal tadpoles of *Rana pipiens*. (*A*) at the beginning and end of premetamorphic (growth) period. (*B*) in metamorphic climax. (*a*) Day before emergence of foreleg (E − 1), (*b*) E + 0, (*c*) E + 1, (*d*) E + 2, (*e*) E + 4. (Metric Scale).

These newly evolved tadpole structures are broken down at metamorphosis and the development of a carnivorous feeding mechanism activated. The segmented muscles of the fish-like stegocephalians are retained as the primary locomotor apparatus in the tadpole. As in fish, they extend into the large tail and enable the animal to swim by undulatory movements. This primitive apparatus is destroyed at metamorphosis and replaced by the complex saltatory apparatus of the frog.

To transform the larval locomotor system to that of the adult frog requires that the legs with associated girdles and muscles be induced to grow rapidly and attain adult proportions *before* the tail is resorbed. Consequently in anurans we find the metamorphic period comprises two

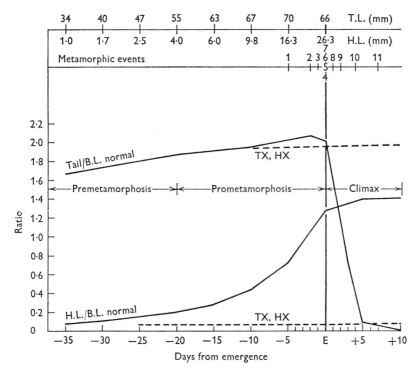

Fig. 2. Pattern of metamorphosis in *Rana pipiens*. Data from one batch of normal animals raised at 23 ± 1 °C shown by solid line. Comparable data for thyroidectomized animals (TX) and hypophysectomized animals (HX) are shown by broken lines. The metamorphic events indicated in the figure by numbers are as follows: (1), anal canal piece, first definite reduction. 2, Anal canal piece reduction completed. 3, Skin window for the forelegs clearly apparent. 4, Loss of second (both) beaks. In experimental animals this is the most satisfactory criterion for the beginning of climax. 5, Emergence of first foreleg to appear. 6, Emergence of second foreleg (E). 7, Mouth widened to level of nostril. Events 4, 5, 6, 7 usually occur within a period of 24 h. 8, Mouth widened to level between nostril and eye. 9, Mouth widened to level of anterior edge of eye. 10, Mouth widened past level of middle of eye. 11, Tympanum definitely recognizable. TL, Total body length (mm.) HL, Hind-limb length (mm) BL, body length.

distinct phases, the first or prometamorphic period is characterized by rapid leg growth and certain other growth processes such as that of skin glands. In the common American species (*Rana pipiens*) this lasts about 3 weeks at 25 °C. This is followed by a shorter period of about 1 week called metamorphic climax, during which the tail and gills are resorbed and the feeding apparatus transformed (Figs. 1 and 2). The length of the larval period is very variable. Some species (i.e. common bullfrogs) remain in a stable tadpole stage, increasing in size but not differentiation, for several seasons. Some species (American toads) pass through the larval stage in a few weeks. On the other hand, the periods of prometamorphosis and of climax are of comparable lengths in different species, varying only in proportion to the size of the animal. There are, of course, many other types of life-histories to be found among amphibians, varying from the suppression of the larval stage to the contrary suppression of the land phase, but we shall not be concerned with these except where they help in the experimental analysis.

Looked at from the viewpoint of the experimentalist, the typical anuran life-history presents many challenging questions. What is the 'clock' mechanism that breaks the stable tadpole premetamorphic state and initiates metamorphic transformation? What is the timing mechanism that determines the correct sequence and scheduling of the changes during prometamorphosis and climax?

When we consider the complexity of the evolutionary and developmental changes involved in anuran metamorphosis it is all the more astonishing that the primary endocrine mechanism involved in its regulation should be as simple as it is. The varied tissue changes are induced by thyroid hormone (thyroxine, T_4, or triiodothyronine, T_3). The thyroid itself is under control of the thyrotropic hormone (TSH) produced by the anterior pituitary. These conclusions were drawn from experiments done in the first quarter of this century and substantiated repeatedly since. I shall not review this literature since numerous adequate reviews are available (Bounhiol, 1942; Lynn & Wachowski, 1952; Kollros, 1961; Etkin, 1963, 1968). Rather we will ask the fundamental questions mentioned above in terms of the pituitary–thyroid (PT) axis. Let us first consider how the different events, all responding to the thyroid stimulus, are spaced out during prometamorphosis and climax.

An obvious possibility for the patterning of metamorphic change is that of a rising concentration of thyroid hormone. Early studies gave clear morphological evidence that thyroid activity was low before metamorphosis and is progressively increased during prometamorphosis to reach an extremely high level at the beginning of climax (for recent discussions see Fox (1966), Coleman, Evennett & Dodd (1968). The picture that emerged from such morphological studies is that premetamorphic thyroid activity is extremely low. Prometamorphosis is characterized by increasing hormone synthesis accompanied by storage and some release of hormone. At the beginning of climax the high cellular activity accompanied in some species by partial evacuation of colloid suggests an explosive activation of

the gland in synthesis and release of hormone. At the conclusion of metamorphosis the thyroid returns to the inactive condition in most anurans.
More recent work with biochemical criteria (radio-iodine and PBI) fully
confirm this concept at the qualitative level but have been disappointing
in not providing reasonable quantification of the activity levels at different

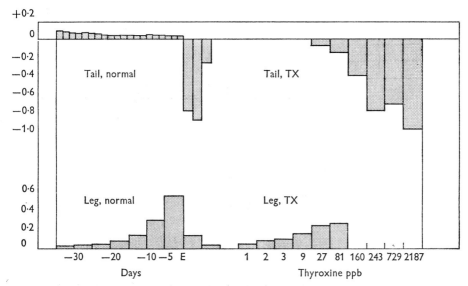

Fig. 3. Results of an experiment (right) to test the effect of different concentrations of
thyroxine (TX) on hind-leg growth and tail resorption in half-grown normal tadpoles.
For comparison the rates of change of leg and tail in normal metamorphosis are shown at
the left. The rates of change are expressed as proportions of body length to make the data
of smaller experimental animals comparable to controls. Each bar for the tail shows the
ratio change over a 2-day period in normal animals and for days 3–5 after exposure to T_4
for the experimentals. Leg ratios are shown for 5-day periods in normals and for days 3–8
in experimentals. It may be noted that in the normals the legs grow most rapidly in the
5-day period before climax begins and tail reduction begins after emergence of the forelegs and drops by 0·8 of the body length in each of the next two 2-day periods. In the
experimental animals it can be seen that the rate of leg growth produced by 27 ppb
thyroxine is about maximum. Animals in concentrations higher than 81 ppb do not survive long enough to provide adequate data. The tail does not show maximal rate of
reduction until the concentration of 243 ppb is reached. It may be concluded that maximal
leg growth such as occurs during late prometamorphosis requires a T_4 concentration
about one-tenth that necessary for the maximal rate of tail resorption.

stages of metamorphosis. Such quantification, however, has been provided
by studies of the responsiveness of tissues to immersion in different concentrations of T_4. It has long been known that early prometamorphic
change (leg growth) responds to low concentrations of hormone and climax
requires 10–20 times higher concentration to achieve normal rates of
response (Fig. 3). Recent studies have used isolated tail tissue *in vitro*
(Derby, 1968). These demonstrate a dramatic increase in endogenous

levels of hormone at the beginning of climax (day of emergence of the forelegs, marked E on the charts). Tissues isolated after E prove to be already programmed for self-destruction, whereas those isolated 1 or 2 days before require the addition of exogenous hormone for this process (Fig. 4). The evidence thus indicates that the scheduling of metamorphic events during prometamorphosis and climax is effected by (1) a rising concentration of thyroid hormone during prometamorphosis and (2) an explosive release of hormone at the beginning of climax (see Fig. 10) It

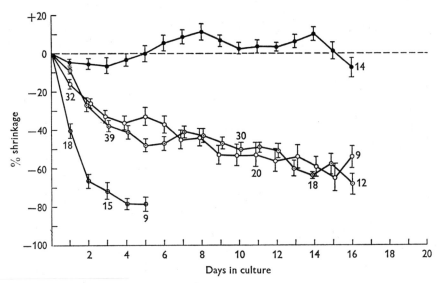

Fig. 4. Rate of shrinkage of tail fin disks in organ culture without exposure to exogenous thyroxine. The prometamorphic disks show no shrinkage but those cut at the beginning of climax show a partial response and those cut at E+1 show complete response, indicating a sudden rise in endogenous hormone at day of E. (From Derby, 1968.) Untreated disks. ●——●, Early prometamorphosis; ○——○, E−1; ⊙——⊙, E+o; ⊕——⊕, E+1).

may be noted that this interpretation indicates that the exact regulation of the rate of rise of hormone concentration during prometamorphosis is essential for the success of the process but that the exact level of hormone action at climax is irrelevant provided the lowest level for maximal response is attained. Thus in artificially induced metamorphosis it is easy to induce climax events of a normal character by simply using strong doses, whereas the induction of a normal prometamorphosis requires the precise regulation of a schedule of increasing dosage (Plate 1).

The dependence of thyroid activity upon pituitary TSH in the tadpole was clearly established by the results of extirpation and injection experiments during the 1920s and 1930s. Cytological studies of the pituitary in relation to metamorphosis have generally been interpreted as supporting the concept of increased activity during metamorphosis (see, for example, van Oordt, 1966, Kerr, 1966). Similarly, work with TSH preparations is in general agreement with the concept of control of thyroid by pituitary TSH, since injections of TSH preparations promote metamorphic change in proportion to the amount of injected hormone. Thus despite admitted lacunae in the evidence (principally lack of evidence of changes in hormone levels in the blood) we are led to the inference that the pattern of metamorphic change (Fig. 2) is governed by the pattern of activation of the PT axis. This axis shows increasing activity during prometamorphosis and culminates in a high peak at the beginning of climax. This concept brings us to the problem of the control of the PT axis during the animal's development.

The concept of hypothalamic regulation of the PT axis as developed for vertebrates generally in the 1950s was soon extended to the tadpole. Autotransplantation of the primordium of the pituitary to the tail of the frog embryo yields an animal which grows more rapidly than the normal and is excessively dark in pigmentation (due to hyperactivity of the pars intermedia) but which fails to go through metamorphic climax. Some individuals show a late development of a slow prometamorphosis but many show no metamorphic change though growing far beyond the normal size or time of metamorphosis (Fig. 5). Similarly the insertion of a barrier between the hypothalamus and the pituitary in a salamander prevents metamorphosis provided the barrier is not by-passed by regenerating blood vessels. Isolation of the pituitary by destruction of the posterior hypothalamus permits a delayed prometamorphosis but not climax (Hanaoka, 1967). Destruction of the pre-optic nucleus stops metamorphosis (Goos, 1969). These experiments indicate that a hypothalamic thyrotropin releasing factor (TRF) passing through the pituitary portal vessels activates pituitary TSH production at metamorphosis. In the absence of portal transfer, as in tadpoles with transplanted glands or with the posterior hypothalamus destroyed, sufficient TRF may sometimes reach the pituitary through the general circulation to elicit a slow and delayed prometamorphosis but not enough to induce the high level of PT activity necessary for climax changes. A study of the normal development of the hypothalamic–pituitary region supported this concept for it showed that the median eminence region of the hypothalamus differentiates its vascular area during prometamorphosis producing a well-defined portal circulation by the beginning of climax (Fig. 6). The implication of this finding is that the initiation and patterning of metamorphic change is a function of hypothalamic development. The metamorphic 'clock' in other words is in the hypothalamus.

W. ETKIN

It was thought that this hypothalamic 'clock' could be dramatically displayed by its differentiation in the thyroidectomized tadpole in which all other tissues remain in the larval condition. To our surprise the hypothalamic–pituitary area in giant (non-metamorphosing) thyroidectomized animals proved to be like that of the premetamorphic animal rather than

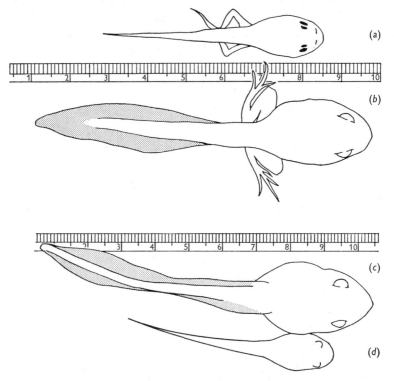

Fig. 5. Effect of heterotopic transplantation of the adenohypophyseal primordium to the tail in embryos of *Rana pipiens*. (*a*) Normal control in late prometamorphosis; (*b*) experimental in stasis in corresponding metamorphic stage; (*c*) experimental, showing no metamorphosis despite large size; (*d*) normal control at maximal premetamorphic size. (Metric Scale)

like the metamorphosed controls (Plate 2). It would appear therefore that the development and activity of the median eminence region of the hypothalamus cannot be a controlling event initiating metamorphosis but is itself dependent upon the metamorphic stimulus. Is it possible it is under thyroid control? By treating giant thyroidectomized tadpoles with appropriately graded concentrations of exogenous thyroxine they were brought into metamorphic climax. Examination of the brains of such animals showed that the median eminence had indeed been transformed to climax morphology in complete parallelism with the changes induced in the other

PLATE I

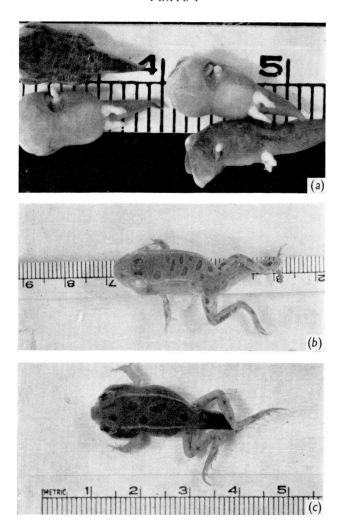

(a) Abnormal metamorphosis resulting from treating tadpoles by immersion in strong thyroxine (243 ppb). Normal-type metamorphosis induced in a large thyroidectomized tadpole (b) and hypophysectomized animal (c) after treatment with a sequence of increasing concentrations of thyroxine, using low concentrations (3–27 ppb) until legs are fully grown, then strong concentration (243 ppb) to bring the tail down at about the normal rate.

PLATE 2

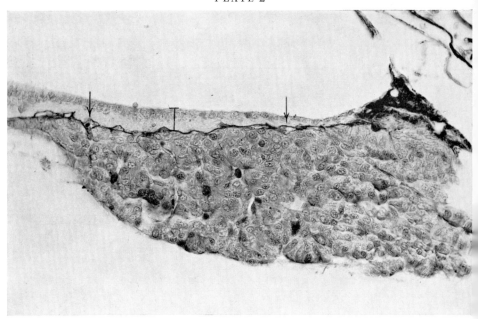

(a)

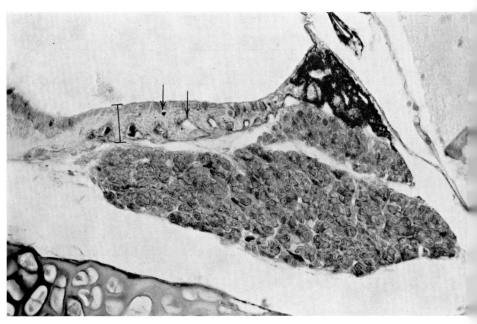

(b)

(a) Median sagittal section of pituitary and hypothalamic floor in a large thyroidectomized tadpol showing failure of median eminence to differentiate. (b) Similar section in an animal brought t midclimax by exogenous thyroxine showing the differentiation of the median eminence and it vascular zone.

tissues (Plate 2). This result led us to postulate a positive feedback mechanism acting between the thyroid and the hypothalamus as follows. Thyroid
hormone stimulates the maturation of the TRF mechanism (shown morphologically in the development of the vascular zone of the median
eminence). Before metamorphosis the thyroid activity is extremely low,

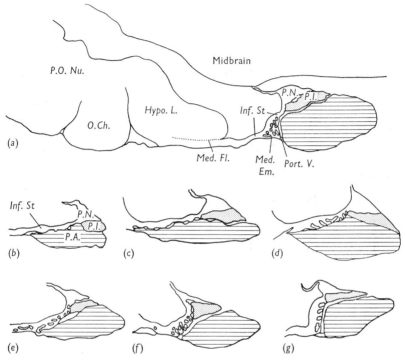

Fig. 6. Line drawings of the hypothalamic–pituitary region of *R. pipiens* as seen in medial
sagittal section. (*a*) adult, (*b*) premetamorphosis, (*c*) early prometamorphosis, (*d*) late prometamorphosis, (*e*) beginning of climax, (*f*) midclimax, (*g*) post-climax. *P.A.*, Pars
anterior; *P.I.*, pars intermedia; *P.N.*, pars nervosa; *Port V.*, portal vein; *Med. Em.*,
median eminence showing primary capillary bed; *Inf. St.*, infundibular stalk; *Med. Fl.*,
median floor of hypothalamus; *Hypo. L.*, hypothalamic lobes or posterior hypothalamus;
O.Ch., optic chiasma; *P.O. Nu.*, preoptic nucleus. Note the increasing vascularity of the
median eminence towards climax.

thus acting only imperceptibly on the tissues, including the hypothalamus.
However, even this low level of activity acting in positive feedback manner
eventually builds up to the point of initiating prometamorphic change.
Like other positive feedback systems, once it reaches an effective level it
accelerates quickly to total activation thus producing the explosive hormone
production of the thyroid that brings on metamorphic climax. In this view
the metamorphic 'clock' is not in the hypothalamus nor in the PT axis but
in the positive feedback interaction between the two.

According to this theory low levels of T_4 applied to the early tadpole should activate the circular feedback mechanism precociously and induce an early metamorphosis of normal pattern. Experiments, however, showed that this was not the case. The possibility was therefore suggested that the hypothalamus of the early larva is not sensitive to T_4 feedback but only becomes so shortly before the normal time of beginning of prometamorphosis. (It will be realized that our previous feedback experiments were all done on giant thyroidectomized tadpoles). Investigation showed that

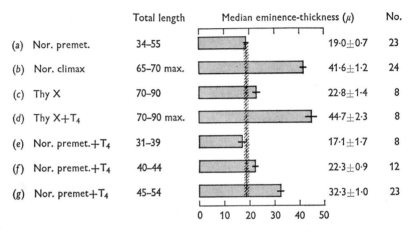

Fig. 7. Thickness of median eminence regions of tadpoles. (*a*) Normal premetamorphic animals, (*b*) normal climax animals, (*c*) large thyroidectomized animals, (*d*) large thyroidectomized animals metamorphosed by exogenous thyroxine, (*e*) smallest group of premetamorphic animals metamorphosed precociously by exogenous hormone, (*f*) medium-sized treated group, and (*g*) group so treated they reached maximal size of normal premetamorphic animals. These date demonstrate that thyroxine is effective in inducing differentiation of the median eminence only in animals at the end of premetamorphosis or later. (Total length measurements in mm.)

this was indeed true. Early tadpoles brought to metamorphic climax by an appropriate pattern of exogenous T_4 do not show median eminence maturation but those treated just before normal prometamorphosis do (Fig. 7). Our concept of the endocrine mechanism determining metamorphosis then assumes the following form (Fig. 8). The pituitary–thyroid axis operates at an extremely low level during the premetamorphic (tadpole) period. The initiation of prometamorphosis is signalled by the acquisition of sensitivity to positive thyroid feedback by the hypothalamic TRF mechanism. Different species differ greatly in the time of acquisition of this sensitivity, thereby differing in the length of the larval period as pointed out above. Prometamorphosis is the period of positive feedback build-up of the HPT (hypothalamic–pituitary–thyroid) axis in all species.

It is therefore of much the same duration in large and small species. The culmination of the feedback operation comes at metamorphic climax, which also is of comparable duration in all species. In this theory the metamorphic 'clock' is *turned on* by the genetically timed change in sensitivity in the hypothalamus. However, the rate at which the 'clock' *runs* through metamorphosis is a function of the parameters of the positive feedback system.

In order to avoid verbal complications, I have not discussed negative feedback of thyroid hormone to the pituitary TSH cells. This concept is

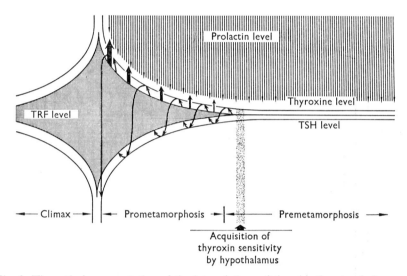

Fig. 8. Theoretical representation of the interrelations of thyroid, thyrotropic hormone (TSH), thyrotropin releasing factor (TRF) and prolactin levels in life-history of the tadpole. Increased activity is indicated by deviation from the midline either above or below. Thin arrows indicate positive feedback relations, heavy arrows indicate inhibition.

well documented for vertebrates generally and comparable evidence is available for tadpoles. For example, goitrogens not only inhibit metamorphosis but lead to hypertrophy of the thyroid gland and to pituitary enlargement. We can fit this negative feedback concept into our picture of the mechanism of metamorphosis by assuming that negative feedback between thyroid and pituitary keeps the PT axis at a very low level during the tadpole stage. The function of the hypothalamic TRF would then be to desensitize the pituitary to this restraint and permit the 'thyrostat' mechanism to rise as described above.

Interesting evidence for this concept was provided by an experiment done 10 years ago but never published since no reasonable explanation of the results could be offered at that time. We transplanted thyroids from

normal tadpoles into giant thyroidectomized hosts. We reasoned that the
TSH level in the hosts would be high in the absence of negative feedback
of thyroid to the pituitary. We expected that the activation of the transplanted thyroid would then lead to a precipitous metamorphosis comparable to that produced by an injection of T_4. Nothing of the kind happened.

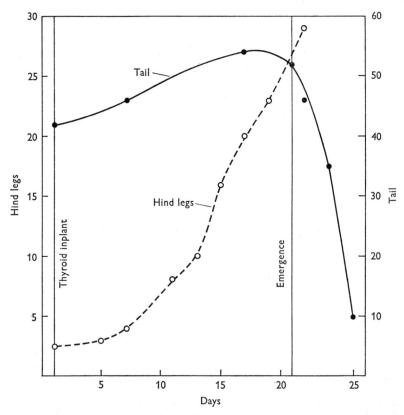

Fig. 9. Effect of implantation of one thyroid lobe from a late prometamorphic donor into
a large thyroidectomized host. The host's hind legs had not developed beyond the 3 mm
stage at the time of implantation. About a week later they began to grow rapidly and the
animal went through a normal prometamorphosis followed by normal climax as shown
by rapid tail reduction after foreleg emergence. (Length measured in mm.)

Instead, the successful hosts showed no immediate metamorphosis but
after 1 week or so underwent an entirely normal prometamorphosis
followed in due time by metamorphic climax (Fig. 9). These results are
readily understood in terms of our present concept. The introduction of
the thyroid gland into the thyroidectomized host led to immediate inhibition of TSH production by negative feedback to the pituitary. At the
same time the thyroid hormone initiated positive feedback to the hypo-

thalamus. The progressive activation of the HPT system then led to a normal pattern of metamorphosis.

As we have indicated in our discussion of the evolution of metamorphosis, the larval period of the anuran life-history has followed its own evolutionary pathway, which has increasingly separated it as a distinct phase from the adult stage. We may therefore inquire into the endocrine balance by which the stability of the tadpole stage is maintained. We have already indicated two points regarding this balance in the tadpole. The first is that the PT axis is maintained at a minimal level by negative feedback of T_4 to the highly sensitive TSH cells of the pituitary. A second was hinted at by the evidence that a tadpole with its own pituitary transplanted to its tail shows a greater than normal rate of growth. Since it has long been known that the pituitary of the tadpole produces a growth factor this last bit of evidence suggests that this factor, like prolactin in mammals, is under inhibitory control by the hypothalamus. Recent evidence from a number of laboratories, indeed, suggests that a prolactin-like hormone is the effective growth factor in tadpoles (for literature see Bern & Nicoll, 1968). Furthermore, it appears that prolactin inhibits the responsiveness of the tissues to T_4 both *in vivo* and *in vitro* (Berman, Bern, Nicoll & Strohman, 1964; Derby, 1968; Etkin, Derby & Gona, 1969). In addition, at high levels prolactin acts as a powerful and non-toxic goitrogen (Etkin & Gona 1967). Since pituitary grafts act like prolactin injections the prolactin–thyroid antagonism appears to be physiological, not pharmacological.

We are thus led to conceive of the tadpole stage of frog development as being predominantly under the influence of prolactin which is being produced autonomously by the pituitary in the absence of hypothalamic control. Prolactin stimulates growth while counteracting any thyroid influence that may be present. Thus the tadpole is stabilized in the larval condition. The activation of the hypothalamic mechanism at the beginning of prometamorphosis not only raises the thyroid hormone level thereby promoting metamorphosis but it cuts back on the action of prolactin thereby removing this brake to metamorphic change (Fig. 8). Whether the same hypothalamic factor which activates the PT axis (TRF) also inhibits prolactin activity (PIF) in the tadpole is unknown.

Although we cannot here delve into the many variations of the metamorphic process to be found among amphibians a few comments must be made. In urodeles generally the mode of life of the larva does not differ as sharply from that of the adult as in anurans. Correspondingly the metamorphic change is not as great. In particular the legs are not thyroid-dependent in their development. As a sequence there is no clearly delimited

prometamorphosis. The commonly recognized 'metamorphosis' of salamanders corresponds to anuran climax. The endocrine pattern appears to
be fundamentally the same. Some differences, however, do appear. The
activation of the thyroid is often even more dramatic than in frogs. In some
species the follicles empty themselves completely. The morphological
differentiation of the median eminence is, on the contrary, much less,
and of course not correlated with leg growth. Perhaps most significant the
'turning off' of thyroid activity at the end of metamorphosis, so apparent
in anurans, seems to be absent in some urodeles. For example, after
thyroid-induced metamorphosis in the axolotl, the thyroid gland is active
in iodine uptake (Prahalad, 1968). Apparently in this animal the thyroid
treatment has a predominantly positive feedback effect on the hypothalamus. This persistence of thyroid activity in the postmetamorphic
urodele may be general. In the American newt the animal shows continued
high thyroid activity after its transformation from an aquatic larva to the
land phase. A second metamorphic transformation eventually sends the
animal back to breed in the water. This second metamorphosis is induced
by prolactin and inhibited by thyroid (Grant, 1961; Grant & Cooper, 1965).
We may infer that this reciprocity between thyroid and prolactin in urodeles (Gona, 1967) is governed by the changes in hypothalamic activity
comparable to that described above for the frog although how the balance
becomes shifted again to favour prolactin in the newt is not known.

Returning to our original consideration of evolution we see that behind
the scenes of the evolutionary drama an extensive backstage machinery
has evolved. Central to this machinery have been evolutionary changes in
the sensitivity of target organs to endocrine messengers and the acquisition
of tissue responses which are specialized parts of the animal's adaptation
to its environment. This is another example of the familiar dogma of
comparative endocrinologists that it is tissue responsiveness not endocrine
messengers which change in evolution. Despite some important limitations
to this dogma (see Heller & Bentley, 1965; Bern & Nicoll, 1968) we think
it is essential to recognize its fundamental validity for the implications of
the idea are wide. For example, in our interpretation, the mechanism of
metamorphosis discussed here is one which has arisen only in the amphibian line of evolution. It is not therefore to be expected that the specific
tissue interactions involved at either the endocrine or peripheral levels
would be the same as those of higher vertebrates. For example, we found
that treatment of the developing rat in the perinatal period with either T_4
or propylthiouracil did not affect the development of the median eminence
(Etkin & Kikuyama, 1964). The metabolic rate response to thyroid so
characteristic of the mammal appears to be absent in the amphibian

(Fletcher & Myant, 1959; Funkhouser & Mills, 1969). We found no goitrogenic effect of prolactin in the rat (Gona & Etkin, unpublished data) and Olivereau (1968) found a different prolactin–thyroid relation in the eel. In short, the specific interactions in the mechanism of metamorphosis are not part of the evolutionary inheritance of higher vertebrates. These animals may have analogous developmental changes such as that of the prolongation of childhood and the delay in reaching puberty in man. But these cannot be expected to be based on the same endocrine interrelations as those governing amphibian metamorphosis. Perhaps, however, by suggesting general modes of action such as positive feedback, stabilizing hormone balances, etc., the study of metamorphosis may be helpful in the analysis of the endocrine mechanisms of other developmental patterns.

Evolutionists are fond of emphasizing that the evolutionary process is, to use Simpson's apt phrase, 'opportunistic'. That is to say, when selective pressure pushes evolution of a species in a particular direction the change is effected by using whatever materials happen to be available at the moment rather than by the invention of new devices *de novo*. Thus in the evolution of the ear ossicles in terrestrial vertebrates, bones which happen to lie in the appropriate area and which were being released from their previous function were used. So we may visualize the evolution of the endocrine mechanism of amphibian metamorphosis. The hormones that were pressed into service to achieve the adaptation of the amphibian life-history, namely the thyroid hormones and prolactin, were clearly present in the Crossopterygian ancestor. Apparently they became available as messengers because, whatever function they were performing in the fish, it was not so vital that it could not be surrendered. In fact endocrinologists find it hard to pin down the functional importance of either of these hormones in fish. Present evidence seems to relate prolactin to the control of salt and water balance in some fish (Bern & Nicoll, 1968) and thyroid to migration between fresh and salt water in others (Dodd & Matty, 1964). These, of course, like the attachment of the hyomandibular to the otic capsule may be considered to be a kind of preadaptation to a possible future role. Yet it must be recognized that in essence the explanation of why tissue response in metamorphosis has been tied to thyroid hormone and inhibition to prolactin (rather than to steroids, for example), must be regarded as an 'accident' of evolution. They were the 'messengers' which happened to be available because they could be taken off their previous roles while the steroids could not since they were vitally involved elsewhere.

This view of the role of endocrines in evolution emphasizes the specificity of each mechanism rather than functional continuity. It stresses the subtlety with which the evolutionary process can use the common group of messengers—hypothalamus, pituitary, thyroid—to code different operations and thus accomplish the specific adaptation of each organism to a particular ecological niche. Above all it throws the emphasis upon the problem of the nature of the receptor substances. It is their differential distribution and mode of action in different cells which appears to lie at the crux of evolutionary change in the use of hormonal messengers. Perhaps this analysis of the metamorphic mechanism will be useful to counteract the tendency of investigators to look for universal 'common denominators', etc. in hormone action. After all, hormones are merely messengers. The nature of the message is not closely related to the nature of the messenger. Or is it?

The research reported here has been supported by National Science Foundation grant G.B.12353.

REFERENCES

BERMAN, R., BERN, H. A., NICOLL, C. S. & STROHMAN, R. C. (1964). Growth-promoting effects of mammalian prolactin and growth hormone in tadpoles of *Rana catesbeiana*. *J. exp. Zool.* **156**, 353–60.

BERN, H. & NICOLL, C. S. (1968). Comparative endocrinology of prolactin. *Recent Prog. Horm. Res.* **24**, 681–720.

BOUNHIOL, J. J. (1942). *Le déterminisme des métamorphoses chez les amphibiens.* Paris: Hermann.

COLEMAN, R., EVENNETT, P. J. & DODD, J. M. (1968). Ultrastructural observations on the thyroid gland of *Xenopus laevis* Daudin throughout metamorphosis. *Gen. comp. Endocr.* **10**, 34–46.

DERBY, A. (1968). An *in vitro* quantitative analysis of the response of tadpole tissue to thyroxine. *J. exp. Zool.* **168**, 147–56.

DODD, J. M. & MATTY, A. J. (1964). Comparative aspects of thyroid function. In *The Thyroid Gland*, pp. 303–56. (eds. R. Pitt-Rivers and W. R. Trotter). London: Butterworths.

ETKIN, W. (1963). The metamorphosis activating system of the frog. *Science, N.Y.* **139**, 810–14.

ETKIN, W. (1968). Hormonal control of amphibian metamorphosis. In *Metamorphosis, a Problem in Developmental Biology* (eds. W. Etkin and L. Gilbert). New York: Appleton-Century-Crofts.

ETKIN, W., DERBY, A. & GONA, A. G. (1969). Prolactin-like antithyroid action of pituitary grafts in tadpoles. *Gen. comp. Endocr.* Suppl. 2, pp. 253–9.

ETKIN, W. & GONA, A. G. (1967). Antagonism between prolactin and thyroid hormone in amphibian development. *J. exp. Zool.* **165**, 249–58.

ETKIN, W. & KIKUYAMA, S. (1964). Thyroid influence on the development of the hypothalamic–hypophysial system in the rat. *Am. Zool.* **4**, 3.

FLETCHER, K. & MYANT, N. B. (1959). Oxygen consumption of tadpoles during metamorphosis. *J. Physiol., Lond.* **145**, 353–68.

FOX, H. (1966). Thyroid growth and its relationship to metamorphosis in *Rana temporaria*. *J. Embryol. exp. Morph.* **16**, 487–96.

FUNKHOUSER, A. & MILLS, K. S. (1969). Oxygen consumption during spontaneous amphibian metamorphosis. *Physiol. Zool.* **42**, 15–22.

GONA, A. (1967). Prolactin as a goitrogenic agent in amphibia. *Endocrinology* **81**, 748–54.

GOOS, H. J. TH. (1969). Hypothalamic neurosecretion and metamorphosis in *Xenopus laevis*. IV. The effect of extirpation of the presumed TRF cells and of a subsequent PTU treatment. *Z. Zellforsch. mikrosk. Anat.* **97**, 449–58.

GRANT, W. C. (1961). Special aspects of the metamorphic process. Second metamorphosis. *Am. Zool.* **1**, 163–71.

GRANT, W. C. & COOPER, G. (1965). Behavioral and integumentary changes associated with metamorphosis in *Diemyctylus*. *Biol. Bull. mar. biol. Lab.*, *Woods Hole* **129**, 510–22.

HANAOKA, Y. (1967). The effects of posterior hypothalectomy upon the growth and metamorphosis of the tadpole of *Rana pipiens*. *Gen. comp. Endocr.* **8**, 417–31.

HELLER, H. & BENTLEY, P. J. (1965). Phylogenetic distribution of the effects of neurohypophysial hormones on water and sodium metabolism. *Gen. comp. Endocr.* **5**, 96–108.

KERR, T. (1966). The development of the pituitary in *Xenopus laevis* Daudin. *Gen. comp. Endocr.* **6**, 303–11.

KOLLROS, J. (1961). Mechanisms of amphibian metamorphosis: hormones. *Am. Zool.* **1**, 107–14.

LYNN, W. G. & WACHOWSKI, H. E. (1952). The thyroid gland and its function in cold blooded vertebrates. *Q. Rev. Biol.* **26**, 123–68.

OLIVEREAU, M. (1968). Action de la prolactine chez l'Anguille. IV. Métabolisme thyroidien. *Z. vergl. Physiol.* **61**, 246–58.

VAN OORDT, P. G. (1966). Changes in the pituitary of the common toad, *Bufo bufo*, during metamorphosis, and the identification of the thyrotropic cell. *Z. Zellforsch. mikrosk. Anat.* **75**, 47–56.

PRAHALAD, K. (1968). Induced metamorphosis rectification of a genetic disability by thyroxine hormone in the mexican axolotl (*Siredon mexicanum*). *Gen. comp. Endocr.* **11**, 21–30.

ROMER, A. S. (1966). *Vertebrate Paleontology*. Chicago University Press.

SCHMALHAUSEN, I. I. (1968). *The Origin of Terrestrial Vertebrates*. New York and London: Academic Press.

SZARSKI, H. (1962). The origin of the Amphibia. *Q. Rev. Biol.* **37**, 189–241.

DISCUSSION

VAN OORDT: The results of extirpation experiments and PTU treatment led Goos (*Z. Zellforsch. mikrosk. Anat.* **97**, 449, 1969) to conclude that TRF is formed in cells in the dorsal preoptic nucleus in larvae of *Xenopus laevis*. The thyroid hormones were found to have a negative feedback influence on TRF excretion by the cells and at the same time stimulated development of the preoptic nucleus, median eminence and the portal vessels. This stimulation of hypothalamic development is supposed to be part of a non-specific morphogenetic effect of thyroid hormones.

It is also believed that during larval development this morphogenetic effect causes a gradual increase in TRF–TSH–thyroid hormone activity, whereas when development has been completed—that is, immediately after metamorphic climax—the negative-feedback effect of the thyroid hormones remains, and thus the TRF–TSH–thyroid hormone level will drop considerably.

ETKIN: Well, as you are aware, we have had occasion to discuss this at some length with Dr Goos and yourself. I think there has been a clarification of this aspect and I don't disagree essentially with the experimental points which you make since you too find an activation of the TRF mechanism. We agree on that. You regard this as part of a general activation phenomenon that accompanies thyroid activity. This is a theoretical point with which I would very strongly disagree. I prefer to discuss this from an evolutionary viewpoint, and it seems to me that the adaptation of the animal has been attained by each tissue developing its own particular response to thyroid stimulation. I do not believe that there is such a thing as a general activation of morphogenesis by thyroid hormone. This is the kind of generalization we carry over from mammalian thinking where we think of thyroid as activating basal metabolism. This may be true of the mammal but it is not true of the amphibian. I think that experimentally this has been well established even though practically every year someone repeats the experiment of subjecting tadpoles or other amphibia to thyroid and measuring metabolic rates. It simply does not go up.

The evolutionary process has been highly selective and each tissue responds in its own way. The legs, for example, of our tadpole respond by growth, but in the salamander the legs are not responsive to thyroid because the salamander has a different mechanism regulating development of the legs. In the toad, the legs develop about half-way before they become thyroid-sensitive. I think the crux of the problem of the use of hormones in evolution comes down to the specificity of the receptors and specificity of the tissue response, and we cannot rely on general terms. Except for the use of this general terminology, I think that we do not differ in fundamentals. There may be also a negative feedback involved to the hypothalamus, but I don't know that the evidence is conclusive on that point.

VAN OORDT: I quite agree that all tissues have their own sensitivity to thyroid hormones, and that their reactions to these hormones depends on tissue-specific factors. What I want to emphasize, however, is that with regard to the hypothalamus, thyroid hormones stimulate its development as well as that of the median eminence, and at the same time inhibit the secretory activity of the TRF in the dorsal preoptic area. The former seems to be part of a general morphogenetic function of the thyroid hormones, whereas the latter is identical to the negative feedback existing in post-metamorphic and adult amphibia. There seems to be no need for the postulation of a positive feedback upon TRF secretion to explain the increase in

TSH and thyroid hormone levels during prometamorphosis and metamorphic climax.

Dodd: May I first make the general point that theories concerning the physiological control of amphibian metamorphosis are very difficult to test because of the difficulties associated with measuring circulating titres of the hormones involved. For example, if we knew the circulating levels of thyroxine and prolactin the existence and nature of the feedback mechanisms that have been postulated could be identified more readily. We have recently been able to measure the amounts of thyroid stimulating hormone in the pituitary glands of *Xenopus* tadpoles in certain of the later stages of metamorphosis and have found that the TSH content reaches a peak at about stage 61. We have also found that the most active stage of the thyroid gland, measured by histology and uptake of radioiodine, is stage 63. It seems possible that TSH is released from the pituitary at stage 61 more actively than it is synthesized and this activates the thyroid, accounting for the high activity in stage 63. I should mention that at stage 63 the body has almost completed its metamorphosis but the tail is still virtually intact. We believe, for several reasons, that tail resorption requires more thyroxine than any other event in metamorphosis.

Professor Etkin has described work on the effects of prolactin on amphibian metamorphosis and we have done similar work on *Xenopus*. We have found that hypophysectomized tadpoles, if injected with 10 μg of prolactin per day, show only minimal signs of metamorphosis, and in particular, the feeding mechanism and tail remain unresorbed and the tadpoles grow to a weight at least twice that of the controls. When thyroxine and prolactin are injected together into hypophysectomized *Xenopus* tadpoles the latter appears to antagonize the metamorphic action of thyroxine, and whereas we can produce complete and tolerably normal metamorphosis by injecting 7 ng of thyroxine per day until metamorphosis is complete, it is necessary to inject four times this amount of thyroxine to produce the same effect if prolactin is injected at the same time. Furthermore, we have shown that prolactin acts as a growth hormone in the completely metamorphosed young adults as well as in the tadpoles.

KIDNEY FUNCTION IN DESERT
VERTEBRATES

By WILLIAM H. DANTZLER

INTRODUCTION

The role of the kidney in the adaptation of vertebrates to a hot, arid environment is variable. It is related both to the vertebrate class involved and to other physiological and behavioural adaptations for meeting the major problems of survival in a desert. These problems result from a high environmental temperature, a rapid evaporative water loss, and a scarcity of water. The contribution of the kidney to survival in the desert lies in its ability to help conserve body water. Water conservation is also achieved by a variety of other means, including avoidance of extreme desert heat and reduction of cutaneous, respiratory, and gastrointestinal water losses. The interrelationships of some of these adaptations have been covered in a number of recent reviews (Cade, 1964; Dawson & K. Schmidt-Nielsen, 1964; Hudson, 1964; K. Schmidt-Nielsen, 1964a; K. Schmidt-Nielsen, 1964b; K. Schmidt-Nielsen & Dawson, 1964; Bentley, 1966a). In the present paper, I shall confine my discussion to the mechanisms by which the kidney helps achieve water conservation. Other mechanisms of water conservation will be mentioned only as necessary for the discussion of renal mechanisms.

PRODUCTION OF URINE HYPEROSMOTIC
TO PLASMA

Production of a urine hyperosmotic to the plasma enables an animal to eliminate ions and nitrogenous waste while conserving water. The greater the osmolality of the urine relative to that of the plasma the greater the advantage to an animal that needs to conserve water. Among the vertebrates, only the mammals and the birds can produce a urine hyperosmotic to the plasma. This has long been correlated with the possession of long loops of Henle by nephrons of these two vertebrate classes. It is now generally accepted that these structures function as countercurrent multipliers. This enables the kidneys of these vertebrates to produce a concentrated urine without active transport of water. The manner in which this concentrating mechanism is presently felt to function in the mammalian kidney has recently been reviewed (Berliner & Bennett, 1967). However,

[157]

there are a number of aspects of the renal concentrating mechanism of desert mammals and birds which raise additional questions about its detailed operation and regulation.

Mammals

As can be seen from Table 1, a number of desert rodents produce a highly concentrated urine compared with some common non-desert species. The concentrating ability of two species of Australian hopping mice (*Notomys alexis* and *Leggadina hermannsburgensis*) recently described by MacMillen & Lee (1967) far exceeds that recorded for any other mammalian species. Thus, in these forms renal concentrating ability plays an important role in the conservation of water in an arid environment. However, other common desert mammals, such as the pack rat and the camel, do not produce a urine more concentrated than that of the common cat or rat. Clearly, in these desert species other mechanisms of water conservation are more important that the renal concentrating ability. The nature of these mechanisms has been discussed in detail by K. Schmidt-Nielsen in his recent monograph on desert animals (1964a).

Table 1. *Maximal urine osmolalities and osmolal urine-to-plasma ratios for a number of mammals*

Species	Urine osmolality (m-osmoles)	Osmolal U/P
Man	1,430	4·2
*Pack rat (Neotoma albigula)	2,700	7·0
White Rat	2,900	8·9
Cat	3,250	9·9
*Camel	2,800	8·0
*Ground squirrel (Citellus leucurus)	3,900	9·5
*Kangaroo rat (Dipodomys merriami)	5,500	14·0
*Gerbil (Gerbillus gerbillus)	5,500	14·0
*Notomys cervinus	4,920	14·2
*Jerboa (Jaculus jaculus)	6,500	16·0
*Sand rat (Psammomys obesus)	6,340	17·0
*Notomys alexis	9,370	24·6
*Leggadina hermannsburgensis	8,970	26·8

* Desert mammals. Values for *N. cervinus*, *N. alexis* and *L. hermannsburgensis* were obtained from MacMillen & Lee (1967). All other values were taken from K. Schmidt-Nielsen (1964a).

The concentrating ability of a countercurrent multiplier system varies directly with the length of the system and the gradient produced between the two limbs of the loop, and inversely with the diameter of the limbs and the rate of flow through the system. The thickness of the renal medulla gives a measure of the length of the loops of Henle and, therefore, of the countercurrent multiplier system. If other factors remained constant, the longer the countercurrent multiplier system, the greater would be the gradient produced from cortex to papilla tip, and the greater would be the concentration of the urine. However, B. Schmidt-Nielsen & O'Dell (1961), extending the classical observations of Sperber (1944), showed that concentrating ability is related to the relative, not the absolute, thickness of the renal medulla (Table 2). Clearly, other factors do not remain constant.

Table 2. *Relationship of concentrating ability of a number of mammals to kidney size, percentage long-looped nephrons, and medullary thickness (modified from B. Schmidt-Nielsen & O'Dell, 1961).*

Species	Kidney size (mm)	Long-looped nephrons (%)	Medullary thickness (mm)	Relative medullary thickness	Maximal osmolal U/P	Maximal urine osmolality
Man	64·0	14	19·0	3·0	4·2	1430
White rat	14·0	28	8·0	5·8	8·9	2900
Cat	24·0	100	14·5	4·8	9·9	3250
Kangaroo rat (*Dipodomys merriami*)	5·9	27	5·0	8·5	14·0	5500
Jerboa (*Jaculus jaculus*)	4·5	33	7·4	9·3	16·0	6500
Sand rat (*Psammomys obesus*)	13·0	100	13·9	10·7	17·0	6340

Data on kidney size, percentage long-looped nephrons, medullary thickness, and relative medullary thickness are taken from Sperber (1944). Values for maximal osmolal U/P ratios and maximal urine osmolalities are from K. Schmidt-Nielsen (1964a).

If they did, the species with the thickest medulla would produce the most concentrated urine. B. Schmidt-Nielsen & O'Dell (1961) chose the relative medullary thickness because the diameter of the nephrons tends to increase with kidney size. Presumably, the larger tubular diameter in the larger kidneys offsets the greater length of the loops in the production of a concentrated urine. It is also possible that the flow rate through the larger nephrons is greater than in the smaller ones. Although such concepts agree with the basic theory of countercurrent multiplication, they have yet to be examined directly.

Of more interest with regard to desert mammals, however, are the interspecific differences in concentrating ability that are not related directly to relative medullary thickness. These differences are apparent in Fig. 1. For example, among the desert rodents, the North American kangaroo rat, *Dipodomys merriami*, produces a urine more concentrated than that produced by the Australian hopping mouse, *Notomys cervinus*, despite a relative medullary thickness only two-thirds as great. The kangaroo rat can also concentrate its urine nearly as well as the African sand rat, *Psammomys obesus*, although it has relatively shorter loops of Henle.

The African jerboa, *Jaculus jaculus*, which resembles the kangaroo rat, can produce a more concentrated urine than either *Psammomys obesus* or *Notomys cervinus*, although it too has shorter loops of Henle than these species. The two species of *Notomys* have approximately equal relative medullary thickness but *N. alexis* can produce urine nearly twice as

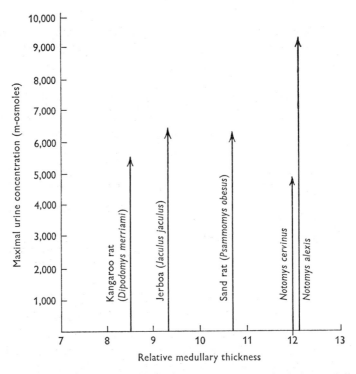

Fig. 1. Relationship between maximal urine concentration and relative medullary thick-ness for a number of desert rodents. Maximal urine concentrations for *Notomys cervinus* and *N. alexis* are from MacMillen & Lee (1967); for the others, from K. Schmidt-Nielsen (1964a). Values for the relative medullary thickness for the kangaroo rat, jerboa, and sand rat are from Sperber (1944). Those for *N. cervinus* and *N. alexis* were computed by the method of Sperber (1944) from the histological sections shown by MacMillen & Lee (1969).

concentrated as *N. cervinus*. Although it is difficult to determine with certainty a maximally concentrated urine, these differences are sufficiently marked to suggest that factors other than the length of the loops of Henle influence the concentrating ability of different desert mammals.

One factor which has been suggested as playing a role is the relative number of long- and short-looped nephrons (B. Schmidt-Nielsen, 1964). The kangaroo rat and jerboa have long, slender renal papillae with only about 30% long-looped nephrons (Table 2). These animals are granivorous with no access to free drinking water and must excrete urea and electrolytes with a minimum of water. *N. alexis* exists under similar conditions and, although the number of long-looped nephrons has not been measured, has a very long, rather slender papilla (MacMillen & Lee, 1969). In contrast, *Psammomys*, which eats succulent plants with a high salt content, must excrete a large volume of urine with a high electrolyte concentration. The kidney in this animal has a long, broad papilla with 100% long-looped nephrons (Table 2). Presumably the larger number of loops in the papilla makes possible the concentration of a larger volume of urine in the collecting ducts. However, the larger number of long-looped nephrons does not explain the failure of *Psammomys* to produce a urine more concentrated than that of the kangaroo rat or jerboa, animals with shorter loops of Henle. Most of the difference results from the failure of *Psammomys* to concentrate urea as well as those other species. It is also possible, despite the effective concentrating of electrolytes, that the driving force behind the countercurrent multiplier system, the active transport of sodium by the ascending limb of the loop of Henle, is less effective than in other species.

The latter possibility may be more important in explaining the concentrating ability of *Notomys cervinus* relative to other desert rodents. This desert rodent is largely granivorous, but, like *Psammomys*, it may also obtain water by eating halophytic plants (MacMillen & Lee, 1969). Also, like *Psammomys*, it appears to be less independent of exogenous water than the kangaroo rat, the jerboa, or *N. alexis* (MacMillen & Lee, 1969). Although its renal papilla appears slightly broader than that of the kangaroo rat and *N. alexis* (MacMillen & Lee, 1969), nothing is known about the relative number of long-looped nephrons. However, as already noted, this animal does not produce as concentrated a urine as desert rodents with the same or less relative medullary thickness (Fig. 1). The data of MacMillen & Lee (1969) again suggest that part of the difference may result from a reduced ability to concentrate urea. But the concentrating ability of *N. cervinus* is so low for its great relative medullary thickness that it suggests that the osmotic gradient produced between the ascending and descending limbs of the loop of Henle (the gradient against which the active sodium transport system can work) may be lower in this species than in these other desert rodents. At present, there is no direct evidence of possible interspecific differences in the osmotic gradient produced between the two limbs of the loop.

Birds

The ability of bird kidneys to produce a concentrated urine, although more variable than once thought, is much more limited than that of mammals. Only one species of those studied is capable of producing a maximum urine-to-plasma osmolality ratio (osmolal U/P ratio) significantly greater than 2·0 (Table 3).

Table 3. *Maximal concentrating ability of some birds*

Species	Osmolal U/P
Domestic fowl (*Gallus gallus*)	2·0
House finch (*Carpodacus mexicanus*)	2·3
Savannah sparrow (*Passerculus sandwichensis brooksi*)	3·2
Salt marsh savannah sparrow (*Passerculus sandwichensis beldingi*)	5·8
Bobwhite (*Colinus virginianus*)	1·6
California quail (*Lophortyx californicus*)	1·7
*Gambel's quail (*Lophortyx gambelii*)	2·5
*Roadrunner (*Geococcyx californianus*)	2·1

* Desert birds. Value for domestic fowl is from Skadhauge & B. Schmidt-Nielsen (1967*a*). Values for house finch and savannah sparrows are from Poulson & Bartholomew (1962). Values for the three species of quail are estimated from the data of McNabb (1969). Value for the road-runner is estimated from the data of Calder & Bentley (1967). Values for the domestic fowl, the three species of quail, and the roadrunner were obtained during dehydration. Values for the others were obtained during drinking of saline solutions.

The anatomy of the bird kidney suggests that the concentrating ability would not be marked since the medulla is less well developed than in the mammalian kidney. In fact, the bird kidney suggests a mixture of reptilian and mammalian nephrons. Most nephrons in each lobule resemble reptilian nephrons with convoluted tubules without loops of Henle emptying at right angles into collecting ducts. However, some nephrons in each lobule have loops of Henle which, together with vasa recta and parallel collecting ducts, form medullary cones (Feldotto, 1929; Sperber, 1949, 1960; Poulson, 1965). The only direct evidence that these structures do function as countercurrent multipliers has been obtained in domestic fowl (Skadhauge & B. Schmidt-Nielsen, 1967*b*). In these animals the osmolality of the medullary cones exceeds that of the cortex by about 30–90 m-osmoles during periods of dehydration and salt-loading, and a small osmotic gradient exists from the base to the tip of the cones. Differences in sodium and chloride concentrations account for most of the observed osmotic differences. Urea, which is a minor end-product of nitrogen metabolism in birds, does not play the significant role in the osmotic concentration of the medullary tissue that it does in mammals.

The one bird capable of producing a urine significantly more concentrated than the plasma (*Passerculus sandwichensis beldingi*, a sub-species of savannah sparrow; Table 3) is a granivorous passerine inhabiting salt marshes (Poulson & Bartholomew, 1962). However, McNabb (1969) has recently observed a slight gradation in maximal osmolal U/P ratios with dehydration among three species of quail, the values increasing with increasing aridity of habitat (Table 3). Thus, there is a tendency for the kidneys of some desert birds to produce a more concentrated urine than those of closely related non-desert species. There is a good correlation between concentrating ability and the number of medullary cones per cross-sectional area of kidney tissue for those taxa studied (Fig. 2).

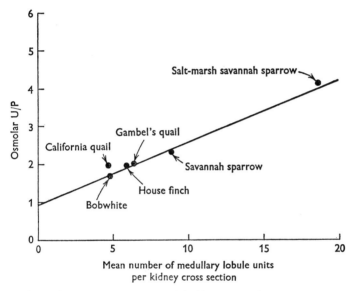

Fig. 2. Relationship between mean osmolal U/P ratios and mean number of medullary lobule units per kidney cross-section in a number of birds (re-drawn after McNabb, 1969). Data for the three species of quail are from McNabb (1969); those for the house finch and savannah sparrows are from Poulson (1965).

Poulson argues that the number of medullary cones per cross-sectional area reflects the number of loops of Henle. His measurements of the length of the loops of Henle (based on layers of loops observed in cross-sectional areas of cones at various positions from base to tip) do not correlate with concentrating ability in the two races of savannah sparrow and the house finch which he studied (Poulson, 1965). If a larger number of loops in bird kidneys makes possible the production of a relatively large volume of concentrated urine as suggested for *Psammomys*, this would seem suited to the high salt intake of the salt-marsh savannah sparrow. However, a greater tendency to have more loops rather than longer loops with increasing concentrating ability, if it resulted in a larger volume of concentrated urine, might also aid in the excretion of uric acid without mechanical obstruction. More detailed measurements of medullary concentration gradients as well as the number and length of Henle's loops must still be made for non-domestic species.

Although there is some increase in concentrating ability of desert quail compared with non-desert species, this ability is still not significantly greater than that of domestic fowl (Table 3). The roadrunner (*Geococcyx californianus*), a desert carnivore which spends most of its time on the ground, also does not produce a urine more concentrated than that of domestic fowl (Table 3). It appears that in many desert birds other adaptations for conservation of water are utilized for survival in deserts. However, few species have been studied, and others may show a concentrating ability as marked as that of the savannah sparrow.

REGULATION OF GLOMERULAR FILTRATION, TUBULAR REABSORPTION, AND BLADDER OR CLOACAL REABSORPTION

Changes in glomerular filtration rate (GFR), tubular reabsorption, and bladder or cloacal reabsorption can also function to regulate excretion of water. In mammals the permeability of the distal tubule and collecting duct to water is regulated by antidiuretic hormone. Such regulation plays an integral role in the ability of the kidney to dilute or concentrate the urine (Berliner & Bennett, 1967). There is no evidence that the response of the distal nephron in desert mammals differs from that in non-desert species. The mammalian bladder plays no role in the reabsorption of water, and changes in filtration rate with states of hydration occur only as an extreme form of response.

Among animals that are unable to produce a highly concentrated urine, however, changes in GFR, tubular reabsorption, and bladder or cloacal reabsorption assume an important role in the regulation of water excretion. The regulation at any one of these three levels may be related to regulation at the other two, to the habitat of the animal, to the end-products of nitrogen metabolism, and to the tolerance of changes in plasma osmolality. The data on desert vertebrates are scarce and I shall be able to consider only a few aspects of this regulation.

Reptiles

More data on renal regulation are available for desert reptiles than for any other non-mammalian class of desert vertebrates. Table 4 shows data on GFR and osmolal U/P ratios for species of each of the three groups of reptiles inhabiting arid regions—the chelonians, the ophidians, and the saurians—during various states of hydration. Similar data are shown for species from each of these groups living in non-arid environments.

Table 4. *Renal function in some reptiles during normal hydration, dehydration, salt loading and water loading*

Species	Condition	GFR ml/kg/h	Osmolal U/P	Osmolal U/P (range for all conditions)
		A. Chelonians		
		(*Gopherus agassizii*)		
1. Desert tortoise	Normal	4·74±0·60 (9)	0·36±0·02 (9)	} 0·3–0·7
	Salt Load	2·94±0·91 (20)	0·57±0·05 (20)	
	No urine flow when plasma osmolality increased 100 m-osmoles			
	Water Load	15·12±6·64 (17)	0·61±0·03 (17)	
		(*Pseudemys scripta*)		
2. Freshwater turtle	Normal	4·73±0·69 (40)	0·62±0·03 (40)	} 0·3–1·0
	Salt Load	2·77±0·90 (10)	0·84±0·06 (10)	
	No urine flow when plasma osmolality increased 20 m-osmoles			
	Water Load	10·27±2·00 (25)	0·60±0·03 (25)	
		B. Ophidians		
		(*Pituophis melanoleucus*)		
1. Arid environment	Salt Load	16·08±1·06 (5)	0·72±0·05 (5)	} 0·5–1·0
	Water Load	10·96±1·07 (5)	0·73±0·09 (5)	
		(*Natrix sipedon*)		
2. Freshwater snakes	Salt Load	13·12±1·26 (11)	0·58 (3)	} 0·1–1·0
	Water Load	22·84±1·75 (12)	0·27 (3)	
	No urine flow when plasma osmolality increased 50 m-osmoles			
		C. Saurians		
		(*Phrynosoma cornutum*		
1. Horned lizard; arid environment	Normal	3·54±0·32 (23)	0·93±0·04 (22)	
	Dehydration	2·14±0·20 (14)	0·97±0·02 (14)	
	Salt Load	1·73±0·40 (10)	1·00±0·01 (10)	
	Water Load	5·52±0·54 (19)	0·90±0·04 (19)	
		(*Tropidurus* sp.)		
2. Galapagos lizards; arid environment	Normal	3·62±0·36 (17)	0·96±0·02 (17)	
	Dehydration	1·23±0·23 (5)	0·97±0·02 (5)	
	Salt Load	2·42 (1)	1·01 (1)	
	Water Load	4·53 (3)	0·99 (3)	
		(*Hemidactylus* sp.)		
3. Puerto Rican Gecko; moist environment	Normal	10·4±0·77 (14)	0·64±0·02(12)	
	Dehydration	3·33±0·37 (6)	0·74±0·02 (6)	
	Salt Load	11·01±2·18 (4)	0·80±0·02 (4)	
	Water Load	24·3±1·67 (12)	0·74±0·02 (12)	

Values are means ± s.e. Numbers in parentheses indicate number of determinations. No standard errors are given when there were fewer than four determinations. Approximate range for osmolal U/P ratio observed under all conditions is given for each species for which it was available. Data on chelonians were compiled from Dantzler & B. Schmidt-Nielsen (1966); on ophidians, from Dantzler (1967a, 1968) and Komadina & Solomon (1970); on saurians, from Roberts & B. Schmidt-Nielsen (1966).

Chelonian reptiles

Renal function in the desert tortoise of the south-western United States (*Gopherus agassizii*) has been examined closely and compared with that of a species of freshwater turtle (*Pseudemys scripta*) (Dantzler & B. Schmidt-Nielsen, 1966). Unfortunately, in the desert tortoise the effects of dehydration on renal function could not be easily studied over periods of a few hours as they could be in the freshwater turtle. Under natural conditions these animals can live without eating or drinking for prolonged periods of time, and dehydration under laboratory conditions caused inconsistent changes in plasma osmolality. Such differences between these species probably resulted from the fact that respiratory and cutaneous water losses in the desert tortoise are about one-tenth of those in the freshwater turtle (K. Schmidt-Nielsen & Bentley, 1966; Bentley & K. Schmidt-Nielsen, 1966). However, it was found that in the freshwater turtle the effects of dehydration on renal function could be mimicked by the intravenous administration of a hyperosmotic (1 M) sodium chloride solution. The changes in renal function with a given increase in plasma osmolality were similar whether the increase was produced by dehydration or salt-loading. For this reason, the effect of increased plasma osmolality induced by a salt load was studied and compared in both species. As we noted at the time, this method has the advantage that the osmolality of the blood can be controlled, but the disadvantage that the salt load also expands the extracellular fluid volume. This may then affect volume regulation in the animal. It should be noted that neither the freshwater turtle nor the desert tortoise has an extrarenal route for salt excretion.

As can be seen from Table 4, renal function in the desert tortoise ceased when the plasma osmolality was raised about 100 m-osmoles. Renal function in these animals at plasma osmolalities below this level is shown in more detail in Fig. 3. Increases in plasma osmolality of less than 50 m-osmoles had very little effect on renal function in *Gopherus*. Glomerular filtration rate and tubular function remained remarkably constant. The urine remained hypo-osmotic to the blood and the osmolal U/P ratio only varied from about 0·3 to 0·6. In no experiments with desert tortoises was the osmolal U/P ratio (based on ureteral urine) greater than 0·7. The free water clearance also did not fall below about 10% of the filtration rate and averaged about 15% of the filtration rate under most circumstances in this study.

These observations on the desert tortoise were in marked contrast to those on the freshwater turtle (Fig. 3; Table 4). In this species a rise in plasma osmolality of between 10 and 20 m-osmoles, whether produced by dehydration or salt-loading, caused a decrease in urine flow, which was a result of a variable decrease in GFR and a consistent increase in tubular reabsorption of water. The urine osmolality rose toward that of the blood and the free-water clearance decreased from 11 % of the filtration rate to only 2 %, When the plasma osmolality was raised 20 m-osmoles or more, renal function ceased.

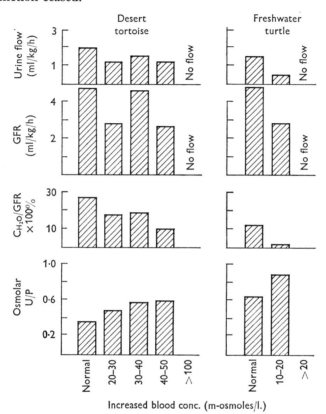

Fig. 3. Effect of salt-loading on renal function in the desert tortoise and the freshwater turtle (modified from Dantzler & B. Schmidt-Nielsen, 1966). Urine flow ceased in the desert tortoise when the osmolality of the blood was increased more than 100 m-osmoles. Urine flow ceased in the freshwater turtle when the osmolality of the blood was raised more than 20 m-osmoles. The results represent a total of 29 clearance periods in the desert tortoise and 50 clearance periods in the freshwater turtle.

In both the desert tortoise and the freshwater turtle an intravenous water load (administered as 5 % dextrose solution) produced a marked increase in urine flow and GFR (Table 4). There was less change in tubular reabsorption of water as demonstrated by the osmolal U/P ratios. In part this may simply reflect the relatively short period of time over which the effects of a water load were observed. However, as already noted, the osmolal U/P ratio for the desert tortoise did not vary greatly throughout this study.

Variations in GFR in both species appear to reflect alterations in the number of functioning glomeruli. This is suggested by the observations that the tubular maximum (T_m) for para-aminohippurate (PAH) varied directly with GFR. This correlation agrees rather well with direct observations of glomerular function in frogs (Sawyer, 1951) and thus appears to give a reasonable indication of glomerular activity. Moreover, recent microscopic work on a number of reptile species (B. Schmidt-Nielsen & Davis, 1968). shows that the ratio of the number of open to closed tubules correlates roughly with the filtration rate, supporting the concept that changes in filtration rate result from changes in the number of functioning nephrons. This type of regulation of GFR appears reasonable in these non-mammalian vertebrates where nephrons do not function together in a countercurrent multiplier system.

The bladder function of the desert tortoise, as well as the glomerular and tubular function, differs from that of the freshwater turtle. The bladder of the desert tortoise is highly permeable to water. Urine entering the bladder rapidly becomes isosmotic with the plasma. At the same time, urates precipitate in the bladder. We found no evidence of a change in permeability of the bladder with the state of hydration of the animal. The bladder of the freshwater turtle, on the other hand, can be less permeable. When the bladders of well-hydrated *Pseudemys scripta* were catheterized, the initial collections were always markedly hypo-osmotic to the blood, whereas those from *Gopherus agassizii* under the same conditions were isosmotic with the blood. *In vitro* studies of bladders of freshwater turtles by a number of investigators (Brodsky & Schlib, 1960, 1965, 1966; Gonzalez, Shamoo & Brodsky, 1967) have shown that sodium and chloride are transported actively and water passively from the mucosal to the serosal side. However, no effects of neurohypophysial hormones on bladder permeability have been observed (Brodsky & Schlib, 1960).

The responses of these desert and non-desert chelonian reptiles on all three levels—glomerular, tubular and bladder—appear related to their habitat, their tolerance of increased plasma osmolality, and the final excretory products of their nitrogen metabolism. The differences in the sensitivity of the renal function in the freshwater turtle and the desert tortoise to changes in plasma osmolality may be related in part to differences in tolerance of increases in plasma osmolality. The desert tortoise tolerated a 200 m-osmoles increase in plasma osmolality during severe dehydration with no apparent ill-effects, while some freshwater turtles died with an increase of only 60 m-osmoles and none survived with an increase of more than 95 m-osmoles. The ability of desert reptiles to survive and, indeed, to function well with marked increases in plasma osmolality was first clearly demonstrated by Bentley (1959). It should be noted that most of the increase in plasma osmolality in the desert tortoise resulted from an increase in the concentration of urea and that vertebrate tissues are less sensitive to high concentrations of urea than to high concentrations of electrolytes (Thesleff & K. Schmidt-Nielsen, 1962; McClanahan, 1964). Although freshwater turtles are predominantly ureotelic (Table 5), a marked increase in plasma urea concentration only occurred shortly before death from dehydration.

Table 5. *Nitrogenous excretory products*

	Ammonia	Urea	Uric acid
Mammals	Minor (2–8)	Major (80–90)	Minor (1–2)
Birds	Minor (3–4)	Minor (10)	Major (87)
Reptiles			
Chelonians			
Desert tortoise (*Gopherus agassizii*)	Minor (3–8)	Minor (15–50)	Major (20–50)
Freshwater turtle (*Pseudemys scripta*)	Minor (4–44)	Major (45–95)	Minor (1–24)
Saurians	Minor	Minor	Major (98)
Ophidians	Minor	Minor (0–2)	Major (98)
Amphibians			
Anurans			
Semi-aquatic and terrestrial	Minor	Major	—
Rana catesbiana	(12)	(84)	—
Bufo bufo bufo	(10–15)	(85–90)	—
Aquatic	Major	Minor	—
Xenopus laevis	(80)	(20)	—

Approximate percentage of total urinary nitrogen appearing as ammonia, urea, and uric acid is given in parentheses when available. Where variations in nitrogen excretion occur among orders of a vertebrate class, only data for those orders with desert representatives are given. Comparable data for non-desert species of the same class or order are shown where differences exist. Data on chelonian reptiles are from Dantzler & B. Schmidt-Nielsen (1966); on saurian and ophidian reptiles, from B. Schmidt-Nielsen (1964) and Khalil (1948a, b); on most amphibians, from Munro (1953); on all other groups, from Prosser & Brown (1961).

Although the regulation of glomerular and tubular function may be partly related to the tolerance of increased plasma osmolality, it appears more closely related to bladder function and the end-products of nitrogen metabolism. The desert tortoise excretes urea, but about half of its excretory nitrogen is in the form of uric acid (Table 5). Renal function does not cease with moderate increases in plasma osmolality, and ureteral urine remains hypo-osmotic to the plasma. This reduces the possibility of precipitation and accumulation of urates in the renal tubules. In the bladder, however, urates precipitate and the urine equilibrates with the blood. This leads to temporary storage of nitrogenous waste within the bladder. Thus, continued renal function during dehydration enables the predominantly uricotelic desert tortoise to get rid of nitrogenous waste. At the same time, water reabsorption from the bladder makes such excretion virtually equivalent to cessation of renal function in the conservation of water. Although bladder function does not vary, the bladder is separated from the ureters by a muscular sphincter and ureteral urine can by-pass the bladder when hydration is adequate.

Such function differs from that of the predominantly ureotelic freshwater turtle. Complete cessation of renal function occurs with only moderate dehydration. Although the turtle bladder does not respond to anti-diuretic hormone, water moves passively out of the bladder and continued reabsorption of sodium and chloride will lead to additional reabsorption of water. This may be important when renal function ceases. Most of the nitrogenous waste is in the form of soluble urea or ammonia (Table 5) which does not precipitate and penetrates the bladder wall. Thus continued production of urine would be of no advantage since the urinary constituents would be reabsorbed from the bladder.

Although there is no definite information on hormonal regulation of GFR or of distal tubular reabsorption of water in the desert tortoise, a few preliminary experiments on freshwater turtles with turtle posterior-pituitary extract suggest that small doses of neurohypophysial peptides increase tubular permeability to water while large doses also lead to a decrease in GFR (Dantzler & B. Schmidt-Nielsen, 1966). It seems likely that those changes observed in GFR and tubular permeability to water in the desert tortoise are also mediated by a neurohypophysial peptide, presumably arginine vasotocin (AVT) (Sawyer, Munsick & van Dyke, 1959). However, the limited change in distal tubular permeability to water makes the nature of the interaction between peptide and membrane of particular interest.

Ophidian reptiles

The pattern of glomerular and tubular responses observed in ophidian reptiles resembles in some ways that seen in chelonian reptiles (Table 4). Although no data on the effects of dehydration on renal function are available for desert-dwelling snakes, it is again possible to compare the effects of a water load and a salt load on this function in arid-living and freshwater forms. No extra-renal route for salt excretion has been identified in terrestrial snakes.

Pituophis melanoleucus obtained from the deserts of Arizona and New Mexico had a lower GFR during a control diuresis (produced by the intravenous infusion of 1·25% mannitol) (Komadina & Solomon, 1970) than freshwater *Natrix sipedon* (Dantzler, 1967a, 1968) (Table 4). As in the desert tortoise, no decrease in GFR occurred in the desert snakes during the intravenous administration of a 5% sodium chloride solution. Instead, the filtration rate increased markedly (Table 4; Komadina & Solomon, 1970). Also, as in the desert tortoise, there was little change in distal tubular reabsorption of water, the average value for the osmolal U/P ratio remaining remarkably constant during both a water load and a salt load (Table 4; Komadina & Solomon, 1970). However, the values for urine and plasma osmolalities from individual animals in the study by Komadina & Solomon (1970) reveal that the osmolal U/P ratio in these snakes can vary at least from about 0·5 to 1·0. Thus, although the average values during salt-loading and water-loading did not change significantly, the animals are capable of producing a hypo-osmotic or an isosmotic urine.

The response of these desert snakes to a salt load contrasted with the response of the freshwater snakes. The latter animals responded to a salt load in the same fashion as the freshwater turtles with a marked decrease in filtration rate and an increase in tubular permeability to water (Table 4; Dantzler, 1968; Komadina & Solomon, 1970). Moreover, renal function appeared to cease when the plasma osmolality had increased by about 50 m-osmoles (Dantzler, 1968). Studies on the renal effects of AVT, the naturally occurring anti-diuretic hormone (Munsick, 1966), in *Natrix sipedon* (Dantzler, 1967a) suggest that these changes in filtration rate and tubular permeability to water are mediated by this peptide. Tubular permeability to water appears to be more sensitive to this hormone than does GFR. During these studies, the osmolal U/P ratio varied from 0·1 in maximal water diuresis to 1·0 with maximal response to AVT. Finally, both studies of variations in the tubular maximum for PAH (Dantzler, 1967a) and microscopic studies (B. Schmidt-Nielsen & Davis, 1968) suggest that changes in GFR in ophidian reptiles also result from changes in the number of functioning nephrons.

All snakes are uricotelic (Table 5). Thus differences in the major end-products of nitrogen metabolism cannot be related to differences in glomerular and tubular function. Snakes do not possess bladders, but differences in the function of the cloaca or distal intestine may be associated with differences in glomerular and tubular function in desert and freshwater forms. Although isosmotic urine can be produced by the renal tubules of freshwater ophidian reptiles (Dantzler, 1967a) and precipitated urates have been observed in the collecting ducts during antidiuresis (unpublished observations), it appears likely that under many circumstances urate is precipitated in some region distal to the kidneys. In freshwater snakes, crystalline urate has been consistently observed in the distal large intestine (Dantzler, 1968) which opens into the cloaca ventral to the ureteral openings. Experiments with dye that is filtered by the kidneys indicate that urine does enter the terminal intestine (Dantzler, unpublished observations).

Active reabsorption of sodium probably occurs in this region (Junqueira, Malnic & Monge, 1966), and might be expected to influence the ability of the kidney to continue filtering during dehydration. However, the details of such a reabsorptive process are not known. If reabsorption from the distal intestine of freshwater snakes were only hyperosmotic even during dehydration or salt-loading, it would create a problem in solute excretion and water conservation in animals without an extra-renal route for solute excretion. Under these circumstances, storage of urate in the distal intestine might not be of sufficient importance for filtration to continue.

It is possible that the function of the cloaca or distal intestine in desert-dwelling snakes more nearly resembles bladder function in the desert tortoise than cloacal function in freshwater snakes. If water is reabsorbed passively from a dilute urine in this region as urate is precipitated, it may be of benefit to these animals to continue producing a hypo-osmotic ureteral urine during dehydration. Data on cloacal or distal intestine function in desert species have yet to be obtained.

It should be noted that both freshwater and desert snakes were subjected to a sodium cloride load and not dehydration. Thus the renal responses observed may also have been related to the expansion of extra-cellular volume or to the quantity of sodium infused.

Saurian reptiles

Roberts & B. Schmidt-Nielsen (1966) compared some aspects of renal function and structure in terrestrial lizards from arid regions (*Phrynosoma cornutum*, a horned lizard from the western United States, and *Tropidurus* sp. from the Galapagos Islands) and from a moist, tropical environment (*Hemidactylus* sp., from Puerto Rico). Data on GFR and osmolal U/P ratios under various conditions are shown for these animals in Table 4. As in the other studies on reptiles reported here, these values were obtained on ureteral urine. The data for the two arid-living species are remarkably similar. The control GFR in these animals was only about one-third of that in the tropical gecko. Although the GFR was reduced to somewhat lower levels during dehydration in these animals than in the gecko, the percentage decrease was not greater than in the tropical form.

A water load produced a marked increase in GFR in the tropical gecko, the value being more than twice the control level. However, there was only a modest increase in GFR in the horned toad and the Galapagos lizard. It is more difficult to interpret the effect of an intraperitoneal sodium chloride load since it could only be administered following hydration. Nevertheless, a similar load produced a decrease in GFR from the control level in the two arid-living forms, but no change in the tropical gecko.

It should be noted that the correlation between the number of open and closed tubules observed histologically and the GFR in another species of terrestrial lizard (B. Schmidt-Nielsen & Davis, 1968) has suggested that changes in GFR in saurian reptiles also result from changes in the number of functioning nephrons. Although no definite information is available on the effect of anti-diuretic hormone on glomerular function in these species, it seems likely from the action of pitressin on urine flow in the Australian lizard, *Trachysaurus rugosus*, (Bentley, 1959) and the effect of AVT on GFR in water snakes (Dantzler, 1967a) that neurohypophysial peptides control GFR. In summary, it appears that the filtration rate in the tropical gecko is greater and more variable than in either the horned lizard or the Galapagos lizard.

The tubular function in these lizards showed even less variability than that of the desert chelonian and ophidian reptiles. In every state of hydration the horned toad and the Galapagos lizard produced a urine nearly isosmotic with the plasma. On the other hand, the tropical gecko always produced a urine hypo-osmotic to the plasma regardless of the state of hydration. Moreover the distal tubule did not appear to regulate the degree of urine dilution since the osmola U/P averaged 0·7 during both dehydration and water-loading (Table 4). This failure of the distal tubule in many species of lizards to regulate the urine osmolality with changes in hydration has been supported by recent micropuncture studies of H. Stolte & B. Schmidt-Nielsen (personal communication) on *Sceloporus cyanogenys*, a terrestrial lizard from a non-desert environment.

They found that the osmolal U/P ratio was normally 0·5–0·6 in this animal and that AVT had no effect on the permeability of the distal tubule to water. However, in this lizard (B. Schmidt-Nielsen, personal communication) and in all others studied (Roberts & B. Schmidt-Nielsen, 1966; Seshadri, 1956) uric acid accumulated and the urine became isosmotic in the cloaca. In *Sceleporus cyanogenys* the urine in the cloaca varied from hypo-osmotic during a water load to isosmotic during dehydration, suggesting that antidiuretic hormone might influence this membrane (B. Schmidt-Nielsen, personal communication). The possible advantage of a failure of the distal tubule of the desert-dwelling lizards to dilute the urine is obscure. Perhaps, with more detailed study the cloaca would be found capable of diluting the urine by reabsorption of a hyperosmotic solution.

There are fine structural differences between the distal tubules of the arid-living lizards incapable of producing a dilute urine (*Phrynosoma* and *Tropidurus*) and those of the tropical gecko (*Hemidactylus*) (Roberts & B. Schmidt-Nielsen, 1966). The distal tubular cells of the latter have marked basal infoldings and numerous, elongated mitochondria; those of the former have few basal infoldings and few mitochondria. Whether these basal infoldings, which are in fact lateral interdigitations between adjacent cells, are similar to the lateral spaces between distal tubular cells described for a number of marine, freshwater, and terrestrial reptiles from each of the three major orders (Davis & B. Schmidt-Nielsen, 1967; B. Schmidt-Nielsen & Davis, 1968) is not yet clear. Nor is it yet known with certainty whether such basal infoldings or lateral spaces play a role in transport similar to that postulated for lateral spaces in the mammalian gall bladder (Diamond & Tormey, 1966; Tormey & Diamond, 1967; Kaye, Wheeler, Whitlock & Lane, 1966). However, the few mitochondria in the distal tubular cells of the horned toad and the Galapagos lizard may relate to the fact that no osmotic gradient is created across these cells and that only a small fraction of filtrate is reabsorbed in the distal tubule (Roberts & B. Schmidt-Nielsen, 1966). The larger number of mitochondria in the distal tubular cells of the gecko would then appear to be related to the re-absorption of a larger volume of hyperosmotic fluid in this region.

Although none of the lizards in the study by Roberts & B. Schmidt-Nielsen (1966) appears to have an extrarenal route of salt excretion, other terrestrial lizards from both desert and non-desert environments, including *S. cyanogenus*, have functional salt glands (K. Schmidt-Nielsen, Borut, Lee & Crawford, 1963; Norris & Dawson, 1964; Templeton, 1964, 1966). If reabsorption of water from the cloaca were dependent upon sodium reabsorption in these species, the presence of a salt gland might permit filtration to continue during dehydration while water was reabsorbed and urate stored in the cloaca (K. Schmidt-Nielsen *et al.* 1963).

Amphibians

Although a number of anuran amphibians inhabit hot, arid regions, no detailed data are available on the renal function of these species. However, some inferences about glomerular and tubular function in desert amphibians can be made from data available on non-desert forms. Amphibians cannot produce a urine hyperosmotic to the plasma, and the kidneys of anurans stop forming urine when the animals are sufficiently dehydrated (Adolph, 1927). In *Rana clamitans*, dehydration leads first to an increase in tubular permeability to water, the urine approaching the osmolality of the blood, and then to a decrease in GFR (B. Schmidt-Nielsen & Forster, 1954). Indirect studies involving the tubular maximum for transport of glucose or PAH (Forster, 1942; B. Schmidt-Nielsen & Forster, 1954) and direct visualization of glomeruli during neurohypophysial hormone administration (Sawyer, 1951) have again suggested that changes in GFR in frogs result from changes in the number of functioning nephrons.

AVT has been identified in the pituitaries of all those amphibians examined (Sawyer et al. 1959; Munsick, 1966) and is considered to function as the natural antidiuretic hormone (Sawyer, 1960a; Uranga & Sawyer, 1960; Jard & Morel, 1963; Bentley, 1969). Small doses of exogenous AVT caused both a decrease in GFR and an increase in tubular permeability to water in frogs (Jard & Morel 1963). The effect of AVT on tubular permeability to water appeared more marked than that on GFR (Jard & Morel, 1963). These data suggest that low levels of circulating AVT promote increased tubular permeability to water while higher levels also cause a decrease in GFR. This would explain the pattern of glomerular and tubular responses to increasing dehydration observed in frogs by B. Schmidt-Nielsen & Forster (1954). In this respect, too, amphibians resemble water snakes (Dantzler, 1967a) and, possibly, freshwater turtles (Dantzler & B. Schmidt-Nielsen, 1966). Although these patterns of renal responses to dehydration and AVT may be shared by desert anurans, this must be determined by further study.

Regulation of renal function in desert amphibians, as in reptiles, may be related to the ability to tolerate increases in plasma osmolality, the end-products of nitrogen metabolism and the function of the bladder. Bentley, in his recent review on adaptations of amphibians to arid environments (1966a), emphasized the remarkable tolerance of amphibians to increased plasma osmolality and noted that some desert species appeared more tolerant than similar aquatic or semi-aquatic species. Thus, complete cessation of renal function is less harmful than might be expected.

The tolerance of increased plasma osmolality among terrestrial amphibians also appears related to the fact that urea is the major nitrogenous excretory product (Table 5). As already noted in the discussion on reptiles, vertebrate tissues are less sensitive to high concentrations of urea than to high concentrations of electrolytes. The recent studies by McClanahan (1967) on the desert spadefoot toad, Scaphiopus couchi, show that the plasma osmolality in these animals doubles to about 600 m-osmoles during prolonged periods of aestivation. Urea accounts for about half of the plasma osmolality. Thus, reduced urine output may be related to the tolerance of high plasma urea concentrations.

The regulation of glomerular and tubular function may also be related to the function of the urinary bladder in amphibians. The anuran urinary bladder functions for the storage and later utilization of water (Bentley, 1966 a, b; Ruibal, 1962; Shoemaker, 1964; McClanahan, 1967). Both dehydration and neurohypophysial extracts stimulate water reabsorption from the bladder (Steen, 1929; Ewer, 1952; Sawyer, 1955; Sawyer & Schisgall, 1956), and it now appears well documented that AVT is the active agent in producing this response (Sawyer, 1960 a, b; Bentley, 1969).

The physiology of the amphibian urinary bladder and the nature of the action of some of the neurohypophysial hormones upon it have been reviewed recently (Bentley, 1966 b; Leaf, 1967; Orloff & Handler, 1967). Bentley (1966 a) has also discussed the integration of hormone action in his review on adaptations of amphibians to the desert. The important point for this discussion is the way in which the storage of water in the amphibian bladder and its reabsorption under the influence of AVT may be related to glomerular and tubular function in desert amphibians. AVT increases the permeability of the bladder (Leaf & Hays, 1962) to urea as well as water. If indeed AVT controls glomerular function in desert amphibians, as seems most likely, then it would seem quite reasonable for glomerular function to cease since any urea and water entering the bladder would be reabsorbed. However, the integration of these mechanisms is still not clear. Presumably the bladder is less sensitive to AVT than is glomerular function, for dilute urine remains in the bladder long after renal function appears to have ceased (McClanahan, 1967). The simultaneous studies of renal and bladder function during dehydration or AVT administration, which would be necessary to examine these points, have not been undertaken. It would be necessary to know the relative sensitivities of GFR, tubular permeability, and bladder permeability to AVT in the intact animal.

Birds

There are few data available on glomerular and tubular function in desert birds, and, again, much must be inferred from data on domestic animals. However, as already discussed, both desert birds and domestic birds are capable of producing a dilute urine and a urine hyperosmotic to the plasma. If a hyperosmotic urine is produced by the action of a countercurrent multiplier system, as appears to be the case, the permeability of the collecting ducts and possibly the distal convoluted tubule must vary. AVT has been identified in the neurohypophysis of birds (Munsick, Sawyer & van Dyke, 1960) and has been demonstrated to have a definite anti-diuretic effect with an increase in urine osmolality in domestic fowl (Skadhauge, 1964).

As already pointed out, glomerular filtration rate in mammals does not change with hydration except when dehydration is severe. Moreover, mammals, unlike many non-mammalian vertebrates, do not respond to a salt load with a reduction in filtration rate (B. Schmidt-Nielsen, 1964). Since mammals can produce a highly concentrated urine, they can eliminate a sodium chloride load in this manner. Also, since all mammalian nephrons have loops of Henle and function together in the production of a concentrated urine, it would not seem reasonable for the number of functioning nephrons to vary widely as appears to be the case in reptiles and amphibians. However, birds cannot produce a highly concentrated urine, most terrestrial birds do not have an extrarenal route for sodium chloride excretion, and their kidneys have nephrons resembling those of both reptiles and mammals. Thus, the glomerular response of birds to different states of hydration and sodium chloride loading is of particular interest.

Skadhauge & B. Schmidt-Nielsen (1967a) found that hydration of domestic fowl following dehydration caused only about a 23% increase in GFR. Most of the increase in urine flow observed resulted from a decrease in tubular permeability to water. This increase in filtration rate is very modest compared with that seen in many reptiles (Table 4) and amphibians (B. Schmidt-Nielsen & Forster, 1954). Skadhauge & B. Schmidt-Nielsen (1967a) also found that a modest intravenous sodium chloride load (12–15 m-equiv/kg body weight) did not lead to a decrease in GFR even in a dehydrated bird. However, I found that a constant intravenous infusion of 6% sodium chloride in normally hydrated domestic fowl did lead to a fall in filtration rate (Dantzler, 1967a). When the plasma osmolality was raised by about 150 m-osmoles, the GFR decreased by about 60%. During these studies, the tubular maximum for transport of both glucose and PAH varied directly with the filtration rate. These data indicate that the decrease in filtration rate, as in reptiles and amphibians, resulted from a decrease in the number of functioning nephrons. Thus, it appears that domestic fowl without an extrarenal route for salt excretion respond to marked increases in plasma osmolality with a reduction in the number of functioning nephrons.

This response raises some intriguing questions concerning renal function in desert birds. First, in those birds which might be subject to severe dehydration, does the filtration rate decrease more readily than in domestic birds? Secondly, when a decrease in the number of functioning nephrons occurs, does this involve nephrons of both reptilian and mammalian types? If only nephrons of the reptilian type ceased functioning, nephrons of the mammalian type could continue to function together, allowing maintenance of concentrating ability. Moreover, normal flow through nephrons of the mammalian type would reduce the possibility of urate deposits in the medullary cones. Third, how is the number of functioning nephrons regulated? Even a sustained infusion of large amounts of arginine vasotocin (total dose: 5,000–9,500 mU) produced no consistent depression of filtration rate in domestic fowl (Dantzler, 1966). Thus, it appears unlikely that avian glomerular filtration rate is controlled primarily by AVT. These questions remain to be answered by future studies.

Urine from the ureters enters the cloaca in birds. In domestic birds, at least, it moves retrogradely into the large intestine and even into the caecum (Skadhauge, 1968). As in reptiles and amphibians, cloacal and intestinal reabsorption might be expected to influence renal function. Sodium chloride and water are reabsorbed from the intestine of domestic fowl (Skadhauge, 1967, 1968) and desert quail (McNabb, 1969) even when the solution is hyperosmotic to the plasma. This sodium reabsorption is not influenced by the presence of AVT (Skadhauge, 1967). However, during hydration, urine flow is so great that ureteral urine does not spend sufficient time in contact with the reabsorptive epithelium to be altered significantly (Skadhauge, 1968).

Skadhauge (1967) demonstrated that solute-linked transport of water may be carried out in the fowl intestine against an osmotic concentration difference of about 65 m-osmoles. As noted by Skadhauge (1968), such reabsorption would be useful during dehydration. Some water must accompany uric acid and the sodium urate is more soluble than crystalline uric acid. If this sodium and water can be reabsorbed in the intestine as uric accumulates, it would be useful for filtration to continue. Thus, reduction in filtration, as in some uricotelic desert reptiles (Dantzler & B. Schmidt-Nielsen, 1966; Roberts & B. Schmidt-Nielsen, 1966) might be expected to occur only with marked osmotic stress. The terrestrial birds discussed in this paper, in which renal and cloacal function have been studied in some detail, lack extrarenal routes for salt excretion. However, some terrestrial birds do have functional salt glands (K. Schmidt-Nielsen et al. 1963; Cade & Greenwald, 1966). As in the case of the saurian reptiles where such an extrarenal route of ion excretion occurs, it might be expected to influence glomerular filtration and cloacal reabsorption during dehydration.

NITROGEN EXCRETION

That the end-products of nitrogen metabolism vary with habitat has long been recognized, and this relationship has been well reviewed by others (for example, B. Schmidt-Nielsen, 1964). The major and some of the minor nitrogenous excretory products of tetrapod vertebrates are shown in Table 5. Some aspects of nitrogen excretion have already been discussed in relation to regulation of glomerular and tubular function in desert vertebrates. Consideration will now be given to those aspects of the renal handling of urea and uric acid which are important for desert vertebrates and are the subject of current research.

Urea

All mammals, regardless of habitat, excrete urea as the major end-product of nitrogen metabolism. This is possible because the countercurrent mechanism for urine concentration makes possible the excretion of highly soluble urea with a minimum of water. In general, urea moves passively across the mammalian renal tubular membranes, but recent evidence indicates that some active reabsorption occurs, probably in the region of the collecting duct (Ullrich, Rumrich, & B. Schmidt-Nielsen, 1967; Lassiter, Mylle & Gottschalk, 1966). However, as already noted, some desert mammals with long loops of Henle (e.g. *Psammomys*) fail to concentrate urea as well as others with relatively shorter loops. The factors accounting for these differences are not understood, but may involve differences in countercurrent exchange for urea (B. Schmidt-Nielsen & Pagel, 1969). Studies on such differences will require detailed evaluation of factors affecting urea movement, e.g. active transport, facilitated diffusion, and membrane permeability.

As already pointed out, those semi-aquatic and terrestrial amphibians studied excrete mainly urea (Table 5). Although urea is highly soluble and amphibians cannot produce a concentrated urine, it has a low toxicity and can be maintained at high concentrations in body fluids. The complete pattern of nitrogen excretion is not available for such desert anurans as *Scaphiopus couchi*, but it seems unlikely that highly toxic ammonia plays any significant role. The ability of the renal tubule of semi-aquatic anurans to secrete urea has been well documented (Marshall, 1932; B. Schmidt-Nielsen & Forster, 1954). This secretion is limited by a tubular maximum for transport (B. Schmidt-Nielsen & Forster, 1954) and is inhibited by 2,4-dinitrophenol and probenecid (Forster, 1954). The detailed nature of this transport is not understood. However, the ability of the tubules to secrete urea and the presence of renal arginase appear simultaneously during metamorphosis (Forster, B. Schmidt-Nielsen & Goldstein, 1963).

It has been suggested that arginase could facilitate transport of urea across the tubular cells by catalysing at the luminal side, the hydrolysis of arginine, which had previously been formed from urea by an energy-requiring, DNP-sensitive process at the contra-luminal side (Robinson &

B. Schmidt-Nielsen, 1963). The tubular transport of urea has not been evaluated in desert amphibians, but it seems most likely that active tubular transport would aid these animals in the elimination of urea.

Among the chelonian reptiles, the end-products of nitrogen metabolism clearly vary with habitat—freshwater turtles being ammono-ureotelic and desert tortoises being predominantly uricotelic (Table 5). In the latter, however, a significant fraction of the excretory nitrogen also appears in the form of urea. In neither freshwater nor desert chelonians was there evidence of active tubular secretion of urea (Dantzler & B. Schmidt-Nielsen, 1966). Khalil & Haggag (1955) reported that the relative amounts of uric acid and urea excreted by a single desert tortoise (genus *Testudo*) varied. They suggested that these animals could shift from uricotelism to ureotelism, perhaps depending upon hydration. We found no such shift when ureteral urine was collected from the desert tortoise, *Gopherus agassizii*, regardless of hydration. However, spontaneous urine collections did reveal shifts in the relative amounts of uric acid and urea excreted since large amounts of crystalline urate could be spontaneously released from storage in the bladder.

Uric acid

Since uric acid is highly insoluble in water, it exerts a very low osmotic pressure and can be excreted in the urine in a solid form. Thus, it is advantageous as an end-product of nitrogen metabolism in animals that must conserve water and cannot produce a concentrated urine. Desert-dwelling chelonian reptiles are primarily uricotelic, but even freshwater chelonians excrete some uric acid. In both freshwater turtles and desert tortoises urate is actively secreted by the renal tubules (Dantzler & B. Schmidt-Nielsen, 1966). Factors regulating urate transport in desert chelonians have not been evaluated.

All birds, ophidian reptiles, and saurian reptiles excrete primarily uric acid (Table 5). The complete urea cycle enzymes do not appear to be present in the livers of birds and ophidian reptiles (Brown & Cohen, 1960), and these animals excrete only a small percentage of nitrogen as urea. Most of the detailed data available on the renal handling of urate in these animals were obtained from non-desert species. Recently, however, I have been able to obtain data on some aspects of urate transport from desert saurian reptiles for comparison with data from non-desert forms. In those species on which clearance data have been obtained, urate is secreted by the renal tubules (Dantzler, 1967b, 1968; Shannon, 1938). Tubular urate transport may be bi-directional in all these animals, but stop-flow studies on conscious water snakes of the genus *Natrix* gave no evidence of tubular reabsorption of urate (Dantzler, 1967b).

Studies on the effects of acute metabolic alkalosis and acidosis, sodium chloride infusion, and acetazolamide administration on tubular secretion of urate in conscious water snakes (Dantzler, 1968) revealed that the relative urate clearance (C_{urate}/GFR) increased during an acute alkalosis (produced by intravenous infusion of sodium bicarbonate) but did not change during an acute acidosis (produced by intravenous infusion of hydrochloric acid) (Fig. 4). An infusion of sodium chloride which produced a decrease in the number of functioning nephrons equivalent to that produced by the sodium bicarbonate infusion also had no effect on the secretion of

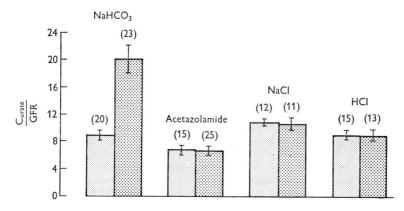

Fig. 4. Relative urate clearance (C_{urate}/GFR) in freshwater snakes (*Natrix* sp.) during sodium bicarbonate, sodium chloride, and hydrochloric acid infusions and acetazolamide administration. Lightly stippled bars, control periods. Darkly stippled bars, experimental periods. Bars represent mean values. Vertical lines represent standard errors. The number of determinations are given in parentheses. Data are from Dantzler (1968).

urate by individual renal tubules (expressed as C_{urate}/ GFR) (Fig. 4). The increase in urate secretion with alkalosis appeared to be correlated with an increase in blood pH and not with urine pH since the administration of sufficient acetazolamide to produce an alkaline urine without a change in blood pH did not affect urate secretion (Fig. 4). Since almost all the uric acid is in the form of the sodium or potassium salt, an increase in tubular secretion during a metabolic alkalosis would eliminate base in the same fashion as an increased excretion of sodium bicarbonate. Although urate secretion has not been studied in desert snakes, it would seem that this response to an alkalosis would also be advantageous.

An increased excretion of uric acid during acidosis, had it occurred, could also have been advantageous. But C_{urate}/GFR did not change. Since the urine pH in this study did not fall to the level of the pKa for uric acid, most of the uric acid would still have been in the form of the sodium or potassium salt. Therefore, the failure of an increase in urate secretion during acidosis seems quite appropriate. However, with a relatively high urine pH, a decrease in urate secretion would seem to have been an even more appropriate response. The failure of tubular urate secretion to decrease may have been related to cloacal function and may be very important for desert reptiles during dehydration when renal function continues. Uric acid is stored in the cloaca, where sodium is probably actively reabsorbed. Further acidification of the urine during acidosis might occur in this region through the secretion of hydrogen ions in exchange for sodium ions. If hydrogen ions normally exchanged for sodium ions of sodium urate in the cloaca during acidosis, this would permit the continued excretion of nitrogenous waste during acidosis while sodium and water were being conserved and acid eliminated.

Other factors which might influence the transport of urate have been evaluated with kidney slices *in vitro*. The uptake of uric acid by kidney slices from desert-dwelling saurian reptiles exceeded that by slices from non-arid ophidian reptiles and domestic fowl (Dantzler, 1969; Dantzler, in preparation). The steady-state urate slice-to-medium ratio (S/M ratio) for slices from the desert spiny lizard (*Sceloporus magister*) was about twice that observed with slices from domestic fowl or garter snakes (*Thamnophis* sp) with the same concentration of urate (3×10^{-7}M) in the incubation medium (Table 6). The addition of energy sources, such as acetate, to the incubation medium did not influence the uptake of urate by slices from these animals. This contrasts with previous studies on organic acid uptake by mammalian slices (Cross & Taggart, 1950). It can only be assumed that sufficient acetate or similar substrate for maximal urate transport is already present in avian and reptilian slices.

The uptake of urate showed a definite requirement for potassium. However, the requirement varied among these animals. The preparation and incubation of slices in potassium-free medium reduced the uptake of urate (Table 6). Although the depletion of tissue potassium was comparable in kidney slices from chickens and both reptilian species, the effect on urate uptake was most marked with avian kidney slices. The active uptake of urate by chicken slices was almost completely eliminated by the absence of potassium from the medium (Table 6). Inhibition was less marked with slices from garter-snake kidneys and least marked with slices from desert spiny-lizard kidneys. The effect of reduction in sodium concentration on urate uptake by chicken and garter snake kidney slices was also evaluated. A significant depression of urate uptake by snake kidney slices occurred only when the medium concentration was reduced to 25 mM while a significant depression of uptake by chicken slices occurred when the

Table 6. *Effects of variations in medium potassium on urate accumulation by kidney slices from reptiles and birds*

Species	Standard ringer	0 mM-K ringer	40 mM-K ringer
Domestic fowl (*Gallus gallus*)	1·94 ± 0·07 (33)	1·19 ± 0·06 (8)	2·38 ± 0·06 (8)
Terrestrial snakes, non-arid environment (*Thamnophis* sp.)	1·67 ± 0·07 (50)	1·35 ± 0·05 (16)	1·60 ± 0·04 (5)
Terrestrial lizards arid environment (*Sceloporus magister*)	3·69 ± 0·38 (18)	2·13 ± 0·18 (6)	5·70 ± 0·28 (18)

Values are mean slice-to-medium concentration ratios (S/M ratios) ± s.e. Figures in parentheses indicate number of experiments. Values for domestic fowl and snakes are from Dantzler (1969).

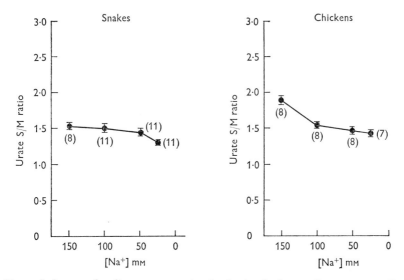

Fig. 5. Influence of sodium concentration in the incubation medium on urate slice-to-medium concentration ratios (S/M ratios) for snake (*Thamnophis* sp.) and chicken kidney slices. Values are means ± s.e. Figures in parentheses indicate number of experiments. Data are from Dantzler (1969).

medium concentration was reduced to 50 mM (Fig. 5). These differences in the sensitivity of urate transport to decreases in sodium and potassium concentrations may be related to the normal variability in plasma ion concentrations. The plasma sodium and potassium concentrations of reptiles are much more variable than those of birds or mammals. Since urate is the major end-product of nitrogen metabolism in reptiles, it appears important that its tubular secretion not be greatly reduced by natural decreases in plasma potassium or sodium.

However, high medium potassium concentrations, which led to similar increases in tissue potassium concentrations in slices from these three species, resulted in a marked increase in urate uptake by *Sceloporus* slices, much less increase by chicken slices, and no increase by *Thamnophis* slices (Table 6). The physiological meaning of these differences, if any, is not clear. But it would seem advantageous for any uricotelic vertebrate to show an increase in tubular urate secretion.

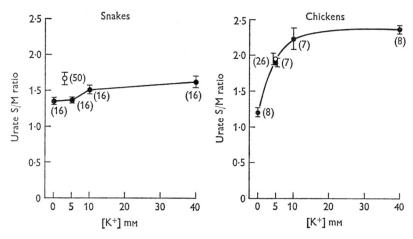

Fig. 6. Influence of potassium concentration in the incubation medium on urate slice-to-medium concentration ratios (S/M ratios) for snake (*Thamnophis* sp.) and chicken kidney slices. Values are means ± s.e. Closed circles represent slices prepared in potassium-free medium. Open circles represent control slices prepared in standard potassium-containing medium appropriate for each species. Figures in parentheses indicate number of experiments. Figure is taken from Dantzler (1969).

Certain differences between the responses of potassium-depleted chicken and snake slices to increasing medium potassium concentrations may have some bearing on the mechanisms involved in urate transport. The urate S/M ratio for potassium-depleted chicken slices returned to control levels with 5 mM potassium in the incubation medium and increased with increasing concentrations of potassium in the medium, reaching a maximum at a potassium concentration between 10 and 40 mM (Fig. 6). However, the tissue potassium concentration was only restored with incubation in 40 mM potassium. This pattern suggests that uptake of urate by avian renal tissue is not closely related to the intracellular concentration of potassium. Potassium in the medium may stimulate urate uptake in some fashion or urate uptake may be more closely related to the process of active accumulation of potassium. On the other hand, the uptake of urate by potassium-depleted snake slices did not reach control levels until the medium potassium concentration reached 40 mM (Fig. 6), but the tissue potassium concentration was restored when the medium concentration was 3 mM. Thus, it appears that severe depletion of potassium impairs some portion of the urate transport system which can only be restored by increased tissue potassium. Preliminary work indicates that, in this respect at least, uptake of urate by slices from desert spiny lizards resembles that by slices from garter snakes.

Finally, studies on urate accumulation by slices from the medullary cones of chicken kidneys were performed (Dantzler, 1969). The uptake was far below that observed with slices from chicken kidney cortex. Since the medullary cones contain the loops of Henle and collecting ducts, the above findings are compatible with the concept that these structures do not contribute significantly to the transport of urate. Since the tubular fluid becomes concentrated as it passes through the loops of Henle and collecting ducts and precipitated urate might be expected to interfere with flow in the loops, it is possible that urate secretion does not occur at all in those nephrons possessed of loops. Thus, urate secretion might occur only in nephrons of the reptilian type where urine does not become concentrated until it reaches the collecting ducts and obstruction would be less of a problem.

SUMMARY

Renal conservation of water by desert vertebrates may involve the production of urine hyperosmotic to the plasma, and, particularly in those animals in which this is not possible, the regulation of glomerular filtration and tubular, bladder or cloacal reabsorption of water. The nature of the nitrogenous excretory products and the methods for their renal excretion may also aid in water conservation and be related to the regulation of other renal functions. Many of the details of the regulatory mechanisms and the precise nature of the interrelationships among them are not yet well understood and await future study.

This work was supported by National Science Foundation Grant GB-7469.

REFERENCES

ADOLPH, E. F. (1927). The skin and kidneys as regulators of the body volume of frogs. *J. exp. Zool.* **47**, 1–30.
BENTLEY, P. J. (1959). Studies on the water and electrolyte metabolism of the lizard *Trachysaurus rugosus* (Gray). *J. Physiol., Lond.* **145**, 37–47.
BENTLEY, P. J. (1966a). Adaptations of amphibia to arid environments. *Science, N.Y.* **152**, 619–23.
BENTLEY, P. J. (1966b). The physiology of the urinary bladder of Amphibia. *Biol. Rev.* **41**, 275–316.

BENTLEY, P. J. (1969). Neurohypophysial function in amphibia: hormone activity in the plasma. *J. Endocr.* **43**, 359–69.

BENTLEY, P. J. & SCHMIDT-NIELSEN, K. (1966). Cutaneous water loss in reptiles. *Science, N.Y.* **151**, 1547–9.

BERLINER, R. W. & BENNETT, C. M. (1967). Concentration of urine in the mammalian kidney. *Am. J. Med.* **42**, 777–89.

BRODSKY, W. A. & SCHILB, T. P. (1960). Electrical and osmotic characteristics of the isolated turtle bladder. *J. clin. Invest.* **39**, 974.

BRODSKY, W. A. & SCHILB, T. P. (1965). Osmotic properties of isolated turtle bladder. *Am. J. Physiol.* **208**, 46–57.

BRODSKY, W. A. & SCHILB, T. P. (1966). Ionic mechanisms for sodium and chloride transport across turtle bladders. *Am. J. Physiol.* **210**, 987–96.

BROWN, G. W. & COHEN, P. P. (1960). Comparative biochemistry of urea synthesis. 3. Activities of urea-cycle enzymes in various higher and lower vertebrates. *Biochem. J.* **75**, 82–91.

CADE, T. J. (1964). Water and salt balance in granivorous birds. *Proc. First Int. Symp. on Thirst in the Regulation of Body Water*, pp. 237–56 (ed. M. J. Wayner). Oxford: Pergamon.

CADE, T. J. & GREENWALD, L. (1966). Nasal salt secretion in falconiform birds. *Condor* **68**, 338–50.

CALDER, W. A. & BENTLEY, P. J. (1967). Urine concentrations of two carnivorous birds, the white pelican and the roadrunner. *Comp. Biochem. Physiol.* **22**, 607–9.

CROSS, R. J. & TAGGART, J. V. (1950). Renal tubular transport: accumulation of *p*-aminohippurate by rabbit kidney slices. *Am. J. Physiol.* **161**, 181–90.

DANTZLER, W. H. (1966). Renal response of chickens to infusion of hyperosmotic sodium chloride solution. *Am. J. Physiol.* **210**, 640–6.

DANTZLER, W. H. (1967a). Glomerular and tubular effects of arginine vasotocin in water snakes (*Natrix sipedon*). *Am. J. Physiol.* **212**, 83–91.

DANTZLER, W. H. (1967b). Stop-flow study of renal function in conscious water snakes (*Natrix sipedon*). *Comp. Biochem. Physiol.* **22**, 131–40.

DANTZLER, W. H. (1968). Effect of metabolic alkalosis and acidosis on tubular urate secretion in water snakes. *Am. J. Physiol.* **215**, 747–51.

DANTZLER, W. H. (1969). Effects of K, Na, and ouabain on urate and PAH uptake by snake and chicken kidney slices. *Am. J. Physiol.* **217**, 1810–19.

DANTZLER, W. H. & SCHMIDT-NIELSEN, B. (1966). Excretion in freshwater turtle (*Pseudemys scripta*) and desert tortoise (*Gopherus agassizii*). *Am. J. Physiol.* **210**, 198–210.

DAVIS, L. E. & SCHMIDT-NIELSEN, B. (1967). Ultrastructure of the crocodile kidney (*Crocodylus acutus*) with special reference to electrolyte and fluid transport. *J. Morph.* **121**, 255–76.

DAWSON, W. R. & SCHMIDT-NIELSEN, K. (1964). Terrestrial animals in dry heat: desert birds. In *Handbook of Physiology, Adaptation to the Environment*, sect. 4, vol. I, ch. 31, pp. 481–92; Washington, D.C.: American Physiological Society.

DIAMOND, J. M. & TORMEY, J. McD. (1966). Role of long extracellular channels in fluid transport across epithelia. *Nature, Lond.* **210**, 817–20.

EWER, R. F. (1952). The effect of pituitrin on fluid distribution in *Bufo regularis* Reuss. *J. exp. Biol.* **29**, 173–7.

FELDOTTO, A. (1929). Die Harnkanälchen des Huhnes. *Z. mikrosk.-anat. Forsch.* **17**, 353–70.

FORSTER, R. P. (1942). The nature of the glucose reabsorptive process in the frog renal tubule. Evidence for intermittency of glomerular function in the intact animal. *J. cell. comp. Physiol.* **20**, 55–69.

FORSTER, R. P. (1954). Active cellular transport of urea by frog renal tubules. *Am. J. Physiol.* **179**, 372–7.

FORSTER, R. P., SCHMIDT-NIELSEN, B. & GOLDSTEIN, L. (1963). Relation of renal tubular transport of urea to its biosynthesis in metamorphosing tadpoles. *J. cell comp. Physiol.* **61**, 239–44.

GONZALEZ, C. F., SHAMOO, Y. E. & BRODSKY, W. A. (1967). Electrical nature of active chloride transport across short-circuited turtle bladders. *Am. J. Physiol.* **212**, 641–50.

HUDSON, J. W. (1964). Water metabolism in desert mammals. *Proc. First Int. Symp. on Thirst in the Regulation of Body Water*, pp. 211–35 (ed. M. J. Wayner). Oxford: Pergamon.

JARD, S. & MOREL, F. (1963). Actions of vasotocin and some of its analogues on salt and water excretion by the frog. *Am. J. Physiol.* **204**, 222–6.

JUNQUEIRA, L. C. U., MALNIC, G. & MONGE, C. (1966). Reabsorptive function of the ophidian cloaca and large intestine. *Physiol. Zoöl.* **39**, 151–9.

KAYE, G. I., WHEELER, H. O., WHITLOCK, R. T. & LANE, N. (1966). Fluid transport in the rabbit gallbladder. *J. Cell Biol.* **30**, 237–68.

KHALIL, F. (1948a). Excretion in reptiles. II. Nitrogen constituents of the urinary concretions of the oviparous snake *Zamenis diadema*, Schlegel. *J. biol. Chem.* **172**, 101–3.

KHALIL, F. (1948b). Excretion in reptiles. III. Nitrogen constituents of the urinary concretions of the viviparous snake *Eryx thebaicus*, Reuss. *J. biol. Chem.* **172**, 105–6.

KHALIL, F. & HAGGAG, G. (1955). Ureotelism and uricotelism in tortoises. *J. exp. Zool.* **130**, 423–32.

KOMADINA, S. & SOLOMON, S. (1970). Comparison of renal function of bull and water snakes (*Pituophis melanoleucus* and *Natrix sipedon*). *Comp. Biochem. Physiol.* (in the Press).

LASSITER, W. E., MYLLE, M. & GOTTSCHALK, C. W. (1966). Micropuncture study of urea transport in rat renal medulla. *Am. J. Physiol.* **210**, 965–70.

LEAF, A. (1967). Membrane effects of antidiuretic hormone. *Am. J. Med.* **42**, 745–56.

LEAF, A. & HAYS, R. M. (1962). Permeability of the isolated toad bladder to solutes and its modification by vasopressin. *J. gen. Physiol.* **45**, 921–32.

MACMILLEN, R. E. & LEE, A. K. (1967). Australian desert mice: independence of exogenous water. *Science, N.Y.* **158**, 383–5.

MACMILLEN, R. E. & LEE, A. K. (1969). Water metabolism of Austalian hopping mice. *Comp. Biochem. Physiol.* **28**, 493–514.

MCCLANAHAN, L. (1964). Osmotic tolerance of two desert inhabiting toads, *Bufo cognatus* and *Scaphiopus couchi. Comp. Biochem. Physiol.* **12**, 501–8.

MCCLANAHAN, L. (1967). Adaptations of the spadefoot toad, *Scaphiopus couchi*, to desert environments. *Comp. Biochem. Physiol.* **20**, 73–99.

MCNABB, F. M. A. (1969). A comparative study of water balance in three species of quail. II. Utilization of saline drinking solutions. *Comp. Biochem. Physiol.* **28**, 1059–74.

MARSHALL, E. K. (1932). The secretion of urea in the frog. *J. cell. comp. Physiol.* **2**, 349–53.

MUNRO, A. F. (1953). The ammonia and urea excretion of different species of amphibia during their development and metamorphosis. *Biochem J.* **54**, 29–36.

MUNSICK, R. A. (1966). Chromatographic and pharmacologic characterization of the neurohypophysial hormones of an amphibian and a reptile. *Endocrinology* **78**, 591–9.

MUNSICK, R. A., SAWYER, W. H. & VAN DYKE, H. B. (1960). Avian neurohypophysial hormones: pharmacological properties and tentative identification. *Endocrinology* 66, 860–71.

NORRIS, K. S. & DAWSON, W. R. (1964). Observations on the water economy and electrolyte excretion of chuckwallas (Lacertilia, *Sauromalus*). *Copeia*, pp. 638–46.

ORLOFF, J. & HANDLER, J. (1967). The role of adenosine 3′,5′-phosphate in the action of antidiuretic hormone. *Am. J. Med.* 42, 757–68.

POULSON, T. L. (1965). Countercurrent multipliers in avian kidneys. *Science, N.Y.* 148, 389–91.

POULSON, T. L. & BARTHOLOMEW, G. A. (1962). Salt balance in the savannah sparrow. *Physiol. Zoöl.* 35, 109–19.

PROSSER, C. L. & BROWN, F. A. (1961). *Comparative Animal Physiology*, 688 pp. Philadelphia: Saunders.

ROBERTS, J. S. & SCHMIDT-NIELSEN, B. (1966). Renal ultrastructure and excretion of salt and water by three terrestrial lizards. *Am. J. Physiol.* 211, 476–86.

ROBINSON, R. R. & SCHMIDT-NIELSEN, B. (1963). Distribution of arginase within the kidneys of several vertebrate species. *J. cell. comp. Physiol.* 62, 147–57.

RUIBAL, R. (1962). The adaptive value of bladder water in the toad, *Bufo cognatus*. *Physiol. Zoöl.* 35, 218–33.

SAYYER, W. H. (1951). Effect of posterior pituitary extracts on urine formation and glomerular circulation in the frog. *Am. J. Physiol.* 164, 457–64.

SAWYER, W. H. (1955). The hormonal control of water and salt-electrolyte metabolism with special reference to the Amphibia. *Mem. Soc. Endocr.* 5, 44–59.

SAWYER, W. H. (1960a). Evidence for the identity of natriferin the frog water-balance principle, and arginine vasotocin. *Nature, Lond.* 187, 1030–1.

SAWYER, W. H. (1960b). Increased water permeability of the bullfrog (*Rana catesbiana*) bladder *in vitro* in response to synthetic oxytocin and arginine vasotocin and to neurohypophysial extracts from non-mammalian vertebrates. *Endocrinology* 66, 112–20.

SAWYER, W. H., MUNSICK, R. A. & VAN DYKE, H. B. (1959). Pharmacological evidence for the presence of arginine vasotocin and oxytocin in neurohypophysial extracts from cold-blooded vertebrates. *Nature, Lond.* 184, 1464.

SAWYER, W. H. & SCHISGALL, R. M. (1956). Increased permeability of the frog bladder to water in response to dehydration and neurohypophysial extracts. *Am. J. Physiol.* 187, 312–14.

SCHMIDT-NIELSEN, B. (1964). Organ systems in adaptation: the excretory system. In *Handbook of Physiology. Adaptation to the Environment*, sect. 4, vol. 1, ch. 13, pp. 215–44. Washington, D.C.: American Physiological Society.

SCHMIDT-NIELSEN, B. & DAVIS, L. E. (1968). Fluid transport and tubular intercellular spaces in reptilian kidneys. *Science, N.Y.* 159, 1105–8.

SCHMIDT-NIELSEN, B. & FORSTER, R. P. (1954). The effect of dehydration and low temperature on renal function in the bullfrog. *J. cell. comp. Physiol.* 44, 233–46.

SCHMIDT-NIELSEN, B. & O'DELL, R. (1961). Structure and concentrating mechanism in the mammalian kidney. *Am. J. Physiol.* 200, 1119–24.

SCHMIDT-NIELSEN, B. & PAGEL, H. D. (1969). Renal countercurrent exchange for urea and water in rat and gerbil. *Fedn Proc. Fedn. Am. Socs exp. Biol.* 28, 523.

SCHMIDT-NIELSEN, K. (1964a). *Desert Animals. Physiological Problems of Heat and Water*. Oxford: Clarendon. Press.

SCHMIDT-NIELSEN, K. (1964b). Terrestrial animals in dry heat: desert rodents. In *Handbook of Physiology. Adaptation to the environment*, sect. 4, vol. 1, ch. 32, pp. 493–507. Washington ,D.C.: American Physiological Society.

SCHMIDT-NIELSEN, K. & BENTLEY, P. J. (1966). Desert tortoise *Gopherus agassizii*: cutaneous water loss. *Science, N.Y.* **154**, 911.

SCHMIDT-NIELSEN, K., BORUT, A., LEE, P. & CRAWFORD, E. (1963). Nasal salt excretion and the possible function of the cloaca in water conservation. *Science, N.Y.* **142**, 1300–1.

SCHMIDT-NEILSEN, K. & DAWSON, W. R. (1964). Terrestrial animals in dry heat: desert reptiles. In *Handbook of Physiology, Adaptation to the environment*, sect. 4, vol. 1. ch. 30. pp. 467–80. Washington D.C.: American Physiological Society.

SESHADRI, C. (1956). Urinary excretion in the Indian house lizard *Hemidactylus flavivividis* (Rüppel). *J. zool. Soc. India* **8**, 63–78.

SHANNON, J. A. (1938). The excretion of uric acid by the chicken. *J. cell. comp. Physiol.* **11**, 135–48.

SHOEMAKER, V. H. (1964). The effects of dehydration on electrolyte concentrations in a toad, *Bufo marinus. Comp. Biochem. Physiol.* **13**, 261–71.

SKADHAUGE, E. (1964). Effects of unilateral infusion of arginine-vasotocin into the portal circulation of the avian kidney. *Acta endocr., Copenh.* **47**, 321–30.

SKADHAUGE, E. (1967). *In vivo* perfusion studies of the cloacal water and electrolyte resorption in the fowl (*Gallus domesticus*). *Comp. Biochem. Physiol.* **23**, 483–501.

SKADHAUGE, E. (1968). The cloacal storage of urine in the rooster. *Comp. Biochem. Physiol.* **24**, 7–18.

SKADHAUGE, E. & SCHMIDT-NIELSEN, B. (1967a). Renal function in domestic fowl. *Am. J. Physiol.* **212**, 793–8.

SKADHAUGE, E. & SCHMIDT-NIELSEN, B. (1967b). Renal medullary electrolyte and urea gradient in chickens and turkeys. *Am. J. Physiol.* **212**, 1313–18.

SPERBER, I. (1944). Studies on the mammalian kidney. *Zool. Bidr. Upps.* **22**, 249–432.

SPERBER, I. (1949). Investigations on the circulatory system of the avian kidney. *Zool. Bidr. Upps.* **27**, 429–48.

SPERBER, I. (1960). Excretion. In *Biology and Physiology of Birds*, pp. 469–92. (ed. A. J. Marshall). New York: Academic Press.

STEEN, W. B. (1929). On the permeability of the frog's bladder to water. *Anat. Rec.* **43**, 215–20.

TEMPLETON, J. R. (1964). Nasal salt excretion in terrestrial lizards. *Comp. Biochem. Physiol.* **11**, 223–9.

TEMPLETON, J. R. (1966). Responses of the lizard nasal salt gland to chronic hypersalemia. *Comp. Biochem. Physiol.* **18**, 563–72.

THESLEFF, S. & SCHMIDT-NIELSEN, K. (1962). Osmotic tolerance of the muscles of the crab-eating frog. *Rana cancrivora. J. cell. comp. Physiol.* **59**, 31–4.

TORMEY, J. McD. & DIAMOND, J. T. (1967). The ultrastructural route of fluid transport in rabbit gall bladder. *J. gen. Physiol.* **50**, 2031–60.

ULLRICH, K. J., RUMRICH, G. & SCHMIDT-NIELSEN, B. (1967). Urea transport in the collecting duct of rats on normal and low protein diet. *Pflügers Arch. ges. Physiol.* **295**, 147–56.

URANGA, J. & SAWYER, W. H. (1960). Renal responses of the bullfrog to oxytocin, arginine vasotocin and frog neurohypophysial extract. *Am. J. Physiol.* **198**, 1287–90.

DISCUSSION

HENDERSON: With reference to the data on the turtle where you showed total cessation of urine flow with increasing plasma osmolality, I wonder if you would like to say a few words about whether this is dependent upon vasotocin or not.

DANTZLER: I am really not sure. We have only done a few preliminary experiments with turtle neurohypophysial extract on that and we do have some evidence, as I said, that there is a decrease in filtration rate and an increase in tubular permeability to water as one would expect. Apparently the tubular permeability to water is more sensitive than the filtration rate to neurohypophysial extract. I don't know about absolute cessation.

GREEP: Dr Dantzler, I was fascinated by your discussion of the physiology of the bladder, which appears to put this organ in a wholly new light. I had come to regard the bladder as an adaptation of convenience for the terrestrial vertebrates. Your discussion emphasized its importance in the excretory process. Apparently the bladder might be regarded as a part of the kidney, but independent of control by the neurohypophysis.

I am curious about the evolution of this organ and I wonder if you would make a few Darwinian comments relative to its origin?

DANTZLER: I think I'll let somebody else try that one. However, I do consider the bladder of many non-mammalian vertebrates as part of the kidney from my point of view.

BERN: I would like to extend part of Professor Greep's comments to ascertain how critical the experimental data are relevant to the effect of neurohypophysial factors on surfaces such as ureters, bladders and cloacal membranes in vertebrates 'above' the amphibia.

DANTZLER: Well, as far as the reptiles are concerned, the *in vitro* studies on the freshwater-turtle bladders carried out by Brodsky and a number of others have shown no effect of neurohypophysial peptides, as far as I know, on water permeability. We also found none with isolated desert-tortoise bladders. We also found no changes doing *in vivo* studies on desert tortoises. They don't appear to respond in either case to neurohypophysial extracts.

MALNUTRITION AND THE ENDOCRINE
SYSTEM IN MAN

By R. D. G. MILNER

The relationship between man and his environment is nowhere more dramatically illustrated than in man's inability to find enough food. Approximately half the population of the world has survived a period of serious deficiency in childhood (Graham, 1967), and today over 300 million children are suffering from varying degrees of malnutrition. The severity of this problem is emphasised by evidence that nutritional deprivation in infancy may stunt growth in later life (Graham, 1968; Thomson, 1968) and, more important, impair development of the brain (Winick, 1969).

Recognition of widespread malnutrition has led to the establishment in various parts of the world of clinical research centres in which all aspects of the problem are studied. Despite intensive investigation of the metabolic adaptation accompanying undernutrition and subsequent recovery, a study of the associated hormonal changes has received only intermittent attention. Animal models have, of course, been used to answer questions which cannot be asked of man (Munro, 1964) but, as ever, problems of extrapolation arise (Kirsch, Saunders & Brock, 1968; Widdowson, 1968) and the importance of the human problem fully justifies its direct study.

Comprehensive studies of malnutrition in adult man, such as the 'Minnesota Experiment' (Keys, Brozek, Henschel, Mickelsen & Taylor, 1950) did not lay emphasis on endocrine changes. More recent studies of the metabolic and hormonal interrelationships during fasting have been concerned with prolonged fasts in obese adults (Beck, Koumans, Winterling, Stein, Daughaday & Kipnis, 1964; Owen, Felig, Morgan, Wahren & Cahill, 1969; Felig, Owen, Wahren & Cahill, 1969), or relatively short fasts in normal adults (Unger, Eisentraut & Madison, 1963; Cahill, Herrera, Morgan, Soeldner, Steinke, Levy, Reichard & Kipnis, 1966). Investigations of this kind cannot be performed in children and we have to capitalize on an experiment of nature to make analagous observations.

Understanding the endocrinology of malnutrition may be of help in the practical treatment of the condition. There is also another important reason for such work. The endocrinology of normal human growth has been studied primarily by observations on children with abnormalities such as endocrine or metabolic disorders and their correction. Malnutrition, on the other hand, is an environmental disorder and evaluation of the endocrine adaptation to both malnutrition and the subsequent 'catch up' growth will form an important facet of our understanding of the control of normal growth.

Infantile malnutrition is a continuous spectrum of deficiencies in calories and/or protein. Thus malnutrition may be of any degree of severity. If the deficiency is primarily of protein, the clinical condition called kwashiorkor results, whereas if both protein and calories are inadequate, the baby becomes marasmic (Waterlow, 1948). It may be anticipated that there will be, likewise, a continuous spectrum in the quantitative endocrine response to malnutrition. Also, in some instances the hormonal response will be the same whether the child has marasmus or kwashiorkor; in other cases there will be a qualitatively different response to a deficiency of protein or to lack of protein and calories.

The endocrine glands studied in malnutrition have been, naturally, those known to have anabolic or catabolic functions. Reviews of the hormonal changes characteristic of undernutrition in the adult (Keys *et al.* 1950; Gillman & Gillman, 1951; Zubiran & Gomez-Mont, 1953) were in agreement principally on the basis of clinical and histological evidence, that there was hypofunction of most glands. A central idea at that time was that malnutrition caused the occurrence of 'pseudohypophysectomy' with secondary hypofunction of glands under pituitary control. This had been introduced first by Mulinos & Pomerantz (1940) on the basis of experiments with rats. The concept of endocrine hypofunction and in particular of pituitary hypofunction was carried over and applied to malnourished children (Mönkeberg, Donoso, Oxman, Pak & Meneghello, 1963; Dean, 1965; Mönkeberg, 1968). Because of refinements in biochemical analyses and hormone assays, more direct evidence is now available and in many instances the concept of endocrine hypofunction has needed revision.

Recent studies in children have attempted to elucidate mainly the roles of the adrenal cortex, the thyroid, growth hormone (GH) and insulin in malnutrition. Little attention has been paid to the gonads or to anterior-pituitary function other than growth hormone secretion.

ADRENAL CORTEX

The adrenal cortex, by most accounts, is atrophied in fatal infantile malnutrition (Lucien, 1908*a*; Marfan, 1921; Gillman & Gillman, 1951; Trowell, Davies & Dean, 1954), but the histological picture is sufficiently varied to make Gillman & Gillman (1951) suggest that hyperfunction might precede adrenal exhaustion and atrophy. The assessment of adrenal function by measurement of urinary steroid excretion has also given varied results. Castellanos & Arroyave (1961) found that marasmic infants had a high excretion of glucocorticoids whilst in kwashiorkor the excretion was low. Lurie & Jackson (1962) also observed low 17-ketosteroid and 17-hydroxysteroid excretion in kwashiorkor, both on admission and during recovery, when compared with normal children. They also demonstrated that steroid excretion could be increased by injection of ACTH. However, measurements of urinary steroids can be unreliable indices of adrenal function (Cope & Pearson, 1965), especially when, as in malnutrition, there may be impairment of renal function (Alleyne, 1967).

More recently, adrenocortical function has been clarified by measurements of the plasma levels, production rate and half-life of cortisol. Alleyne & Young (1967) showed that fasting plasma levels of cortisol were high in malnourished children. With recovery the level fell. In both sick and recovered children the production rates were similar, but the half-life of cortisol in the sick infants was prolonged, thus accounting for their higher fasting plasma levels. In both kwashiorkor and marasmus there is a similar rise in plasma cortisol levels after the administration of Synacthen (β^{1-24} corticotrophin) when the infants are sick or recovered (Alleyne & Young, 1967; Rao, Srikantia & Gopalan, 1968). Also, plasma cortisol can be suppressed by the administration of dexamethasone (Alleyne & Young, 1967). Although circulating levels of corticotrophin have not yet been measured, the cortisol response to Synacthen and dexamethasone suggests some preservation of the pituitary-adrenal axis.

The data on adrenocortical function can be unified by proposing that when the malnourished child is first seen he has high levels of cortisol caused by decreased catabolism. Associated with this is decreased urinary steroid excretion and, it is postulated, normal or hyperplastic adrenals. If the condition is fatal, death is preceded by adrenal atrophy and hypofunction. The metabolic consequences of high circulating levels of cortisol are complex but among them may be included the fatty liver (Waterlow, 1948) and the impaired glucose tolerance characteristic of infantile malnutrition.

THYROID

Keys *et al.* (1950) state firmly that 'the evidence is quite consistent that the human thyroid atrophies during starvation', and they cite comprehensive evidence to support this viewpoint. As with the other endocrine glands there is great variation in the histology of the thyroid in different reports, but early workers agreed that in infantile malnutrition there was atrophy, reduction of colloid and histological evidence of thyroid hypofunction (Thompson, 1907; Lucien, 1908*b*; Marfan, 1921; Nicolaeff, 1923). Direct measurement of thyroid function was made many years ago by Benedict, Miles, Roth & Smith (1919), who showed a fall in basal metabolic rate (BMR) in the adult on reduction of the caloric intake from 4,000 to 1,400 calories per day. Early measurements of the BMR of undernourished infants gave varied results (Benedict & Talbot, 1914, 1915), due in part to the many different disease states studied and in part to the conditions under which the measurements were made.

A more direct approach to the study of thyroid function is now possible and from it has come clear evidence of diminished thyroid activity in marasmus. Uptake of radioactive iodine by the thyroid is reduced (Mönke-berg, Barzellato, Beas & Waissbluth, 1957). Plasma levels of protein-bound iodine (PBI) and butanol extractable iodine (BEI) are low (Valledor, Lavernia, Borbolla, Satanowsky, Costales, Prieto & Bardelas, 1959; Lifshitz, Chavarria, Cravioto, Frenk & Morales, 1962). Interpretations of measurements of BMR are complicated by variations in body composition (Montgomery, 1962*a*; Beas, Mönkeberg & Horwitz, 1966) but there is agreement that an untreated malnourished child or one who is not growing has a diminished oxygen consumption per unit of lean body mass.

In an attempt to analyse further the cause of diminished thyroid function, Beas *et al.* (1966) gave thyroid-stimulating hormone (TSH) to malnourished children and control subjects. In both groups TSH caused a rise in the BEI, ^{131}I uptake by the thyroid gland and, in the marasmic infants, a rise in the BMR. In every variable measured, the level of activity in the mal-nourished child, before or after TSH, was lower than in the control subjects. The results were interpreted to indicate that the reduced thyroid activity in marasmus was in part intrinsic, but mainly due to a decrease of TSH secretion. Varga & Mess (1968) have recently produced direct evidence of a reduction in both plasma and pituitary levels of TSH in both marasmus and kwashiorkor.

Mönkeberg (1968) has contrasted these findings with those in infants with kwashiorkor, in whom oxygen consumption is not significantly reduced (Mönkeberg, 1966). The difference between the groups is explained by the fact that there is an adequate caloric intake in kwashiorkor but not in marasmus.

A further illustration of the lability of the thyroid response to mal-nutrition arises from the study of malnourished Jamaican children (Montgomery, 1962*a*). The BMR was found to be subnormal in relation to the calculated surface area, but approximately normal in terms of body solid mass. Further evidence cited in favour of normal thyroid function in these children was the normal excretion of a radioactive iodine load (Montgomery, 1962b) and normal thyroid histology (Stirling, 1962). This apparent divergence from the findings of workers in other centres may be reconciled since Alvarado (personal communication) has shown recently that there is a uniform depression of metabolic rate in the children in Jamaica on the day after admission, but that this rapidly returns to normal as treatment commences. The patients thought to have normal thyroid function were studied initially at varying intervals, ranging from 1 to 24 days after admission (Montgomery, 1962c).

INSULIN AND GROWTH HORMONE

Consideration of insulin and GH together is justified by the fact that these hormones have a synergistic action in promoting protein synthesis and growth (Manchester & Young, 1961; Knobil & Hotchkiss, 1964). Furthermore, insulin and GH have antagonistic actions in the control of carbohydrate metabolism (Weil, 1965), which is known to be abnormal in infantile malnutrition (Alleyne & Scullard, 1969).

A clear-cut difference in glucose tolerance between children with kwashiorkor and those with marasmus has been reported on two occasions (Bowie, 1964; Hadden, 1967). Kwashiorkor is characterized by a reduced disappearance of an intravenous glucose load, whereas, in marasmus, glucose tolerance is normal. The data for kwashiorkor have been corroborated (Baig & Edozien, 1965) but Oxman, Maccioni, Zuniga, Spada & Mönkeberg (1968) have found that marasmics also have decreased glucose tolerance. Plasma insulin determinations have been made in oral and intravenous glucose-tolerance tests. The fasting level is low and the insulin response to an intravenous glucose load is blunted (Baig & Edozien, 1965; Hadden, 1967; James, 1968a). Of other variables known to affect glucose tolerance only free fatty acid (FFA) levels have been measured; these were high in kwashiorkor and low in marasmus and, in both, became normal following treatment (Hadden, 1967).

The use of glucose as a tool to investigate hormonal and metabolic interrelations in malnutrition has revealed different metabolic responses in the two clinical subdivisions—kwashiorkor and marasmus—and differences between marasmics in different parts of the world. These differences spotlight one of the recurrent problems facing workers in this field: that of choosing a homogeneous sample of a malnourished population for study. We believe that if a classification into categories called kwashiorkor and marasmus is to be retained, it is reasonable to base it on two objective criteria only: the presence or absence of oedema and whether or not the child is less than 60% of his expected weight for age. The criterion of normal weight is chosen empirically as the 50th percentile of the Boston standards (Nelson, 1959), since these are widely available. The majority of malnourished children seen in Jamaica are less than 60% of their expected weight and have oedema; they may therefore be described as having 'marasmic-kwashiorkor' (Waterlow, 1948; Jelliffe, Bras & Stuart, 1954).

Information about GH and malnutrition comes entirely from reports of the histology of the pituitary and measurements of plasma GH levels. Pituitary morphology in starvation, like that of the adrenal cortex and thyroid, is variable. In adults there may be an appreciable incidence (29–72%) of eosinophilia in the adenohypophysis (Lamy, Lamotte & Lamotte-Barillon, 1946; Uehlinger, 1947), whilst Gillman (1942) and Vint (1949) have reported the development of cysts and pituitary atrophy. In malnourished children information is more scanty, but Tejada & Russfield (1957) observed no evidence of atrophy and a variable increase of some secretory cells in a small group of children who had died from kwashiorkor or marasmus.

Measurements of plasma GH suggest also that this part of anterior-pituitary function is normal or increased. Fasting levels are high in both untreated marasmus and kwashiorkor and fall towards normal as the children recover (Pimstone, Wittman, Hansen & Murray, 1966; Pimstone, Barbezat, Hansen & Murray, 1967, 1968). Pimstone and his colleagues (1968) showed also that the fall in GH with recovery was associated with the rise in plasma albumin concentration, but that acute elevation of the plasma albumin by infusions in the sick child had no effect on the raised plasma GH level (Hansen, personal communication). Glucose infusions caused a fall in plasma GH which was greater and occurred earlier in the recovered than in the sick child (Pimstone et al. 1967).

The apparently normal or hyperplastic appearance of the islets of Langerhans (Tejada & Russfield, 1957) belies the uniform evidence of impaired β-cell function in infantile malnutrition and is probably more a reflection of the profound exocrine pancreatic atrophy which occurs (Waterlow, 1948; Trowell et al. 1954). Atrophy also occurs in the small-bowel mucosa (Trowell et al. 1954) and may result in impaired absorption of glucose (James, 1968b). Variable glucose absorption coupled with the possibility of impaired secretion of enteric hormones known to affect insulin release (Dupré, Curtis, Unger, Waddell & Beck, 1969) reduces the value of oral glucose loads as a test of β-cell function in malnutrition.

In my own studies on the role of insulin and GH in the adaptation of a child to undernutrition and his subsequent 'catch-up' growth I felt that it was important to measure both hormones and certain metabolites in the same test. The use of other stimuli in addition to glucose also seemed desirable. These investigations were carried out in Jamaica where, as stated above, most malnourished children have oedema and are 60% underweight for their age. On admission to our Unit their ages commonly range between 6 and 18 months. During recovery their growth rate is approximately 15 times that of a normal child of the same age or five times that of a normal child of the same weight or height (Ashworth, 1969). By studying a child shortly after admission and again some 6–12 weeks later it is possible to assess his adaptation to malnutrition and to the period of accelerated growth. Also, variability of results due to differences between individuals is avoided. By the use of microtechniques it is possible to estimate glucose, FFA, α-amino nitrogen, insulin and GH in duplicate on 300 μl plasma. It is thus feasible to make repeated determinations of these metabolites and hormones on one child on one day.

In the first study (Milner, 1970) ten malnourished children were given intravenous glucose shortly after admission. Blood samples were collected before the glucose injection and at 3, 10, 30 and 60 min afterwards. The test was repeated, on the same children, several weeks later when they had recovered clinically. The infants had an impaired glucose tolerance which improved but did not become normal on recovery (Fig. 1). The initial tests were performed on the first or second day after admission when FFA levels were high, but FFA could not be incriminated in the poor glucose tolerance on recovery, as the levels had fallen to normal (Fig. 1). There was no significant insulin response to intravenous glucose in the sick children, but a rise at 3 min when they had recovered, which returned to fasting levels later in the test despite persistent hyperglycaemia (Fig. 1). Plasma GH levels were very high on admission and fell with recovery, but in both groups rose following intravenous glucose (Fig. 1).

Glucagon was used as another approach to the same problem (Milner, 1970). Glucagon is known to cause hyperglycaemia in malnourished children (Alleyne & Scullard, 1969) and to stimulate insulin secretion independently of its glycogenolytic action (Samols, Marri & Marks, 1965). In the new-born, glucagon causes a rise in plasma GH levels (Milner & Wright, 1967) and it was of interest to see if these older, yet small, children would resemble the normal new-born in this respect. It was argued that the direct action of glucagon plus hyperglycaemia might prove a more effective stimulus of insulin secretion than glucose alone.

Intravenous glucagon caused a rise in plasma glucose levels, in nine children on the second to sixth day of admission, which was maximal at 30 min and remained steady at 60 min. In the recovered children the rise was also maximal at 30 min but was greater than in the sick children and there was a fall in plasma glucose levels between 30 and 60 min (Fig. 2). Since glyogenolysis is normal in both the sick and recovered children the difference in the rise in plasma glucose was thought to be due to the

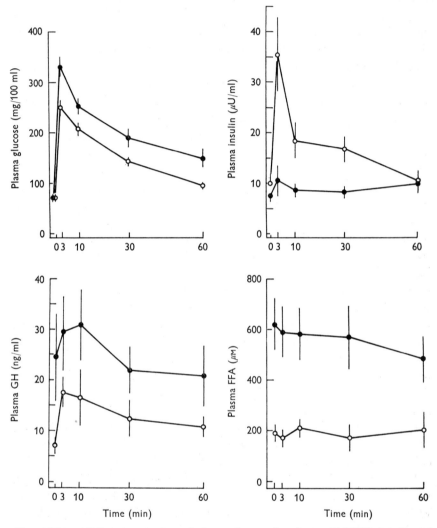

Fig. 1. Mean±S.E. concentrations of plasma glucose, free fatty acid (FFA), insulin and growth hormone (GH) in ten malnourished infants in response to intravenous glucose (0·5 g/kg body weight). Measurements were made on the first or second day after admission (solid circles) and 6–12 weeks later (open circles).

decreased hepatic glycogen of the sick child (Alleyne & Scullard, 1969). Plasma FFA, in both groups, rose in the first 10 min but the changes were not significant. If the rise were real it could be due to the direct lipolytic effect of the hormone (Lefebvre, 1966). The fall in plasma FFA between 10 and 60 min was significant and was interpreted as being due to inhibition of lipolysis by hyperglycaemia (Randle, Garland, Hales & Newsholme, 1963). In neither group was there a significant change in plasma

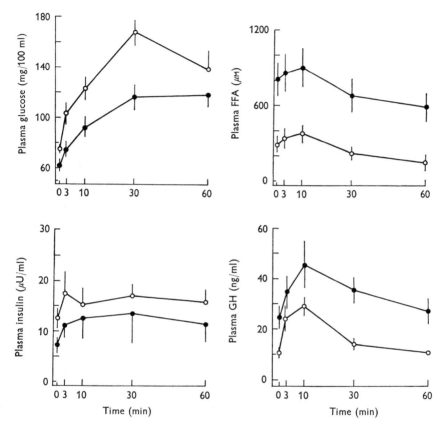

Fig. 2. Mean ± s.e. concentrations of plasma glucose, free fatty acid (FFA), insulin and growth hormone (GH) in nine malnourished infants in response to intravenous glucagon (0·1 mg/kg body weight). Measurements were made on the second to sixth day after admission (solid circles) and 6–12 weeks later (open circles).

insulin levels (Fig. 2). Because of the complicated nature of the stimulus to insulin secretion, variation in the time of the maximum insulin response was anticipated, but testing either the sum of the increases or the maximum rise in plasma insulin in each group still failed to reveal a significant change from the fasting level.

Plasma GH levels appeared to rise following intravenous glucagon, as they had after glucose. Before a causal relationship was invoked, the possibility was considered that the rise was a non-specific consequence of the stress of five venepunctures in 1 h (Helge, Weber & Quabbe, 1969). Plasma GH levels were measured at the same time intervals in six recovered children who received no injection, and the change from the fasting level was compared with that in the children receiving glucose or glucagon

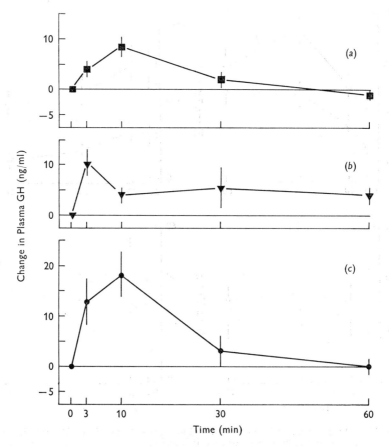

Fig. 3. Mean ± s.e. change from fasting levels of plasma growth hormone (GH) in three groups of infants who had recovered from malnutrition: (a) six infants who had five vene-punctures but no injection, (b) ten infants who received intravenous glucose, (c) nine infants who received intravenous glucagon.

(Fig. 3). The stress of venepuncture did cause a rise in plasma GH which was similar to that seen following the injection of glucose. The rise after glucagon was greater, however, and it was concluded that glucagon had caused a real rise in plasma GH levels, as it did in the new-born (Milner & Wright, 1967). This rise was considered to be due to increased secretion of GH for the following reasons: GH release from rat pituitary slices is stimulated by activation of the adenyl cyclase system (Steiner, Peake,

Utiger & Kipnis, 1969) and glucagon causes a 30-fold rise *in vivo* of circulating levels of cyclic $3',5'$-adenosine monophosphate (cyclic AMP) (Kaminsky, Broadus, Hardman, Ginn, Sutherland & Liddle, 1969). It seems possible that the GH-releasing action of glucagon as well as its stimulation of insulin secretion and glycogenolytic and lipolytic effects may be mediated by cyclic AMP.

A third way in which insulin and GH may be studied is by the administration of protein or amino acids. Oral protein or the intravenous infusion of amino acids causes a rise in plasma insulin (Floyd, Fajans, Conn, Knopf & Rull, 1966*a*, *b*) and GH levels (Knopf, Conn, Fajans, Floyd, Guntsche & Rull, 1965; Merimee, Lillicrap & Rabinowitz, 1965). Both Knopf *et al.* (1965) and Rabinowitz, Merimee, Maffezolli & Burgess (1966) have suggested that the sequential rise of insulin and GH following oral protein leads to synergism of the anabolic roles of these hormones. This type of stimulus therefore is most appropriate for a study of insulin and growth hormone in malnutrition.

Because we did not know how the infants would respond to large intravenous doses of amino acids and since there is known to be impaired digestion and absorption of foodstuffs, a compromise was made. The children were given oral loads of a mixture of the ten essential amino acids (Floyd *et al.* 1966*b*) and blood samples were collected before and at intervals up to 3 h afterwards. The amino acid load caused a similar rise in plasma α-amino nitrogen, in both the sick and recovered children, which was maximal 60 min after ingestion of the load (Fig. 4). All the other variables measured showed similar changes in the sick and recovered groups. There were reciprocal changes in glucose and FFA levels; glucose rose slightly in the first hour and then fell, while FFA fell in the first hour and then rose. In neither group was there any significant change in the plasma insulin level. Surprisingly, amino acid ingestion caused a fall in plasma GH levels (Fig. 4). This response was qualitatively different to that expected.

By combining the results of different tests some generalizations on the role of GH and insulin in malnutrition may be made. In agreement with other workers, malnourished infants in Jamaica have been found to have high fasting FFA and GH levels which fall with recovery. The fasting plasma insulin on the other hand is low and rises with recovery (Table 1). No change is seen in plasma glucose or α-amino nitrogen concentrations. The high fasting levels of GH are similar to those reported previously (Pimstone *et al.* 1966, 1967, 1968). From the scattergram of fasting plasma GH levels and duration of stay in hospital (Fig. 5) it is seen that there is no clear-cut sex difference in fasting GH levels, that the rate of fall of GH

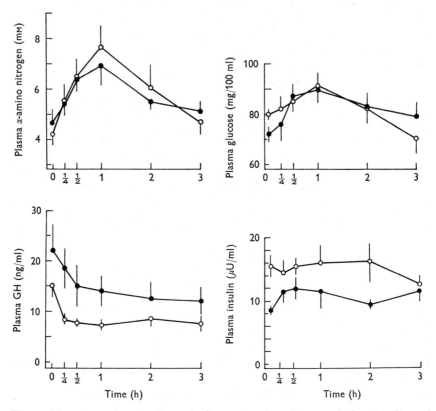

Fig. 4. Mean±s.e. concentrations of plasma α-amino nitrogen, glucose, insulin and growth hormone (GH) in two groups of malnourished children in response to an oral mixture of essential amino acids (0·5 g/kg body weight). One group of eight infants were studied between the first and sixteenth day after admission (solid circles). The other group of seven infants was studied after 6–12 weeks treatment (open circles).

Table 1. *Mean ± s.e. fasting plasma levels of certain metabolites and hormones in infantile malnutrition*

Metabolite or hormone	Clinical state*	
	Sick	Recovered
Glucose (mg/100 ml)	67±4	73±3
FFA (μM)	707±91	238±39†
α-amino nitrogen (mM)	2·6±0·2	3·2±0·2
Insulin (μU/ml)	7·3±0·9	11·9±1·2†
GH (ng/ml)	24·5±4·6	8·7±1·2†

* Measurements were made on 15 infants 1–6 days after admission (sick) and 6–12 weeks later (recovered).

† Statistical analysis by Student's *t* test shows *P* < 0·01.

cannot be accurately assessed because of the infrequent observations between days 10 and 30, and that after the first month the levels reach a steady range. Fasting plasma GH levels fall in the first 8 weeks of life (Cornblath *et al.* 1965) and in older normal children (Stimmler & Brown, 1967; Root, Saenz-Rodriguez, Bongiovanni & Eberlein, 1969) are lower than in the recovered children of this study. It is not possible to state if the

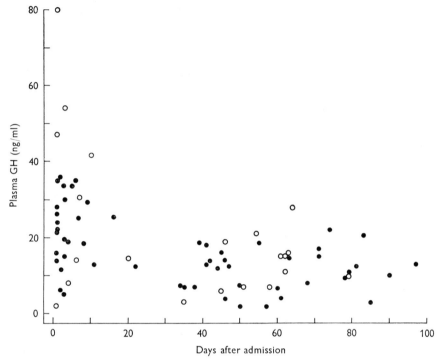

Fig. 5. Scattergram showing fasting levels of plasma growth hormone (GH) and length of stay in hospital of malnourished infants. Solid circles depict boys, and open circles girls.

mean level at this time is normal since the fasting GH level of normal infants aged one year is not known.

That malnutrition should cause a rise in plasma GH is suggested by other data in man. Fasting (Roth, Glick, Yalow & Berson, 1964), anorexia nervosa (Marks, Haworth & Greenwood, 1965), renal failure (Wright, Lowy, Frazer, Spitz, Rubenstein & Berschn, 1968) and protein-losing enteropathy due to giardiasis (Pimstone, Bank & Buchanan-Lee, 1968) are all associated with raised circulating GH levels. Of more direct relevance to infantile malnutrition may be the fact that babies born small for their gestational age, who are thought to have been malnourished *in utero*, have high GH levels (Laron, Mannheimer, Nitzan & Goldmann, 1967) while babies born to diabetic mothers, who may have been over-nourished *in utero*, have low GH levels at birth (Westphal, 1968). The present study demonstrated that elevated fasting plasma GH levels could be raised further by stress or by the administration of glucagon. In this respect also there is a similarity between the malnourished infant and the new-born (Milner & Wright, 1967). If the amino acid load had a causal relationship to the fall in plasma GH levels, it is possible that the circulating level of an amino acid or acids may have a direct feedback control on GH release as has been postulated for glucose in the adult (see Randle, Ashcroft & Gill, 1968).

Before it was appreciated that circulating GH levels were high in malnutrition, children were given exogenous GH to determine if this was a limiting factor in their failure to grow as had been originally suggested by Mulinos & Pomerantz (1940). Marasmic infants given GH retained nitrogen (Mönkeberg *et al.* 1963) but the degree of nitrogen retention was similar to that seen in normal children given GH (Vest, Girard & Buhler, 1963; Prader, Illig, Szeky & Wagner, 1964) and did not support the deduction that the marasmic infants were GH-deficient (Mönkeberg *et al.* 1963; Mönkeberg, 1968). No significant metabolic effect of exogenous GH was noted in kwashiorkor (Hadden & Rutishauser, 1967).

What are the metabolic consequences of raised GH levels? James & Hay (1968) have shown that there is a decreased rate of catabolism of albumin in malnourished infants. Others have shown, in the adult, that exogenous GH prolongs the half-life of albumin (Gabuzda, Jick & Chalmers, 1963; Jeejeebhoy, Boucher & Hartog, 1967). The same mechanism may operate in infancy. The high FFA levels are not directly the consequence of raised GH levels as the plasma FFA falls rapidly with the resumption of a normal caloric intake (Lewis, Hansen, Wittman, Krut & Stewart, 1964), whereas GH levels fall more slowly. The high GH levels in recovered children may be responsible in part for the persistence of an abnormal glucose tolerance.

Diminished secretion of insulin is now a well-established feature of infantile malnutrition. The present study confirms earlier work (James, 1968a) in showing that although glucose will stimulate insulin release in the recovered child the response is still less than that of normal children. The failure of these stimuli may be interconnected since the mechanism by which amino acids cause insulin release may involve the secretion of glucagon and other enteric hormones (Dupré *et al.* 1969; Milner, 1969a, b). A fall in insulin secretion fits well with the hypothesis that this is the primary hormonal regulator of energy supply in starvation (Cahill *et al.*

1966), but failure of β-cell function to return to normal after treatment is less easy to explain. No study of the long-term effects of infantile malnutrition on endocrine pancreatic function has yet been made and the important question of whether these infants will subsequently have an increased tendency to develop diabetes mellitus remains unanswered.

COMMENT

A simple view of the hormonal control of growth is embodied in the concept that GH is responsible for growth by increase in cell number, and insulin for growth by increase in cell size (Cheek & Graystone, 1969). Although this must be an over-simplification of a more complex control mechanism, it fits the data at present available for malnourished infants. Montgomery (1962d) found that the sartorius muscle of Jamaican infants dying from malnutrition was reduced to a size comparable to that of a normal 31-week foetus. There was slight loss of cell number but the major loss was in cell size. Similar children have a deficiency of insulin secretion and high circulating GH levels. Analysis of the hormonal adaptation to malnutrition and recovery has shown that the endocrines play a complicated and, as yet, poorly understood role in man's adaptation to this environmental change. It is important to obtain direct evidence of the secretory status of other pituitary trophic hormones such as TSH and ACTH as well as their target glands. Since malnutrition is associated with decreased catabolic rates of some proteins, it becomes important, in analysing the significance of plasma concentrations, to consider the half-lives of hormonal polypeptides. The half-life of other proteins, such as those which bind steroids or thyroid hormones, is of equal relevance. Does the decreased binding of cortisol in the plasma of malnourished infants (Leonard & MacWilliam, 1964), for example, indicate a greater biological effectiveness for a given plasma cortisol concentration? The study of infantile malnutrition offers the chance of answering this and many other questions relevant to the welfare of the starving child and the control of growth in man.

I wish to thank Professor J. C. Waterlow, Director of the M.R.C. Tropical Metabolism Research Unit, for his encouragement throughout these investigations and Miss M. Ceballos for technical assistance.

REFERENCES

ALLEYNE, G. A. O. (1967). The effect of severe protein calorie malnutrition on the renal function of Jamaican children. *Pediatrics* **39**, 400–11.

ALLEYNE, G. A. O. & SCULLARD, G. H. (1969). Alterations in carbohydrate metabolism in Jamaican children with severe malnutrition. *Clin. Sci.* (in the Press).

ALLEYNE, G. A. O. & YOUNG, V. H. (1967). Adrenocortical function in children with severe protein-calorie malnutrition. *Clin. Sci.* **33**, 189–200.

ASHWORTH, A. (1969). Growth rates in children recovering from protein-calorie malnutrition. *Brit. J. Nutr.* (in the Press).

BAIG, H. A. & EDOZIEN, J. C. (1965). Carbohydrate metabolism in kwashiorkor. *Lancet* ii, 662–5.

BEAS, F., MÖNKEBERG, F. & HORWITZ, I. (1966). The response of the thyroid gland to thyroid-stimulating hormone (TSH) in infants with malnutrition. *Pediatrics* **38**, 1003–8.

BECK, P., KOUMANS, J. H. T., WINTERLING, C. A., STEIN, M. F., DAUGHADAY, W. H. & KIPNIS, D. M. (1964). Studies of insulin and growth hormone secretion in human obesity. *J. Lab. clin. Med.* **64**, 654–67.

BENEDICT, E. G., MILES, W. R., ROTH, P. & SMITH, H. M. (1919). Human vitality and efficiency under prolonged restricted diet. *Publs Carnegie Instn*, no. 280.

BENEDICT, F. G. & TALBOT, F. B. (1914). The gaseous metabolism of infants with special reference to its relation to pulse rate and muscular activity. *Publs Carnegie Instn*, no. 201.

BENEDICT, F. G. & TALBOT, F. B. (1915). The physiology of the new-born infant. Character and amount of the katabolism. *Publs Carnegie Instn*, no. 233.

BOWIE, M. D. (1964). Intravenous glucose tolerance in kwashiorkor and marasmus. *S.A. med. J.* **38**, 328–9.

CAHILL G. F., JR., HERRERA, M. G., MORGAN, A. P., SOELDNER, J. S., STEINKE, J., LEVY, P. L., REICHARD, G. A. JR, & KIPNIS, D. M. (1966). Hormone–fuel interrelationships during fasting. *J. clin. Invest.* **45**, 1751–69.

CASTELLANOS, H. & ARROYAVE, G. (1961). Role of the adrenal cortical system in the response of children to severe protein malnutrition. *Am. J. clin. Nutr.* **9**, 186–95.

CHEEK, D. B. & GRAYSTONE, J. E. (1969). The action of insulin, growth hormone and epinephrine on cell growth in liver, muscle, and brain of the hypophysectomized rat. *Pediat. Res.* **3**, 77–88.

COPE, C. L. & PEARSON, J. (1965). Clinical value of the cortisol secretion rate. *J. clin. Path.* **18**, 82–7.

CORNBLATH, M., PARKER, M. L., REISNER, S. H., FORBES, A. E. & DAUGHADAY, W. H. (1965). Secretion and metabolism of growth hormone in premature and full-term infants. *J. clin. Endocr. Metab.* **25**, 209–18.

DEAN, R. F. A. (1965). Kwashiorkor. In *Recent Advances in Paediatrics*, pp. 234–65 (ed. D. M. Gairdner) London: Churchill.

DUPRÉ, J., CURTIS, J. D., UNGER, R. H., WADDELL, R. W. & BECK, J. C. (1969). Effects of secretin pancreozymin or gastrin on the response of the endocrine pancreas to administration of glucose or arginine in man. *J. clin. Invest.* **48**, 745–58.

FELIG, P., OWEN, O. E., WAHREN, J. & CAHILL, G. F. JR, (1969). Amino acid metabolism during prolonged starvation. *J. clin. Invest.* **48**, 584–94.

FLOYD, J. C. JR, FAJANS, S. S., CONN, J. W., KNOPF, R. F. & RULL, J. (1966a). Insulin secretion in response to protein ingestion. *J. clin. Invest.* **45**, 1479–86.

FLOYD, J. C. JR, FAJANS, S. S., CONN, J. W., KNOPF, R. F. & RULL, J. (1966b). Stimulation of insulin secretion by amino acids. *J. clin. Invest.* **45**, 1487–502.

GABUZDA, T. G., JICK, H. & CHALMERS, T. C. (1963). Human growth hormone and albumin metabolism in patients with cirrhosis. *Metabolism* 12, 1–10.

GILLMAN, T. (1942). The cytology of the anterior lobe of the human (Bantu) pituitary gland. M.Sc. thesis, Rand, cited by Gillman & Gillman (1951), p. 438. See below.

GILLMAN, J. & GILLMAN, T. (1951). *Perspectives in Human Malnutrition*, pp. 584. New York: Grune and Stratton.

GRAHAM, G. G. (1967). The effect of infantile malnutrition on growth. *Fedn Proc. Fedn. Am. Socs exp. Biol.* 26, 139–43.

GRAHAM, G. G. (1968). The later growth of malnourished infants; effects of age, severity and subsequent diet. In *Calorie Deficiencies and Protein Deficiencies*, pp. 301–16. (eds. R. A. McCance and E. M. Widdowson). London: Churchill.

HADDEN, D. R. (1967). Glucose, free fatty acid, and insulin interrelations in kwashiorkor and marasmus. *Lancet* ii, 589–93.

HADDEN, D. R. & RUTISHAUSER, I. H. E. (1967). Effect of human growth hormone in kwashiorkor and marasmus. *Archs Dis. Childh.* 42, 29–33.

HELGE, H., WEBER, B. & QUABBE, H. J. (1969). Growth hormone release and venepuncture. *Lancet* i, 204.

JAMES, W. P. T. (1968*a*) cited by Waterlow, J. C. (1968). Observations on the mechanism of adaptation to low protein intakes. *Lancet* ii, 1091–7.

JAMES, W. P. T. (1968*b*). Intestinal absorption in protein-calorie malnutrition. *Lancet* i, 333–5.

JAMES, W. P. T. & HAY, A. M. (1968). Albumin metabolism: effect of the nutritional state and the dietary protein intake. *J. clin. Invest.* 47, 1958–72.

JEEJEEBHOY, K. N., BOUCHER, B. J. & HARTOG, M. (1967). The effect of growth hormone on albumin turnover in man. *Metabolism* 14, 67–74.

JELLIFFE, D. B., BRAS, G. & STUART, K. L. (1954). Kwashiorkor and marasmus in Jamaican infants. *W. Indian med. J.* 3, 43–55.

KAMINSKY, N. I., BROADUS, A. E., HARDMAN, J. G., GINN, H. E., SUTHERLAND, E. W. & LIDDLE, G. W. (1969). Effects of glucagon and parathyroid hormone on plasma and urinary 3′,5′-adenosine monophosphate in man. *J. clin. Invest.* 48, 42 a.

KEYS, A., BROZEK, J., HENSCHEL, A., MICKELSEN, O. & TAYLOR, H. L. (1950). *The Biology of Human Starvation*. University of Minnesota Press.

KIRSCH, R. E., SAUNDERS, S. J. & BROCK, J. F. (1968). Animal models and human protein calorie malnutrition. *Am. J. clin. Nutr.* 21, 1225–8.

KNOBIL, E. & HOTCHKISS, J. (1964). Growth hormone. *Am. Rev. Physiol.* 26, 47–74.

KNOPF, R. F., CONN, J. W., FAJANS, S. S., FLOYD, J. C. JR, GUNTSCHE, E. M. & RULL, J. A. (1965). Plasma growth hormone response to intravenous administration of amino acids. *J. clin. Endocr.* 25, 1140–4.

LAMY, M., LAMOTTE, M. & LAMOTTE-BARILLON, S. (1946). Etude anatomique des états de dénutrition. *Bull. Mém. Soc. méd. Hôp, Paris* 62, 435–9.

LARON, Z., MANNHEIMER, S., NITZAN, M. & GOLDMANN, J. (1967). Growth hormone, glucose and free fatty acid levels in mother and infant in normal, diabetic and toxaemic pregnancies. *Archs Dis. Childh.* 42, 24–8.

LEFEBVRE, P. (1966). The physiological effect of glucagon on fat mobilization. *Diabetologia* 2, 130–2.

LEONARD, P. J. & MACWILLIAM, K. M. (1964). Cortisol binding in the serum in kwashiokor. *J. Endocr.* 29, 273–6.

LEWIS, B., HANSEN, J. D. L., WITTMAN, W., KRUT, L. H. & STEWART, F. (1964). Plasma free fatty acids in kwashiorkor and the pathogenesis of the fatty liver. *Am. J. Clin. Nutr.* 15, 161–8.

LIFSHITZ, F., CHAVARRIA, L., CRAVIOTO, J., FRENK, S. & MORALES, M. (1962). Iodo hormonal en la desnutricion avanzada delrino. *Bol. Hosp. Inf. Mex.* **19**, 319–26.

LUCIEN, M. (1908 a). Capsules surrénales et arthrepsie. *C. r. Séanc. Soc. Biol.* **64**, 462–4.

LUCIEN, M. (1908 b). Considerations anatomo-pathologiques sur l'arthrepsie. *C. r. Séanc. Soc. Biol.* **64**, 236–8.

LURIE, A. O. & JACKSON, W. P. U. (1962). Adrenal function in kwashiorkor and marasmus. *Clin. Sci.* **22**, 259–68.

MANCHESTER, K. L. & YOUNG, F. G. (1961). Insulin and protein metabolism. *Vitams Horm. Lpz.* **19**, 95–132.

MARFAN, A. B. (1921). Les états de dénutrition dans la première enfance. Description de la hypothrepsie et athrepsie. *Nourrison* **9**, 65–86.

MARKS, V., HAWORTH, N. & GREENWOOD, F. C. (1965). Plasma growth-hormone levels in chronic starvation in man. *Nature, Lond.* **208**, 686–7.

MERIMEE, T. F., LILLICRAP, D. A. & RABINOWITZ, D. (1965). Effect of arginine on serum-levels of human growth hormone. *Lancet* ii, 668–70.

MILNER, R. D. G. (1969 a). Stimulation of insulin secretion *in vitro* by essential amino acids. *Lancet* i, 1075–6.

MILNER, R. D. G. (1969 b). The mechanism by which leucine and arginine stimulate insulin secretion *in vitro*. *Biochim. biophys. Acta* (in the Press).

MILNER, R. D. G. (1970). Metabolic and hormonal responses to glucose and glucagon in infantile malnutrition. *Pediat. Res.* (in the Press).

MILNER, R. D. G. & WRIGHT, A. D. (1967). Plasma glucose, non-esterified fatty acid, insulin and growth hormone response to glucagon in the newborn. *Clin. Sci.* **32**, 249–55.

MÖNKEBERG, F. (1966). Alteracions bioquimicas en la desnutriccion infantil. *Nutr. Bromatol. Toxicol.* **5**, 31–42.

MÖNKEBERG, F. (1968). Adaptation to caloric and protein restriction in infants. In *Calorie Deficiencies and Protein Deficiencies*, pp. 91–108 (eds. R. A. McCance and E. M. Widdowson). London: Churchill.

MÖNKEBERG, F., BARZELATTO, J., BEAS, F. & WAISSBLUTH, H. (1957). Captacion de yodo radioactivo por el tiroide en el lactante distrofico. *Rev. Chile Pediat.* **28**, 173–5.

MÖNKEBERG, F., DONOSO, G., OXMAN, S., PAK, N. & MENEGHELLO, J. (1963). Human growth hormone and infant malnutrition. *Pediatrics* **31**, 58–64.

MONTGOMERY, R. D. (1962 a). Changes in the basal metabolic rate of the malnourished infant and their relation to body composition. *J. clin. Inv.* **41**, 1653–63.

MONTGOMERY, R. D. (1962 b). Urinary radio-iodine excretion in the malnourished infant. *Archs Dis. Childh.* **37**, 383–6.

MONTGOMERY, R. D. (1962 c). Changes in the basal metabolic rate of the malnourished infant and their relation to body composition and to thyroid activity. M.D. Thesis, University of London, Pp. 158 + xvii.

MONTGOMERY, R. D. (1962 d). Muscle morphology in infantile protein malnutrition. *J. clin. Path.* **15**, 511–21.

MULINOS, M. G. & POMERANTZ, L. (1940). Pseudohypophysectomy: a condition resembling hypophysectomy produced by malnutrition. *J. Nutr.* **19**, 493–504.

MUNRO, N. H. (1964). General aspects of the regulation of protein metabolism by diet and by hormones. In *Mammalian Protein Metabolism*, pp. 382–482 (eds. H. N. Munro and J. B. Allison). London: Academic Press.

NELSON, W. E. (1959). *Textbook of Pediatrics*, 7th edn., pp. 50–61. Philadelphia: W. B. Saunders.

NICOLAEFF, L. (1923). Influence de l'inanition sur la morphologie des organes infantiles. *Presse med.* ii, 1007–9.

OWEN, O. E., FELIG, P., MORGAN, A. P., WAHREN, J. & CAHILL, G. F. JR, (1969). Liver and kidney metabolism during prolonged starvation. *J. clin. Invest.* **48**, 574–83.

OXMAN, S. V., MACCIONI, A. S., ZUNIGA, A. C., SPADA, R. G. & MÖNKEBERG, F. B. (1968). Distribution of carbohydrate metabolism in infantile marasmus. *Am. J. Clin. Nutr.* **21**, 1285–90.

PIMSTONE, B., BANK, S. & BUCHANAN-LEE, B. (1968). Growth hormone in protein losing enteropathy. *Lancet* ii, 1246–7.

PIMSTONE, B. L., BARBEZAT, G., HANSEN, J. D. L. & MURRAY, P. (1967). Growth hormone and protein-calorie malnutrition, impaired suppression during induced hyperglycaemia. *Lancet* ii, 1333–4.

PIMSTONE, B. L., BARBEZAT, G., HANSEN, J. D. L. & MURRAY, P. (1968). Studies on growth hormone secretion in protein-calorie malnutrition. *Am. J. Clin. Nutr.* **21**, 482–7.

PIMSTONE, B. L., WITTMAN, W., HANSEN, J. D. L. & MURRAY, P. (1966). Growth hormone and kwashiorkor. Role of protein in growth hormone homeostasis. *Lancet* ii, 779–80.

PRADER, A., ILLIG, R., SZEKY, J. & WAGNER, H. (1964). The effect of human growth hormone on hypopituitary dwarfism. *Archs Dis. Childh.* **39**, 535–44.

RABINOWITZ, D., MERIMEE, T. J., MAFFEZOLLI, R. & BURGESS, J. A. (1966). Patterns of hormonal release after glucose, protein and glucose plus protein. *Lancet* ii, 454–7.

RANDLE, P. J., ASHCROFT, S. J. H. & GILL, J. R. (1968). Carbohydrate metabolism and release of hormones. In *Carbohydrate Metabolism and Its Disorders*, pp. 427–47. (eds. F. Dickens, P. J. Randle and W. J. Whelan). London University Press.

RANDLE, P. J., GARLAND, P. B., HALES, C. N. & NEWSHOLME, E. A. (1963). The glucose fatty acid cycle and its role in insulin sensitivity and the metabolic disturbances of diabetes mellitus. *Lancet* i, 785–9.

RAO, K. S. J., SRIKANTIA, S. G. & GOPALAN, C. (1968). Plasma cortisol levels in protein-calorie malnutrition. *Archs Dis. Childh.* **43**, 365–7.

ROOT, A. W., SAENZ-RODRIGUEZ, C., BONGIOVANNI, A. M. & EBERLEIN, W. R. (1969). The effect of arginine infusion in plasma growth hormone and insulin in children. *J. Pediat.* **74**, 187–97.

ROTH, J., GLICK, S. M., YALOW, R. S. & BERSON, S. A. (1964). The influence of blood glucose on the plasma concentration of growth hormone. *Diabetes* **13**, 355–61.

SAMOLS, E., MARRI, G. & MARKS, V. (1965). Promotion of insulin secretion by glucagon. *Lancet* ii, 415–6.

STEINER, A. L., PEAKE, G. T., UTIGER, R. & KIPNIS, D. (1969). Median eminence stimulation of growth hormone (GH) and thyrotrophin (TSH) secretion and the pituitary adenyl cyclase system. *J. Clin. Invest.* **48**, 80a.

STIMMLER, L. & BROWN, G. A. (1967). Growth hormone secretion provoked by insulin induced hypoglycaemia in children of short stature. *Archs Dis. Childh.* **42**, 232–8.

STIRLING, G. A. (1962). The thyroid in malnutrition. *Archs Dis. Childh.* **37**, 99–102.

TEJADA, C. & RUSSFIELD, A. B. (1957). A preliminary report in the pathology of the pituitary gland in children with malnutrition. *Archs Dis. Childh.* **33**, 343–6.

THOMPSON, R. L. (1907). Atrophy of the parathyroid glands and other glandular structures in infantile atrophy. *Am. J. Med. Sci.* **134**, 562–76.

210   R. D. G. MILNER

THOMSON, A. M. (1968). The later results in man of malnutrition in early life. In *Calorie Deficiencies and Protein Deficiencies*, pp. 289–99 (eds. R. A. McCance and E. M. Widdowson). London: Churchill.
TROWELL, H. C., DAVIES, J. N. P. & DEAN, R. F. A. (1954). *Kwashiorkor*, Pp. 308. London: Edward Arnold.
UEHLINGER, E. (1947). Die hypophyse bei inanition. *Schweiz. Z. Path. Bakt.* **10**, 144–58.
UNGER, R. H., EISENTRAUT, A. M. & MADISON, L. L. (1963). The effect of total starvation upon the levels of circulating glucagon and insulin in man. *J. Clin. Invest.* **42**, 1031–9.
VALLEDOR, Y., LAVERNIA, F., BORBOLLA, L., SATANOWSKY, C., COSTALES, F., PRIETO, E. & BARDELAS, A. (1959). Thyroid function disturbances in malnourished infants and small children. *Rev. Cuba. Pediat.* **31**, 533–40.
VARGA, F. & MESS, B. (1968). Serum thyrotrophin in semistarvation. *Acta paediat. hung.* **9**, 197–203.
VEST, M., GIRARD, J. & BUHLER, U. (1963). Metabolic effects of short term administration of human growth hormone in infancy and early childhood. *Acta endocr., Copenh.* **44**, 613–24.
VINT, F. W. (1949). Post-mortem findings in the natives of Kenya. *E. Afr. med. J.* **13**, 332–40.
WATERLOW, J. C. (1948). Fatty liver disease in infants in the British West Indies. *Spec. Rep. Ser. med. Res. Coun.* no. 263.
WEIL, R. (1965). Pituitary growth hormone and intermediary metabolism. I. The hormonal effect on the metabolism of fat and carbohydrate. *Acta endocr., Copenh.* (Suppl.) **98**, 1–92.
WESTPHAL, O. (1968). Growth hormone—a methodological and clinical study. *Acta paediat., Stokh.* (Suppl.) **182**, 1–81.
WIDDOWSON, E. M. (1968). The place of experimental animals in the study of human malnutrition. In *Calorie Deficiencies and Protein Deficiencies*, pp. 225–36 (eds. R. A. McCance and E. M. Widdowson). London: Churchill.
WINICK, M. (1969). Malnutrition and brain development. *J. Pediat.* **74**, 667–79.
WRIGHT, A. D., LOWY, C., FRASER, T. R., SPITZ, I. M., RUBENSTEIN, A. H. & BERSCHN, I. (1968). Serum growth hormone and glucose intolerance in renal failure. *Lancet* i, 798–801.
ZUBIRAN, S. & GOMEZ-MONT, F. (1953). Endocrine disturbances in chronic human malnutrition. *Vitam. Horm.* **11**, 97–132.

DISCUSSION

BERN: Is there evidence in infants that supposedly have increased growth-hormone levels of anything suggestive of increased prolactin effects, the witches' milk phenomenon, for example.

MILNER: No, they do not show any clinical manifestations such as the one you described.

BERN: I was thinking of the inherent prolactin activity of human growth hormone.

MILNER: As I said, there is no clinical evidence.

MATTY: I wonder if Dr Milner would care to comment on how he sees his investigations on growth hormone and insulin levels of children suffering from malnutrition in the light of work recently reported by Luft and his co-workers. Luft and Cerasi (*Diabetologia* **4**, 1, 1968) have recently suggested that HGH, although decreasing the peripheral utilization of glucose, is not a primary diabetogenic factor.

MILNER: I think, if we are speaking teleogically, the role that growth hormone is playing in these infants is probably quite different from that which it plays in the adult, in whom Cerasi and Luft have made most of their observations. High plasma GH levels may be contributing towards the impaired glucose tolerance in these children, but further than that I wouldn't like to comment, apart from saying that GH is not there primarily to stimulate growth. It is also apparent from this work that normal β-cell function is not necessary for extremely rapid growth.

QURESHI: You said that malnutrition has an effect on the β cells and that they are atrophied during malnutrition. You did not get the change in plasma insulin levels after glucagon injections—which as you mentioned has a stimulatory effect on insulin secretion. In this connection it probably becomes important to know how the α-cells of pancreatic islets, known to secrete glucagon, behave during malnutrition. I will appreciate your opinion on this.

MILNER: Reports of the histology of the endocrine pancreas are sparse in contrast to the reports on the exocrine pancreatic changes. As far as I know, I don't think this information is in the literature.

QURESHI: There is some evidence that D-cells of pancreatic islets probably secrete gastrin; it is also known that gastrin stimulates insulin secretion from the beta cells. Do you agree that it would be worth ascertaining the role of gastrin in malnutrition?

MILNER: This is not known, but it may be relevant to refer to the interesting hypothesis proposed by Henderson (*Lancet* ii, 469–70, 1969); not only do the enteric hormones of the upper gut influence endocrine pancreatic function but the endocrine pancreas may influence the normal function of the exocrine pancreas and, in fact, we may be seeing the interrelationship of the two very well brought out in malnutrition.

CLEGG: Although these children were retarded as far as their gross size was concerned, did you in fact see whether their skeletal development was impaired?

MILNER: No. we did not, but other workers have demonstrated that it is.

OLIVEREAU: I am sorry to compare these poor children to fish. However, eels are able to withstand complete starvation for 18 months; they lose more than 60% body weight. In the pituitary gland the growth-hormone secreting cells exhibit all signs of a strong stimulation; they are more numerous, and enlarged, with a well-developed endoplasmic reticulum and appear completely degranulated. It seems that they are secreting continuously a large amount of growth hormone. This observation fits quite well with Dr Milner's.

LEATHEM: I just insert a question here. Is it possible that these elevated levels of growth hormone in malnutrition are also in part due to disuse atrophy in that you have circulating levels of hormone which simply cannot be utilized? For example, if you have a threonine deficiency you negate the action of growth hormone, and the classic ten essential amino acids may not be the classic amino acids for growth-hormone responsiveness. In perfusion experiments it has actually been shown that you could increase growth-hormone action well above normal by having the right amino acids in circulation at the right time so the two obviously would have to be playing a concert of the intertwining facets to allow the growth hormone to express itself.

SESSION III
THE INTEGUMENT IN THE TERRESTRIAL ENVIRONMENT

THE CONTROL OF
THE MAMMALIAN MOULT

By F. J. EBLING and PATRICIA A. HALE

INTRODUCTION

The integument is a major organ by which the animal maintains a homeostatic relationship with its environment. In particular, by moulting a mammal can alter the colour or the insulating properties of its coat, which depend on the length and density per unit area of the hairs (Tregear, 1966), to accommodate itself to alterations in camouflage requirements and in temperature with the changing seasons.

Mammalian hair and its derivatives also function in sex display, defence (e.g. the spines of the hedgehog or porcupine) and tactile sensitivity. In addition, various skin glands are concerned with temperature control or with external chemical communication—a property dealt with in another contribution to this Symposium (Strauss & Ebling, 1970).

THE PELAGE AND ITS REPLACEMENT

Hairs are produced by follicles which are downgrowths of the epidermis into the dermal layer of the skin. In most species the follicle population is complete within a few days of birth, and thereafter in the general body skin there is no neogenesis of hair follicles. The only proved exception to this rule is the skin or 'velvet' covering the antlers of deer.

Typically there are two main categories of hairs in the coat—overhairs and underhairs. The long stiff overhairs or guard hairs bear the brunt of the wear and tear of contact with the surroundings, can be erected to increase the thickness of the coat, and serve to entrap and shed water droplets, keeping the deeper fur layer dry. These hairs are also sensitive organs of touch, each follicle being surrounded by a network of sensory neurones. The finer, shorter underhairs or fur form the main insulating layer of the coat.

A mammal's coat is not produced once and for all at the beginning of its life. Hair follicles go through cycles. At the end of each active period or *anagen* (Fig. 1 d) the follicle passes through a transition phase or *catagen* (Fig. 1 e) to a resting stage or *telogen* (Fig. 1 a, f). During this process the base of the hair becomes expanded and keratinized to form a 'club' which anchors the completed hair in the skin, and the matrix of the follicle regresses to a small 'secondary germ'. When the follicle again becomes

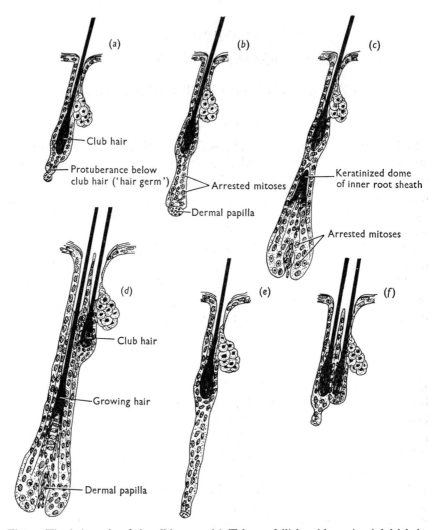

Fig. 1. The hair cycle of the albino rat. (a) Telogen follicle with resting 'club' hair.
(b) Very early anagen (active) stage; the first mitoses are visible. (c) Later anagen; keratinization is beginning. (d) Full anagen; the new hair has erupted beside the old club hair.
(e) Catagen; the base of the follicle is regressing. (f) Telogen follicle with two club hairs.
(From Johnson, 1958a).

active, the secondary germ grows inwards (downwards) to reconstitute the
matrix and a new hair starts to form, which eventually erupts by the side
of the old club hair (Fig. 1b, c). The latter is subsequently moulted, but
not until the new hair is at least partly grown.

In the growth of the first coat, follicular activity may be more or less
synchronized, all follicles being in the same stage of the cycle at the same

time. On the other hand, follicular activity may appear completely asynchronous, producing a mosaic pattern of hair replacement such as is found in the adult human scalp. This pattern has long been believed to be characteristic of the guinea-pig, but recent work by Jackson (unpublished) has shown that there is some degree of synchrony within each follicle type, although the cycles of follicles producing different types of hairs are out of phase.

The most common type of hair replacement involves the initiation of follicular activity in a specific region of the skin with neighbouring follicles in synchrony. The area of activity then expands in a characteristic pattern

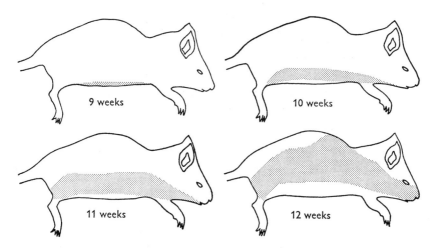

Fig. 2. The passage of a wave of hair growth in the rat. Stippling indicates the area in which follicles are active.

so that eventually it passes like a wave over the entire skin (Fig. 2). This situation has been described for various small rodents such as *Onchomys* (Ruffer, 1965) and the golden mouse *Ochrotomys* (Linzey & Linzey, 1967), for weasels and stoats, (Rothschild, 1944; Rust, 1962), in laboratory mammals such as the mouse (Dry, 1926; Nay & Fraser, 1954) and rat (Butcher, 1934; Johnson, 1958a), and in species whose fur or skin is commercially valuable such as the chinchilla (Lyne, 1965), mink (Bassett & Llewellyn, 1949) and cattle (Dowling & Nay, 1960). Much experimental work on sheep has been carried out in Australia and New Zealand, where wool production is a major industry, but the situation in domesticated varieties is somewhat different from other species in that the active phase of the follicular cycle may extend over several years, and the resting phase is very short. However, wool does not grow at a constant rate, but shows seasonal peaks in the weight produced per unit area per unit time. Ryder & Stephenson (1968) have recently published a comprehensive review of work on wool growth.

INFLUENCE OF THE ENVIRONMENT

(a) *Role of the photoperiod*

There is much experimental evidence from various species indicating that alterations in the photoperiod are of prime importance in controlling seasonal changes in the mammalian coat. This was first demonstrated by Bissonnette (1935) working on the ferret, and later on the mink (Bissonnette & Wilson, 1939) and weasel (Bissonnette & Bailey, 1944). The ferret was more intensively investigated by Harvey & Macfarlane (1958), who compared control animals kept in normal seasonal temperature and lighting conditions with an experimental group subjected to a complete and exaggerated reversal of the seasonal change in photoperiod (Fig. 3). In both groups, with a decreasing photoperiod both sexes became sexually inactive, there was extensive shedding of old hair and the follicles became active, producing a dense new coat. With an increasing daylength reproductive activity was resumed, hair growth ceased and there was some shedding of fine underhairs. They concluded that in autumn a new winter coat is grown while the old summer coat is entirely shed, but in spring only part of the underhair or fur layer is lost, leaving a less dense coat better suited to summer conditions.

Yeates (1955) also reversed the seasonal changes in daylength and thus induced cattle to adopt a 'summer' type of coat in winter and vice versa, without reference to the ambient temperature. Heat-tolerance tests confirmed that animals with a 'summer' coat withstood warm conditions better. He also investigated the adaptation of European breeds of cattle to tropical conditions (Yeates, 1957). The response to a fixed photoperiod of $12\frac{1}{2}$L:$11\frac{1}{2}$D varied with the time of year the experiment began, but eventually all the cattle lost the seasonal cycle of hair shedding and replacement characteristic of temperate zones, and grew a coat intermediate in type between the long furry heat-retaining winter coat and the short glossy summer coat. There was diffuse hair shedding at all seasons.

In commercial breeds of sheep there is a seasonal peak of wool production (wt./unit area/unit time) in summer and autumn. This peak, which could be achieved by an increase in either the diameter or the rate of growth in length of the wool fibres, can be influenced both in its amplitude and its timing by alterations of photoperiod, while the temperature conforms to the natural environmental cycle (Hart, 1961). Morris (1961)

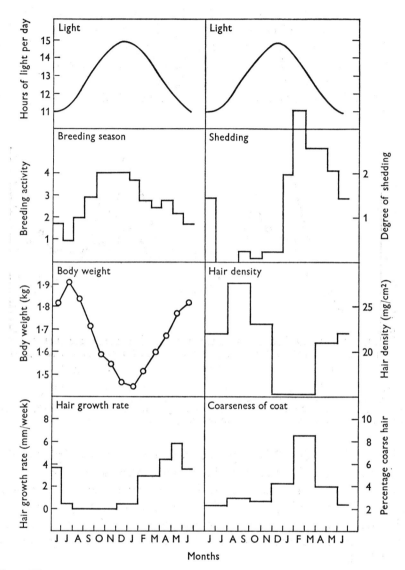

Fig. 3. The correlation of changes in daylength with breeding season, body weight, hair growth, hair shedding, hair density and coarseness of the coat in normal male ferrets (after Harvey & Macfarlane, 1958).

found that wool growth was not affected immediately by experimental alterations of the photoperiod, but required some time to become entrained to a new rhythm, although Hart, Bennett, Hutchinson & Wodzicka-Tomaszewska (1963) found that the effect was immediate. They believed that their different results might be due to interaction with different temperature conditions. Hart (1961) found that hooding the eyes of sheep eliminated the effect of changes in photoperiod, so that wool production remained almost constant throughout the year (Fig. 4). His experiments also suggest that there is a threshold of light intensity below which sheep do not respond, for when periods of high and low light intensity were alternated, the results were similar to those obtained with alternating light and dark periods.

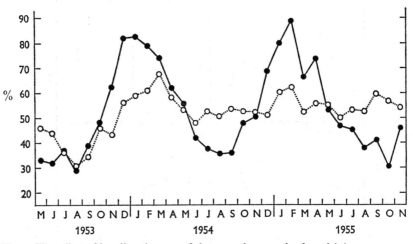

Fig. 4. The effect of hooding the eyes of sheep on the growth of wool (g/10 cm square, as a percentage of the average growth during the previous December, January and February). ●——●, Control (natural photoperiod and temperature range); ○ – – – ○, hooded (8L:16D photoperiod, natural temperature range). (After Hart, 1961.)

Recently Pinter (1968) has reported on the effects of altered photoperiod on hair growth and replacement in the vole, *Microtus montanus*. Under long day conditions (18L:6D) the juvenile and subadult moults were completed more rapidly than under short days (6L:18D). Both guard hairs and underfur were longer under short photoperiods than long, but whether this was due to faster growth or a longer anagen stage was not ascertained.

(b) Influence of temperature

Since the coat is concerned with heat insulation, it might be expected that changes in environmental temperature would be of prime importance in controlling the activity of hair follicles. However, under natural conditions the annual cycle of temperature change closely parallels that of change in the photoperiod, so that it is difficult to assess their effects separately.

From field studies of *Lepus timidus* in Scotland both Hewson (1958) and Watson (1963) have concluded that the progress of the spring moult can be modified by temperature. Watson observed that in cold springs the change from white to brown was later than in mild ones, and in addition, hares living at high altitudes did not complete the spring moult until later than those at lower altitudes (Fig. 5). Rothschild (1942) also noted an effect of environmental temperature on the autumn moult of stoats. In this case, as might be expected, cold conditions led to a more rapid assumption of the winter coat.

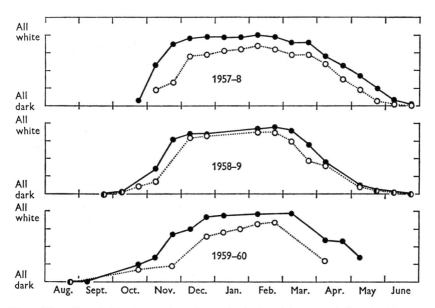

Fig. 5. The effect of altitude on colour change in the Scottish mountain hare in Glen Esk. The winter 1958/9 was more severe than in the other two seasons. ●——●, High area; ○ – – – ○, low area. (After Watson, 1963.)

Berman & Volcani (1961) compared the coats of cattle kept in three districts in Israel—the Jordan Valley (hot conditions), the Coastal Plain (warm) and the Jerusalem region (cool). In all three regions they found seasonal coat cycles, but the hair mass produced per unit area and the coat thickness were significantly less in the Jordan Valley than in the other regions. The authors claim that these differences were due to the differences in environmental temperature, but they might equally well be accounted for, in part at least, by differences in time of shedding or of fibre diameter, or by inter-strain differences, for which they made no attempt to compensate.

Sheep's wool grows faster in summer than in winter, and attempts have been made to increase wool production by raising the winter temperature. Coop & Hart (1953) found that a rise of 7 °F had no effect, and Hutchinson & Wodzicka-Tomaszewska (1961) quote Canadian work in which it was raised 29 °F without effect. However, Ferguson (quoted from Ryder & Stephenson, 1968) found that keeping Merinos at a constant 27 °C for 18 months abolished the seasonal rhythm of wool production, suggesting that at the latitude at which he was working (34 °S), seasonal temperature variations control wool production. Morris (1961) and Bennett, Hutchinson & Wodzicka-Tomaszewska (1962) concluded that seasonal temperature variations might have a modifying effect although the main factor influencing wool growth was the seasonal change in photoperiod. The latter group did find evidence of a local effect of lowered temperature on wool follicles, for in sheep kept in a constant daylength but reversed temperature seasons, clipped patches fitted with an insulating cover behaved like those on control animals (which were also in a constant daylength but experiencing a natural temperature cycle), but uncovered patches responded to the experimental cycle of temperature change.

Hutchinson (1965) has also investigated the interaction of temperature and photoperiod on shedding of hairs from the legs of sheep. He found that under equatorial daylength but reversed temperature conditions there was an autonomous cycle which decreased in amplitude over 3 years, but with normal environmental temperature conditions and reversed photoperiods the peak of shedding came into phase with the altered photoperiodic cycle within 6 months.

(c) Correlation of moult with breeding cycles

It is now well established that seasonal breeding cycles are regulated by the natural change in photoperiod, so in view of the results reviewed in the preceding section it is not surprising to find that in many species there is a close correlation between cycles of hair growth and the reproductive cycle.

In many seasonal breeders, e.g. hares (Watson, 1963) and ponies (Burkhardt, 1947), resumption of sexual activity coincides with the onset of the spring moult. Pregnancy and lactation delay, or may prevent, the completion of a moult in mice (Nay & Fraser, 1955), bats (Dwyer, 1963), pocket mice (Speth, 1969) and elephant seals (Ling, 1965), although Hewson (1963) believes that moult cycles are not affected in either the brown or the mountain hare. Shedding and replacement of the ferret's coat are initiated during pregnancy and continue during lactation (Harvey & Macfarlane, 1958), while in this species and the rabbit, club hairs are shed late in pregnancy and used for lining the nest (Bissonnette, 1935; Ebling, 1965 b). In ewes, wool growth is reduced during pregnancy and lactation even though food intake is sufficient to maintain the weight of maternal tissues (Ferguson, Wallace & Lindner, 1965).

MECHANISMS OF PHYSIOLOGICAL CONTROL

In the foregoing sections two important features of the moult have emerged. First, it does not usually occur synchronously over the whole body but progresses as a wave. Secondly, in mature animals it is influenced by environmental changes of which alteration in the photoperiod is much the most important but probably not the only factor. How is this influence exercised?

There seems to be little doubt that hair follicles have an intrinsic rhythm which is characteristic of their site. Though in many species activity starts in one region of the body and spreads in a definite pattern, it is unlikely to be propagated. If a rectangular piece of skin is cut from the flank of a rat and then stitched back into position, passage of the moult from venter to dorsum is delayed on the grafted area as compared to the adjacent body. Nevertheless, eruption of new hair dorsal to the graft is not contingent on the arrival of the advancing front on the graft. When skin grafts are rotated or translocated the periodicity of the hair follicles appears to remain characteristic of their site of origin (Ebling & Johnson, 1959).

Such intrinsic rhythms are, however, slowly modified by systemic influences. When grafts are exchanged between rats of different ages whose hair cycles are out of phase, the first eruption after grafting takes place at about the same time as that on an autograft on a symmetrical site on the donor rat, but the subsequent eruptions on the homografts tend to be advanced or retarded to become synchronized with the host's autografts (Ebling & Johnson, 1961). Moult cycles are also brought gradually into phase in rats of different ages parabiotically joined (Ebling & Hervey, 1964). Some evidence of site influence was also shown in experiments in which grafts of dorsal flank skin were placed in ventral sites on hosts of a different age and vice versa (Ebling, 1965a). Both local and systemic factors can, however, be overridden by experimental interference. If club hairs are plucked from resting follicles a new cycle of activity is induced and new hair erupts from the skin surface about 12 days afterwards (Johnson & Ebling, 1964), even when the next spontaneous eruption is not due until much later.

In summary it seems that activity of the hair follicles involves an inherent rhythm which is slowly modified by systemic factors which are themselves under environmental influences. Thus it is reasonable to consider the possible role of the endocrine system.

INFLUENCE OF ENDOCRINE ORGANS

The passage of the moult wave is affected by several endocrine systems. In general terms, it is accelerated by thyroid hormone and retarded by steroid hormones whether of gonadal or adrenocortical origin. The effects appear to be produced by alterations in the relative lengths of the resting phase of the follicular cycle in different regions, the duration of anagen remaining unchanged. Hormones may also affect other aspects of hair growth; for

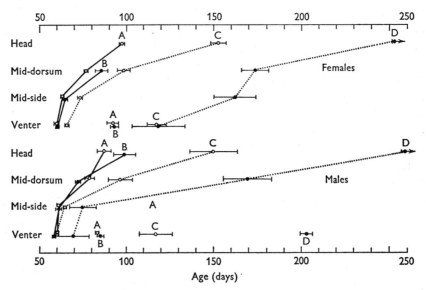

Fig. 6. The effect of thyroid hormone on age at eruption of the third hair generation in the rat, and the beginning of the fourth generation. The limits represent 1·5 × S.E.; when they do not overlap they are significantly different with $P < 0.05$. A, 100 μg thyroxine/ day; B, 20 μg thyroxine/day; C, normal controls; D, propylthiouracil, 0·02 % in drinking water. (After Ebling & Johnson, 1964.)

example, the rate of hair growth and the rate of shedding of club hairs can be altered, or the type of hair produced may be influenced; that hypophysectomized dogs retain 'puppy' fur was shown by Houssay in 1918.

The effect of thyroxine on hair growth of shaved rats was first noted by Chang (1926). Mohn (1958) showed that thyroxine accelerated, and inhibition of the thyroid with propylthiouracil retarded, the replacement of hair in plucked follicles. Spontaneous moult cycles in the rat are accelerated by administration of thyroxine (Ebling & Johnson, 1964); eruption of hair on the venter occurs earlier than in controls and the moult moves more rapidly (Fig. 6). Hair length is also significantly increased in intact females but not in males.

In sheep, thyroxine administration increases the weight of wool grown/ unit area/unit time (Ferguson *et al.* 1965), mainly by increasing the rate of elongation (Thierez & Rougeot, 1962), although Downes & Wallace (1965) found a small local increase in fibre diameter after intradermal injections of thyroxine.

Ovariectomy accelerates the passage of the hair growth wave in the mouse (Fraser & Nay, 1953), and Pinter (1968) has observed that in voles the juvenile and subadult moults are initiated later and completed more slowly in young females than in males, and the hairs of females are longer. In the rat, oestradiol delays initiation of follicular activity, slows the rate of growth of hairs, reduces their ultimate length, and also delays the shedding of old club hairs (Mohn, 1958; Johnson, 1958*b*). Oestradiol and diethylstilboestrol both reduce the rate of length growth of sheep's wool (Slen & Connell, 1958), although these authors believed this to be an effect of feedback to the pituitary resulting in a depression of thyroid activity and stimulation of the adrenal cortex.

Progesterone appears not to affect hair growth in the rat (Mohn, 1958) or sheep (Slen & Connell, 1958).

Testosterone has been shown to delay the passage of the growth wave in the rat (Johnson, 1958*b*), though it did not affect hair length or delay shedding. Houssay, Epper & Pazo (1965) similarly state that in mice testosterone delays the initiation of hair growth. Dwyer (1963), working on bats, and Pinter (1968) on voles both observed that males passed through the moult faster than females, but of course this could be due to absence of oestrogenic inhibition rather than to the presence of testosterone. Pinter (1968) found also that the hairs of castrate voles were longer than those of intact males.

A further effect of testosterone is to coarsen the fibre, as shown in the rat (Mohn, 1958) and in the sheep (Ryder & Stephenson, 1968). Finally it must be mentioned that in a number of species hair growth on certain regions of the skin is androgen-dependent; for example, the lateral gland region of hamsters (Hamilton & Montagna, 1950) and the ventral gland of gerbils (Mitchell & Butcher, 1966). In the latter species castration precipitates active hair follicles in the gland region into catagen and telogen within a week, although it has no effect on the rest of the pelage. Mitotic activity within these follicles recommences within 48 h of testosterone administration.

Adrenocorticoids inhibit hair growth. In the rat, they delay initiation of follicular activity and retard the passage of the growth wave but do not affect hair length or shedding (Johnson, 1958*c*; Mohn, 1958). Cortisone retards the initiation of hair growth induced by plucking in the rat (Mohn, 1958) and prevents it in the mink (Rust, Shackelford & Meyer, 1965), where it also inhibits the shedding of club hairs. The administration of ACTH or cortisone has been shown to reduce wool growth and the diameter of the fibres in sheep (Lindner & Ferguson, 1956; Downes & Wallace, 1965). Hair growth is stimulated by adrenalectomy in the rabbit (Whiteley, 1958) and the mouse, where replacement may be diffuse instead of in a definite pattern (Houssay et al. 1965).

Since the thyroid, gonads and adrenal cortex are all involved in the control of hair cycles, it is axiomatic that the pituitary must be concerned in their regulation. This has been experimentally investigated in many species. However, in view of the many possible ways, direct or indirect, in which the pituitary might act, and of the several different effects which might be produced, experimental results are often difficult to interpret.

Two results of hypophysectomy have been clearly demonstrated in rodents. First, removal of the pituitary advances the initiation and spread of follicular activity, as shown in the rat by Mohn (1958) and Ebling & Johnson (1964). Similarly, in hypophysectomized mice the wave pattern is said to be lost and hair growth becomes diffuse (Houssay et al. 1965), and the pituitary must be present for a normal moult and pattern of hair replacement to occur in *Mustela* and in the mink (Rust, 1965; Rust et al. 1965). Secondly, hypophysectomy of young animals causes the hair to remain of the 'puppy' type.

The effect of hypophysectomy in accelerating the hair growth wave may be mediated solely or partly through other endocrine organs. If this is so, it is clear that the reduction of gonadal and adrenocortical steroids consequent on hypophysectomy must produce an effect which totally overrides that of reduced thyroid activity. Whether hypophysial hormones play any direct role in the process is not known. The evidence is further complicated by the finding that hypophysectomy of the sheep completely suppresses wool growth. Ferguson, Wallace & Lindner (1965) found that this suppression could be partly overcome by thyroxine treatment, but not by prolactin or growth hormone, though the latter hormone stimulated wool growth in intact sheep. From experiments with fractionated extracts of sheep pituitary they concluded that an unidentified factor (which is not the thyroid stimulating hormone (TSH), the follicle stimulating hormone (FSH), growth hormone (GH), prolactin or adrenocorticotrophic hormone (ACTH)) stimulates wool growth.

It is generally believed that the maturation of hair, the change from the infantile to the adult pelage, is due to a direct action of hypophysial hormones. Rennels & Callahan (1959) published evidence that prolactin—rather than growth hormone—was responsible for this change in the rat, though Mohn (1958) had believed it to have no effect. In sheep, Downes

& Wallace (1965) obtained conflicting local responses to various concentrations of prolactin administered intradermally but concluded the hormone could produce a decrease in rate of growth and an increase in wool fibre diameter. They found that growth hormone was without effect after maturity but was necessary for growth of an adult type of fleece.

INTERACTION OF LOCAL AND SYSTEMIC FACTORS

If we accept that moult patterns are dependent on inherent local cycles of activity in the hair follicles which are gradually influenced by systemic hormonal conditions, the problem remains how these systems interact. For example, if the period of the inherent cycle is normally influenced to some extent by systemic hormonal conditions, we might expect the reaction of the follicle to local interference, such as plucking the hair, to be similarly affected.

We have recently investigated the interaction of the response to plucking with various hormonal treatments which are known to affect spontaneous hair cycles (Hale & Ebling, unpublished). Groups of six female litter-mate rats were randomized at weaning (21 days) and assigned to one of the following treatments: (i) intact control, (ii) spayed, (iii) spayed + oestradiol, (iv) spayed + propylthiouracil, (v) spayed + thyroxine, (vi) spayed + oestradiol + thyroxine. A strip $1-1\frac{1}{2}$ cm wide extending from mid-ventral to mid-dorsal via the left flank was plucked free of hair at a known age ranging from 25 to 60 days. The right flank served as a control. This range of ages at plucking ensured that groups of intact animals were plucked before the eruption of G_2 hairs (nomenclature after Dry, 1926) and other groups at various times during and after the growth of G_2. All six litter-mates were plucked on the same day, which meant that their follicular cycles were out of phase as a result of the treatments they had been receiving.

Our observations of eruption on the plucked flank of intact female rats as compared with a similar level on the contralateral control flank (Fig. 7) largely bear out those of Johnson & Ebling (1964). Plucking resting hairs up to 9 days before eruption occurred on the control mid-flank induced eruption after a constant interval of 9 days, while plucking resting hairs after mitotic activity had recommenced in the hair bulb (about 5 days before eruption) had no effect. (In the present series of experiments we have found no evidence of a delay of eruption if resting hairs were plucked between 9 and 5 days before eruption of new hairs, such as that observed by Johnson & Ebling, 1964.) It thus seems that the follicle's course of action is irrevocably determined late in the telogen stage of the cycle, several days before mitotic figures can be identified in the hair bulb.

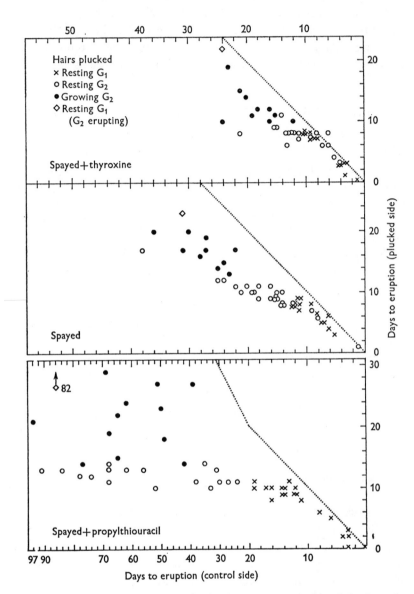

Fig. 7. The effect of thyroid hormone on the local response to plucking hairs from the mid-flank of the rat. Thyroxine treatment: rats weighing under 100 g, 50 μg/day I.P. in 0·9 % NaCl; rats over 100 g, 100 μg/day. Propylthiouracil treatment: 0·02 % in drinking water. Symbols would lie on the broken diagonal line if plucking had no effect on the subsequent activity of the follicles.

Plucking growing G_2 hairs also advances the next eruption, but the interval to eruption is longer than that after resting hairs are plucked.

Plucking induces the same basic response in hormone-treated as in intact rats, but the time-scale is altered—hairs erupt sooner after plucking on rats undergoing treatments which accelerate spontaneous hair cycles, and vice versa. Thus eruption follows about 8 days after plucking the mid-flank of thyroxine-treated rats, but not until 10–12 days afterwards in rats receiving propylthiouracil, i.e. in these animals the time taken to respond to the 'trigger' at plucking (or at spontaneous determination of follicular activity) is increased.

Ebling (1965a) has shown that over several cycles follicular activity on rat skin grafts tends towards synchrony with that on the surrounding host skin. For this reason it seemed possible that follicles from which hair had been plucked would do likewise, in spite of a contrary finding in the mouse by David (1934). Accordingly we continued our observations over several subsequent hair cycles. We found that the difference in age at eruption on the plucked and control sides was maintained, and even enhanced in many groups of animals, in the later hair generations.

THE SIGNIFICANCE AND CONTROL OF COAT COLOUR

Many mammalian species have characteristic patterns of coat colour which serve as camouflage. Thus within the Felidae the tiger's striped fur helps to conceal it amongst the parallel grass stems and reeds of its habitat, the arboreal African leopards have spotted or streaked coats which disguise them in the dappled pattern of light and shade among trees, while the snow leopard which lives in the treeless highlands of Central Asia, often above the snow-line, has long white fur (Cott, 1940). The giraffe's outline is broken up by its reticulated coat pattern, and this is probably also the function of the zebra's stripes.

Some species change colour seasonally. Cott suggests that such pelage changes are of importance in a deciduous forest habitat. In summer, when the leaves scatter sunlight into flecks, a spotted livery provides camouflage, but in winter, after the leaves have fallen, it would be conspicuous. Both the European fallow deer and the Japanese deer have white spots in summer, but become a uniform greyish brown in winter. The most striking examples of colour change are the mountain and prairie hares, arctic fox and some species of stoats and weasels, all of which are dark in summer but become white in winter. The usual explanation of this change is that it is an adaptation for concealment from either predators or prey when the ground is snow-covered (Cott, 1940), but many species which live in the same high latitudes—for example, the rabbit, reindeer, moose and musk-ox—never change colour in any part of their range. This is probably because concealment is unnecessary for the large herbivores because of their size and gregariousness, for rabbits because of their burrowing habits, and for species such as the pine marten because they are arboreal However, Hadwen (quoted from Severaid, 1945) believes that a white pelage subserves a different function in that it prevents radiation of heat and so conserves body warmth.

Change of colour is associated with the moult cycle. Since a mammalian hair is a dead appendage in which pigment is incorporated during its growth, to effect a major colour change a mammal must shed its old coat and replace it by a new one of different pigmentation.

Hewson (1958) and Watson (1963) observed the moulting of the mountain hare, *Lepus timidus scoticus*, in Scotland, where it usually turns white in winter, although the Irish race *Lepus timidus hibernicus* does not do so. They believe that the progress of at least the spring moult in any year can be adapted to the prevailing climatic conditions, since the completion of the colour change from white to brown was later in cold springs and at higher altitudes. Rothschild (1942) noted that keeping English *Mustela erminea* in cold conditions induced an autumn moult to white in a district where wild stoats remained brown in winter, while Rust (1962) found that cold delayed the onset of the spring moult (white to brown) of American *M. erminea*. Subsequently he has shown that the ability to change the colour of the pelage depends on the presence of the intact pituitary, for after hypophysectomy at any season the hair grown is white (Rust, 1965). Replacement experiments indicate that the pituitary factor essential for normal hair pigmentation is melanocyte-stimulating hormone (MSH), although adrenocorticotrophic hormone (ACTH) has some activity, probably due to its chemical similarity to MSH. The central nervous system appears to inhibit synthesis and/or release of pituitary MSH directly, for when pituitaries were transplanted under the kidney capsule the animals grew brown coats even though they were exposed to a photoperiod which induced intact controls to grow a white winter coat (Rust & Meyer, 1968).

α-MSH also influences hair pigmentation in mice (Geschwind, 1966), while Bronson, Eleftheriou & Dezell (1969) have implicated both the pituitary and the adrenals in the control of coat colour in deermice (*Peromyscus maniculatus bairdii*). In this species adrenalectomy causes darkening, correlated with raised levels of plasma MSH, but these effects do not occur in the absence of the pituitary, indicating a feedback control of MSH production. This system seems not to operate in pigmented strains of mice, where adrenalectomy does not affect coat colour.

GENERAL CONCLUSIONS

As targets for hormonal action the hair follicles are undoubtedly more difficult to understand than the reproductive organs. Reproductive cycles have genetically programmed indigenous components which are to a greater or lesser extent influenced by the environment, and Bullough (1965) has postulated that the cyclic activity of hair follicles is similarly controlled. However, hair follicles show the additional complication that their phase varies from region to region, with the different regions related to each other in a patterned sequence. We have cited evidence that follicles have their own characteristic rhythms, but that these rhythms can be gradually influenced by systemic factors. It must be emphasized that our knowledge of possible factors is based on experiments on domesticated laboratory animals such as the rat which show so-called spontaneous moult cycles when kept under constant conditions. There is incontrovertible evidence that in this and other laboratory species several endocrine systems contribute to the maintenance of normal hair cycles. Endocrine control of various aspects of hair growth can also be inferred from observations of various wild mammals in the field.

Normal cycles of hair growth and moulting appear to depend on the maintenance of a balance between the inhibitory effects of the steroids produced by the gonads and adrenals, and stimulation by the thyroid. These glands are controlled by the hypothalamo-hypophysial system, and that the presence of the pituitary is necessary for normal hair growth cycles to continue has been shown. There is also clear evidence that the natural changes in the photoperiod, which are known to control alterations in pituitary activity governing seasonal sexual cycles (e.g. the work of Bissonnette (1935) on the ferret), initiate seasonal moults or variations in hair growth (see, for example, the work of Yeates (1955, 1957) on cattle, Hart (1961) and Hutchinson (1965) on sheep and Rust (1965) on *Mustela*). Harvey & Macfarlane (1958), who also studied the ferret, have shown that hair cycles are linked both to gonadal cycles and to alterations in photoperiod. The response of the ferret pituitary to alterations in the photoperiod is mediated via the eyes, being prevented by section of the optic nerves (Donovan & Harris, 1954). Confirmation of the role of the eyes comes from the work of Hart (1961), who found that covering the eyes of sheep eliminated the seasonal peak of wool growth.

From the evidence outlined above it is reasonable to conclude that seasonal cycles of hair growth are maintained by the natural change in photoperiod, acting through the eyes and mediated via the hypothalamus, pituitary, thyroid, gonads and adrenals. Environmental temperature may have a modifying influence.

In recent years evidence has accumulated that the pineal organ is involved in the control of cyclic phenomena linked to changes in photoperiod (for example, see Quay, 1963). Pineal enzyme levels and melatonin secretion change in step with alterations in photoperiod, and its presence is essential for changes in pituitary MSH level with altering photoperiod to occur in the rat (Howe & Thoday, 1969). Moore and his associates have shown that in both the rat (Wurtman, Axelrod, Sedvall & Moore, 1967; Moore, Heller, Wurtman & Axelrod, 1968) and the rhesus monkey (Moore, 1969) signals from the retina to the pineal travel via the accessory optic tracts, which remain uncrossed at the optic chiasma and are not essential for the maintenance of visual discrimination. The majority of incoming nerve fibres to the pineal are sympathetic (Kappers, 1965), and it has been found that superior cervical ganglionectomy results in pineal neural degeneration. The link between the nerve fibres of the accessory optic tracts and the superior cervical ganglia has not yet been identified.

Herbert (1967) has shown that the response of the ferret gonad to added lighting is lost following pinealectomy, and in view of the work of Bissonnette (1935) and Harvey & Macfarlane (1958) it appears likely that this procedure would also influence the hair cycles. So far only Houssay *et al.* (1966), working on the mouse, seem to have produced evidence that the pineal is implicated in the control of hair cycles, but whether this is entirely mediated via the pituitary and thence the thyroid, adrenals and gonads has not yet been established.

REFERENCES

BASSETT, C. F. & LLEWELLYN, L. M. (1949). The molting and fur growth pattern in the adult mink. *Am. Midl. Nat.* **42**, 751–6.

BENNETT, J. W., HUTCHINSON, J. C. D. & WODZICKA-TOMASZEWSKA, M. (1962). Annual rhythm of wool growth. *Nature, Lond.* **194**, 651–2.

BERMAN, A. & VOLCANI, R. (1961). Seasonal and regional variations in coat characteristics of dairy cattle. *Aust. J. agric. Res.* **12**, 528–38.

BISSONNETTE, T. H. (1935). Relation of hair cycles in ferrets to changes in the anterior hypophysis and to light cycles. *Anat. Rec.* **63**, 159–68.

BISSONNETTE, T. H. & BAILEY, E. E. (1944). Experimental modification and control of moults and changes of coat colour in weasels by controlled lighting. *Ann. N.Y. Acad. Sci.* **45**, 221–60.

BISSONNETTE, T. H. & WILSON, E. (1939). Shortening daylight periods between May 15 and September 12 and the pelt cycle of the mink. *Science, N.Y.* **89**, 418–19.

BRONSON, F. H., ELEFTHERIOU, B. E. & DEZELL, H. E. (1969). Melanocyte-stimulating activity following adrenalectomy in deermice. *Proc. Soc. exp. Biol. Med.* **130**, 527–9.

BULLOUGH, W. S. (1965). Mitotic and functional homeostasis: a speculative review. *Cancer Res.* **25**, 1683–727.

BURKHARDT, J. (1947). Transition from anoestrus in the mare and the effects of artificial lighting. *J. agric. Sci., Camb.* **37**, 64–8.

BUTCHER, E. O. (1934). The hair cycles in the albino rat. *Anat. Rec.* **61**, 5–19.

CHANG, H. C. (1926). Specific influence of the thyroid gland on hair growth. *Am. J. Physiol.* **77**, 562–7.

COOP, J. E. & HART, D. S. (1953). Environmental factors affecting wool growth. *Proc. N.Z. Soc. Anim. Prod.* **13**, 113–19.

COTT, H. B. (1940). *Adaptive Coloration in Animals.* London: Methuen.

DAVID, L. THIGPEN (1934). Modification of hair direction and slope on mice and rats (*Mus musculus* and *Mus norvegicus albinus*). *J. exp. Zool.* **68**, 519–28.

DONOVAN, B. T. & HARRIS, G. W. (1954). Effect of pituitary stalk section on light-induced oestrus in the ferret. *Nature, Lond.* **174**, 503–4.

DOWLING, D. F. & NAY, T. (1960). Cyclic changes in the follicles and hair coat in cattle. *Aust. J. agric. Res.* **11**, 1064–71.

DOWNES, A. M. & WALLACE, A. L. C. (1965). Local effects on wool growth of intradermal injections of hormones. In *Biology of the Skin and Hair Growth*, ch. 42, pp. 679–703 (eds. A. G. Lyne and B. F. Short). Sydney and London: Angus and Robertson.

DRY, F. W. (1926). The coat of the mouse (*Mus musculus*). *J. Genet.* **16**, 287–340.

DWYER, P. D. (1963). Seasonal changes in pelage of *Miniopterus schreibersi blepotis* (Chiroptera) in North-eastern New South Wales. *Aust. J. Zool.* **11**, 290–300.

EBLING, F. J. (1965*a*). Systemic factors affecting the periodicity of hair follicles. In *Biology of the Skin and Hair Growth*, ch. 31, pp. 507–24. (eds. A. G. Lyne and B. F. Short). Sydney and London: Angus and Robertson.

EBLING, F. J. (1965*b*). Comparative and evolutionary aspects of hair replacement. In *Comparative Physiology and Pathology of the Skin*, pp. 87–102 (eds. A. J. Rook and G. S. Walton). Oxford: Blackwell.

EBLING, F. J. & HERVEY, G. R. (1964). The activity of hair follicles in parabiotic rats. *J. Embryol. exp. Morph.* **12**, 425–38.

EBLING, F. J. & JOHNSON, E. (1959). Hair growth and its relation to vascular supply in rotated skin grafts and transposed flaps in the albino rat. *J. Embryol. exp. Morph.* **7**, 417–30.

EBLING, F. J. & JOHNSON, E. (1961). Systemic influence on activity of hair follicles in skin homografts. *J. Embryol. exp. Morph.* **9**, 285–93.

EBLING, F. J. & JOHNSON, E. (1964). The action of hormones on spontaneous hair growth cycles in the rat. *J. Endocr.* **29**, 193–201.

FERGUSON, K. A., WALLACE, A. L. C. & LINDNER, H. R. (1965). Hormonal regulation of wool growth. In *Biology of the Skin and Hair Growth*, ch. 41, pp. 655–77. (eds. A. G. Lyne and B. F. Short). Sydney and London: Angus and Robertson.

FRASER, A. S. & NAY, T. (1953). Growth of the mouse coat. II. Effect of sex and pregnancy. *Aust. J. biol. Sci.* **6**, 645–56.

GESCHWIND, I. I. (1966). Change in hair colour in mice induced by injection of α-MSH. *Endocrinology* **79**, 1165–7.

HAMILTON, J. B. & MONTAGNA, W. (1950). The sebaceous glands of the hamster. I. Morphological effects of androgens on integumentary structures. *Am. J. Anat.* **86**, 191–233.

HART, D. S. (1961). The effect of light–dark sequences on wool growth. *J. agric. Sci., Camb.* **56**, 235–42.

HART, D. S., BENNETT, J. W., HUTCHINSON, J. C. D. & WODZICKA-TOMASZEWSKA, M. (1963). Reversed photoperiodic seasons and wool growth. *Nature, Lond.* **198**, 310–11.

HARVEY, N. E. & MACFARLANE, W. V. (1958). The effects of day length on the coat-shedding cycles, body weight, and reproduction of the ferret. *Aust. J. biol. Sci.* **11**, 187–99.

HERBERT, J. (1967). The effect of removal or sympathetic denervation of the pineal upon light-induced oestrus in the ferret. *Acta endocr., Copenh.* (Suppl.) **119**, 46.

HEWSON, R. (1958). Moults and winter whitening in the mountain hare, *Lepus timidus scoticus*, Hilzheimer. *Proc. zool. Soc. Lond.* **131**, 99–108.

HEWSON, R. (1963). Moults and pelages in the brown hare *Lepus europaeus occidentalis* de Winton. *Proc. zool. Soc. Lond.* **141**, 677–87.

HOUSSAY, B. A. (1918). Extirpacíon de la hipófisis en el perro. *Endocrinology* **2**, 497–8.

HOUSSAY, A. B., EPPER, C. E. & PAZO, J. H. (1965). Neurohormonal regulation of the hair cycles in rats and mice. In *Biology of the Skin and Hair Growth*, ch. 40, pp. 641–54. (eds. A. G. Lyne and B. F. Short). Sydney and London: Angus and Robertson.

HOUSSAY, A. B., PAZO, J. H. & EPPER, C. E. (1966). Effects of the pineal gland upon the hair cycles in mice. *J. invest. Derm.* **47**, 230–4.

HOWE, A. & THODAY, A. J. (1969). Post-natal development of rat pituitary melanocyte-stimulating hormone and the effect of light. *Nature, Lond.* **222**, 781.

HUTCHINSON, J. C. D. (1965). Photoperiodic control of the annual rhythm of wool growth. In *Biology of the Skin and Hair Growth*, ch. 34, pp. 565–73 (eds. A. G. Lyne and B. F. Short). Sydney and London: Angus and Robertson.

HUTCHINSON, J. C. D. & WODZICKA-TOMASZEWSKA, M. (1961). Climate physiology of sheep. *Anim. Breed. Abstr.* **29**, 1–14.

JOHNSON, E. (1958a). Quantitative studies of hair growth in the albino rat. I. Normal males and females. *J. Endocr.* **16**, 337–50.

JOHNSON, E. (1958b). Quantitative studies of hair growth in the albino rat. II. The effect of sex hormones. *J. Endocr.* **16**, 351–9.

JOHNSON, E. (1958c). Quantitative studies of hair growth in the albino rat. III. The role of the adrenal glands. *J. Endocr.* **16**, 360–8.

JOHNSON, E. & EBLING, F. J. (1964). The effect of plucking hairs during different phases of the follicular cycle. *J. Embryol. exp. Morph.* **12**, 465–74.

KAPPERS, J. A. (1965). Survey of the innervation of the epiphysis cerebri and the accessory pineal organs of vertebrates. *Prog. Brain. Res.* **10**, 87–151.

LINDNER, H. R. & FERGUSON, K. A. (1956). The influence of the adrenal cortex on wool growth and its relation to 'break' and 'tenderness' of the fleece. *Nature, Lond.* **177**, 188–9.

LING, J. K. (1965). Hair growth and moulting in the southern elephant seal, *Mirounga leonina* (Linn.). In *Biology of the Skin and Hair Growth*, ch. 32, pp. 525–44 (eds. A. G. Lyne and B. F. Short). Sydney and London: Angus and Robertson.

LINZEY, D. W. & LINZEY, A. V. (1967). Maturational and seasonal molts in the golden mouse, *Ochrotomys nuttalli*. *J. Mammal.* **48**, 236–41.

LYNE, A. G. (1965). The hair cycle in the chinchilla. In *Biology of the Skin and Hair Growth*, ch. 29, pp. 467–89 (eds. A. G. Lyne and B. F. Short). Sydney and London: Angus and Robertson.

MITCHELL, O. G. & BUTCHER, E. O. (1966). Growth of hair in the ventral glands of castrate gerbils following testosterone administration. *Anat. Rec.* **156**, 11–18.

MOHN, M. P. (1958). The effects of different hormonal states on the growth of hair in rats. In *The Biology of Hair Growth*, ch. 15, pp. 336–99. (eds. W. Montagna and R. A. Ellis). New York and London: Academic Press.

MOORE, R. Y. (1969). Pineal response to light: mediation by the accessory optic system in the monkey. *Nature, Lond.* **222**, 781–2.

MOORE, R. Y., HELLER, A., WURTMAN, R. J. & AXELROD, J. (1968). Visual pathway mediating pineal response to environmental light. *Science, N.Y.* **155**, 220–3.

MORRIS, L. R. (1961). Photoperiodicity of seasonal rhythm of wool growth in sheep. *Nature, Lond.* **190**, 102–3.

NAY, T. & FRASER, A. S. (1954). Growth of the mouse coat. III. Patterns of hair growth. *Aust. J. biol. Sci.* **7**, 361–7.

NAY, T. & FRASER, A. S. (1955). Growth of the mouse coat. V. Effects of pregnancy and lactation. *Aust. J. biol. Sci.* **8**, 428–34.

PINTER, A. J. (1968). Hair growth responses to nutrition and photoperiod in the vole, *Microtus montanus. Am. J. Physiol.* **215**, 828–32.

QUAY, W. B. (1963). Circadian rhythm in rat pineal serotonin and its modifications by estrous cycle and photoperiod. *Gen. comp. Endocr.* **3**, 473–9.

RENNELS, E. G. & CALLAHAN, W. P. (1959). The hormonal basis for pubertal maturation of hair in the albino rat. *Anat. Rec.* **135**, 21–7.

ROTHSCHILD, M. (1942). Change of pelage in the stoat, *Mustela erminea* L. *Nature, Lond.* **149**, 78.

ROTHSCHILD, M. (1944). Pelage change of the stoat, *Mustela erminea* L. *Nature, Lond.* **154**, 180–1.

RUFFER, D. G. (1965). Juvenile molt of *Onchomys leucogaster. J. Mammal.* **46**, 338–9.

RUST, C. C. (1962). Temperature as a modifying factor in the spring pelage change of short-tailed weasels. *J. Mammal.* **43**, 323–8.

RUST, C. C. (1965). Hormonal control of pelage cycles in the short tailed weasel (*Mustela erminea bangsi*). *Gen. comp. Endocr.* **5**, 222–31.

RUST, C. C. & MEYER, R. K. (1968). Effect of pituitary autografts on hair color in the short-tailed weasel. *Gen. comp. Endocr.* **11**, 548–51.

RUST, C. C., SHACKELFORD, R. M. & MEYER, R. K. (1965). Hormonal control of pelage cycles in the mink. *J. Mammal.* **46**, 549–65.

RYDER, M. L. & STEPHENSON, S. K. (1968). *Wool Growth.* New York and London: Academic Press.

SEVERAID, J. H. S. (1945). Pelage changes in the snowshoe hare, *Lepus americanus struthopus* Bangs. *J. Mammal.* **26**, 41–63.

SLEN, S. B. & CONNELL, R. (1958). Wool growth in sheep as affected by the administration of certain sex hormones. *Can. J. Anim. Sci.* **38**, 38–47.

SPETH, R. L. (1969). Patterns and sequences of molts in the Great Basin pocket mouse, *Perognathus parvus. J. Mammal.* **50**, 284–9.

STRAUSS, J. S. & EBLING, F. J. (1970). Control and function of skin glands in mammals. *Mem. Soc. Endocr.* **18**, 341–68.

THIERIEZ, C. & ROUGEOT, J. (1962). Action des hormones thyroïdiennes sur la croissance en longueur du brin de laine. *Annls Biol. anim. Biochim. Biophys.* **2**, 5–11.

TREGEAR, R. T. (1966). *Physical Functions of Skin.* New York and London: Academic Press.

WATSON, A. (1963). The effect of climate on the colour changes of mountain hares in Scotland. *Proc. zool. Soc. Lond.* **141**, 823–35.

WHITELEY, H. J. (1958). The effect of adrenalectomy and adrenocortical hormones on the hair growth cycle in the rabbit and rat. *J. Endocr.* **17**, 167–76.

WURTMAN, R. J., AXELROD, J., SEDVALL, G. & MOORE, R. Y. (1967). Photic and neural control of the 24-hour norepinephrine rhythm in the rat pineal gland. *J. Pharmac. exp. Ther.* **157**, 487–92.

YEATES, N. T. M. (1955). Photoperiodicity in cattle. I. Seasonal changes in coat character and their importance in heat regulation. *Aust. J. agric. Res.* **6**, 891–902.

YEATES, N. T. M. (1957). Photoperiodicity in cattle. II. The equatorial light environment and its effect on the coat of European cattle. *Aust. J. agric. Res.* **8**, 733–9.

DISCUSSION

JOHNSON: We have been looking into the factors which affect hair growth in *Microtus agrestis*, the short-tailed field vole. *Microtus* is a seasonal breeder, as well as showing evidence of seasonal changes in the other endocrine glands, and it also has a seasonal change of coat. In winter the coat has more hairs per unit area than in summer. In summer, the guard hairs are coarser than in winter. Thus there are fewer hairs grown in summer, but a greater secretion of hairs by the individual follicles. Subjecting animals to long days in autumn and winter resulted in the maintenance of their sexual activity, and they grew a summer-type coat. It thus seemed likely that the effects on hair were mediated through the endocrine system in the way postulated by Professor Ebling. Our evidence suggests that the male sex hormone inhibits activity of the hair follicles; the reduced amount of sex hormone in autumn allows more hair to grow. On the other hand, thyroid hormone affects the type of hair produced; the increased secretion in spring results in coarser hair.

QUAY: I wish first of all to compliment Dr Ebling on his very interesting paper and its logical development. The question I wish to pose concerns age changes in the moult pattern. In young rodents there is a regular 'wave-like' pattern of moult but with age this becomes progressively patchy and unpredictable. Is this due to systemic or local factors? Is it a matter of increasingly asynchronous local cycles in the follicle or are there other local factors involved?

EBLING: It is certainly true that the patterns become progressively more difficult to follow. Most studies, apart from the beautiful work which Dr Johnson has just reported, have been on domesticated mammals in which the environmental link may have become loosened. In rats kept under constant conditions, cycles seem to be spontaneous, not only when the animals are growing, but even subsequently. Patterns appear to break up as the animals get older. In the guinea-pig, activity of hair follicles appears synchronous at first, but the pattern breaks up to such an extent that it resembles a mosaic. The scalp of man may be similar. I quite agree with you that the local mechanisms must be of overriding importance, but any hypotheses which have so far been proposed about such local mechanisms are too crude to be of much value.

QUAY: This asynchrony seems to occur in wild rodents as well as in laboratory animals; that is, wild species collected in the field and examined in the laboratory *post mortem.*

STRAUSS: In the experiments that you have described in which light-related variations in wool growth in sheep were eliminated by hooding of the animals, were there adequate controls of other possible contributing factors such as grazing habits, i.e. food intake? I would assume that hooding could alter such factors, thereby influencing wool growth independently of any oculo-humoral pathway.

EBLING: Dr Hale, was not this work done at the CSIRO Wool Biology Laboratory at Prospect? The experiments would have been very carefully controlled in that case.

HALE: These sheep were kept in controlled conditions at Christchurch, New Zealand, and were not free to graze outside. Both groups received a standard diet which maintained a constant live weight.

ENVIRONMENTAL STIMULI AND
THE MAMMARY GLAND

By J. S. TINDAL and G. S. KNAGGS

The mammary gland, in addition to being a highly specialized part of the mammalian integument, is a remarkable organ in that when it is fully developed, as it is at parturition, it is almost entirely dependent on external stimuli for normal functioning to occur. Of the several sensory modalities which must be considered, the most important and indeed the most widely studied is the tactile stimulus initiated by the young suckling the teat.

THE SUCKLING STIMULUS

The control exerted over mammary function by the suckling stimulus is mediated by a neuroendocrine circuit involving ascending nervous pathways from teat to spinal cord, brain and pituitary body, and a descending link in the vascular system which conveys endocrine secretions to the mammary gland. This external stimulus, which occurs as an integral part of the act of suckling, has a dual role. It exerts a short-term effect by evoking a transient release of oxytocin which contracts the myoepithelial cells surrounding the mammary alveoli and ducts, causing milk to be expelled from the alveoli to the large ducts and gland sinuses or udder cisterns. This process is termed the milk-ejection reflex, and in most species its occurrence is essential if the young are to obtain milk from the gland (see Denamur, 1965). However, the suckling stimulus also exerts a long-term effect by causing release of pituitary trophic hormones including prolactin (Selye, 1934; Grosvenor & Turner, 1957), adrenocorticotrophin (Denamur, Stoliaroff & Desclin, 1965; Voogt, Sar & Meites, 1969), somatotrophin (Grosvenor, Krulich & McCann, 1968) and also melanophore-stimulating hormone (Taleisnik & Orías, 1966), most of which, especially prolactin, are necessary to ensure the continued secretion of milk by the mammary glands (see Cowie, 1969). The train of events associated with the removal of milk by the young, therefore, represents a rather neat example of supply being regulated to meet demand, as well as a triggering mechanism to flush the secretory product of the mammary gland when it is required, all brought about by an environmental stimulus (Fig. 1).

The pattern of suckling differs widely throughout the mammals. In the monotremes, it was thought until recently that the milk exuded from

[239]

the mammary glands and the young merely licked the milk from the mammary hairs. However, at least in the echidnas, it has been noted that the young sucks vigorously and audibly at the areolar region, and it seems likely that the milk-ejection reflex occurs in these animals, since injection of oxytocin causes the mammary lobules to change shape and milk to exude from the areolae (Griffiths, 1968). In the marsupials the young remain permanently attached to the teat for a time after birth, while the

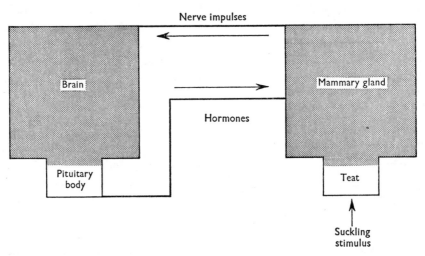

Fig. 1. Diagram to indicate the relationship between the mammary gland and the brain. The process of suckling triggers nerve impulses which pass from teat to spinal cord and then to brain. Impulses passing through the hypothalamic paraventricular nucleus to the neurohypophysis evoke release of oxytocin into the blood. On reaching the mammary gland, the hormone causes contraction of the mammary myoepithelial cells, which produces milk ejection, thus making the alveolar milk available to the suckling or milker. Impulses which activate other hypothalamic mechanisms controlling the adenohypophysis cause release of trophic hormones, including prolactin, which are essential for the continued secretion of milk by the mammary gland.

placental mammals exhibit a variety of types of suckling behaviour, ranging from very brief periods at the teat in such species as the pig and rabbit, to the more prolonged attachment seen in the rat, where the young may even fall asleep at the teat. The aquatic mammals present a specialized case; here, the milk is said to be pumped out of the mammary gland into the mouth of the suckling young by muscular action (see Turner, 1939). However, this interpretation should be treated with caution, since an analogous mechanism was originally proposed for the marsupial and has now been disproved (Enders, 1966).

There is considerable variation too in the degree of functional importance attached to the suckling stimulus in different species. The stimulus appears to be essential for the maintenance of lactation in the rat (Eayrs & Baddeley, 1956) and cat (Beyer, Mena, Pacheco & Alcaraz, 1962) since lactation was arrested after spinal cord section and was not restored in the cat even when oxytocin was administered to ensure milk ejection (Beyer *et al.* 1962). Although lactation has been reported to continue at a reduced level in the cord-sectioned rat if the operation is performed at mid-lactation when the pups are vigorous and if oxytocin is administered (Grosvenor, 1964), this has been attributed to incomplete denervation of the mammary glands, since after detailed studies of the innervation of the mammary glands in the rat, severance of the appropriate dorsal spinal roots resulted in the complete failure of both milk ejection and milk secretion (Edwardson & Eayrs, 1967; J. A. Edwardson, unpublished results, quoted by Edwardson & Eayrs, 1967).

By way of contrast, in the sheep and goat not only is the occurrence of the milk-ejection reflex not vital for the removal of milk, although it may occur under normal circumstances, but lactation can continue in the absence of any nervous connections between mammary gland and central nervous system. This was demonstrated initially in French and Russian laboratories by studies involving transection of the spinal cord or denervation of the udder (see Denamur, 1965) and confirmed later by transplantation of the goat's udder to the neck (Linzell, 1963). A direct demonstration of the non-essential nature of the milk-ejection reflex in the goat was provided by Folley & Knaggs (1966) since, in the majority of cases, they found no detectable release of oxytocin during hand-milking, although milk yields were normal. The reason why the reflex is unimportant in this species is not entirely clear, although the goat udder does contain a large proportion of cisternal spaces and large ducts which might be emptied fairly readily by a vigorous milking procedure. Moreover, it is possible that milk removal might be assisted by local contraction of myo-epithelial cells in response to direct mechanical stimulation of the udder, as has been reported for the mammary gland of the rabbit (Cross, 1954). Although it has been known for some years that the milking stimulus may not be necessary for anterior pituitary activation in the cow (Mielke & Brabant, 1963), the occurrence of the milk-ejection reflex has generally been considered to be essential for efficient milk removal in this species. However, recent work on the cow suggests that this may not be entirely true (see Folley, 1969).

The situation in the rabbit appears to lie between the two extremes outlined above, since, although the milk-ejection reflex is essential for removal of milk from the alveoli, the suckling stimulus may not be vital for anterior pituitary activation (Tindal, Beyer & Sawyer, 1963). It has been found that whereas spinal cord section in the rabbit will cause the cessation of lactation even if oxytocin is given before each suckling period, if, in addition, either prolactin or adrenocorticotrophin are injected for a few days, lactation is restored and is maintained even after withdrawal of trophic hormone (Mena & Beyer, 1963). A discussion of the possible mechanisms which would account for the apparent functional autonomy of the lactating mammary gland of some species when isolated from the central nervous system would be out of place here. Several hypotheses have been proposed and are discussed elsewhere (Tindal *et al.* 1963; Cowie & Tindal, 1965; Tindal, 1967).

ASCENDING PATHWAYS FOR TACTILE INFORMATION

Sensory receptors

The sensory receptors which are activated by the suckling stimulus are thought to be confined essentially to the teat, since local anaesthesia of the teat is effective in blocking the milk-ejection reflex, at least in the rabbit (Findlay, 1968). Histochemical studies in this species have not revealed any innervation of the mammary parenchyma and have shown that while extensive innervation occurs in the teat, it is much more attenuated in the surrounding skin (Ballantyne & Bunch, 1966). Electron-microscope observations have confirmed this and have shown that the sensory receptors consist of unmyelinated nerve fibres (Cross & Findlay, 1969).

Studies on the cutaneous innervation of the human breast have shown that the nipple and areola are specialized areas when compared with the remainder of the mammary skin; indeed the nipple has been considered to be one of the most highly innervated tissues of the body (Cathcart, Gairns & Garven, 1948). Although the hair follicles are small, the breast skin peripheral to the areola is essentially hair-covered skin and is basically innervated by free fibres and by the circular and palisade fibres associated with hair follicles. Follicles are sparse in the areola and absent in the nipple. In the nipple and areola the most frequently occurring nerve terminals are free-fibre endings concentrated in the deeper portions of the dermis. There is little innervation associated with the superficial dermis or the epidermis, unlike the mammary skin peripheral to the areola where such innervation occurs (Miller & Kasahara, 1959).

Pathways in the spinal cord

The exact pathway traversed by suckling- or milking-induced impulses within the spinal cord is still very much a matter for speculation, and from the data which are available at the present time it seems possible that the spinal pathway in the small ruminant may differ from that in the rodent or lagomorph. Thus, in the goat, section of the dorsal funiculi blocks the milk-ejection reflex (Tsakhaev, 1953; Popovici, 1963), suggesting the participation of either the dorsal column-medial lemniscal, or the spino-cervico-thalamic systems. In contrast to this, in the rat and rabbit, section of the lateral or ventrolateral funiculi, respectively, blocks the reflex (Eayrs & Baddeley, 1956; Mena & Beyer, 1968a), which suggests that some component of the spinothalamic system is involved.

In all three species so far investigated, the pathway has been reported to be ipsilateral to the side being suckled or milked (Tsakhaev, 1953; Popovici, 1963; Mena & Beyer, 1968a) although in the rat there is also a minor contralateral component (Eayrs & Baddeley, 1956). It would appear that a more precise identification of the ascending path of the reflex must await electrical stimulation studies to pin-point the region of the cord which conveys neuroendocrine, as against purely tactile, sensory information from the mammary glands.

Pathways in the brain

Most studies on the ascending path of the suckling stimulus in the brain have, up to now, been concerned with detection of the release of oxytocin by monitoring intramammary pressure in the lactating animal. However, although various reports indicated that oxytocin release could be evoked by stimulation of a number of sites in the midbrain and forebrain of the rabbit (see Cross, 1966), no definable pattern or pathway emerged from these studies. We decided to base our own observations on the guinea-pig and it is perhaps worth recording that our initial discovery of the pathway occurred as a result of one of those strokes of good fortune which happen all too rarely.

On one particular occasion we had unknowingly made a faulty insertion of the stereotaxic ear bars, resulting in the guinea-pig's head being displaced forward a few millimetres and the electrode penetrating not the diencephalon, as intended, but the lateral wall of the mesencephalon, a region we had not explored at that time. Stimulation at a particular point in this track evoked a large milk-ejection response. Further investigation confirmed this result and showed that the afferent path of the milk-ejection reflex in the midbrain of the guinea-pig is compact and lies in the lateral wall of the tegmentum. Further forward, at the meso-diencephalic boundary, the pathway lies just medio-ventral to the medial geniculate body in a region with ill-defined boundaries, the posterior thalamic complex, but as it passes forward in the diencephalon, this single pathway on each side of the brain divides into two. One portion, which we have termed the dorsal path, moves medially to pass forwards close to the rostral central grey matter and periventricular region, while the other portion, the ventral path, passes medio-ventrally into the subthalamus. At the level of the posterior hypothalamus the two pathways merge again on each side of the brain and pass forwards to more rostral levels of the hypothalamus.

The position of the pathway in the midbrain led us to propose that it was the spino-thalamic system of fibres which was responsible for conveying impulses triggered by the suckling stimulus (see Tindal, 1967; Tindal, Knaggs & Turvey, 1967b). This system of fibres relays with the ventrobasal and posterior thalamic complexes (see Poggio & Mountcastle, 1960; Whitlock & Perl, 1961) for further onward routing to the telencephalon, but it also has collateral branches which relay with other ascending systems to the diencephalon, both periventricular (Nauta, 1960) and subthalamic (Scheibel & Scheibel, 1967), which appear to be represented in our dorsal and ventral pathways respectively.

Although it was tempting to generalize, and assume that what applied to the guinea-pig applied to other mammals, we thought it prudent, indeed essential, to investigate another species. We have now found that the same ascending reflex pathways in the brain occur in the rabbit as in the guinea-pig, and also, in both species, that the paths are concerned with the preferential release of oxytocin from the neurohypophysis. Milk-ejection responses obtained by stimulation of these paths were found to be caused entirely by oxytocin, and vasopressin release could not be detected by the pressor response, even though the preparations themselves responded to exogenous vasopressin (Tindal, Knaggs, & Turvey 1968, 1969). Although the possibility is not excluded that some vasopressin may have been released, it is of interest that in the rabbit significant vasopressin release did occur, but only after stimulation of sites outside the confines of the afferent path of the milk-ejection reflex (Tindal et al. 1969).

Within the hypothalamus, using a simple electrical stimulation technique it was not possible to distinguish between the release of oxytocin caused by stimulation of pathways ascending from the mammary gland to the hypothalamus and that caused by stimulation of other pathways converging on the hypothalamus. By combining stimulation of the pathway in the midbrain with the use of a special brain knife to undercut the paraventricular nuclei, it has been found that the afferent path of the milk-ejection reflex in the guinea-pig does, in fact, pass to the paraventricular nuclei and then down to the neurohypophysis. It was also shown by ablation experiments in the guinea-pig that the cerebral cortex, hippocampus and amygdala, as well as structures rostral to the paraventricular nuclei, are not involved in the ascending path of the reflex (Tindal & Knaggs, un-published results). This is not to deny, of course, that other forebrain structures may be involved in release of oxytocin by means other than the suckling stimulus.

After studying the ascending path for oxytocin release in the brain of the rabbit, we were then in a position to make parallel observations to determine whether a similar ascending path was utilized for the release of prolactin from the adenohypophysis. The rabbit is a convenient species for such work, because release of prolactin can be detected by the lacto-genic response of the mammary glands in the pseudopregnant animal. By using a technique of electrical stimulation through permanently implanted electrodes it was found that in the midbrain the ascending path for release of prolactin is the same as for the release of oxytocin. Further forwards, however, whereas the oxytocin-release pathway divides into dorsal and ventral paths, that for prolactin utilizes only the dorsal and not the ventral path to enter the hypothalamus (Tindal & Knaggs, 1969). It seems likely, therefore, that this is the path activated by the suckling stimulus to cause release of prolactin and perhaps other trophic hormones from the adeno-hypophysis. Since, in the pregnant rat, stimulation of the mammary glands by self-licking of the nipple line contributes to the degree of mammary development at parturition (Roth & Rosenblatt, 1966, 1968), the same ascending pathway might be involved in trophic hormone release before parturition.

The hypothalamic site of action of the suckling stimulus in causing release of prolactin has yet to be determined. It may act directly at the level of inhibiting or releasing factors in the hypophysiotrophic area in the basal hypothalamus, yet it could equally well act elsewhere if, for instance, there is a tonic neuronal mechanism in the forebrain which must be either suppressed or activated if prolactin release is to occur. Indeed, it is already known that in the rabbit at least, the basal part of the amygdaloid complex (Tindal, Knaggs & Turvey, 1967a; Mena & Beyer, 1968b) and the telencephalon (Beyer & Mena, 1965) have a role to play in the mecha-nisms governing the secretion and release of prolactin, and the identifica-tion of the cerebral switch which turns this process on or off is awaited with interest.

NON-TACTILE STIMULI

In the preceding section it was tacitly assumed that the act of suckling involves a stimulus which is purely tactile. A moment's reflection will reveal that this may be an oversimplification, since other factors, such as the body heat of the young, may exert some additional facilitatory effect on the receptors in the teat. Be that as it may, we chose to consider thermal factors as being at least associated with the tactile stimulus of suckling. However, there are other sensory modalities which must be considered as affecting the mammary gland, not directly, but indirectly, entering the body through the organs of special sense, reaching the hypothalamus and causing release of hormones, a process which is normally effected by the act of suckling (see Fig. 2).

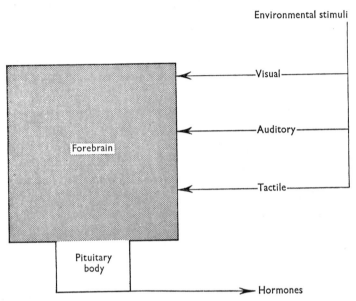

Fig. 2. Diagram to illustrate the fact that the ascending pathway for impulses triggered by the tactile stimulus of suckling is but one of several routes for environmental stimuli to reach the hypothalamus and pituitary body. Under appropriate conditions, release of the pituitary hormones which are necessary both for milk ejection and milk secretion can occur in response to auditory or visual stimuli.

Recent evidence for the participation of auditory stimuli in the milk-ejection mechanism came from the observation that when lactating rats were isolated from their litters for 9 h and were then placed within sound of other rats nursing litters for 15 min before being allowed to suckle their own young, the amount of milk obtained by the young was significantly greater than that obtained by control litters whose mothers were not exposed to the exteroceptive stimulus (Deis, 1968). The exteroceptive stimulation before nursing could be mimicked by injection of oxytocin, which indicates a pre-nursing, in addition to the normal nursing-induced release of oxytocin. The effect was mediated by auditory and not by visual or olfactory means since it did not occur in experimentally deafened rats. A little less physiological perhaps, but none the less dramatic, was the release of oxytocin in the rat caused by playing a recording of a thunderstorm or of a 150 Hz sound at 100 dB (Ogle & Lockett, 1966; Ogle, 1967).

When it is realized that some auditory fibres do converge on the posterior thalamic complex (Poggio & Mountcastle, 1960), that auditory potentials have been recorded from this region (Perl & Whitlock, 1961; Hotta & Kameda, 1963; Calma, 1965) and that oxytocin release can occur after stimulation of the tectum (Tindal *et al.* 1967*b*) it becomes a little easier to understand how auditory stimuli can evoke release of pituitary hormones. Since the posterior thalamic complex is believed to lie on or close to the ascending path of the suckling stimulus (Tindal, 1967) this would be the logical site for directing auditory stimuli onwards to the hypothalamus (Tindal *et al.* 1967*b*).

In the case of prolactin, the suckling of a lactating rat by her litter normally evokes a significant drop in the pituitary content of this hormone (Grosvenor & Turner, 1957), yet if the mother has been isolated for several hours from her 14-day-old pups and is then placed over them, but contact between them is prevented, the presence of the litter will induce a fall in pituitary prolactin content, just as if the mother had been suckled (Grosvenor, 1965). This effect does not occur, however, when the rat pups are only 7 days old (unpublished observations, quoted by Grosvenor & Mena, 1967). More recently, Moltz, Levin & Leon (1969) have claimed that, in the post-partum rat whose nipples have been removed before pregnancy, the mere physical presence of the pups will cause release of prolactin, assessed by the deciduoma reponse and the arrest of the oestrous cycle. However, the experimental plan did not exclude the possibility of post-surgical irritation of the proximal ends of the severed mammary nerves, since there were no observations on thelectomized rats deprived of pups. Physical irritation of ascending paths in spinal cord and brain, normally activated by the suckling stimulus, can evoke release of prolactin (Relkin, 1967; Tindal & Knaggs, 1969) and it should be borne in mind that a similar mechanism may have been operative in the experiments reported by Moltz *et al.* (1969).

It is a little difficult to draw the dividing line between these phenomena and conditioning of hormone release, which is known to apply to the milk-ejection reflex, at least in some species. The situation in the human is well known, where sight or sound of the child or some accustomed mechanical task will trigger the reflex (see Newton, 1961). In the cow, also, it is a widespread belief that certain external stimuli other than the milking stimulus will trigger milk ejection (see Zaks, 1962), but until recently direct evidence for release of oxytocin has been lacking. This has now been demonstrated for the cow in our laboratory, where a release of oxytocin accompanied by a rise in udder pressure was found to occur when the milker entered the milking parlour (Cleverley & Folley, 1970).

The main stumbling block in interpreting conditioned release of hormones concerned with lactation is a general lack of knowledge of brain function. Although it seems likely that memory plays some part in the conditioned release of hormones, the neural basis of memory is not fully understood at the present time. One fact which may be of significance is that in the human, degeneration of the mammillary bodies in the posterior hypothalamus or of the hippocampus which sometimes occurs as a consequence of alcoholism—the Korsakoff syndrome—results in complete loss of recent memory and the ability to memorize (see Magoun, 1963; Campbell, 1965). These structures are also known to be part of a neural circuit in the brain, involving the mammillo-thalamic tract, the anterior thalamic nuclei, the cingulate cortex and the hippocampus, which then projects by way of the fornix back to the hypothalamus and mammillary bodies. This circuit, which has side-branches to allow ingress and egress of information, was recognized by Papez (1937) and bears his name. It was originally proposed as a circuit of emotion, but may also be concerned with memory processing, and a part of it, the hippocampus, has been implicated in the acquisition of learned behaviour (Adey, Walter & Hendrix, 1961). Also, stimulation of several sites in the forebrain, including parts of the Papez circuit, results in the release of oxytocin and Cross (1966) has suggested that such structures may be concerned with conditioning of reflex responses. This merely serves to emphasize the fact that the ascending pathway traversed by impulses triggered by the suckling stimulus represents but one of several routes to the diencephalon.

EVOLUTION OF TACTILE SENSORY SYSTEMS

Up to now we have been considering sensory pathways as they appear in the highly specialized brain of the mammal, where the general picture is of long ascending pathways from spinal cord to thalamus, from where impulses are directed to their destinations within the forebrain. Since the very complexity of the mammalian brain makes it difficult to view the sensory mechanisms associated with lactation in broad perspective, it would be appropriate at this point to have a brief look at the phylogenetic development of these pathways.

The pathways which feed sensory information to the diencephalon in the lower vertebrates are thought to be represented by the central grey matter, and, alongside it in the tegmentum, the central tegmental fasciculus. These are phylogenetically ancient sensory paths within the brainstem which receive afferent fibres from the spinal cord only after extensive synaptic relays in the hindbrain and midbrain (see Mehler, 1957, 1966; Clezy, Dennis & Kerr, 1961; Dennis & Kerr, 1961). However, even in the lowest tetrapods, the Amphibia, the rudiments of more direct ascending systems make their appearance. This trend is continued in the Reptilia, and involves a progressive by-passing of the reticular neuropil of the brainstem (see Noback & Shriver, 1966). This means that there will be fewer synaptic relays between the periphery and the primitive thalamus, and hence the possibility of more precise information reaching the diencephalon. By the time the Marsupialia are reached on the evolutionary scale, these direct pathways have undergone further development and are recognizable as spinothalamic and medial lemniscal fibres extending from spinal cord to diencephalon. However, they are still sparse and there is still a dominance of the medial, indirect sensory pathways to the medial thalamus (Mehler, 1957; Dennis & Kerr, 1961). The marsupials thus represent a stage between the reptiles, where such indirect paths predominate, and the higher mammals, where the full development of lemniscal and spinothalamic systems has occurred.

Hand-in-hand with the development of new sensory pathways, the thalamus is also undergoing rapid evolutionary development. Part of it becomes enlarged to form the ventroposterior thalamic nucleus, which also develops projections to the rapidly expanding cerebral cortex. Thus, in the marsupial, and also in the lower placental mammal, such as the hedgehog, the ventro-posterior nucleus represents an interesting evolutionary stage, since it is a region of sensory convergence responding to both tactile and auditory stimuli, owing to the overlapping distribution of the brachium of the inferior colliculus and the medial lemniscal and spinothalamic systems. Furthermore, it cannot be distinguished readily from the neighbouring medial geniculate body, and has little topographic organization (Dennis & Kerr, 1961; Erickson, Hall, Jane, Snyder & Diamond, 1967; Jane, Yashon & Diamond, 1968).

As the mammalian scale is ascended, modality-specific nuclei emerge from this multi-modal complex. The medial geniculate body becomes a distinct entity to deal exclusively with auditory information, and part of the ventroposterior nucleus gives rise to the ventroposterior lateral and ventroposterior medial thalamic nuclei, which receive point-to-point information from the body surface, mainly via the medial lemniscus. However, the caudal part of the primitive ventroposterior nucleus and the medial magnocellular portion of the medial geniculate body lying adjacent to it remain largely unchanged as an ill-defined region, the posterior thalamic complex, which receives tactile information mainly from spinothalamic fibres, and also some auditory information, thus retaining much of the character of the primitive ventroposterior nucleus (see Perl & Whitlock, 1961; Erickson *et al.* 1967). Seen in this context, therefore, although the posterior thalamic complex and spinothalamic fibres are a relatively

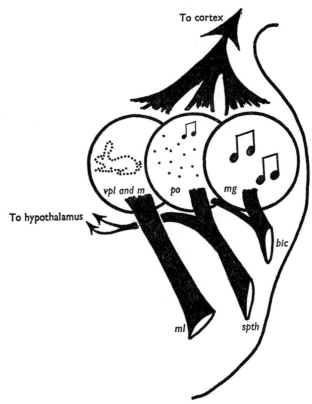

Fig. 3. A simplified diagram of a transverse section through the brain at the level of the caudal diencephalon, to illustrate the relationship between the ascending path for impulses triggered by the suckling stimulus and the sensory thalamic nuclei. Note the topographic representation conveyed to the ventroposterior lateral and ventroposterior medial thalamic nuclei by the medial lemniscus, the purely auditory input to the medial geniculate body, and, between them, an ill-defined and fairly primitive region, the posterior thalamic complex, in which there is little or no topographic representation of the body surface and which receives a small auditory input from the brachium of the inferior colliculus and its major tactile input from the spinothalamic system. In general, the latter system is activated by more powerful and abrupt stimuli than the medial lemniscal system, which can respond to such delicate stimuli as movement of hairs. It is now known that the spinothalamic system also provides the ascending path in the brain for impulses triggered by the suckling stimulus which are destined to activate neuroendocrine mechanisms, and that within the diencephalon the spinothalamic path relays with other ascending fibre systems to reach the hypothalamus. It is not known whether the posterior thalamic complex lies directly on the ascending path of the suckling stimulus or whether this structure is by-passed *en route* to the hypothalamus. *bic*, Brachium of inferior colliculus; *mg*, medial geniculate body; *ml*, medial lemniscus; *po*, posterior thalamic complex; *spth*, spinothalamic fibres; *vpl*, *m*, ventroposterior lateral and ventroposterior medial thalamic nuclei. For reasons of simplicity, facial representation is shown as being conveyed by medial lemniscus.

recent development when the vertebrates are considered as a whole, nevertheless functionally, they represent a primitive somaesthetic system as far as the mammals are concerned, and, at least in the guinea-pig and rabbit, the spinothalamic system appears to provide the afferent path of the milk-ejection reflex. At the present time it is not known whether the posterior thalamic complex lies directly on the ascending reflex path, or whether the impulses triggered by the suckling stimulus by-pass this structure on their way to hypothalamus and pituitary.

SPECIFICITY IN A NON-SPECIFIC SYSTEM

Our reasons for proposing that the spinothalamic system carries the messages which are destined to activate hormone-releasing mechanisms in the forebrain were initially neuroanatomical, but when neurophysiological considerations were also borne in mind this choice became more compelling. Consider the situation in the lactating animal. When it moves or lies down, the surface of its body, including the mammary glands, will receive tactile stimulation from the environment, from objects such as bedding, or the young brushing against their mother, or, if domesticated, from the structure of its cage. However, stimuli of this type will not normally trigger the milk-ejection reflex, yet the animal will be aware of the type of stimulation and whereabouts on the body it is being touched.

Such stimuli as these are thought to be conveyed by the lemniscal system, involving the dorsal spinal funiculi and medial lemnisci, which project first to the ventrobasal thalamus in a topographic or point-to-point manner and thence to the somatosensory cortex, giving an exquisite localization of sensation. In contrast to this, although the spinothalamic system may make some topographic sensory contribution to the ventrobasal thalamus, its contribution to the posterior thalamic complex, now thought to lie on or close to the milk-ejection reflex pathway, is essentially of a non-specific nature, giving little or no localization of sensation from stimuli impinging on the body surface. The two systems also differ in the quality of stimulus necessary for their activation, since the extra-lemniscal system requires that a tactile stimulus should be of an alerting nature, having the properties of abruptness, rapidity of onset, and of being in general more forceful than the delicate stimuli, such as movement of hairs, which can activate the lemniscal system (for fuller discussion see Perl & Whitlock, 1961; Albe-Fessard, 1967; Horridge, 1968).

These differences between the two major tactile sensory systems of the body are also reflected in the pattern of peripheral innervation. Whereas the rich innervation of the teat is confined essentially to the dermis

(Cathcart *et al.* 1948; Miller & Kasahara, 1959), in areas of the body where sensory discrimination is highly developed nerve endings are concentrated in the epidermis and superficial dermis (Miller, Ralston & Kasahara, 1958). In this context it is extremely pertinent that nearly 40 years ago Wood-Jones & Turner (1931) demonstrated that in the human the areolar and nipple skin has only limited powers of sensory discrimination. Light touch is not well perceived, the quality of objects cannot be determined, and two-point discrimination is very poor.

Therefore, although tactile stimulation of the mammary gland may activate both the lemniscal and spinothalamic pathways in spinal cord and brain, it is the functional characteristics of the receptors in the teat and of the spinothalamic system which can best explain why general tactile stimuli are ineffective and why the suckling stimulus is effective in triggering neurohumoral mechanisms. This system would give the animal the ability to select automatically from the barrage of environmental tactile stimuli the one stimulus which happens to satisfy the requirements for activation of the one ascending pathway which is now known to project to the hormone-releasing mechanisms of the hypothalamus (Tindal, 1967; Tindal *et al.* 1967*b*).

Indeed, if we consider the suckling stimulus to be an alerting stimulus, or even a mild, presumably pleasurable, form of stress, this will come as no surprise to workers in the field of lactation, who will be aware of the vigorous, if not brutal, assaults made by young mammals on the mother's teats, until they have achieved the triggering of the milk-ejection reflex, and the subsequent period of comparative calm as they empty the mammary glands. The reason why activation of spinothalamic fibres and the triggering of the reflex does not normally occur after vigorous tactile stimulation of other parts of the integument may be a combination of the fact that sensory nerve endings are concentrated in the teat (Ballantyne & Bunch 1966; Cross & Findlay, 1969) and the, perhaps naïve, observation that the physical projection of the teat from the body happens to present a convenient handle with which to achieve maximum stimulation of these sensory nerve endings. This would seem a reasonable supposition since, in species with multiple pairs of mammary glands, spatial summation occurs in the suckling-induced release of oxytocin in the rabbit (Fuchs & Wagner, 1963) and of prolactin in the rat (Mena & Grosvenor, 1968), indicating that the total size of the sensory barrage is important.

In conclusion, the spinothalamic system appears to have a suitable pedigree to qualify as message-carrier for the suckling stimulus, and it would seem entirely appropriate that a primitive mammalian system of sensibility should subserve what is, after all, a primitive mammalian neuroendocrine reflex. The picture presented in this brief review is inevitably oversimplified, limited by the present state of knowledge concerning the integration of information by the central nervous system. There must be many checks and balances and sophisticated control mechanisms for processing data arising from environmental stimuli of which we are unaware, or even if they are recognized are little understood. Perhaps it is best to allow our appreciation of the complexity of control of mammary gland function by the environment to spur us on to decipher these mechanisms, since, in the absence of a functional mammary gland, few mammals could survive.

Grateful acknowledgement is made to the Ford Foundation and the Rockefeller Foundation for generous financial support of the recent studies described in this paper.

REFERENCES

ADEY, W. R., WALTER, D. O. & HENDRIX, C. E. (1961). Computer techniques in correlation and spectral analyses of cerebral slow waves during discriminative behaviour. *Expl Neurol.* **3**, 501–24.

ALBE-FESSARD, D. (1967). Organization of central somatic projections. In *Contributions to Sensory Physiology*, vol. 2, pp. 101–67 (ed. W. D. Neff). New York and London: Academic Press.

BALLANTYNE, B. & BUNCH, G. A. (1966). The neurohistology of quiescent mammary tissue in *Lepus albus*. *J. comp. Neurol.* **127**, 471–87.

BEYER, C. & MENA, F. (1965). Induction of milk secretion in the rabbit by removal of the telencephalon. *Am. J. Physiol.* **208**, 289–92.

BEYER, C., MENA, F., PACHECO, P. & ALCARAZ, M. (1962). Effect of central nervous system lesions on lactation in the cat. *Fedn Proc. Fedn Am. Socs exp. Biol.* **21**, 353.

CALMA, I. (1965). The activity of the posterior group of thalamic nuclei in the cat. *J. Physiol., Lond.* **180**, 350–70.

CAMPBELL, H. J. (1965). *Correlative Physiology of the Nervous System*. London and New York: Academic Press.

CATHCART, E. P., GAIRNS, F. W. & GARVEN, H. S. D. (1948). The innervation of the human quiescent nipple, with notes on pigmentation, erection, and hyperneury. *Trans. R. Soc. Edinb.* **61**, 699–717.

CLEVERLEY, J. D. & FOLLEY, S. J. (1970). The blood levels of oxytocin during machine milking in cows with some observations on its half-life in the circulation. *J. Endocr.* **46**, 347–61.

CLEZY, J. K. A., DENNIS, B. J. & KERR, D. I. B. (1961). A degeneration study of the somaesthetic afferent systems in the marsupial phalanger, *Trichosurus vulpecula*. *Aust. J. exp. Biol. med. Sci.* **39**, 19–28.

COWIE, A. T. (1969). General hormonal factors involved in lactogenesis. In *Lactogenesis, the Initiation of Milk Secretion* (eds. M. Reynolds and S. J. Folley). Philadelphia: University of Pennsylvania Press (in the Press).

COWIE, A. T. & TINDAL, J. S. (1965). Some aspects of the neuro-endocrine control of lactation. *Proc. 2nd Int. Congr. Endocr.*, London, 1964. Excerpta Medica Int. Congr. Series, no. 83, pp. 646–54.

CROSS, B. A. (1954). Milk ejection resulting from mechanical stimulation of mammary myoepithelium in the rabbit. *Nature, Lond.* **173**, 450.

CROSS, B. A. (1966). Neural control of oxytocin secretion. In *Neuroendocrinology*, vol. 1, ch. 7, pp. 217–59 (eds. L. Martini and F. Ganong). New York and London: Academic Press.

CROSS, B. A. & FINDLAY, A. L. R. (1969). Comparative and sensory aspects of milk ejection. In *Lactogenesis, the Initiation of Milk Secretion* (eds. M. Reynolds and S. J. Folley). Philadelphia: University of Pennsylvania Press (in the Press).

DEIS, R. P. (1968). The effect of an exteroceptive stimulus on milk-ejection in lactating rats. *J. Physiol., Lond.* **197**, 37–46.

DENAMUR, R. (1965). The hypothalamo-neurohypophysial system and the milk-ejection reflex. *Dairy Sci. Abstr.* **27**, 193–224, 263–80.

DENAMUR, R., STOLIAROFF, M. & DESCLIN, J. (1965). Effets de la traite sur l'activité corticotrope hypophysaire des petits ruminants en lactation. *C. r. hebd. Séanc. Acad. Sci., Paris* **260**, 3175–8.

DENNIS, B. J. & KERR, D. I. B. (1961). Somaesthetic pathways in the marsupial phalanger, *Trichosurus vulpecula. Aust. J. exp. Biol. med. Sci.* **39**, 29–42.

EAYRS, J. T. & BADDELEY, R. M. (1956). Neural pathways in lactation. *J. Anat.* **90**, 161–71.

EDWARDSON, J. A. & EAYRS, J. T. (1967). Neural factors in the maintenance of lactation in the rat. *J. Endocr.* **38**, 51–9.

ENDERS, R. K. (1966). Attachment, nursing and survival of young in some didelphids. *Symp. zool. Soc. Lond.* **15**, 195–203.

ERICKSON, R. P., HALL, W. C., JANE, J. A., SNYDER, M. & DIAMOND, I. T. (1967). Organization of the posterior dorsal thalamus of the hedgehog. *J. comp. Neurol.* **131**, 103–30.

FINDLAY, A. L. R. (1968). The effect of teat anaesthesia on the milk-ejection reflex in the rabbit. *J. Endocr.* **40**, 127–8.

FOLLEY, S. J. (1969). The milk-ejection reflex: a neuroendocrine theme in biology, myth and art. *J. Endocr.* **44**, x–xx.

FOLLEY, S. J. & KNAGGS, G. S. (1966). Milk-ejection activity (oxytocin) in the external jugular vein blood of the cow, goat and sow, in relation to the stimulus of milking or suckling. *J. Endocr.* **34**, 197–214.

FUCHS, A. R. & WAGNER, G. (1963). Quantitative aspects of release of oxytocin by suckling in unanaesthetized rabbits. *Acta endocr., Copenh.* **44**, 581–92.

GRIFFITHS, M. (1968). *Echidnas. International Series of Monographs in Pure and Applied Biology*. Zoology Division, vol. 38 (ed. G. A. Kerkut). Oxford: Pergamon Press.

GROSVENOR, C. E. (1964). Lactation in rat mammary glands after spinal cord section. *Endocrinology* **74**, 548–53.

GROSVENOR, C. E. (1965). Evidence that exteroceptive stimuli can release prolactin from the pituitary gland of the lactating rat. *Endocrinology* **76**, 340–2.

GROSVENOR, C. E., KRULICH, L. & McCANN, S. M. (1968). Depletion of pituitary concentration of growth hormone as a result of suckling in the lactating rat. *Endocrinology* **82**, 617–19.

GROSVENOR, C. E. & MENA, F. (1967). Effect of auditory, olfactory and optic stimuli upon milk ejection and suckling-induced release of prolactin in lactating rats. *Endocrinology* **80**, 840–6.

GROSVENOR, C. E. & TURNER, C. W. (1957). Release and restoration of pituitary lactogen in response to nursing stimuli in lactating rats. *Proc. Soc. exp. Biol. Med.* **96**, 723–5.

HORRIDGE, G. A. (1968). *Interneurons*. London and San Francisco: Freeman.

HOTTA, T. & KAMEDA, K. (1963). Interaction between somatic and visual or auditory responses in the thalamus of the cat. *Expl Neurol.* **8**, 1–13.

JANE, J. A., YASHON, D. & DIAMOND, I. T. (1968). An anatomic basis for multi-modal thalamic units. *Expl Neurol.* **22**, 464–71.

LINZELL, J. L. (1963). Some effects of denervating and transplanting mammary glands. *Q. Jl exp. Physiol.* **48**, 34–60.

MAGOUN, H. W. (1963). *The Waking Brain*, 2nd ed. Springfield: Thomas.

MEHLER, W. R. (1957). The mammalian 'pain tract' in phylogeny. *Anat. Rec.* **127**, 332.

MEHLER, W. R. (1966). Some observations on secondary ascending afferent systems in the central nervous system. In *Pain*, pp. 11–32 (eds. R. S. Knighton and P. R. Dumke). Boston: Little, Brown and Co.

MENA, F. & BEYER, C. (1963). Effect of high spinal section on established lactation in the rabbit. *Am. J. Physiol.* **205**, 313–16.

MENA, F. & BEYER, C. (1968*a*). Effect of spinal cord lesions on milk ejection in the rabbit. *Endocrinology* **83**, 615–17.

MENA, F. & BEYER, C. (1968*b*). Induction of milk secretion in the rabbit by lesions in the temporal lobe. *Endocrinology* **83**, 618–20.

MENA, F. & GROSVENOR, C. E. (1968). Effect of number of pups upon suckling-induced fall in pituitary prolactin concentration and milk ejection in the rat. *Endocrinology* **82**, 623–6.

MIELKE, H. & BRABANT, W. (1963). Laktogenese und Galactopoese beim Rind ohne Saug-, Melk- oder andere exogene, zur Milchejektion führende Euterreize. *Arch. exp. VetMed.* **16**, 909–19.

MILLER, M. R. & KASAHARA, M. (1959). The cutaneous innervation of the human female breast. *Anat. Rec.* **135**, 153–67.

MILLER, M. R., RALSTON, H. J. III & KASAHARA, M. (1958). The pattern of cutaneous innervation of the human hand. *Am. J. Anat.* **102**, 183–217.

MOLTZ, H., LEVIN, R. & LEON, M. (1969). Prolactin in the postpartum rat: synthesis and release in the absence of suckling stimulation. *Science, N.Y.* **163**, 1083–4.

NAUTA, W. J. H. (1960). Some neural pathways related to the limbic system. In *Electrical Studies on the Unanesthetized Brain*, ch. 1, pp. 1–16 (eds. E. R. Ramey and D. S. O'Doherty). New York: Hoeber.

NEWTON, M. (1961). Human lactation. In *Milk: the Mammary Gland and Its Secretion*, pp. 281–320 (eds. S. K. Kon and A. T. Cowie). New York and London: Academic Press.

NOBACK, C. R. & SHRIVER, J. E. (1966). Phylogenetic and ontogenetic aspects of the lemniscal systems and the pyramidal system. In *Evolution of the Forebrain*, pp. 316–25. (eds. R. Hassler and H. Stephan). Stuttgart: Georg Thieme.

OGLE, C. W. (1967). Low frequency sound and oxytocic activity of plasma in rats. *Nature, Lond.* **214**, 1112–13.

OGLE, C. W. & LOCKETT, M. F. (1966). The release of neurohypophysial hormone by sound. *J. Endocr.* **36**, 281–90.

PAPEZ, J. W. (1937). A proposed mechanism of emotion. *Archs Neurol. Psychiat., Chicago* **38**, 725–45.

PERL, E. R. & WHITLOCK, D. G. (1961). Somatic stimuli exciting spinothalamic projections to thalamic neurons in cat and monkey. *Expl Neurol.*, **3**, 256–96.

POGGIO, G. F. & MOUNTCASTLE, V. B. (1960). A study of the functional contributions of the lemniscal and spinothalamic systems to somatic sensibility. *Bull. Johns Hopkins Hosp.* **106**, 266–316.

POPOVICI, D. G. (1963). Recherches neurophysiologiques sur le réflexe d'évacuation du lait. *Rev. Biol., Bucarest* **8**, 75–81.

RELKIN, R. (1967). Neurologic pathways involved in lactation. *Dis. nerv. Syst.* **28**, 94–7.

ROTH, L. L. & ROSENBLATT, J. S. (1966). Mammary glands of pregnant rats; development stimulated by licking. *Science, N.Y.* **151**, 1403–4.

ROTH, L. L. & ROSENBLATT, J. S. (1968). Self-licking and mammary development during pregnancy in the rat. *J. Endocr.* **42**, 363–78.

SCHEIBEL, M. E. & SCHEIBEL, A. B. (1967). Structural organization of nonspecific thalamic nuclei and their projection toward cortex. *Brain Res.* **6**, 60–94.

SELYE, H. (1934). On the nervous control of lactation. *Am. J. Physiol.* **107**, 535–8.

TALEISNIK, S. & ORÍAS, R. (1966). Pituitary melanocyte-stimulating hormone (MSH) after suckling stimulus. *Endocrinology* **78**, 522–6.

TINDAL, J. S. (1967). Studies on the neuroendocrine control of lactation. In *Reproduction in the Female Mammal*, pp. 79–109. *Proc. 13th Easter Sch. agric. Sci. Univ. Nott.*, 1966 (eds. G. E. Lamming and E. C. Amoroso). London: Butterworth.

TINDAL, J. S., BEYER, C. & SAWYER, C. H. (1963). Milk ejection reflex and maintenance of lactation in the rabbit. *Endocrinology*, **72**, 720–4.

TINDAL, J. S. & KNAGGS, G. S. (1969). An ascending pathway for release of prolactin in the brain of the rabbit. *J. Endocr.* **45**, 111–20.

TINDAL, J. S., KNAGGS, G. S. & TURVEY, A. (1967*a*). Central nervous control of prolactin secretion in the rabbit: effect of local oestrogen implants in the amygdaloid complex. *J. Endocr.* **37**, 279–87.

TINDAL, J. S., KNAGGS, G. S. & TURVEY, A. (1967*b*). The afferent path of the milk-ejection reflex in the brain of the guinea-pig. *J. Endocr.* **38**, 337–49.

TINDAL, J. S., KNAGGS, G. S. & TURVEY, A. (1968). Preferential release of oxytocin from the neurohypophysis after electrical stimulation of the afferent path of the milk-ejection reflex in the brain of the guinea-pig. *J. Endocr.* **40**, 205–14.

TINDAL, J. S., KNAGGS, G. S. & TURVEY, A. (1969). The afferent path of the milk-ejection reflex in the brain of the rabbit. *J. Endocr.* **43**, 663–71.

TSAKHAEV, G. A. (1953). O prirode afferentnykh puteĭ refleksa molokootdachi. *Dokl. Akad. Nauk SSSR* **93**, 941–4.

TURNER, C. W. (1939). *The Comparative Anatomy of the Mammary Glands.* Columbia, Missouri: University Cooperative Store.

VOOGT, J. L., SAR, M. & MEITES, J. (1969). Influence of cycling, pregnancy, labor, and suckling on corticosterone-ACTH levels. *Am. J. Physiol.* **216**, 655–8.

WHITLOCK, D. G. & PERL, E. R. (1961). Thalamic projections of spinothalamic pathways in monkey. *Expl Neurol.* **3**, 240–55.

WOOD-JONES, F. & TURNER, J. B. (1931). A note on the sensory characters of the nipple and areola. *Med. J. Aust.* **1**, 778–9.

ZAKS, M. G. (1962). *The Motor Apparatus of the Mammary Gland*, 1st English edn. (ed. A. T. Cowie), Edinburgh and London: Oliver and Boyd.

DISCUSSION

FOLLEY: You mention early in your paper the fact that the milk-ejection reflex is not essential for milking-out ruminants. Recently, as you no doubt know, our colleague Alan McNeilly has obtained evidence in the goat that with regard to the release of oxytocin into the jugular blood as measured by the lactating guinea-pig assay, the nature of the stimulus seems to be important. In previous work in goats in our laboratory, it was found by G. S. Knaggs and J. D. Cleverley that oxytocin release could be detected in about 30% of experiments in goats which were hand-milked. In

McNeilly's recent experiments in which goats suckled by kids were compared with goats which were hand-milked oxytocin was detected in the jugular blood in almost 90% of the experiments on the suckled animals. I wonder if you would like to comment on this in relation to the concluding remarks in your paper.

TINDAL: There must be many sensory modalities involved here. Even things which one cannot measure, such as the maternal attitude to the young, must play a facilitatory role in these mechanisms. Perhaps the answer lies in the role of the central inhibition of oxytocin release. When we know more about these central inhibitory mechanisms, we will be in a better position to understand why some stimuli are more effective than others in causing a release of this particular hormone.

SMITH: The stimulation of discrete afferent pathways leading to the release of oxytocin, which you have demonstrated so elegantly, also causes the release of some vasopressin. How efficient is the neurohypophysis in responding selectively to different incoming stimuli? In particular, have you any information concerning the relative amounts of oxytocin and vasopressin released when the dorsal rather than the ventral afferent pathway to the paraventricular nucleus is stimulated?

TINDAL: Time did not permit me to discuss the relative release of oxytocin and vasopressin. We have studied this aspect fully. The pathways I described are pathways for the preferential release of oxytocin and not of vasopressin. A little vasopressin may possibly be released but if so it is far less than the release of oxytocin. We can obtain vasopressin release by stimulating other regions of the brain which I did not mention, but the tactile pathway in question is a selective pathway for oxytocin release. We have measured simultaneously intramammary pressure and arterial blood pressure and have followed this through a vast number of experiments both in guinea-pigs and in the rabbit, and we are absolutely certain that you do not get an equal release of the two hormones or even a measurable release of vasopressin. Of course, when you stimulate further on in the pathway, in the pituitary stalk, that is a different matter; you will then get both hormones released, but certainly until you reach the paraventricular nucleus it is the release of oxytocin which occurs predominantly.

LEATHEM: As a matter of clarification, is it not true that the cord contains ventral and lateral spinothalamic tracts and do these tracts coalesce in the midbrain? Slusher (*Expl Brain Res.* **1**, 184–94) has suggested some relationship between midbrain and ACTH release. Would you comment on these data?

TINDAL: As far as is known, the two spinothalamic pathways at the cord level are thought to coalesce at the level of the midbrain, but, as you will be aware, they are very poorly myelinated, difficult to see, and not a lot of work has been done on this aspect. They do appear, however, to be one entity at the level of the midbrain. Regarding the work of Slusher, I think, if I remember correctly, her effective sites in the midbrain were in the reticular formation, which is a more diffuse phenomenon than the one we have been investigating. It is interesting that we have, in fact, obtained vasopressin release when stimulating such regions, but not oxytocin release.

HUTCHINSON: Olfactory stimuli seem to be important in the recognition and acceptance of the young by the mother. Do such stimuli play a role in milk secretion?

TINDAL: The short answer is that I can't reply to this because I am not aware of any work in this field. The work of Deis (see References) certainly showed that the effective exteroceptive stimulus was auditory. He deafened his rats experimentally, which I think was done by injecting alcohol into the ear of the rats when young, and by this means ruled out olfactory or visual stimuli.

BERN: I wonder particularly in so far as the neuroendocrine control of prolactin release is concerned what one might infer from similar studies in birds regarding the development of the brood patch, and also whether tactile stimuli and other stimuli, visual and auditory, might be involved therein. It would be interesting to know whether analagous nervous pathways have evolved in birds.

TINDAL: Yes, I think that this would be a very rewarding study.

BRIGGS: There is very high demand in Western Society for primary suppression of lactation and for the moment the popular clinical procedure is to administer potent doses of oestrogens, but the possible side-effects are now making this method less acceptable. Do you feel that the work you have been describing offers any insights into possible alternate methods for lactation suppression?

TINDAL: This question is rather difficult to answer because when lactation really gets going it is like trying to stop a runaway train. If it is caught early enough then these various methods employing steroids of one sort or another may be effective. The eternal problem is that if you remove milk to relieve engorgement, you will then stimulate further release of hormones; if you don't remove milk in these circumstances, you may finish up with severe pathological troubles associated with distension of the gland. There are proprietary steroids available now which are being subjected to clinical trials which have had mixed results, but there seems to be no clear answer. I would like to end on a rather plaintive note. It seems a shame that one has to think in terms of suppression of lactation, which is such a natural phenomenon, and it is saddening to me that mothers do not wish to breast-feed their own children.

ENDOCRINE–EPIDERMAL RELATIONSHIPS IN SQUAMATE REPTILES

BY P. F. A. MADERSON, K. W. CHIU AND J. G. PHILLIPS

A. INTRODUCTION

There is nothing unusual about the fact that from time to time some vertebrates 'lose their skins'. The production, maturation, and eventual loss to the external environment, of cells from the outer epidermis, or specialized epidermal derivatives e.g. hair or feathers, appears to be a vertebrate characteristic. However, certain forms manifest this characteristic in a rather spectacular fashion. From time to time, many amphibia (urodeles and anurans), as well as some reptiles, lose a relatively large amount of material from the body surface in a form which mirrors exactly the external appearance of the animal. The German word for a shed snake skin *Die Natterhemnd*—literally 'snake-shirt'—precisely describes the phenomenon.

Although Maderson (1968a) has shown that a certain pattern of skin-shedding is a unique characteristic of the subclass Lepidosauria among extant reptiles, there is no information regarding the physiology of the sloughing process in the single surviving rhynchocephalian, the New Zealand tuatara, and we therefore propose to restrict our discussion to the more widely known squamate forms, the snakes and lizards.

B. CYCLIC ACTIVITY IN THE SQUAMATE EPIDERMIS

When a lizard or snake sheds its skin it loses a *complete, mature* 'epidermal generation'. The epidermis is then said to enter the 'resting phase' defined as a period of time during which only a single epidermal generation is represented above the stratum germinativum (Maderson 1967). In the immediate post-slough period the outer epidermal generation consists of a mature β-layer (with its superficial *Oberhautchen*), a mature mesos layer, a partly formed α-layer, and a number of presumptive α-cells. Within a maximum of 5–7 days these latter differentiate further and become incorporated into the mature tissue. The epidermis then enters a period of apparent inactivity during which little if any cell production or differentiation takes place; this is the so-called 'perfect resting condition' (Plate 1 a, c) (Maderson & Licht, 1967).

Eventually, the resting phase ends as new cells are laid down which represent the final components of the outer generation, these are the lacunar and clear layer cells. The outer generation is therefore said to be *complete* at the end of the resting phase in the sense that all its constituent

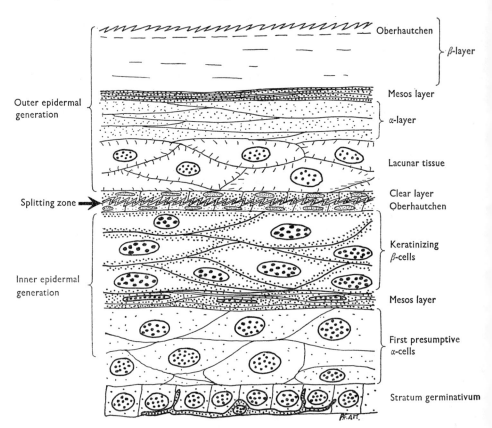

Fig. 1. Diagrammatic representation of a generalized squamate epidermis some 4–8 days before shedding. The outer generation is shed from the body when the constituent cells on its innermost aspect (the clear layer) finally keratinize, by which time the *Oberhautchen*, β-layer, mesos layer, and perhaps a small portion of an α-layer of a new inner generation have also matured. Figure taken from Maderson (1967) and reproduced here by kind permission of the editors of *Copeia*.

parts are present. The ensuing 'renewal phase' is a period during which much of a new inner epidermal generation is laid down and matures, while the lacunar and clear layer components of the outer generation finally mature prior to shedding (Fig. 1). When the inner generation has a mature β-layer (with its superficial *Oberhautchen*), a mature mesos layer and part of an α-layer, the clear layer of the outer generation separates from the new *Oberhautchen* and shedding occurs (Plate 1b). The previous inner generation becomes the new outer generation and the cycle is repeated. Detailed accounts of patterns of histogenesis during epidermal generation formation are presented elsewhere (Lillywhite & Maderson, 1968; Maderson, 1965 a, b, 1966, 1967, 1968 a–c; Maderson, Chiu & Phillips, 1970;

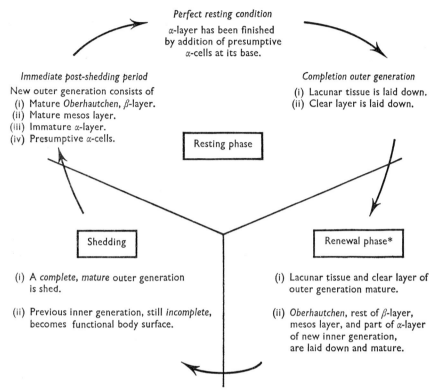

Fig. 2. Summary of the major histogenic events in the epidermal sloughing cycle of squamates. The cycle essentially consists of two parts, the resting phase and the renewal phase, the duration of which is measured in days. The act of shedding usually only occupies 1–12 h, depending on the species, but is indicated here to emphasize the changes which occur in general epidermal topography. The renewal phase (*) has been divided into five arbitrary stages covering various aspects of the differentiation sequence of the inner generation (Lillywhite & Maderson, 1968; Maderson, 1958b, 1966, 1967, 1968a–c; Maderson & Licht, 1967).

Maderson, Flaxman, Roth & Szabo, 1970; Maderson & Licht, 1967; Maderson, Mayhew & Sprague, 1969) and the major events are summarised in Fig. 2.

The physical removal of the mature generation from the entire body rarely takes longer than about 12 h, but nevertheless considerable variation in the degree of development of the α-layer of the new outer generation has been observed at this time (Maderson, 1966; Maderson & Licht, 1967;

Maderson & Chiu, unpublished data). Furthermore, there is a peak of mitotic activity which occurs around the time of shedding which appears to coincide with the laying-down of presumptive α-cells (Maderson, Pang & Roth, unpublished data). The period of time elapsing between the appearance of the first identifiable presumptive lacunar cells at the end of the resting phase (an event also associated with another mitotic peak), and eventual shedding has been found to be 8–14 days in a variety of squamate species so far studied (Chiu & Maderson, unpublished data). The examination of numerous biopsy samples from large numbers of individuals or species which are known to have relatively long shedding cycles (Maderson *et al.* 1969; Maderson, unpublished data) shows a very high percentage of individuals in the 'perfect resting condition'. It is as yet uncertain whether the 'perfect resting condition' represents a period of absolute germinal quiescence. It is clear, however, that very little cell division or maturation takes place during this time. The available evidence suggests that the time elapsing from the laying-down of presumptive lacunar cells through the completion of the α-layer of the next generation is maximally 16–18 days, with the shed occurring around the 8th–14th day of this period. Therefore, when an individual or species has a total interslough period of more than 18 days, the increased time is represented by an extended 'perfect resting condition'. This conclusion is important in assessing the pattern of action of hormones in controlling the cycle length.

C. LITERATURE REVIEW

1. *Introduction*

Over the past 40 years the problem of the endocrine control of skin-shedding in squamates has attracted the attention of a number of workers, and has been specifically reviewed a number of times (Sembrat & Drzewicki, 1936; Eggert, 1935, 1936; Lynn, 1960, 1970; Maderson, 1965 a). Reviewers, and authors of comparative endocrinological texts have always indicated that the situation appears to be extraordinarily complex and that there appear to be a number of inexplicable paradoxes. It is now certain that many of these problems arise from the fact that there are indeed significant differences in endocrine/epidermal relationships between snakes and lizards. For this reason we will not attempt at this stage to consider the two sets of data together. The literature will be reviewed in chronological order and the results and interpretations will be expressed in terms of events which have been defined earlier (pp. 259–262).

2. *Lizards*

Drzewicki (1926) thyroidectomized *Lacerta agilis* and concluded that this operation prevented sloughing, although some members of the small experimental group did slough after the operation. In 1927 Drzewicki presented a histological analysis and noted a characteristic α-layer hyperplasia in animals which did not shed. Further experiments recorded in 1928 showed that these hyperkeratotic symptoms disappeared if and when the animals shed. This suggested that thyroidectomy always *delayed* skin-shedding, but sometimes, for unknown reasons, if an animal survived the operation for long enough, then shedding could occur even in the absence of the thyroid gland. Eggert's widely quoted (1933) paper describing cessation of skin-shedding in three species of *Lacerta* following thyroidectomy, contained no observations on the hyperkeratotic or exophthalmic conditions described by Drzewicki (1927).

Noble & Bradley (1933) hypophysectomized and thyroidectomized the gecko *Hemidactylus brookii*. Reporting on a large number of individuals kept at constant 28 °C, the authors found a decreased shedding periodicity resulting from both operations, but complete inhibition was not found. These authors also reported that thyroxine injections restored normal periodicity in thyroidectomized animals, but similar treatment of normal animals did not result in an increased frequency.

Sembrat & Drzewicki (1935, 1936) reported further extensive investigations of the effects of thyroidectomy on *Lacerta agilis*, followed in some cases by replacement therapy using homoplastic (*Lacerta*) and xenoplastic (shark) thyroid implants. Whereas in the 1935 paper they reported some return to a normal condition when grafting was performed at the first indications of hyperkeratosis, their control and experimental animals showed considerable variation in sloughing periodicity and no clear conclusion can be drawn from the results. Administration of dried thyroid gland tissue also gave equivocal results (Sembrat & Drzewicki, 1936). However, the authors did offer the first explanation of the events which followed thyroidectomy. They suggested that if epidermal differentiation was advanced beyond a certain point at the time of operation, then the imminent slough would occur, but none thereafter. Sembrat & Drzewicki referred extensively to the pioneer study of amphibian moulting by Adams, Kuder & Richards (1932) and apparently concurred with their conclusion that 'the release of thyroid hormone is the fundamental factor at work in the moulting reaction'. During the same period, Eggert published two further studies on the relationship of the thyroid to sloughing in *L. agilis* and *L. vivipara*, reporting the cyclic changes in the gland associated with epidermal changes (Eggert 1935), and analysing the effects of thyroidectomy (Eggert, 1936). He reported pathological changes similar to those observed by Sembrat & Drzewicki (1935, 1936) and also a greater degree of success was recorded in the restoration of the cycle following homoplastic implants. Eggert (1936) indicated that it was the immediate post-slough period which was the critical period when thyroid-ectomy would result in a complete inhibition of the shedding process, but he could offer no explanation of Noble & Bradley's (1933) results.

Adams & Craig (1950) stated that there was no change in the sloughing activity of *Lacerta* following administration of thiourea or thiouracil, but as these goitrogens did not produce a change in thyroid histology, this conclusion should be treated with caution. Wilhoft (1958) showed that maintenance of *Sceloporus occidentalis* at 35 °C under constant light produced hyperkeratosis associated with hyperthyroid activity. Although Licht (1968) suggested that this might well have represented a pathological condition, the experiment has, however, been repeated several times and such epidermal changes have not been observed (Wilhoft & Maderson, unpublished data).

Maderson *et al.* (1969) demonstrated the existence of well-defined annual patterns of sloughing activity in the desert iguanids *Dipsosaurus* and *Uma* but were unable to find any specific correlation with any aspect of the species' ecology. It is particularly significant that although sex differences in the patterns were found in *Uma* (though not in *Dipsosaurus*) there was no indication of any association with sexual or breeding activity of the animals. It therefore seems reasonable to conclude that neither sex hormones nor gonadotrophins are directly involved in sloughing activity in these forms, confirming experimental data from another iguanid, *Anolis carolinensis* (Maderson & Licht, 1967). Analysis of shedding frequency and epidermal morphology in a large experimentally maintained population of the Yucca night lizard, *Xantusia vigilis*, some of which had been pinealectomized, indicated that this operation does not affect the shedding cycle (Maderson & Stebbins, unpublished data).

The results of numerous studies on the Tokay (*Gekko gecko*) and the American chameleon, *Anolis carolinensis*, species which have been most frequently used in recent years, are summarized in Tables 1 and 2 respectively. There is some evidence for stimulatory, depressant and possibly total inhibitory action by hormones in both species, but the available data do not permit a clear understanding of the role hormones play in the events which determine shedding frequency. Broadly speaking, both sets of data strongly suggest that the thyroid/pituitary axis will emerge as a central feature in the over-all scheme when it is finally elucidated.

Table 1. *Summary of results of endocrinological investigations of the sloughing cycle in the tokay* (Gekko gecko)

Histological examination of biopsy samples showed that in all cases where the cycle was increased or decreased in length, only the resting phase was affected: the renewal phase remained the same. In those treatments where subsequent sloughs were inhibited, no renewal phases were ever observed after the beginning of the treatment.

	Treatment		Dosage*	Conclusion†	Reference
1.	Intact plus T_4	i	0·08 μg daily	Enhanced	Chiu *et al.* (1967)
		ii	0·16 μg	Enhanced	Chiu & Phillips (1970 d)
2.	Intact plus TSH		0·04 USP unit	Enhanced	Chiu & Phillips (1970 d)
3.	Intact plus prolactin		0·4 i.u.	Enhanced	Chiu & Phillips (1970 d)
4.	Intact plus prolactin plus T_4		0·4 i.u.	Enhanced‡	Chiu & Phillips (1970 d)
			0·16 μg		
5.	Intact plus corticosterone		0·03 μg daily	No effect	Chiu & Phillips (unpubl.)
6.	Intact plus metopirone		0·06 μg daily	No effect	Chiu & Phillips (unpubl.)
7.	Thyroidectomized	i	—	Decreased	Chiu *et al.* (1967)
		ii	—	Decreased	Chiu & Phillips (1970 b)
8.	Thyroidectomized plus T_4		0·16 μg daily	Enhanced	Chiu *et al.* (1967)
9.	Thyroidectomized plus prolactin		0·4 i.u.	Enhanced	Chiu & Phillips (1970 b)
10.	Hypophysectomized		—	Decreased§	Chiu & Phillips (1970 a)
11.	Hypophysectomized plus TSH		0·02 USP daily	Enhanced	Chiu & Phillips (1970 a)
12.	Hypophysectomized plus prolactin	i	0·26 i.u.	‖	Chiu & Phillips (1970 b)
		ii	0·40 i.u.	Enhanced	Chiu & Phillips (1970 b)
13.	Hypophysectomized plus ACTH		0·02 i.u. daily	Inhibits	Chiu & Phillips (1970 a)
14.	Hypophysectomized and thyroidectomized		—	Inhibits	Chiu & Phillips (1970 c)
15.	Hypophysectomized and thyroidectomized plus prolactin		0·4 i.u.	Enhanced	Chiu & Phillips (1970 c)
16.	Hypophysectomized and thyroidectomized plus TSH		0·04 USP unit	Inhibits	Chiu & Phillips (unpubl.)
17.	Intact plus testosterone (female)		2·00 μg daily	Enhanced	Chiu, Lofts & Tsui (unpubl.)
18.	Hypophysectomized plus testosterone propionate (male)		2·00 μg daily	Inhibits	Chiu, Maderson, Lofts & Tsui (unpubl.)
19.	Hypophysectomized plus testosterone		2·00 μg daily	Inhibits	Chiu, Maderson, Lofts & Tsui (unpubl.)

* Dosage 1 g body weight and on alternate days unless otherwise stated.

† Due to the fact that exact means for control animals vary slightly from experiment to experiment (see text, p. 266), and some of the treatments produce relatively small percentage increases or decreases in shedding frequency, only a qualitative conclusion is reported here; sloughing frequency with reference to the control group in each case was either unaffected (no effect), more frequent (enhanced), less frequent (decreased), or completely stopped (inhibits).

‡ This treatment produced a greater percentage increase in shedding frequency than either T_4 or prolactin given alone at similar dose levels, indicating synergistic action.   § *Some* animals never shed after surgery.

‖ This dosage caused the hypophysectomized animals to shed with the same frequency as intact and sham-operated animals.

Table 2. *Summary of results of endocrinological investigations of the sloughing cycle in the American chameleon* (Anolis carolinensis)

Hormone injections alternate days unless otherwise stated, general results expressed as detailed in Table 1.

	Treatment	Dosage	Conclusion	Reference
1.	Intact plus ACTH	0·1 i.u.	Decreased	Chiu & Lynn (unpubl.)*
2.	Intact plus gonadotropin	0·19 i.u. daily	No effect	Maderson & Licht (1967)*
3.	Intact plus prolactin	0·16 i.u. daily	Enhanced	Maderson & Licht (1967)*
4.	Intact plus prolactin and gonadotropin	0·16 i.u.	Enhanced	Maderson & Licht (1967)*
		0·19 i.u. daily		
5.	Intact plus thiourea			
	(i)	0·33 or 0·67 mg (3 times/week)	Decreased†	Ratzersdorfer, Gordon & Charipper (1949)
	(ii)	0·25–5·00 mg (6 times/week)	Decreased†	Adams & Craig (1951)
6.	Intact plus thiouracil	1·0–5·0 mg (6 times/week)	Decreased†	Adams & Craig (1951)
7.	Thyroidectomy			
	(i)	—	Decreased†	Adams & Craig (1951)
	(ii)	—	Decreased	Chiu & Lynn (unpubl.)*
	(iii)	—	Decreased	Clark & Maderson (unpubl.)*
8.	Thyroidectomy plus ACTH	0·1 i.u.	Inhibits	Chiu & Lynn (unpubl.)*

* These studies included histological analysis and indicated that only the resting phase was affected.

† There may have been some indication of inhibition in these studies, but the data are difficult to interpret (see discussion, Maderson & Licht, 1967).

3. General comments on available data for lizards

Maderson & Licht (1967) and Chiu, Phillips & Maderson (1967) drew attention to the fact that earlier studies contained only bare details regarding the temperatures at which experimental animals were kept, or else quoted figures which now seem to be inadequate in the light of more recent studies on the thermal requirements of various lizards (Licht, 1968). It seemed possible therefore that this parameter (temperature) might be responsible for discrepancies between the different experimental results deriving from studies using the same or different species. While this possibility still exists, recent studies with *Gekko gecko* and *Anolis carolinensis* have provided data which raise important questions regarding thermal factors in experimental situations. It is worthy of note, therefore, that figures for sloughing frequency for the Tokay in different studies (Chiu *et al.* 1967; Chiu & Phillips, 1970*a–d*) show variation in the control groups despite the fact that experimental conditions remained the same. Recent studies with *Anolis* have utilized a variety of light and temperature regimes: natural daylight with solar heat augmented sporadically by radiant heat from a bulb, constant 32 °C with 8 h, or 16 h of light (Maderson, unpublished data). In all cases, the shedding frequencies for adult male animals have been higher than those reported for control animals by Maderson & Licht (1967). While continued observation of these animals may reveal that there are seasonal differences in this species (see Maderson & Licht, 1967), we may tentatively conclude that (*a*) photoperiod length does not appear to influence shedding frequency, and (*b*) as long as a certain daily quantum of heat is available to the animals, variation in maintained temperature throughout the day does not affect the process of periodic shedding. Both of these conclusions are somewhat surprising in view of the fact that tail regeneration—recently demonstrated to be under the influence of the hypophysial–thyroid axis (Licht & Howe, 1969 and personal communication)—in the same animals is differentially affected according to the conditions of maintenance. While the exact geographical origin of the Tokays used in the Hong Kong studies probably varied, thus raising the possibility that shedding frequencies differ in different populations, *Anolis* used in past and present studies come from the same, fairly restricted location. Indeed, the animals used in the recent studies were collected within a 5-mile radius to attempt to demonstrate that the recorded variation in tail regeneration (Maderson & Licht, 1968) was a population-dependent phenomenon: however, exactly the same mean, and approximately the same degree of variation, has been observed in the present work, in sharp contrast to the shedding data. The recent long-term study (125 consecutive days) has also shown that although consecutive inter-slough periods for individual *Anolis* may be remarkably constant as in short-term studies (Maderson & Licht, 1967), variations of 0–100% from the population mean are not uncommon (Maderson, unpublished data). It may well be that variation of this order, thus far observed in shedding frequency, tail regeneration (Maderson & Licht, 1968: Licht, personal communication: Maderson & Salthe, unpublished data) and cutaneous wound-healing (Maderson, unpublished data) may be characteristic of *Anolis*, if not all other lizards, or ectotherms in general. Alternatively, the observed differences may be simple reflections of the animals' physiological state, since it is always difficult to provide a degree of husbandry for captive animals of wild origin to ensure a uniform and healthy population. As Maderson *et al.* (1969) have indicated, the bulk of presently available data on shedding frequency derives from laboratory-maintained animals. Whereas it is customary to demand some degree of constancy of conditions of maintenance for experimental animals, we should not lose sight of the fact that an ectotherm, by definition, lives in a constantly changing environment. Unfortunately, practical considerations limit the amount of data available on such problems as individual shedding frequencies, or rates of tail regeneration in wild populations of lizards.

Bearing in mind the fact that unexplained individual variations frequently occur in experimental series both within and between individual species and genera, it is perhaps safer to restrict our conclusions to a general, qualitative treatment of the available data.

A direct action on the epidermis by prolactin is suggested from the evidence of the stimulatory role of this hormone in the hypophysectomized/thyroidectomized gecko (Chiu & Phillips, 1970c). The evidence from many other available studies on the control of the lizard sloughing cycle indicate a significant role for the thyroid gland. However, the question raised by Maderson & Licht (1967) as to whether prolactin is *the* stimulatory factor in precipitating a renewal phase, or whether it normally acts through the thyroid gland, is still unanswered. We note, however, that the failure of prolactin to activate the thyroid gland in the normal and hypophysectomized gecko supports the thesis of a direct action on the epidermis (Chiu & Phillips, 1970b, d).

It is perhaps permissible to go further than just to speak broadly of a hypophysial/thyroid axis control of the cycle by considering further certain aspects of the Tokay thyroidectomy experiments (Chiu *et al.* 1967). It was found that if the operation was performed more than 6–7 days after the previous shed, that interslough was unaffected, but subsequent ones were (Chiu & Phillips, 1967). Sembrat & Drzewicki (1936) and Eggert (1936) also found a similar situation in *Lacerta* and both papers indicate that this was the explanation for the occurrence of some post-operative sheddings in their respective earlier experiments. Although it is known that the cycle of thyroid-gland activity accompanying the epidermal cycle is slightly different in *Gekko* than in *Lacerta* (see discussion Chiu *et al.* 1967; Maderson, Chiu & Phillips, 1970), it would appear that there is reason to believe that thyroid hormone levels, or perhaps sequences of changes therein, are important throughout the resting phase. The possibility also exists of a synergism between prolactin and thyroxine in their effect on the epidermis of the Tokay (Chiu & Phillips, 1970*d*).

4. Snakes

Schaeffer (1933) gave a brief report of the effects of hypophysectomy and thyroidectomy on the snakes *Thamnophis sirtalis* and *T. radix*, suggesting that both operations caused an increased shedding frequency. Krockert (1941) fed thyroid gland to a python (he did not thyroidectomize the animal as erroneously stated by Gorbman, 1963, and Maderson, 1965*a*) and reported that the animal shed only four times compared with eight for control animals, indicating an inhibitory action.

Halberkann (1953, 1954*a*, *b*) injected a variety of hormones into intact *Natrix natrix* and reported that T_4 (thyroxine), TSH (thyroid stimulating hormone), STH (somatotrophin) and ACTH (adrenocorticotrophic hormone) prevented shedding, while cortisone and desoxycorticisterone had no effect. Goslar (1958*a*) extended these studies using methylthiouracil as a goitrogen and *p*-hydroxy-propriophenon to inhibit TSH synthesis in the pituitary: both treatments increased shedding frequency. Histological examination of epidermal material from animals treated with T_4 and TSH showed α-hyperplasia with an indefinite resting stage. Goslar reported that when T_4 and TSH treatments were stopped, the animals eventually shed. Claiming a 'natural anti-thyroid action' for thymus extract, Goslar (1957, 1958*b*) found that injection or feeding with this material increased sloughing frequency. However, two problems are associated with these observations. First, the effect of repeated treatments was said to become less with successive sloughs so that the animals returned to normal. Second, Goslar's comments on thyroid histology in the treated animals are indicative of an inactive gland, which is contrary to what would be expected with other goitrogenic agents. It is important to note that Goslar reported that all treatments with thyroid stimulants or depressants were effective even if begun quite late on in the resting phase.

Goslar (1958*c*) commented that female animals shed irregularly during the breeding season, and also showed (Goslar, 1958*a*, *d*) that a variety of gonadotrophic and sex hormones, i.e. serum gonadotrophins, chorionic gonadotrophins, oestrone, oestradiol, progesterone, and testosterone all tended to decrease sloughing frequency, and in some cases to cause inhibition. Significant differences were observed from the effects of thyroid stimulation in that treatments with sex hormones were only effective if started soon after shedding, and no hyperkeratotic symptoms were observed.

Notwithstanding the many problems presented by his 1958 observations on the effects of sex hormones on the shedding cycle, Goslar (1964) presented a model of the role of thyroid hormone alone. He suggested that minimal thyroid activity permitted the completion of the outer epidermal generation and the onset of the renewal phase. Increasing thyroid hormone levels towards the time of shedding restored the resting phase. Proof of such thyroid activity in association with snake epidermal cycles has been obtained for *Ptyas korros* (Maderson, Chiu & Phillips, 1970) and for *Naja naja* (Chiu, unpublished). Chiu & Lynn (1970) have shown that thyroidectomy of *Chionactis occipitalis* increases shedding frequency by reducing the length of the resting phase, while T_4 injections inhibited shedding and produced α-hyperplasia.

5. General comments on available data for snakes

The data for snakes are neither as detailed nor as broadbased as they are for lizards, but there is positive indication that the role of thyroid hormones is the reverse of that in lizards. We note therefore that studies of thyroid gland activity and experimental data point to completion of the outer epidermal generation and an ensuing renewal phase occurring only in the presence of a low titre of circulating thyroid hormone. The reported inhibitory action of hypophysial hormones such as STH, ACTH and gonadotrophins as well as the similar role reported for the gonadal hormones is noted, but all require further investigation.

D. THE ENDOCRINE CONTROL OF SQUAMATE SKIN-SHEDDING

1. General comments

If we are to attempt to understand the role that the endocrine glands play in controlling periodicity of skin-shedding it is necessary to consider the significance of the phenomenon in the life of the animals. The extreme variation observed in shedding frequency within some species and particularly between different genera and families is important here. If we were dealing with a phenomenon in any way comparable to physiological regeneration as known in mammals, e.g. periodic renewal of red blood cell population, more consistency would be expected. Studies on a variety of squamate species have shown that the renewal phase lasts about 10 days. Knowing that species such as *Anolis carolinensis* and *Gekko gecko* shed approximately every 20–30 days and that the chances of observing a renewal stage in a random biopsy from these forms are about 30–40%, we can give some indication of sloughing frequency based on museum collections. In fact, this technique tells us that frequent skin-shedding is indeed characteristic of most geckos, but apart from *Anolis*, most iguanids, skinks, xantusids and most snakes shed rather infrequently. The earlier studies of *Lacerta* (Drzewicki, 1926, 1927, 1928; Sembrat & Drzewicki, 1935, 1936; Eggert, 1933, 1935, 1936) claim an approximately 30-day cycle for three species, but as the quoted figures for control animals indicate considerably lower frequencies for some individuals, it is difficult to place these forms in the over-all spectrum

The recent study of the desert-living forms *Dipsosaurus* and *Uma* (Maderson *et al.* 1969) provided interesting data in this context. In an arid environment, percutaneous water loss, or its prevention might so alter the functional status of the epidermal tissues that frequent shedding might be predicted. This would also be expected if indeed skin-shedding represented an extra excretory mechanism with the conservation of water as a physiologically desirable side-effect (Maderson, 1965a). Ultraviolet radiation is known to stimulate epidermal proliferation in man (Szabo & Horkay, 1967) and it might be predicted that in lizards such exposure during thermo-regulatory behaviour (McGinnis & Dickson, 1967) would produce a high shedding frequency. In fact, a variety of large iguanids (desert-living or otherwise) appear to shed rather infrequently (Maderson *et al.* 1969). We must conclude that there is as yet no conclusive evidence relating ecology and shedding frequency.

The possibility that shedding is associated with growth has still to be fully investigated and has been suggested as a possible basis for the observed annual shedding pattern in wild forms (Maderson *et al.* 1969). However, recent studies with adult and juvenile male *Anolis* kept under a variety of conditions have so far suggested that young animals may shed slightly less frequently than the adults (Maderson, unpublished data). These studies have also indicated that the stress of capture, handling or surgery does not appear to precipitate precocious shedding in this species, contrary to a suggestion for gekkonid lizards (Maderson, 1968b).

A possible association between squamate skin-shedding and excretion has not been investigated, but the evidence suggesting a specific association with steroid metabolism (Maderson, 1965a) has been weakened by subsequent studies of lizard epidermal changes. It was suggested that the reports of high 17-ketosteroid content in shed skins (Pesonen, 1954) might be associated with the role of adrenocorticosteroids in stimulating eosinophil release during the early renewal phase (Maderson, 1965 b). However, while subsequent studies have indicated that eosinophil immigration into the epidermis at this time is characteristic of all the major snake families, the phenomenon has only been observed in one lizard *Anolis* (Maderson & Licht, 1967), and definitely does not occur in *Lacerta* (Breyer, 1929; Eggert, 1933, 1935, 1936; Sembrat & Drzewicki, 1936), various geckos (Maderson, 1966, 1967, 1968 b, c, and unpublished data), other iguanids (Maderson *et al.* 1969), scincids or xantusids (Maderson, unpublished data).

We must conclude, therefore, that there is little positive indication of the role of periodic skin-shedding in squamate physiology. It is very difficult to believe that it is a universal physiological regenerative phenomenon and recent studies already cited may tend to support Maderson's (1965 a) suggestion that it is a mere side-effect of other endocrinological activities and not a consummating activity in itself.

2. The mode of action of hormones on epidermal activity

It is generally accepted that snakes arose from lacertilian ancestors during the Cretaceous period (Romer, 1968). Morphological studies of epidermal renewal have indicated that not only is the periodic shedding of epidermal generations a fundamental and unique lepidosaur characteristic (Maderson, 1968 a), but that the details of the associated patterns of germinal proliferation and cell differentiation appear to be identical between lizards and snakes. Any attempt to establish a working hypothesis regarding the role of hormones in controlling shedding periodicity should take into account the known differences in the patterns of control between the two suborders.

If there were only *in vivo* evidence available, it would be necessary to consider the possibility that perhaps the hormonal *milieu* was responsible not only for the *frequency* of skin-shedding, but in fact controlled the patterns of differentiation of cells arising from the stratum germinativum. Therefore, generation formation, which the morphological evidence tells us only results from a sequential vertical distribution of *Oberhautchen*, β-, mesos, α-, lacunar and clear layer cells, would be dependent on a precisely regulated sequence of changes in the hormonal composition of the blood. However, when pieces of intact lizard integument are cultured *in vitro* at any stage of the shedding cycle, generation formation will often continue (Flaxman, Maderson, Szabo & Roth, 1968). As many as three and a half complete epidermal 'shedding cycles' have been observed under these conditions (Plate 1*d*). Given this information, we could then postulate that there is a permanent programme governing generation formation 'built into' the integument, and that this programme is switched on and off by hormones. Dermis, with a superficial layer of epidermal germinal cells can be obtained by trypsin treatment of lizard integument (Plate 1*e*) (Flaxman *et al.* 1968). If such material is cultured, no generations are formed but a stratified squamous epithelium of abnormal form is obtained (Plate 1*f*). Histological and ultrastructural examination of this epithelium show that it consists of but a single cell type. This observation presents two possibilities. First, generation formation is not a result of dermal influences upon the germinal layer as has been demonstrated for hair and feathers.

If the dermis were the controlling factor, then generation formation should occur if the culture is maintained: this does not happen. Second, we see that the germinal layer may show division and give rise to daughter cells, but in the absence of overlying tissues, only a single pathway of differentiation is observed, resulting in the stratified epithelium mentioned above. These results therefore suggest the possibility that the maturation of cells leading to generation formation is dependent upon a feedback mechanism originating from the epidermal elements above the stratum germinativum. Final substantiation of the intrinsic potential of the epidermis to undergo generation formation in the absence of either somatic factors or dermal material was obtained by splitting perfect resting stage epidermis from the dermis with EDTA (Plate 2*a*) (Flaxman *et al.* 1968). Cell proliferation leading to generation formation occurred within 10 days (Plate 2*b*). These *in vitro* data are summarized schematically in Fig. 3.

Ultrastructural studies of the lizard epidermis (Maderson *et al.* 1970) have demonstrated that *all* the constituent cells of an epidermal generation appear morphologically identical when they first arise from the stratum germinativum. It is some time after presumptive *Oberhautchen* and β-cells are laid down that they begin to show indications of their specialized patterns of protein synthesis. Logically, therefore, we would aruge that it must be just prior to this event that the feedback mechanism must be 'switched on'. Although the nature or mode of action of the feedback mechanism will remain in doubt for some time, we can still ask what activates it initially. There are two possible answers to this question. First, the germinal cells could possess a permanent internal programme which is switched on and off from time to time. If this were so, one might expect that presumptive cells of a particular type would show signs of their eventual fate as they leave the germinal population, or at least earlier than they do. The second, and simpler possibility, is that the germinal population consists of mother cells, the products of whose division have a general capacity for the synthesis of fibrous proteins. The specific pathways of synthesis associated with differentiation of clear layer, *Oberhautchen* and β-cells are the result of activation of a feedback mechanism. This activation could be brought about most easily by some change in the pattern of cell

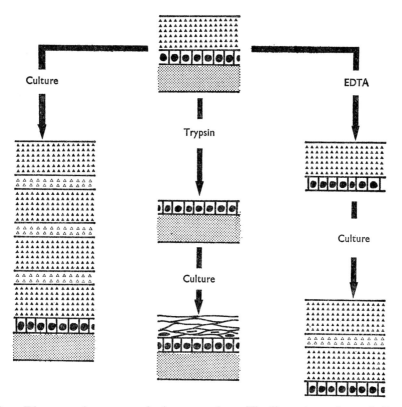

Fig. 3. Diagrammatic summary of culture procedures. The illustration at the top indicates a piece of skin with the epidermis in the perfect resting condition. ▲, *Oberhautchen*, β-, mesos and α-layers of a generation; △, lacunar and clear layer components of a generation; ●, Cells of stratum germinativum; ▦, dermis. For explanation see text.

production, i.e. associated with the mitotic rate. Evidence exists that sharp rises in mitotic activity occur as the outer epidermal generation is completed, and around the time of shedding. There is a possible third peak during the renewal phase as the presumptive mesos population is laid down, but this requires further study (Maderson, Pang & Roth, unpublished data).

As two of the mitotic peaks described above indicate the beginning and the end of the renewal phase, we see that the perfect resting condition must represent a period of low, or perhaps nil, mitotic activity. Is such a condition actively or passively maintained *in vivo*? Unpublished data from organ culture studies suggest an actively maintained inhibitory situation. It has been found that intact lizard skin does not *always* show generation formation *in vitro*. However, whenever explants are put up while the epidermis is in a perfect resting condition, changes *always* occur in the shape of the germinal cells and in the number of layers of immature cells between the germinal layer and the α-layer (Plates 1c, 2c). Long-term maintenance of such cultures indicate continuing low mitotic levels often eventually producing α-hyperplasia, but not generation formation (Plate 2d). These observations can only be explained by suggesting that culture media do not continue an inhibitory action maintaining the perfect resting condition. Apparently, however, they often lack the stimulant required to produce a high enough mitotic level to activate the feedback mechanism. Further experiments are in progress to attempt to identify the postulated stimulant, but it is most probable that it is a hormone.

The significance of the resting phase and levels of mitotic activity therein is indicated by all endocrinological investigations which have involved histological examination of epidermal biopsies. However, it should be noted that differences exist between species as to the way in which this part of the cycle is affected. In the studies of *Lacerta* cited earlier, where extended inter-slough periods, or total inhibitions were recorded, following thyroidectomy, germinal proliferation apparently continued leading to characteristic α-hyperplasia. Here then, apparently, the necessary conditions for lacunar tissue and clear layer formation (prerequisites for completion of the epidermal generation) were not obtained although continued α-layer maturation did take place. Similar results were obtained by Goslar (1958a) with the snake *Natrix* after treatment with T_4 or TSH. Experimental treatments leading to delayed or inhibited skin-shedding with Tokay (*Gekko gecko*) (Chiu & Phillips, 1970a–d; Chiu et al. 1967) are not accompanied by such continued germinal proliferation. In fact, in this species there is evidence that some regions of the epidermis demonstrate an abnormal 'lack' of new daughter cells due to depressed germinal activity. In defining the 'perfect resting stage', Maderson & Licht (1967) indicated that only a single layer of immature cells was detectable between the stratum germinativum and the base of the α-layer over the normal body epidermis of *Anolis* (the 'one-cell condition' (Maderson & Licht, 1967)). In a description of the specialized foot-pad of the same animal, Lillywhite & Maderson (1968) emphasized that such a 'reduced' condition was rarely seen in this position: usually the minimum number of immature cells was 2–5. A similar situation exists in the Tokay (Plate 2e). In hypophysectomized Tokays, whereas biopsies of unspecialized regions of the epidermis may show only a slightly greater flattening of germinal cells than that seen in the normal 'one-cell condition', this condition is also approached over the foot-pads (Plate 2f). This is not a completely abnormal condition which can only result from experimental interference, since such a condition of the foot-pads is occasionally observed in intact *Anolis* (Lillywhite & Maderson, 1968) and *Gekko* (Chiu & Maderson, unpublished data). Sometimes it is seen in individuals which have shown relatively increased inter-slough intervals; sometimes it accompanies cycles of normal length. It is of interest to note that the extreme reduction in the total thickness of the epidermis associated with a single layer of cells beneath the α-layer, and extreme flattening of the germinal cells, is characteristic of animals with long inter-slough periods (Maderson *et al.* 1969).

The dynamic events described and postulated in the preceding pages are shown schematically in Fig. 4. For convenience of discussion, certain terms descriptive of the entire cycle are shown on line *A* and their duration is shown on line *B*. The terms surrounded by cross-hatching indicate that part of the cycle where only whole or part of a *single* epidermal generation is represented above the stratum germinativum. We see from this figure that an animal which showed an inter-slough period of 8 days would have a mature α-layer at the time of shedding, would not therefore show a 'post-shedding resting stage', but would immediately complete the outer generation. A renewal phase would follow immediately. Evidence from a variety of studies suggest that long inter-slough periods, i.e. more than 18 days must be due to a protracted perfect resting condition. The duration of shedding, i.e. the physical loss of the mature generation from the body surface, is usually measured in hours and in those lizards which have been described as shedding 'piecemeal', retarded physical removal, rather than differential regional epidermal maturation, has been observed (Maderson *et al.* 1969). However, by the time the last remnants are removed from the body, the epidermis shows a 'perfect resting condition'.

Lines *C*, *D*, and *E* in Fig. 4 indicate features which have been discussed earlier. The feedback mechanism as indicated is defined as that mechanism which is responsible for triggering the development of the clear layer/*Oberhautchen* complex and subjacent β-layer.

None of the available endocrinological data suggests a specific pathway of hormone involvement in controlling the cycle. However, there is reason to believe that when and where the 'perfect resting condition' exists, it is an actively maintained inhibitory situation. The completion of the outer generation at the end of the resting phase and the switching-on of the

A Descriptive term for part of cycle under discussion	Perfect resting condition	Late resting phase	Renewal phase	Post-shedding resting phase
B Duration	0–n days	2–4 days	6–10 days	0–4 days
C Histogenic events	Probably no cell differentiation, but if any, it would lead to further enhancement of the α-layer of the outer generation	Completion outer generation	Final maturation of outer generation: maturation of part of inner generation	Maturation of α-layer of new outer generation
D Germinal mitotic activity	Nil or minimal	Sudden rise, possible second peak halfway through renewal phase, but definitely a peak around the time of shedding	Sudden rise, possible second peak	Steady drop
E Feedback mechanism	Inactive	Activated	? Deactivated or passively expires	Inactive
F Possible function of hormone milieu	Probably a steady balance which inhibits, or greatly diminishes, ability of germinal cells to divide	Change which stimulates mitotic activity	No known role	Change which may 'establish' the inhibitory milieu of the perfect resting condition

Shedding

Fig. 4. Schematic representation of events occurring in the epidermis throughout the sloughing cycle, indicating the points at which hormones may act. For explanation see text.

feedback mechanism would presumably result from a sudden change in the endocrine *milieu*. Evidence from organ culture studies (Maderson, unpublished data) suggests that the cell proliferation and differentiation patterns which occur during the renewal phase are not subject to hormonal control. If pieces of skin are explanted during the late resting phase, the next renewal phase *always* takes place. However, long-term culture of such pieces sometimes may not produce any further generation formation, although α-hyperplasia may be observed.

It seems likely that the physical act of shedding *in vivo* could provide a source for a feedback of information to the endocrine system leading to the establishment of the 'perfect resting condition'. Bruner (1907) reported raising of blood pressure at the time of shedding and such a phenomenon could act as a trigger. Unfortunately, no other data are available on this point. The possibility that the change in the hormonal *milieu* in the post-shedding period produces a preprogrammed 'perfect resting conditon' of *specific* duration is suggested by the fact that various surgical and/or therapeutic treatments must be begun at this time, if the time of the next shedding is to be affected. There are no data suggesting how such a mechanism might operate.

Although it is unsatisfactory to simply speak in terms of 'changes in', or 'stimulatory action of', a general hormone *milieu*, it is impossible to present the currently available data in any specific fashion. While we note that the snake 'feedback mechanism' is apparently activated by a low activity of thyroid hormone, where in lizards it is a consequence of high activity, this activity is part of a different cycle of action in different lizards. In *Lacerta*, the thyroid gland is maximally active throughout the resting phase, and, as the 'feedback mechanism' is switched on, the activity decreases (Eggert, 1935). On the other hand the thyroid activity of the gecko is low during the resting phase, and there is a sudden peak in activity when the feedback mechanism is activated (Chiu *et al.* 1967). Unfortunately, we do not yet know whether thyroxine can act directly on the epidermis, as prolactin has been shown to do, but then equally we do not know whether (*a*) prolactin usually acts in this fashion *in vivo*, or (*b*) whether it usually acts through or with the thyroid gland, or indeed (*c*) whether there are quite different endocrine relationships of the epidermis in various taxa.

Various aspects of the work reported by the authors in this review have been supported by grants to J.G.P. from The Nuffield Foundation and Sir Shiu-Kin Tang and to P.F.A.M. from N.I.H.

The authors are also grateful to Dr S. I. Roth and Dr B. A. Flaxman of Harvard Medical School for valuable discussions on the culture data.

REFERENCES

ADAMS, A. E. & CRAIG, M. (1950). Observations on normal, thyroidectomized, thiourea and thiouracil-injected *Lacerta agilis*. *Anat. Rec.* **106**, 263.

ADAMS, A. E. & CRAIG, M. (1951). The effects of anti-thyroid compounds on the adult lizard *Anolis carolinensis*. *J. exp. Zool.* **117**, 287–315.

ADAMS, A. E., KUDER, A. & RICHARDS, L. (1932). The endocrine glands and molting in *Triturus viridescens*. *J. exp. Zool.* **63**, 1–55.

BREYER, H. (1929). Über Hautsinnesorgane und Haütung bei Lacertilien. *Zool. Jahb.* (Anat.) **51**, 549–80.

BRUNER, H. L. (1907). On the cephalic veins and sinuses of reptiles, with descriptions of the mechanism for raising the venous blood pressure in the head. *Am. J. Anat.* **2**, 1–117.

CHIU, K. W. & LYNN, W. G. (1970). The role of the thyroid in skin-shedding in the shovel-nosed snake *Chionactis occipitalis*. (In the Press.)

CHIU, K. W. & PHILLIPS, J. G. (1967). Some aspects of the hormonal control of sloughing in the gecko *Gekko gecko* in *Proc. 3rd Asia and Oceania Congress of Endocrinology* Manila Phillippines, **2**, 298–302.

CHIU, K. W. & PHILLIPS, J. G. (1970a). The effects of hypophysectomy and of replacement therapy using TSH and ACTH on the sloughing cycle of the gecko, *Gekko gecko* L. (In the Press).

CHIU, K. W. & PHILLIPS, J. G. (1970b). The role of prolactin in the sloughing cycle of the gecko, *Gekko gecko* L. (In the Press.)

CHIU, K. W. & PHILLIPS, J. G. (1970c). The effect of prolactin on the sloughing cycle in the hypophysectomized/thyroidectomized gecko, *Gekko gecko* L. (In the Press.)

CHIU, K. W. & PHILLIPS, J. G. (1970d). The effect of thyroxine, prolactin, thyrotropin and thyroxine combined with prolactin on the thyroid activity and on the sloughing cycle of the gecko, *Gekko gecko* L. (In the Press.)

CHIU, K. W. PHILLIPS, J. G., & MADERSON, P. F. A. (1967). Observations on the role of the thyroid in the control of the sloughing cycle in the Tokay (*Gekko gecko*, Lacertilia). *J. Endocr.* **39**, 463–72.

DRZEWICKI, S. (1926). L'influence de l'extirpation de la glande thyroïde sur la mue du lézard (*Lacerta agilis*). *C. r. Séanc. Soc. Biol.* **95**, 893–5.

DRZEWICKI, S. (1927). Examen histologique des lézards thyroidectomisés. *C. r. Séanc. Soc. Biol.* **97**, 925–6.

DRZEWICKI, S. (1928). Über den Einfluss der Schilddrüsenextirpation auf die Zauneidechse. *Wilhelm Roux Arch. EntwMech. Org.* **114**, 2/3, 155–76.

EGGERT, B. (1933). Über die histologischen und physiologischen Beziehungen zwischen Schilddrüse und Haütung bei den einheimischen Eidechsen. *Zool. Anz.* **105**, 1–9.

EGGERT, B. (1935). Zur Morphologie und Physiologie der Eidechsen-Schilddrüse. I. Das jahreszeitliche Verhalten der Schilddrüse von *Lacerta agilis* L., *L. vivipara* Jacq. und *L. muralis* Laur. *Z. wiss. Zool.* **147**, 205–62.

EGGERT, B. (1936). Zur Morphologie und Physiologie der Eidechsen-Schilddrüse. III. Über die nach Entfernung der Schilddrüse auftretenden allegemeinen Ausfallserscheinungen und über die Bedeutung der Schilddrüse für die Häutung und für die Kaltestarre. *Zeits. wiss. Zool.* **148**, 221–60.

FLAXMAN, B. A., MADERSON, P. F. A., SZABO, G. & ROTH, S. I., (1968). Control of cell differentiation in lizard epidermis *in vitro*. *Devl. Biol.* **18**, 354–74.

GORBMAN, A. (1963). Thyroid hormones. In *Comparative Endocrinology*, vol. 1, pp. 291–324 (eds. U. S. von Euler and H. Heller). New York and London: Academic Press.

GOSLAR, H. G. (1957). Zur Beinflussung des Hautungsgeschehens bei der Ringelnatter (*Natrix natrix*) durch einen wasserigen Thymusextract. *Arzneimittel-Forsch.* **7**, 399–400.

GOSLAR, H. G. (1958*a*). Die Reptilienhaut als endokrines Testobjekt. *Endokrinologica.* **36**, 279–84.

GOSLAR, H. G. (1958*b*). Über die wirkung eines standardisierten Thymusextraktes auf die Häutungsvorgange und auf einige Organe von *Natrix natrix* L. *Arch. exp. Path. Pharmak.* **233**, 201–5.

GOSLAR, H. G. (1958*c*). Beiträge zum Häutungsvorgang der Schlangen. 1. Mitteilung: Histologische und topochemische Untersuchungen an der Haut von *Natrix natrix* L. während der Phasen des normalen Häutungszyklus. *Acta histochem.* **5**, 182–212.

GOSLAR, H. G. (1958*d*). Über die Wirkung verschiedener Sexualhormone auf die Häutungsvorgango der Ringelnatter (*Natrix natrix* L.). *Derm. Wochschr.* **137**, 139–46.

GOSLAR, H. G. (1964). Beiträge zum Häutungsvorgang der Schlangen. 2. Mitteilung: Studien zur Fermenttopochemie der Keratogenese und Keratolyse am Modell der reptilien Haut. *Acta histochem.* **17**, 1–60.

HALBERKANN, J. (1953). Untersuchungen zur Beeinflussung des Häutungszyklus der Ringelnatter durch Thyroxin. *Arch. Derm. Syph.* **197**, 37–41.

HALBERKANN, J. (1954*a*). Zur hormonalen Beeinflussung des Häutungszyklus der Ringelnatter. *Z. Naturf.* **98**, 77–80.

HALBERKANN, J. (1954*b*). Die Häutungsablauf der Ringelnatter unter Methylthiouracil. *Naturwissenschaften* **41**, 237–8.

KROCKERT, G. (1941). Kontinuierliche Hyperthyreoidisierung und Epiphysierung an *Python bivittaotus. Vitam. Horm.* **1**, 24–31.

LICHT, P. (1968). Response of the thermal preferendum and heat resistance to thermal acclimation under different photoperiods in the lizard *Anolis carolinensis. Am. Mid. Nat.* **79**, 149–58.

LICHT, P. & HOWE, N. R. (1969). Hormonal dependence of tail regeneration in the lizard *Anolis carolinensis. J. exp. Zool.* (In the Press).

LILLYWHITE, H. B. & MADERSON, P. F. A. (1968). Histological changes in the epidermis of the sub-digital lamellae of *Anolis carolinensis* during the shedding cycle. *J. Morph.* **124**, 1–23.

LYNN, W. G. (1960). Structure and functions of the thyroid gland in reptiles. *Am. Mid. Nat.* **64**, 309–26.

LYNN, W. G. (1970). The thyroid. In *The Biology of the Reptilia* (eds. A. d'A. Bellairs, C. Gans and E. E. Williams). New York and London: Academic Press. (In the Press.)

MADERSON, P. F. A. (1965*a*). The structure and development of the squamate epidermis. In *Biology of the Skin and Hair Growth*, pp. 129–53. (eds. A. G. Lyne and B. F. Short). Sydney: Angus and Robertson.

MADERSON, P. F. A. (1965*b*). Histological changes in the epidermis of snakes during the sloughing cycle. *J. Zool.* **146**, 98–113.

MADERSON, P. F. A. (1966). Histological changes in the epidermis of the Tokay (*Gekko gecko*) during the sloughing cycle. *J. Morph.* **119**, 39–50.

MADERSON, P. F. A. (1967). The histology of the escutcheon scales of *Gonatodes* (Gekkonidae) with a comment on the squamate sloughing cycle. *Copeia*, pp. 743–52.

MADERSON, P. F. A. (1968*a*). Observations on the epidermis of the Tuatara (*Sphenodon punctatus*). *J. Anat.* **103**, 311–20.

MADERSON, P. F. A. (1968b). The epidermal glands of *Lygodactylus* (Gekkonidae, Lacertilia), *Breviora* **288**, 1–35.

MADERSON, P. F. A. (1968c). On the presence of 'escutcheon scales' in the Euble-pharine gekkonid *Coleonyx*. *Herpetologica*, **24**, 99–103.

MADERSON, P. F. A., CHIU, K. W. & PHILLIPS, J. G. (1970). Changes in the epidermal histology during the sloughing cycle in the rat snake, *Ptyas korros* Schlegel, with correlated observations on the thyroid gland. (In the Press.)

MADERSON, P. F. A., FLAXMAN, B. A., ROTH, S. I. & SZABO, G. (1970). Ultra-structural contributions to the identification of cell types in the lizard epidermis. (In the Press).

MADERSON, P. F. A. & LICHT, P. (1967). Epidermal morphology and sloughing frequency in normal and prolactin-treated *Anolis carolinensis* (Iguanidae, Lacertilia). *J. Morph.* **123**, 157–72.

MADERSON, P. F. A. & LICHT, P. (1968). Factors influencing rates of tail regeneration in the lizard *Anolis carolinensis*. *Experientia* **24**, 1083–6.

MADERSON, P. F. A., MAYHEW, W. W. & SPRAGUE, G. (1969). Observations on the epidermis of desert-living iguanids. *J. Morph.* (In the Press).

McGINNIS, S. M. & DICKSON, L. L. (1967). Thermo-regulation in the desert iguana, *Dipsosaurus dorsalis*. *Science, N.Y.* **156**, 1757–9.

NOBLE, G. K. & BRADLEY, H. T. (1933). The relation of the thyroid and the hypophysis to the molting process in the lizard *Hemidactylus brookii*. *Biol. Bull. mar. biol. Lab., Woods Hole* **64**, 289–98.

PESONEN, S. (1954). 17-Ketosteroids and keratin production of the skin in verte-brates. *Arch. Soc. Zool. Bot. fenn. Vanamo* **16**, 1–15.

RATZERSDORFER, C., GORDON, A. S. & CHARIPPER, H. A. (1949). The effects of thiourea on the thyroid gland and molting behavior of the lizard, *Anolis carolinensis*. *J. exp. Zool.* **112**, 13–27.

ROMER, A. S. (1968). *Notes and Comments on Vertebrate Paleontology*. University of Chicago Press.

SCHAEFER, W. H. (1933). Hypophysectomy and thyroidectomy of snakes. *Proc. Soc. exp. Biol. Med.* **30**, 1363–5.

SEMBRAT, K. & DRZEWICKI, S. (1935). Influence de la glande thyroïde des Sela-chiens sur la mue des lézards. *C. r. Séanc. Soc. Biol.* **118**, 1599–601.

SEMBRAT, K. & DRZEWICKI, S. (1936). The influence of soelachian thyroid upon the moulting processes of lizards, with some remarks on the skin, the eyes, and the ultimo-branchial body of the thyroidectomized lizards. *Zool. Pol.* **1**, 119–69.

SZABO, E. & HORKAY, I. (1967). Effect of ultra-violet light on the epidermis. *Acta morph. hung.* **15**, 71–80.

WILHOFT, D. C. (1958). The effect of temperature on thyroid histology and survival in the lizard *Sceloporus occidentalis*. *Copeia* pp. 265–76.

PLATE I

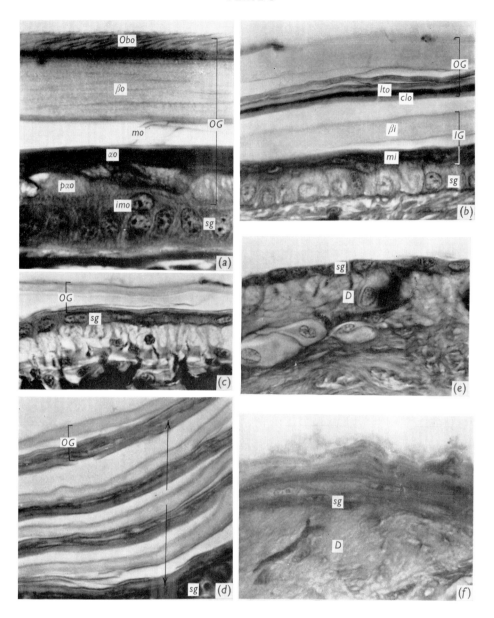

For explanation see p. 283

(facing p. 280)

PLATE 2

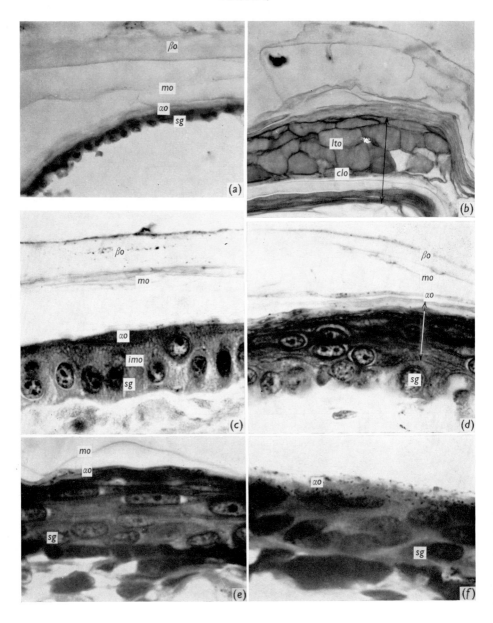

For explanation see p. 283

DISCUSSION

WRIGHT: Have you done any continuation work using hormones on *in vitro* systems and if so with what results?

MADERSON: I have tried this summer, but this is a rather fractious system. In the initial *in vitro* studies, everything went beautifully, the second season was not so good and this year nothing in the way of generation formation was seen, although mitoses continued. If the media conditions can be settled satisfactorily, we hope to try the effects of hormones.

WRIGHT: So you don't really have any information about the effects on the basal cell layer?

MADERSON: Not directly, no.

BERN: I would like to interject that before talking about a prolactin-thyroxine synergism that the importance of considering relative and absolute doses cannot be over emphasized. What may be a synergism at one dose level may be an antagonism if the relations between the two hormones are modified. This is extrapolating from other systems; however, I think Dr Etkin you would agree that this is something one should be concerned about?

ETKIN: The problem we face repeatedly in dealing with hormones is, of course, a question of species specificity and the question of whether we are using prolactin is significant and I think can only be approached by inducing the animals' pituitary to enter into the arrangement on the theory that it would produce an appropriate prolactin. I don't think we worry too much about that with the thyroid.

MADERSON: I think that in that particular experiment, two dose levels of thyroxine alone were tried, which gave quantitatively different results, so that the action with prolactin looks like synergism.

PHILLIPS: As far as that is concerned I think the higher dose level of thyroxine did in fact increase the frequency in the presence of the same dose of prolactin.

ELIZABETH JOHNSON: I was very struck by the similarity between your epidermal system in the reptiles and the hair follicle. In both we have an intrinsic rhythm, which is modified by systemic factors. I was interested that you should suggest that the intrinsic rhythm in the reptilian epidermis may be controlled by an inhibitor mechanism. Our evidence on the hair follicle does not support an inhibitor mechanism. I wonder if you have any direct evidence for an inhibitor in your animals.

MADERSON: If you take perfect resting stage epidermal material it is characterized by a very flat basal cell population, with only one layer of living cells between the stratum germinativum and the base of the mature α-layer. If you put this material into organ culture intact it does not always show generation formation. However, it never maintains that morphological condition, the germinal cells always divide, the cells change in shape and you get two to three layers of living cells where there was only one before. Now I would argue that the fact that *some* change and *some* divisions *always* take place immediately when you put the material *in vitro* after taking it from an *in vivo* situation suggests that an active inhibitor has been removed.

Now the next problem is, if it is showing any cell division at all why doesn't it form a generation. Our thesis at the moment is that generation formation can only take place consequent to a certain rate of mitotic activity. We would suggest that as far as a hormonal control is concerned here, there may be one hormonal *milieu* which is controlling the inhibition, which is replaced by another, which stimulates mitosis. In lizards this might be thyroxine, but no general hypothesis is possible at this time. I would like to take your other point about the hair follicle. There is a fundamental difference here between the pattern of hair growth and this system. To my knowledge no one has ever shown any direct developmental relationship between the mature hair and the new one that is growing. When the follicle 'grows down' ready for the new growth phase it has got nothing directly to do with the old hair, and I think the same system might apply to feathers. However, in this system you must have the mature element above to literally and metaphorically fit the other new system in, so that there is a mechanical and direct tissue relationship. Absence of dermal influences on squamate epidermal activity also suggests that the situation is different from hair and feathers.

BERN: Have you fully eliminated the possible influence of the dermis if one conceives of a possible inhibitory effect that may be coming from the dermis to the epidermis in regards to subsequent generation development?

MADERSON: When I speak of dermal influence I am thinking of its inductive capacity, not the hormonal content of the tissues. I think that even if we only had the intact system growing in culture I would have argued on grounds of pure logic that every other dermal influence which has ever been demonstrated in either an embryonic or adult tissue has always been a one-shot affair. You take a piece of dermis or a little clump of dermal cells—dermal papilla—it can do something, but it only does it once. The moment you have four generations piled up on top of one another, this is a cyclic system. Now when I saw this when we started this work, I said, 'If anyone can suggest in what never-never land in the dermis such a mechanism could reside, you have got to have some sort of pattern of replication of some description and there are no special cells that would suggest that this is true.'

EBLING: I am rather uncertain about the validity of your comment that the dermis is only concerned with 'once for all' inductive processes. There are great changes in the dermis during the cycle of hair growth, and the dermal changes run in parallel with the follicular changes. Which one causes the other we do not know, but there must be cyclic interactions between the dermis and epidermis.

MADERSON: Yes. I take your point.

DESCRIPTION OF PLATES

PLATE 1

(a) The epidermis of the snake *Ptyas korros* just prior to the 'perfect resting condition'. It consists of an *incomplete* outer epidermal generation (OG) above a stratum germinativum (sg). The generation has a serrate *Oberhautchen* (Obo) which is the outermost portion of the β-layer (βo). This latter artifactually splits from the subjacent tissues during preparation providing a clear indication of the strands which comprise the mesos layer (mo). The α-layer (αo) is now well developed and possibly finished, although the chromophobic cells immediately beneath it (pαo) may represent the final constituents of this tissue. The single layer of cells immediately above the stratum germinativum (imo) may contribute to the α-layer or they may form the first component of the lacunar tissue which forms at the end of the resting phase. Haematoxylin and eosin. × 800.

(b) The epidermis of the lizard *Anolis carolinensis* 1–1½ days before shedding. The outer generation (OG) will be lost from the body, and the present inner generation (IG) will take its place. The spinules representing the mature *Oberhautchen* in this species, and the mould-like inter-digitations with the clear layer (clo) can only be detected at high magnifications. Note the characteristic chromophobic, columnar appearance of the germinal cells which accompanies the laying-down and final maturation of the cells of the new mesos layer (mi). Haematoxylin and eosin. (lto, lacunar layer; clo, clear layer; βi, β layer of inner epidermal generation). × 800.

(c) Perfect resting condition of the epidermis of *Anolis carolinensis*. There is an incomplete outer generation (OG) above a stratum germinativum (sg). Note the characteristic flattened form of the germinal cells at this time. Haematoxylin and aniline blue/orange G. × 800.

(d) After 49 days in culture, a piece of intact skin similar to that shown in fig. (c) has continued to show differentiation of generations. The original outer generation is now complete (OG) and three similar units have been formed beneath. All the tissues between the arrow marks, above the stratum germinativum (sg) have resulted from proliferation and differentiation *in vitro*. Haematoxylin and aniline blue/orange G. × 900.

(e) A piece of skin from *Gekko gecko* similar to that shown in fig. (c) has been immersed in trypsin and subjected to mechanical splitting: the outer epidermal generation has been removed, leaving the germinal population (sg) on top of the dermis (D).

(f) After 10 days in organ culture, a piece of skin similar to that shown in fig. (e) has continued to show epidermal proliferation and cell differentiation. However, an epidermal generation has not formed, and the stratified, squamous epithelium has light and electron-microscopic characteristics indicating a similarity to α-tissue. Haematoxylin and eosin. (sg, Germinal population; D, dermis.) × 800.

PLATE 2

(a) A piece of skin from *Iguana iguana* with the epidermis showing a perfect resting condition similar to that shown in Plate 1(c) has been treated with EDTA so that the epidermal tissues could be separated from the subjacent dermis. Abbreviations as in Plate 1(a). Haematoxylin and eosin. × 700.

(*b*) After 10 days in culture, the epidermal strip shown in fig. (*a*) has undergone proliferation and differentiation. The original outer generation has been completed by the laying-down of lacunar (*lto*) and clear layer (*clo*) components, and beneath are the outer parts of a new inner generation. All the cells and tissues between the arrows have been formed *in vitro*. Haematoxylin and aniline blue/orange G. × 500.

(*c*) A piece of skin from *Dipsosaurus dorsalis* which was put up in culture when the epidermis was in a perfect resting condition and resembled Plate 1 (*c*) exactly. After only 3 days in culture there are 2–3 layers of cells (*imo*) above the stratum germinativum (*sg*). The germinal cells are much changed from their previous extreme flattened condition and occasional mitotic figures are seen throughout the biopsy. Haematoxylin and eosin. (*mo*, Mesos layer; *αo*, *βo*, alpha, *β* layer.) × 800.

(*d*) A piece of skin from *Anolis carolinensis* which was similar to that shown in Plate 1 (*c*) after culture for 20 days. Generation formation has not occurred although proliferation and maturation is obvious. The α-layer (*αo*) is much thicker than in the original perfect resting condition and there is a sequence of differentiating cells all of which resemble presumptive α-cells in various stages of maturation (arrow). Haematoxylin and eosin. (*βo*, *β* layer; *mo*, mesos layer; *sg*, stratum germinativum.) × 900.

(*e*) The specialized foot-pad epidermis of an intact *Gekko gecko*, showing the equivalent of a 'perfect resting condition' for this region. Note that the α-layer (*αo*) is fairly thin and there are 5–7 layers of immature cells between it and the stratum germinativum (*sg*). Haematoxylin and eosin. (*mo*, Mesos layer.) × 900.

(*f*) The specialized foot-pad epidermis of a hypophysectomized *Gekko gecko*, biopsy taken at the end of an extended resting phase. The α-layer (*αo*) is very much thicker than in the normal animal, it is heavily pigmented, and there are fewer layers of immature cells between it and the stratum germinativum (*sg*). Haematoxylin and eosin. × 900.

THE CONTROL OF ACTIVITIES IN INSECT EPIDERMAL CELLS

By M. LOCKE

INTRODUCTION

One of the most important characteristics of insects is that they have never evolved the means for co-ordinating the development of masses of cells in three dimensions. Except in the CNS, complexity is attained with epithelial sheets and cylinders only one cell thick, giving an adult insect an organiza-

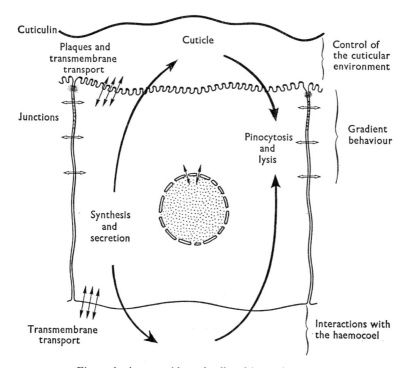

Fig. 1. An insect epidermal cell and its environment.

tion almost as simple as that of a vertebrate gastrula. In particular, the body wall is composed only of the integument; that is, a single layer of epidermal cells and the cuticle which they secrete. The body form is determined almost solely by the epidermis and the study of its development holds a central position in insect biology. In a sense, insects are 'epidermal organisms' in the way that vertebrates are 'mesodermal organisms', but the problems of development have remained those posed by cells in simple epithelia.

Three environments influence an insect epidermal cell (Fig. 1). The basal surface is the port of entry from the haemocoel for hormones and precursors which influence the sequence of events comprising the moult/intermoult cycle. At the apical face the epidermis secretes the cuticle and controls the microenvironment within it bounded by the cuticulin layer. Laterally the epidermal cells interact to give gradients and polarized fields important in determining patterns and in co-ordinating growth.

The role of these three environments in influencing the growth and development of an insect epidermal cell is explored below in experiments on the 5th-stage larva of *Calpodes ethlius* (Lepidoptera, Hesperiidae) and on the blood-sucking bug *Rhodnius prolixus*. In this paper I am concerned to show the need for studying the complete environment of a cell and all of its activities. A cell responding to and secreting molecules at its surface knows no distinction between cell biology and endocrinology.

HORMONES AND THE RESPONSES OF THE EPIDERMIS

The problem

The classical work on insect development (Wigglesworth, 1934; Williams, 1952) has shown that moulting (i.e. the formation of a new cuticle with the resorption and ecdysis of the old one) is initiated by the moulting hormone ecdysone which appears after the prothoracic glands have been activated by a neurosecretion from the brain. The moulting hormone is the cue for the operation of intrinsic factors controlling the precise co-ordination of the sequence of interdependent syntheses comprising the moult in the responding tissue of an insect. There may be secondary cues in particular instances (see, for example, Beck, Shane & Garland, 1969; Cottrell, 1964; Fraenkel & Hsiao, 1965), but in general, moulting only requires initiation for it to proceed to completion. Initiation may be prolonged but moulting is triggered in the sense that it needs no further outside stimulus after initiation. This intrinsic control of the moulting sequence is in marked contrast to the continuous control of events in the intervals between moults which make up most of the preimaginal life. Studies on the epidermis of *Calpodes* show that it is active in secreting cuticle during the intermoult periods as well as at moulting (Locke, 1965*a*, 1967*a*, 1970). The type of synthesis undertaken by the epidermis may be similar during the moult and intermoult, but the intermoult differs in that (1) the activities are not triggered by the moulting hormone, (2) they require the constant presence of a factor from the head as well as the prothoracic glands, and (3) they occur over a protracted period. In the 5th-stage epidermis of *Calpodes* both moult and intermoult are preceded by a phase of cellular preparation, characterized by an elevated rate of RNA synthesis (Locke, Condoulis & Hurshman, 1965).

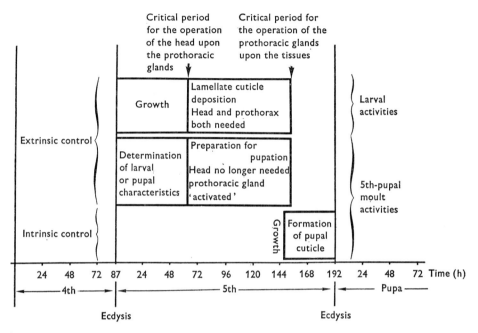

Fig. 2. The main events in the epidermis of *Calpodes* during the last larval stadium. Larval syntheses are concurrent with preparation for the intrinsically controlled sequence of pupal syntheses but are controlled differently. Larval syntheses require the head and prothorax while preparation for pupation ceases to need the head after M + 66 h. Growth refers to periods of elevated RNA synthesis with a change in morphology appropriate for later activity.

The important features of one moult/intermoult in *Calpodes* are shown in Fig. 2, which anticipates some of the results to be presented. The tissues have two sorts of response to the head and prothoracic gland complex. The dichotomy may be introduced by a dual action of the prothoracic gland, by neurosecretions acting directly on the epidermis coincidentally with the moulting hormone, or by a subtle interplay of time and concentration dependence upon ecdysone coupled with changes in the responding system. However, before we can experiment on these more complex control mechanisms, we need to know the occupations of the epidermal cells in more detail. We need to know exactly when they grow and when they are active in synthesis for export relative to critical changes in the hormonal *milieu*.

Critical events in the 5th stadium of Calpodes

Calpodes ethlius Stoll (Lepidoptera, Hesperiidae) has been a useful experimental insect for studying wax secretion (Locke, 1960, 1965 a), oenocytes (Locke, 1969 a) and for experiments on tracheae (Locke, 1958), where the transparent integument facilitated surgery. Although it is a continuous feeder the precise timing of its development makes it particularly suitable for the study of transient phenomena requiring carefully timed sequences. The 5th stadium lasts for 192 ± 13 h. Morphological changes mark the progress from one event to the next within the stadium and increase the accuracy of staging still further. This knowledge of the timing allows precise correlation between changes in the epidermis and the hormonal milieu.

Larvae of various ages were ligated around the head or around the prothorax to determine the time for the action of the brain neurosecretory complex and the prothoracic glands upon moulting. By Moult $(M) + 66$ h the prothoracic glands no longer need the head to ensure pupation later. After $M + 156$ h the prothoracic glands are no longer needed by the other tissues for pupation. These critical times divide the stadium into three phases of development. Hormonal cues are required during the first two phases but from $M + 156$ to $M + 192$ h pupation is controlled intrinsically from within the cell.

The epidermis while under extrinsic control

Changes in the epidermis

Lamellate endocuticle is deposited continuously from ecdysis until the formation of ecdysial droplets just before $M + 156$ h when the cells become independent of the prothoracic glands. The rates of deposition were estimated from electron micrographs of the integument prepared from a series of larvae of all ages through the stadium. Little cuticle is deposited at first, only one lamella forming every 3 h, but at $M + 66$ h there is an abrupt switch to the fast rate of deposition of 1 lamella every 10 min. About 90% of the lamellate cuticle deposited during the 5th stadium is laid down during this central phase of synthesis for export. The switch in rate of synthesis at the time the prothoracic gland is activated (i.e. when it no longer needs the brain in order to initiate pupation 90 h later) is typical of the activities of most tissues, e.g. wax secretion by the wax gland, blood protein accumulation from the fat body (Collins, 1969), lipid synthesis and storage in the fat body, etc., which all have an elevated rate of synthesis for export from $M + 66$ to $M + 156$ (Locke, 1970). This 90 h is a phase for larval syntheses.

Electron-microscope observations and autoradiographic studies of the incorporation of [³H]uridine show that the epidermis prepares for its phase of synthesis by a phase of growth characterized by increased RNA synthesis and the formation of ribosomes. RNA synthesis increases after ecdysis, rises to a peak by about $M + 36$ h and falls again by $M + 66$ h to the low level maintained through the synthetic phase. Although there are numerous ribosomes, the rough-surfaced endoplasmic reticulum (RER) is not extensive and the Golgi complexes are small.

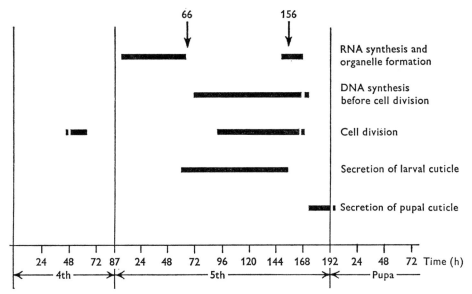

Fig. 3. Summary of the changes in the epidermis secreting surface cuticle in *Calpodes* during the 5th stadium. The secretion of larval cuticle refers to the time when there is an elevated rate of deposition of lamellate larval endocuticle. (after Locke, 1970).

Nuclear DNA synthesis and cell division do not occur during the growth phase but are concurrent with the deposition of lamellae in the synthetic phase. Autoradiographs after the incorporation of [³H]thymidine show that nuclear DNA synthesis begins at $M + 66$ h and reaches a peak by the time the prothoracic glands are no longer needed at $M + 156$ h. Counts of colchicine arrests in Feulgen-stained whole mounts show cell division beginning about 24 h after DNA synthesis and reaching a peak at about $M + 156$ h. Mitoses in the epidermis occur coincidentally with the secretion of larval cuticle. This is a remarkable finding in several ways. In vertebrate tissues the normal sequence of development is for a critical mitosis to precede the formation of a cell type devoted to specific syntheses (Bischoff & Holtzer, 1969; Ishikawa, Bischoff & Holtzer, 1968). This gave

19   BME 18

rise to the dogma that mitosis is usually no longer possible in cells which have differentiated (e.g. Holtzer, 1961). The dogma is clearly inapplicable to insect tissues which undergo repeating cycles of cell division and the synthesis of specific proteins.

Fig. 3 summarizes the changes taking place in the epidermis during the 5th stadium and displays the complexity of the problem of control. How many of these events depend upon the interplay of hormones and how much of the complexity of the response is due to progressive changes in the responding system itself?

The control of larval syntheses

While the brain is acting upon the prothoracic glands from $M + o$ h to $M + 66$ h the epidermis is preparing for intermoult larval syntheses. Intermoult syntheses and cell division begin when once the prothoracic glands have been activated and no longer require the brain to induce pupation. This suggested that growth for intermoult syntheses and the syntheses themselves might be, like pupation, controlled directly by the prothoracic glands. The effects of decapitation after $M + 66$ h show that intermoult syntheses are not triggered like pupation but require the constant presence of the head and prothorax for their continuation. Wax synthesis and lamellate cuticle deposition stop after decapitation although pupation still occurs later. Wax synthesis can be induced in decapitated larvae if the corpus allatum/corpus cardiacum complex is implanted with activated prothoracic glands. Although the prothoracic glands are undoubtedly involved in the control of intermoult syntheses, the epidermis responds in a different way from the induction of moulting.

An ultrastructural study of the prothoracic glands has been made in an attempt to throw light on the dichotomy of this control. The prothoracic glands change markedly in structure after $M + 66$ h when the brain is no longer needed for pupation. An abundant smooth endoplasmic reticulum appears with confronting cisternae around mitochondria deep within the cell and extending to confront plasma membrane pockets closer to the surface (Plates 1–3). The structure is similar to that in silkmoths but microvesicular mitochondria of the sort reported by Beaulaton (1968) have not been seen. This characteristic structure (M–SER–PM, Fig. 4) is maintained from $M + 66$ to $M + 156$ h when there is a phase of autolysis involving isolation membranes like those in the fat body (Locke & Collins, 1965). From what is known of vertebrate steroid synthesizing tissues the M–SER–PM structure would be appropriate for the terminal syntheses in the formation of a strongly hydroxylated steroid like ecdysone. Steroid hydroxylating enzymes are known to occur in mitochondria.

The normal M–SER–PM structure of the prothoracic glands present while preparations for pupation are coincident with intermoult larval activities, has been compared with the structure seen after decapitation at M + 80 h when intermoult syntheses stop but pupation is not blocked. If decapitation results in no qualitative change, we should expect that a factor

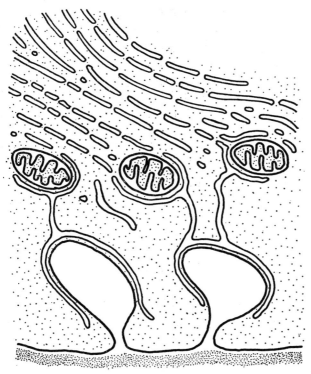

Fig. 4. The characteristic structures of the prothoracic glands during the central phase from M + 66 to M + 156 h when they are believed to be active in influencing moult and intermoult activities. The smooth endoplasmic reticulum confronts mitochondria, and cisternae also confront numerous pockets invaginated from plasma membrane (M–SER–PM complex). Compare with Plates 1, 2 and 3.

from the head might be acting with the prothoracic glands directly on the epidermis to control the intermoult larval syntheses. If, on the other hand, decapitation causes a change in the prothoracic glands, then we might tentatively correlate the lost structures with those controlling intermoult larval syntheses. Although pupation would still have taken place without the head, the prothoracic gland reverted to the structure seen before activation by the brain. Thus although the characteristic M–SER–PM structure of the prothoracic gland would be appropriate for ecdysone synthesis, it is most clearly related to intermoult larval events. A transient growth of SER in time to induce pupation could be overlooked, so that we cannot say that the M–SER–PM structure is not related to moulting, but the results are consistent with it being concerned with intermoult events separately from any involvement in moulting. We may conclude that the control of the dual response of the epidermis is probably present at the level of the prothoracic glands. In this connection it is of particular interest that different naturally occurring ecdysones have been shown to have diverse effects on development (Oberlander, 1969). Another way in which even the same ecdysone could vary its effect would be through the specificity of uptake of a carrier molecule. Vertebrate steroid hormones are known to be attached to proteins (e.g. Mills, 1960). Protein uptake has been shown to take place in nearly all insect tissues (Locke & Collins, 1967). If ecdysones are attached to proteins it could explain the observation of Ohtaki, Milkman & Williams (1968) that the ecdysone deactivation mechanism requires whole cells. There are also indications that the oenocytes could have a role in steroid metabolism. The oenocytes in *Calpodes* are of two sorts; one has a cycle of growth and synthesis of SER appropriate for functioning during the synthetic phase from M + 66 to M + 156 h, the other sort undergoes a cycle of growth as if its activity were relating to moulting. Both have a structure appropriate for sterol metabolism (Locke, 1969a). It could be that oenocytes as well as the prothoracic glands have a hand in changing plant steroids in the diet to moulting hormones.

Preparation for the intrinsically controlled sequence at pupation

The epidermis is engaged in larval syntheses from M + 0 to M + 156 h but this is also the time when moulting and the character of the moult are being determined. The critical period for the operation of the head at M + 66 h sharply separates the first phase during which larval or pupal characteristics are determined from the second phase when cell division determines the number of cells available for forming the pupa. The state of determination of the epidermis towards larval or pupal characteristics has been tested throughout the stadium by causing premature moulting with an injection of ecdysterone (Wang, 1970, unpublished observations). From M + 0 to M + 66 h an appropriate low level of ecdysone causes a larval moult about 24 h later. From M + 66 to M + 156 h a similar injection causes the formation of a pupa. Higher doses cause death before M + 66 h and pupation afterwards. The same results are obtained in isolated abdomens. There is a change in competence of the epidermis to react to ecdysone at M + 66. The epidermis is responding to the absence of juvenile hormone and determining that it will moult to a pupa at the same time as it is preparing for larval syntheses and the brain is acting on the prothoracic glands.

Development is stepwise. The epidermis first decides to moult in the pupal way and only then divides to create the cells needed to form the pupa. It is perhaps significant that the switch from larval–larval moult to larval–pupal moult may only occur when at least some cells have had the opportunity to divide, giving support to the notion of critical mitoses (Ishikawa *et al.* 1968); that is, mitoses which are obligatory before a change in the synthesis of specific proteins becomes possible.

The complexity of the sequence of epidermal activities during the first part of the 5th stadium in *Calpodes* shows us that classical work upon ecdysone and moulting is but a first step in our understanding of insect development. The problem outlined here concerns the control of the dual response of the epidermis in carrying out overt larval syntheses concurrently with less obvious preparations for pupation. Even allowing for changes in competence of the responding system it seems unlikely that the parallel sequence is controlled only by levels of one ecdysone determined by one brain thoracotropic hormone and one juvenile hormone.

The epidermis under intrinsic control

For the last 36 h before ecdysis to the pupa the epidermis undergoes a sequence of changes without further hormonal cues. The transition between extrinsic and intrinsic control is marked by the deposition of ecdysial droplets instead of lamellate cuticle and the separation of the cuticle from the epidermis by the formation of moulting gel. The layers of the new cuticle are then deposited and the old cuticle digested before ecdysis. The details of cuticle structure and deposition have been described elsewhere (Locke, 1966, 1969*b*) but three features are of interest in the context of this essay.

The similarity between some processes under extrinsic and intrinsic control

The secretion of lamellate cuticle occurs as an extrinsically controlled larval activity and during the formation of the pupa as part of an intrinsically controlled sequence. If these activities are identical, then presumably the terminal parts of the chains of command controlling them are also the same. The study of the extrinsic control may then lead us to corresponding links within the cell, or conversely. By such comparisons we may be able to arrive at the full significance of our knowledge of hormone action and speculate upon the evolution of control mechanisms.

The alternation in involvement of the plasma membrane plaques and Golgi complexes in the secretion of cuticle

Although a static view of the classical 'layers' of cuticle makes them appear complicated, ultrastructural studies during development show that they can be resolved as overlapping time sequences of two basic processes, those involving the discharge of vesicles of cuticle precursor and those also connected with sites (plaques) on the plasma membrane concerned with the assembly and/or transmembrane transport of cuticular components. Cuticle composition is controlled through these two processes and their interactions. Any discussion on the hormonal control of cuticle deposition must take into account the different chains of command within the cell which we should expect of the two processes.

One way in which the type of cuticle varies is through qualitative changes in the contents of vesicles discharging at the apical surface. The precursors for the protein epicuticle are clearly different from the ecdysial droplets. The control of vesicle composition would presumably be similar to that in better studied cells involving transcription and ribosomes on the RER.

The other way the cuticle varies is through the activity of the plasma membrane plaques. These are areas of the apical plasma membrane differentiated by their density and are often at the tips of microvilli. Cuticulin arises above the plaques and strands of material connect them to its growing edges. Microfibres making up the lamellae also arise at plaques, although the functional difference should make us suspect differences in details of the structure. The control of the activity of the plaques may be complicated, involving on the one hand their formation and maintenance, and on the other the cuticular precursors traversing the cell sap. Free polysomes and internal protein composition are important, for even at the height of lamellate cuticle deposition the epidermis does not have much RER or well-developed Golgi complexes.

The alternation in dominance between plaques and vesicle discharge in cuticle formation is most clearly seen in the sequence at the end of the 4th stadium, beginning with the deposition of 4th instar larval lamellae (plaques) followed by the discharge of ecdysial droplets (vesicles), deposition of 5th instar cuticulin (plaques), formation of the protein epicuticle (vesicles) and finishing with the deposition of pre-ecdysial lamellate cuticle (plaques). Both processes overlap periods of extrinsic and intrinsic control. The mechanism accounting for the switch from lamellate cuticle formation (plaques) to ecdysial droplet formation (vesicles) without evidence for a qualitative change in hormonal milieu is presumably related to the mechanism for the switches occurring during the intrinsically controlled part of the sequence.

The importance of the cuticulin layer in defining the cuticular environment

The first step in the intrinsically controlled sequence is the demarcation of a new apical extracellular environment to replace that limited by the old cuticle about to be digested. This is accomplished by the secretion of the cuticulin layer, a thick dense membrane which completely invests an insect even to the tips of its tracheoles. Outside the investment digestion of the old cuticle proceeds without damage to the cells. Within it the epidermis can control the environment in a way appropriate for the transport of precursors and the reactions which result in the formation of the new cuticle. One way in which the epidermis can control this environment is the subject of the next section.

THE CONTROL OF THE CUTICULAR ENVIRONMENT

The epidermis is influenced not only from the basal haemocoel face but also responds to changes in the cuticular environment on the apical face. Wigglesworth (1948) found that the epidermis reacted to surface abrasions to the cuticle, which he inferred was 'alive'. The cuticle is also alive in the sense that the epidermis may control the composition of some of its components through their continuous secretion and resorption. One aspect of this control has been studied by following the turnover of a peroxidase occurring naturally in the cuticle.

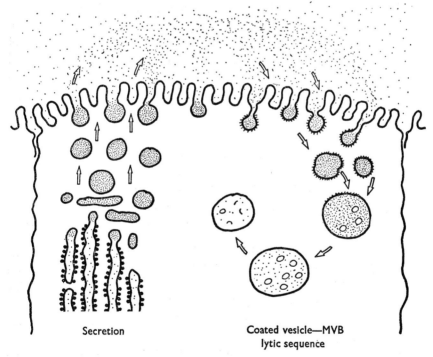

Secretion   Coated vesicle—MVB
                             lytic sequence

Fig. 5. One way in which an epidermal cell controls the composition of the apical environment within the cuticle. MVB, Multivesicular body.

Each spine on the proleg of *Calpodes* is formed by one elongated cell secreting a cylinder of cuticle around itself (Locke, 1969c). At one stage the hardening of the cuticle involves a peroxidase, perhaps in the mechanism put forward by Andersen (1966). At this time the peroxidase can be traced from the cisternae of the RER through Golgi complexes and into secretory vesicles which discharge apically into the cuticle. Peroxidase can also be followed back into the cell through pinocytosis in coated vesicles which carry their contents to the apical multivesicular bodies for lysis. The way peroxidase is introduced to and removed from the cuticle suggests a general mechanism by which the epidermis could control the environment above its apical face with temporal precision. The spatial distribution is also well defined, for macromolecules like peroxidase only spread a few microns laterally in the cuticle above the cell which secretes it. The general hypothesis is summarized in Fig. 5. It may explain the great sensitivity of insects to topically applied reagents such as juvenile hormone and insecticides.

THE RELATIONS BETWEEN EPIDERMAL CELLS

The integument of insects is segmented on a gross scale in the body segments and more finely in the appendages. The epidermis within each segment is arranged in a gradient, the nature of which involves the study of the transport of information from one cell to another through the lateral faces (Locke, 1959, 1967b). The experiments which gave rise to the hypothesis of a segmentally repeating gradient (Fig. 6) involved the interchange

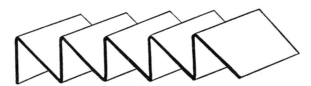

Fig. 6. Conceptual view of the arrangement of cells in the segmented integument of insects determined from their responses to grafts transposed laterally and axially.

of grafts between different positions on the body segments and the appendages and the study of the host graft interactions. In their response to a change in position within a segment the cells differ axially but are like one another transversely with only minor differences between segments. The gradient concept has been confirmed and amplified by Lawrence (1966) and Stumpf (1966).

The gradient properties of the epidermal cells in their responses to one another may be related to the co-ordination of growth. In the tracheal

PLATE I

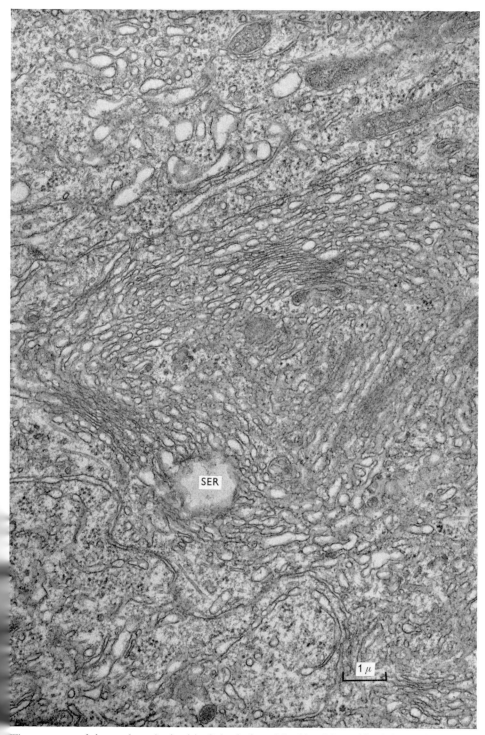

The structure of the prothoracic gland in *Calpodes* from M + 66 to M + 156 h while it is believed to influence intermoult activities and the preparation for pupation. Many swirls of smooth endoplasmic reticulum (SER) develop at this time.

(*facing p.* 296)

PLATE 2

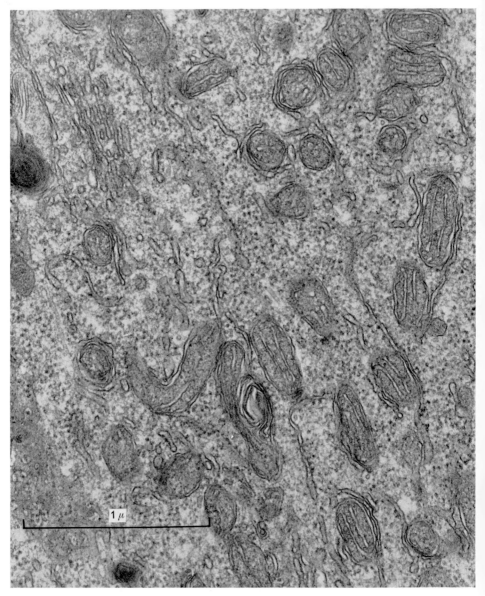

The structure of the prothoracic gland in *Calpodes* from M + 66 to M + 156 h while it is believed to influence intermoult activities and the preparation for pupation. Some of the smooth endoplasmic reticulum is juxtaposed to mitochondria.

PLATE 3

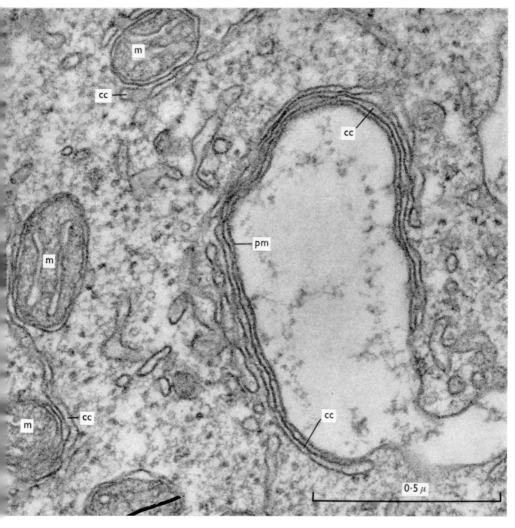

e structure of the prothoracic gland in *Calpodes* from M+66 to M+156 h while it is believed to influence
ermoult activities and the preparation for pupation. At the surface of the cell cisternae of the smooth
ioplasmic reticulum confront the plasma membrane where it invaginates in pockets (M-SER-PM complex).
, Plasma membrane of a pocket leading to the surface; *cc*, confronting cisternae of smooth endoplasmic
culum; *m*, mitochondria.

epidermis there is a polarized conduction of information about how much to grow, and grafts of the surface epidermis apposed to host cells alien in their axial position may fail to grow. In *Galleria* the relative position in the gradient may determine qualitative features of the cuticle (Stumpf, 1968). The mechanism inducing these responses has yet to be studied, but Loewenstein's work on the passage of molecules as large as resorcinol from cell to cell in Dipteran salivary glands may give a clue (Loewenstein, 1968). The dye is presumed to pass through septate desmosomes of the sort also found in epidermal cells (Locke, 1965*b*). The gradient can be thought of as a system of selectively interconnected boxes, each cell being a relay unit with interpretation and perhaps limited amplification of the information it receives. If the gradient does involve the selective movement of information carrying molecules through the septate desmosomes, then its study bridges the gap between the intrinsic and extrinsic controls on epidermal activities already discussed.

CONCLUSION

In this discussion I have outlined the problems posed by the activities of insect epidermal cells and the way they are influenced by their immediate environment. Although insect endocrinologists have been successful in determining the chemistry of two hormones important in development (the juvenile hormone, Roller, Dahm, Sweely & Trost, 1967; ecdysone, Siddal, Cross & Fried, 1966) we are limited in our understanding of their mode of action by our lack of detailed knowledge of the changes in the responding system. The objective is to elucidate the chain of command from an activity such as cuticle formation through the cell to the hormone. The approach I have taken starts with the responses of the cell, what it does and when. The epidermis leads a double life, covertly preparing for pupal syntheses while openly synthesizing larval products. The larval syntheses are controlled separately and are not obligatory for pupation. The converse may not be true; in the experiments so far, larval syntheses have only been maintained under conditions resulting in pupation. While it leads this double life the prothoracic glands are necessary to elevate it to a state of independence after which it can pupate without further hormonal assistance.

The work was generously supported by a grant from the National Institutes of Health, GM 09960.

REFERENCES

ANDERSEN, S. O. (1966). Covalent cross-links in a structural protein, resilin. *Acta. physiol. scand.* **66**, 1–81.

BEAULATON, J. A. (1968). Modifications ultrastructurales des cellules sécrétrices de la glande prothoracique de vers à soie au cours des deux derniers âges larvaires. I. Le chondriome, et ses relations avec le reticulum agranulaire. *J. Cell Biol.* **39**, 501–25.

BECK, S. D., SHANE, J. L. & GARLAND, J. A. (1969). Ammonium-induced termination of diapause in the European corn borer, *Ostriner nubilalis. J. Insect Physiol.* **15**, 945–51.

BISCHOFF, R. & HOLTZER, H. (1969). Mitosis and the processes of differentiation of myogenic cells in vitro. *J. Cell Biol.* **41**, 188–200.

COLLINS, J. V. (1969). The hormonal control of fat body development in *Calpodes ethlius* (Lepidoptera, Hesperiidae). *J. Insect Physiol.* **15**, 341–52.

COTTRELL, C. B. (1964). Insect ecdysis with particular emphasis on cuticular hardening and darkening. In *Advances in Insect Physiology*, vol. 2, 175–218 (eds. J. W. L. Beament, J. E. Treherne and V. B. Wigglesworth). New York and London: Academic Press.

FRAENKEL, G. & HSIAO, C. (1965). Bursicon, a hormone which mediates tanning of the cuticle in the adult fly and other insects. *J. Insect Physiol.* **11**, 513–56.

HOLTZER, H. (1961). Aspects of chondrogenesis and myogenesis. In *Synthesis of Molecular and Cellular Structure*, pp. 35–87 (ed. D. Rudnick). New York: Ronald Press.

ISHIKAWA, H., BISCHOFF, R., & HOLTZER, H. (1968). Mitosis and intermediate-sized filaments in developing skeletal muscle. *J. Cell Biol.* **38**, 538–55.

LAWRENCE, P. A. (1966). The hormonal control of the development of hairs and bristles in the milkweed bug, *Oncopeltus fasciatus*, Dall. *J. exp. Biol.* **44**, 507–22.

LOCKE, M. (1958). The coordination of growth in the tracheal system of insects. *Q. Jl microsc. Sci.* **99**, 373–91.

LOCKE, M. (1959). The cuticular pattern in an insect, *Rhodnius prolixus* Stal. *J. exp. Biol.* **36**, 459–77.

LOCKE, M. (1960). The cuticle and wax secretion in *Calpodes ethlius* (Lepidoptera, Hesperidae). *Q. Jl microsc. Sci.* **101**, 333–8.

LOCKE, M. (1965*a*). The hormonal control of wax secretion in an insect *Calpodes ethlius*, Stoll Lepidoptera, Hesperiidae. *J. Insect Physiol.* **11**, 641–58.

LOCKE, M. (1965*b*). The structure of septate desmosomes. *J. Cell Biol.* **25**, 166–9.

LOCKE, M. (1966). The structure and formation of the cuticulin layer in the epicuticle of an insect, *Calpodes ethlius* (Lepidoptera, Hesperiidae). *J. Morph.* **118**, 461–94.

LOCKE, M. (1967*a*). What every epidermal cell knows. Essay presented to Sir Vincent B. Wigglesworth in *Insects and Physiology*, vol. 7, pp. 69–82 (eds. J. W. L. Beament and J. E. Treherne). Edinburgh: Oliver and Boyd.

LOCKE, M. (1967*b*). The development of patterns in the integument of insects. *Adv. Morph.* **6**, 33–88.

LOCKE, M. (1969*a*). The ultrastructure of the oenocytes in the molt/intermolt cycle of an insect (*Calpodes ethlius* Stoll). *Tissue and Cell* **1**, 103–54.

LOCKE, M. (1969*b*). The structure of an epidermal cell during the formation of the protein epicuticle and the uptake of molting fluid in an insect. *J. Morph.* **127**, 7–40.

LOCKE, M. (1969c). The localization of a peroxidase associated with hard cuticle formation in an insect, *Calpodes ethlius* Stoll, Lepidoptera, Hesperiidae. *Tissue and Cell* **1**, 555–74.

LOCKE, M. (1970). The molt/intermolt cycle in the epidermis and other tissues of an insect, *Calpodes ethlius* (Lepidoptera, Hesperiidae). *Tissue and Cell* **2**, (In the Press).

LOCKE, M. & COLLINS, J. V. (1965). The structure and formation of protein granules in the fat body of an insect. *J. Cell Biol.* **26**, 857–85.

LOCKE, M. & COLLINS, J. V. (1967). Protein uptake in multivesicular bodies in the molt/intermolt cycle of an insect. *Science, N.Y.* **155**, 467–9.

LOCKE, M., CONDOULIS, W. V. & HURSHMAN, L. F. (1965). Molt and intermolt activities in the epidermal cells of an insect. *Science, N.Y.* **149**, 437–8.

LOEWENSTEIN, W. R. (1968). Emergence of order in tissues and organs. In *The Emergence of Order in Developing Systems*. 27th Symposium of the Society for Developmental Biology, pp. 151–82 (ed. M. Locke). New York and London: Academic Press.

MILLS, I. H. (1960). The transport state of steroid hormones. *Mem. Soc. Endocr.* no. 11, pp. 81–9.

OBERLANDER, H. (1969). Effects of ecdysone, ecdysterone, and inokosterone on the *in vitro* initiation of metamorphosis of wing disks of *Galleria mellonella*. *J. Insect Physiol.* **15**, 297–304.

OHTAKI, T., MILKMAN, R. D. & WILLIAMS, C. M. (1968). Dynamics of ecdysone secretion and action in the fleshfly *Sarcophaga peregrina*, *Biol. Bull. mar. biol. Lab., Woods Hole* **135**, 322–34.

ROLLER, H., DAHM, K. H., SWEELY, C. C. & TROST, B. M. (1967). The structure of the juvenile hormone. *Argewandte Chemie* **6**, 179–80.

SIDDALL, J. B., CROSS, A. D. & FRIED, J. H. (1966). Synthetic studies on insect hormones. II. The synthesis of ecdysone. *J. Am. chem. Soc.* **88**, 862.

STUMPF, H. F. (1966). Über gefälleabhangige bildungen des insektensegmentes. *J. Insect Physiol.* **12**, 601–17.

STUMPF, H. F. (1968). Further studies on gradient-dependent diversification in the pupal cuticle of *Galleria mellonella*. *J. exp. Biol.* **49**, 49–60.

WIGGLESWORTH, V. B. (1934). The physiology of ecdysis in *Rhodnius prolixus* (Hemiptera). II. Factors controlling moulting and 'metamorphosis'. *Q. Jl microsc. Sci.* **77**, 191–222.

WIGGLESWORTH, V. B. (1948). The insect cuticle as a living system. *Discussions Faraday Soc.* **3**, 172–7.

WILLIAMS, C. M. (1952). Physiology of insect diapause. IV. The brain and prothoracic glands as an endocrine system in the cecropia silkworm. *Biol. Bull. mar. biol. Lab., Woods Hole* **103**, 120–38.

SESSION IV
THE BIOTIC ENVIRONMENT

AGEING AND ENDOCRINE RESPONSES
TO ENVIRONMENTAL FACTORS: WITH
PARTICULAR REFERENCE TO MAMMALS

By D. BELLAMY

The meaning of the term 'ageing' is open to much dispute when applied to biological studies (Chandler, 1952). Some workers use the term to cover all aspects of development, from fertilization to death; others restrict the ageing phase to the terminal stages of life when there is a rapid decrease in life expectancy. For the purpose of the present discussion ageing is defined as 'a decrease in adaptation as a consequence of loss of tissue and functional reserves' (Albertini, 1952). The definition implies that a major feature of ageing is a decline in the ability of the homeostatic systems of the body to cope with fluctuations in the external world. On this basis ageing begins long before there is a marked increase in mortality. In the human, a steady decline in physiological performance is first noticeable during the third decade of life; the corresponding stage in the laboratory rat (life-span 800 days) occurs at about 200 days; in the fruit fly, *Drosophila* (life-span 27 days), ageing of some systems begins at about 10 days. Thus, it appears that ageing phenomena are noticeable for over half the life-span.

The literature will be reviewed, first in relation to the ability of the ageing, resting organism to maintain a constant composition, and secondly with regard to the ability of the body to combat internal and external pressures. It is concluded that despite large age-changes in the metabolic balance of certain tissues, there are few progressive alterations in cellular and physiological organization and these are often inconsistent with the idea that old age results in a gradual loss of resting function.

The third section will deal with ageing of the endocrine system in as far as this plays a part in the loss of vitality, and the last part contains a discussion of the age-dependent decline in endocrine function in relation to increased mortality.

COMPOSITION OF TISSUES

If the composition of the blood is taken as an indicator of the over-all state of homeostasis there appears to be no impairment of functional capacity with age. The acid-base balance of the blood in resting humans varies little between the ages of 25 and 85 (Shock & Yiengst, 1950). There are no significant changes in the carbon dioxide tension, total carbon dioxide content and bicarbonate concentration. In agreement with this the pH only changes from 7·400 to 7·368 between the second and eighth decades of life. Organic constituents of blood show a similar stability (Horvath, 1946; Praetorius, 1951; Rogers, 1951 (Table 1)). The general constancy of the ionic composition of the body is borne out by studies on individual tissue (Table 2) and subcellular fractions (Griswold & Pace, 1956). No significant age-changes are observed in total water, fat, potassium and the sodium content of liver, whilst the intracellular water, nitrogen, potassium and phosphorus in muscle decrease only by between 5% and 8% (Yiengst, Barrows & Shock, 1959). There is no change in muscle lactic acid and creatine (Horvath, 1946) and analysis of human aortae (Kanabrocki, Fells, Decker & Kaplan, 1963) taken over an age range from 2 to 69 years revealed no significant alterations in either total nitrogen or sulphur. Another study in subjects from 15 to 58 years of age indicates that aortic elastin and creatine also remain relatively constant (Myers & Lang, 1946).

Table 1. *Blood uric acid in humans**

Age range (years)	N	Relative mean concentration	Coefficient of variability[†]
65–70	19	1·00	3·47
71–75	35	1·00	3·93
76–80	25	1·18	4·11
81–91	12	1·02	4·30

* Gertler & Oppenheimer, 1953. † (S.E./$\bar{x}$) × 100.

A constancy of composition was also found to be a feature of the ribonucleic acid in ventricular muscle of the rat from 100 to 1200 days of age (Wulff, Piekielniak & Wayner, 1963). In keeping with this the mean DNA per nucleus and the mean volume of nuclei in rat liver do not change over an age range of 12–27 months (Falzone, Burrows & Shock, 1959).

The failure to detect age-changes indicative of a large-scale deterioration in tissue function impressed the early workers using histological methods. Only moderate signs of ageing are found in the histochemistry of liver and there are no obvious functional deficiencies in the brain. Kidney tissue is marked by having the largest age-alterations but this appears to be a consequence of a large decrease in the relative cell mass without a marked change in composition of the remaining cells (Lowry, Hastings, McCay & Brown, 1946).

Despite the evidence in favour of stability of tissue structure, age-

Table 2. Electrolyte content* of tissues in young† and old† C57 mice§

	Kidney				Muscle				Brain			
	Young		Old		Young		Old		Young		Old	
	M	F	M	F	M	F	M	F	M	F	M	F
N ...	10	10	10	10	10	10	10	10	9	9	9	9
Potassium (m-equiv/g wet wt)	93·00 ±2·14	92·25 ±1·69	88·00 ±2·10	89·73 ±2·30	89·18 ±3·53	88·75 ±6·28	79·25 ±2·24	98·05 ±3·18	98·90 ±1·92	94·75 ±1·81	93·25 ±2·73	95·83 ±1·77
Sodium (m-equiv/g wet wt)	77·75 ±2·25	83·50 ±3·25	88·00 ±7·53	89·18 ±3·20	21·67 ±1·32	23·00 ±0·73	24·50 ±0·82	25·55 ±2·35	49·73 ±1·84	47·25 ±1·12	48·25 ±1·75	50·00 ±1·72
Water (% wet wt)	75·33 ±0·35	74·47 ±0·52	75·75 ±0·30	74·98 ±1·13	71·78 ±3·24	75·79 ±0·51	73·17 ±0·93	72·67 ±1·30	79·28 ±0·7	79·43 ±0·29	78·49 ±0·24	76·97 ±0·48

* Means ± S.E.M. † Old = 18 months. ‡ Young = 2 months. § Bellamy unpublished data.

Table 3. Haematological changes* in the ageing rat
(relative concentrations)

Mean age (days)	N	Haemoglobin		Red cells		White cells		Eosinophils	
		$\bar{x}$	v	$\bar{x}$	v	$\bar{x}$	v	$\bar{x}$	v
329·4	50	1·00	0·49	1·00	0·96	1·00	3·28	1·00	7·08
558·1	50	0·99	0·86	1·00	1·08	1·04	3·79	0·64	9·62
761·5	50	1·07	1·43	1·01	2·14	0·99	4·39	0·47	12·78
963·0	50	1·12	3·22	1·08	2·44	1·09	10·14	0·44	18·47

* Everitt & Webb, 1958. v = coefficient of variability $(\text{S.E.}/(\bar{x})) \times 100$.

changes in composition are frequently observed. This applies to a small number of blood constituents (Ackermann & Kheim, 1964; Das, 1964). Often, age-changes are not consistent between tissues; although there is no variation in the riboflavin content of human brain, heart and skeletal muscle over seven decades (Schaus & Kirk, 1956), aortic tissue is characterized by a 60% fall in the riboflavin content over the same period (Shaus, Kirk & Laursen, 1955). Also, in contrast to ventricular tissue the folia tissue of the cerebellum loses 30% of its ribonucleic acid throughout the life-span (Wulff et al. 1963). Some of the largest alterations in tissue composition have been found in the lipid fraction. Cholesterol in the blood of healthy women increases twofold in concentration from the age of 20 to 60 (Swanson, Leverton, Gram, Roberts & Pesek, 1955). There are also substantial increases in the concentration of tissue elastin, collagen (Hall, Keech, Reed, Saxl, Tunbridge & Wood, 1955; Schaub, 1963), mucopolysaccharides (Kirk & Dyrbye, 1956b) and calcium (Lansing, Rosenthal, & Alex, 1950; Yu & Blumenthal, 1963). These changes, although a general feature of many ageing tissues, are not a universal phenomenon (Streicher, 1958; Bashey, Torii & Angrist, 1967; Sobel & Hewlett, 1967; Sobel, Hrubant & Hewlett, 1968).

Taking the available evidence on the gross composition of tissues it appears that on the whole, age changes are not very marked in those intracellular components, such as water, nitrogen and inorganic ions, which are concerned with fundamental cellular organization. On the other hand, there are marked changes in the chemistry of the extracellular compartment, notably in the ground substance. These changes are probably a reflection of the general tendency for there to be an age-dependent shift in the balance of cell populations. For example, the eosinophil count in rats decreases progressively with age, there being a 50% drop between the ages of 300 and 800 days. Despite this loss there is little change in other blood cells (Everitt & Webb, 1958; Table 3). In ageing cattle it has been observed that a decrease in blood lymphocytes is accompanied by a rise in the proportion of blood neutrophils and a fall in the total leucocyte count (Riegle & Nellor, 1966).

Organ involution is most marked in the thymus. Here the process which results in a progressive general decrease in cellular mass is accompanied by proliferation of epithelial cells which are distinct from the thymocytes (Thung, 1966). Age-involution is also a feature of skeletal muscle, there being a 30% loss of thigh muscle between the first and second year of life in male rats (Yiengst et al. 1959). This process is not so marked in females. Taking total body potassium as a measure of cell mass the number of cells per unit body weight shows a steady decline in humans from the late teens to the age of 80 (Allen, Anderson & Langham, 1960). In another study (Shock, Watkin, Yiengst, Norris, Gaffney, Gregerman & Falzone, 1963), it was found that the extracellular space did not change although the total body water diminished significantly with increasing age. Intracellular water, calculated as the difference between total body water and extra-cellular space, showed a significant age regression. This is interpreted as a reflection of the loss of functioning cells with increasing age. From evidence of this nature it is generally held that involution is one of the most consistent features of ageing.

METABOLISM

In the experiments described at the end of the previous section (Shock *et al.* 1963) the basal oxygen uptake, expressed per unit of intracellular water, did not show a regression with age. From this it is concluded that there is no impairment in the respiratory metabolism in old cells. The basal whole-body oxygen consumption of adult rats actually increases with age between 10% and 13% (Ring, Dupuch & Emeric, 1964). This discrepancy between rats and human subjects has not been resolved but it is likely that the decrease in muscle mass of rats is counteracted by a higher rate of respiration of cells in other tissues. On the other hand, no age-changes have been detected in aerobic metabolism of tissue preparations from rat liver (Barrows, Yiengst & Shock, 1958). Similar experiments with kidney indicate only a small age-decrement in aerobic metabolism, possibly due to a fall in the number of mitochondria per cell (Barrows, Falzone & Shock, 1960). The maintenance of aerobic metabolism at the cellular level is born out by results obtained for guinea-pig tissues (Rafsky, Newman & Horonick, 1952). Further, the specific dynamic action of a standard protein meal expressed as excess oxygen uptake is essentially the same in old (mean age 77 year) and young (mean age 24 year) men, although older subjects show a slower initial response (Tuttle, Horvath, Preson & Daum, 1952).

Little is known about anaerobic metabolism. Indirect evidence is suggestive of a decrease in the demand of the ageing brain for anaerobically produced energy (Enzmann & Pincus, 1934; Samson, Balfour & Dahl, 1958). Some glycolytic enzymes in rat kidney have been observed to decrease with age by between 15% and 20%. However, these changes are not associated with an impairment of glycolysis measured *in vitro* (Zorzoli & Li, 1967).

Ageing rat liver and brain are remarkable for their metabolic stability. Liver does not undergo age involution (Yiengst *et al.* 1959; Kurnick &

Kernen, 1962) and there is no alteration in protein turnover (Fletcher & Sanadi, 1961; Barrows & Roeder, 1961; Beauchene, Roeder & Barrows, 1967). It also appears that the enzyme pattern of liver and brain is largely independent of age (Schmukler & Barrows, 1967; Hollander & Barrows, 1968; Gold, Gee & Strehler, 1968; Shukla & Kanungo, 1969). Despite this, there is a decrease in nitrogen content of liver attributable mainly to losses of mitochondria and other membrane structures, and a decrease in the RNA content per cell (Detwiler & Draper, 1962). In humans there is a rise in the volume ratio of nucleus to cytoplasm (Tauchi & Sato, 1962). The actual quantity of RNA per nucleus is increased and there is an alteration in the base composition. These latter changes may be connected with the marked increase in RNA turnover which is observed in nuclei of a number of tissues of old rats (Wulff, Quastler & Sherman, 1964). Despite an alteration in RNA metabolism, cytophotometric measurements of Feulgen-stained nuclei in neurones of the cerebellum and liver cells indicate that there is no loss of DNA from individual cells during ageing (Enesco, 1967). However, a comparison of this work with a histochemical study of the ageing rat submandibular gland, which revealed a decrease in the activity of several enzymes and a fall in RNA concentration (Bogart, 1967), stresses the lack of a common ageing pattern.

It could be argued that the biochemical features of ageing in the laboratory rat are a consequence of inbreeding. This objection has been overcome with regard to some of the phenomena described above in that they are also found in the ageing wild rat (Barrows, Roeder & Falzone, 1962). The quantitative differences are indicative of a slower rate of ageing in the wild strain (Chvapil & Roth, 1964).

Taken together, the biochemical work suggests that each tissue is characterized by a particular pattern of ageing, but to support this idea it is desirable that comprehensive studies be made on each tissue. So far only one such analysis has been carried out. Kirk and his co-workers have made a detailed study of ageing in human blood vessels and a summary of data on 30 enzymes in the normal aorta is presented in Table 4. The figures represent the mean enzyme activity through seven decades of life relative to that in the first decade. It can be seen that there is a considerable variation in both the extent and pattern of change in enzyme activity. Some enzymes increase in activity (maximum, 3·5-fold rise), others hardly change and some at first increase and then decrease in activity. They may be divided into four groups according to the time when the activity reaches a maximum (Table 5), but there are no obvious similarities in function of enzymes within any group. Enzymes in group 1 show a steady decline in activity from the first decade amounting to about -5% per decade. Those

Table 4. Relative changes in the activity of enzymes in the human aorta

Enzyme		Decade of life							
		1st	2nd	3rd	4th	5th	6th	7th	8th
	Group 1								
Glutamic dehydrogenase (Kirk, 1965)		1·00	0·74	0·89	0·89	0·64	0·68	0·54	0·70
6-phosphogluconate dehydrogenase (Kirk et al. 1959)		1·00	0·92	0·69	0·85	0·85	0·85	0·85	0·85
Leucine amino peptidase (Kirk, 1960a)		1·00	0·93	0·84	0·95	0·95	0·81	0·67	0·99
Aconitase (Kirk, 1961a)		1·00	0·88	0·66	0·73	0·68	0·67	0·70	0·69
Phenylsulphatase (Kirk & Dyrbye, 1956a)		1·00	1·00	0·97	0·75	0·54	0·53	0·43	0·50
	Group 2								
Sorbitol dehydrogenase (Ritz & Kirk, 1967)		1·00	2·40	1·64	2·01	1·82	1·71	1·70	1·57
Inorganic pyrophosphatase (Kirk, 1959a)		1·00	1·93	1·60	1·07	1·27	0·93	0·93	1·27
Creatine phosphokinase (Kirk, 1962d)		1·00	1·83	1·11	0·91	0·79	0·63	0·57	0·41
Glycogen phosphorylase (Kirk, 1962a)		1·00	1·39	0·82	0·95	0·91	0·73	0·59	0·67
Cytochrome c reductase (Kirk, 1962c)		1·00	1·34	0·98	0·98	0·86	0·83	0·64	0·86
Phosphoglucomutase (Kirk, 1966)		1·00	1·17	1·11	0·90	0·90	0·83	0·82	0·61
Glyoxylase (Kirk, 1960a)		1·00	1·15	0·86	0·79	0·83	0·69	0·69	0·64
Phosphomannose isomerase (Kirk, 1966)		1·00	1·12	1·07	0·96	0·97	1·03	1·09	0·86
	Group 3								
Glyceraldehyde-3-phosphate dehydrogenase (Kirk & Ritz, 1967)		1·00	2·69	3·03	2·78	2·70	3·01	2·18	2·53
α-glycerophosphate dehydrogenase (Kirk & Ritz, 1967)		1·00	1·58	2·78	2·50	1·91	1·52	0·98	1·40
Phosphomonoesterase (Kirk, 1959a)		1·00	1·83	2·25	1·92	2·04	1·75	2·00	1·50
Diaphorase (Kirk, 1962c)		1·00	1·49	1·77	1·40	1·17	1·36	1·16	1·13
Adenyl pyrophosphatase (Kirk, 1959a)		1·00	1·18	1·27	0·91	0·95	0·91	0·86	0·91
Malic enzyme (Kirk 1960c)		1·00	1·12	1·15	0·98	1·00	1·13	1·04	0·81
Phosphoglyceric acid mutase (Kirk, 1966)		1·00	0·89	1·02	0·80	0·80	0·81	0·80	0·71
	Group 4								
Glucose-6-phosphate dehydrogenase (Kirk, Wang & Brandstrup, 1959)		1·00	2·73	3·10	3·34	2·71	2·68	2·68	2·53
Enolase (Wang & Kirk, 1959)		1·00	1·44	1·50	1·81	1·81	1·63	1·63	1·25
	Group 5								
5-nucleotidase (Kirk, 1959b)		1·00	1·86	2·63	2·66	3·66	2·65	3·11	2·96
	Group 6								
Isocitric dehydrogenase (Kirk, 1960c)		1·00	1·00	1·03	0·89	1·02	1·05	1·03	0·81
Ribose-5-phosphate isomerase (Kirk, 1959c)		1·00	0·95	0·75	1·65	2·00	2·25	0·70	1·25
	Group 7								
Glutathione reductase (Kirk, 1965)		1·00	1·35	1·24	1·31	1·46	1·35	1·49	1·19
Aldolase (Kirk & Sørensen, 1956)		1·00	0·64	0·95	1·18	1·30	1·20	1·46	1·19
	Group 8								
Phosphofructokinase (Ritz & Kirk, 1967)		1·00	1·88	2·87	1·80	2·14	1·58	3·03	3·46
Cathepsin (Kirk, 1962b)		1·00	1·11	0·75	1·08	1·22	1·04	1·07	1·38
Purine nucleoside phosphorylase (Kirk, 1961b)		1·00	1·23	1·00	1·10	1·05	1·01	1·08	1·27

in group 2 have a peak activity in the second decade (mean 50% increase) then show a decline, which between the third and eighth decade proceeds on average at a rate similar to that for enzymes of group 1. Enzymes in group 3 reach their maximum activity in the third decade (mean, twofold increase) then decline at a rate similar to those in groups 1 and 2. Group 4 contains enzymes that reach their peak between the third and ninth decades. The mean rate of increase to the peak is the same for all groups. This activity pattern may be interpreted as follows. There is a mechanism that tends to increase the activity of all enzymes at a similar rate. This is counteracted by an age-dependent involution process which has a selective effect in that it comes into play at different times during the life-span, depending on the enzyme.

Table 5. *Enzyme activities* of human aorta grouped according to pattern of ageing*

	Group 1	Group 2	Group 3	Groups 4–8	All groups
N ...	5	8	7	10	30
Decade					
2nd	0·932 ±0·001	1·541 ±0·161	1·540 ±0·223	1·465 ±0·204	1·405 ±0·100
3rd	0·810 ±0·054	1·148 ±0·104	1·895 ±0·304	1·754 ±0·296	1·458 ±0·137
4th	0·834 ±0·031	1·071 ±0·134	1·612 ±0·301	1·770 ±0·256	1·378 ±0·126
5th	0·732 ±0·070	1·043 ±0·118	1·510 ±0·268	1·927 ±0·277	1·377 ±0·134
6th	0·708 ±0·054	0·922 ±0·118	1·498 ±0·281	1·710 ±0·219	1·268 ±0·122
7th	0·638 ±0·070	0·880 ±0·134	1·288 ±0·212	1·916 ±0·266	1·258 ±0·134
8th	0·746 ±0·077	0·861 ±0·134	1·284 ±0·234	1·831 ±0·298	1·244 ±0·137

* Activity expressed relative to that in 1st decade = 1.

Taking all enzymes together there is no evidence to support the idea that there is a general deterioration in metabolism with age. By the ninth decade about 50% of the enzymes retain more than 90% of the activity characteristics of the first decade; only one enzyme retains less than 50% of its activity in the first decade.

CAPACITY FOR ADAPTATION

The kind of study that has been discussed so far, although pointing to a number of age-changes in metabolism has not thrown light on the reasons for the increased chances of mortality as time passes. There is no consistent trend that can be definitely interpreted as a deterioration in function, and apart from tissue involution there seems little change in functional organization. This points to the fact that the ageing process is manifest for the most part at a physiological level. Although physiological parameters that depend upon a complete interaction of several organs, such as ion balance and blood pressure, hardly change with age there is an increasing tendency for departures from these norms to take place. Thus, the systolic blood pressure of CBA mice under controlled conditions remains within the range 126 ± 12 mmHg throughout the life-span of $2\frac{1}{2}$ years (Henry, Meehan, Stephens & Santisteban, 1965) but the normal distribution curve for rat blood pressure becomes biphasic with age (Berg & Harmiston, 1955). When the rats pass the age of 600 days about 30 % of males and 20 % of females become hypertensive with marked renal disease. The latter condition is probably the main cause of the hypertension.

A decline in the ability to cope with increased functional demands due to endogenous processes can be clearly seen in the reproductive capacity of rodents. After 14 months the litter-size of hamsters is greatly reduced and the length of gestation increases (Solderwall, Kent, Turbyfill & Britenbaker, 1960). Similarly, data for the mouse suggest that parental age influences the length of the period of fertilization plus gestation, the litter-size and the still-born mortality rate (Roman & Strong, 1961). Resorption during the late phase of gestation is thought to be a contributory factor in the reduction in litter-size in senescent females.

A failure of physiological adaptation is most clearly seen under laboratory conditions in response to specific stresses. Old $C 57 B 2$ female mice have a high mortality rate following exposure to a temperature of 6–7 °C which is not seen in young, mature females. The failure of integrative function which is responsible for the death of older animals appears early in the experimental period. Old mice which survive the first 2 days of cold exposure have a good chance of surviving for the next 12 days in the cold (Grad & Kral, 1956). Mortality of old mice is reduced when they are slowly adapted to the lower temperature, but the adaptive process is less effective than in the young.

A number of the individual physiological responses of rodents to cold such as the increased oxygen uptake, rise in blood sugar, body weight decrease and changes in blood cell counts also become less marked with age (Grad & Kral, 1956). Other studies on humans suggest that the mechanisms for minimizing heat loss deteriorate in old subjects (Krag & Kountz, 1950).

The response of the human body to heat stress has also been examined for possible age-effects (Shattuck & Hilferty, 1932; Friedfeld, 1949). In general, there is an increased degree of discomfort at high temperatures and the physiological responses in the old become more variable (Krag & Kountz, 1952). One relevant feature is that men over the age of 40 have a greatly increased vasodilation in response to heat (Hellon & Lind, 1958). Since in a hot environment heat loss is almost entirely by evaporation, the greater skin blood flow in old subjects would not be expected to have a marked effect on their heat regulation. Age differences in rectal and skin temperatures are not significant in heat-exposed subjects, therefore the view may be taken that the faster skin circulation of older men is unnecessary for heat regulation and places a needless burden on the general circulation. Responses to temperature are intimately linked with the basal metabolic rate. There is a significant fall in basal heat production and carbon dioxide elimination as humans age (Shock *et al.* 1963; Table 6). There is also a larger day-to-day variation in several aspects of respiratory function (Shock & Yiengst, 1955) which may be of significance with regard to the efficiency of adaptation.

Table 6. *Age changes in basal heat production and kidney function*
in humans

Decade	Basal heat production*			Glomerular filtration rate†			Tubular absorption of glucose†		
	N	Relative mean	Coefficient‡ of variation	N	Relative mean	Coefficient‡ of variation	N	Relative mean	Coefficient‡ of variation
3rd	14	1·00	7·98	9	1·00	2·86	9	1·00	5·97
4th	23	1·01	7·40	12	0·93	2·55	12	0·95	4·33
5th	31	1·01	9·85	14	0·87	5·75	14	0·92	5·85
6th	43	0·96	11·29	14	0·76	4·96	14	0·78	6·88
7th	45	0·97	11·88	15	0·71	4·41	15	0·72	5·18
8th	27	0·90	11·65						
9th	5	0·85	19·92						

* Shock *et al.* 1963.   † Miller *et al.* 1952.   ‡ (S.E./$\bar{x}$) × 100.

Starvation, another common stress encountered by animals in the wild, may be examined conveniently in the laboratory. However, there have been few investigations of starvation in relation to age. With regard to rats no significant age differences have been observed in either the rate of weight loss in long-term starvation or the survival-time after food deprivation (Jakubczak, 1967).

With regard to experimental hypoxia there is a marked loss of adaptation in old animals (Fluckiger & Verzár, 1955; Sulkin & Sulkin, 1967). With respect to traumatic injury the data for humans do not support the concept that older individuals have less chance of survival (Collins, 1950). On the other hand, this conclusion relates only to traumatic abdominal injuries. Regeneration of the liver of rats after partial hepatectomy occurs to an equal extent in the young and old but old animals take more time to form the requisite number of new cells (Bucher & Glinos, 1950). Following unilateral nephrectomy senescent rats have the same capacity for compensatory growth as young ones (McKay, McKay & Addis, 1924). However, as for liver regeneration, the kidney of old animals responds slowly (Reiter, McCreight & Sulkin, 1964). There is no evidence that new cells produced in response to unilateral nephrectomy by old rats are biochemically different from those in young rats (Barrows, Roeder & Olewine, 1962). Despite this finding, the chronic hypertrophy of kidneys in rats given a pathogenic diet is associated in old animals with a decreased capacity of mitochondrial oxidative phosphorylation (Fedorčákova, Bachledová, Niederland & Bózner, 1968).

Responses to drugs, anaesthetics and toxic agents have also been used to measure age-changes in adaptation (MacNider, 1946). Increased age is often a factor which is responsible for a loss of drug-responsiveness. For example, patients with arthritis are more responsive to steroid therapy in the third decade than in later years. This difference is not related to the duration of the disease. Hexabarbital anaesthesia is prolonged in old rats compared with young ones. Alongside this age-change there is an increased mortality attributable to the drug (Streicher & Garbus, 1955). The endocrine status is thought to be an important factor underlying this rise in toxicity with age but neurophysiological and neuropathological factors may well be involved. In this connection, the toxic action of uranium salts is manifest locally by specific degenerative changes in the kidney tubule which are more marked in old animals. These histological changes in the epithelial cells go in parallel with the loss of functional capacity of the kidney (MacNider, 1917).

Age-dependent failures of adaptive renal responses have been the subject of much detailed investigation. Studies on p-aminohippurate clearance indicate that renal plasma flow is diminished in old subjects although there is no impairment of the ability of the kidney to extract p-hippurate from the tubules (Miller, McDonald & Shock, 1951). In contrast, the capacity to reabsorb glucose from the glomerular filtrate shows a linear decline with age (Miller, McDonald & Shock, 1952; Table 7), which amounts to about 10% from the third to the ninth decade. Changes in glomerular filtration parallel the changes in reabsorptive capacity (Table 7) which may be of significance in relation to the age-dependent depression of glucose tolerance in humans (Schneeberg & Finestone, 1952) and the failure to induce a reduction in the alkali reserve after treatment of old dogs with anaesthetics (MacNider, 1943). However, there are many physiological functions that deteriorate with age, any one of which may be equated with a loss of vigour (Conrad, 1960).

There is a considerable body of statistical evidence which attests to the fact that on the average the ability to exercise and carry out physical work declines with age (Ruger & Stoessiger, 1927; Burke, Tuttle, Thompson, Janney & Weber, 1953). This may be related to changes in several factors such as oxygen uptake, pulmonary ventilation, diffusing capacity of lungs, work pulse and blood lactate. Between the ages of 30 and 70 there is a twofold increase in the time for recovery from exercise.

Table 7. *Effect of age on glucose tolerance* in humans*

Time after glucose injection (min)	Aged 16–39			Aged 41 and over		
	N	Relative mean glucose concn.	Coefficient[†] of variability	N	Relative mean glucose concn.	Coefficient[†] of variability
0	48	1·00	21·96	49	1·00	20·44
15	48	2·43	13·03	48	2·52	13·95
30	48	1·89	16·92	48	2·00	18·97
60	48	1·07	22·29	49	1·43	22·46
75	48	0·90	17·86	48	1·15	22·87
90	48	0·88	19·16	48	0·99	21·69
120	40	0·89	17·79	16	0·86	21·90

* Schneeberg & Finestone, 1952, † $(\text{s.d.}/\bar{x}) \times 100$.

An important aspect of many adaptive responses is the ability to make rapid neurophysiological adjustments. Thus there is a consistent decline in the capacity to see at low light intensities and an increased limitation upon the extent of dark adaptation (McFarland & Fisher, 1955). There are also age-dependent losses in the ability to move rapidly in response to environmental changes (Birren, 1955; Birren & Botwinick, 1955; Birren & Kay, 1958; Pierson & Montoye, 1958; Hodgkins, 1962).

In long-term responses involving locomotory activities the impairment in ventilation and gas exchange appears to underlie the age deterioration (Norris, Shock & Yiengst, 1955). However, in short-term reflex delays in the aged it is suggested that the increased reaction time is due to decreased excitatory influences from higher levels (Magladery, Teasdall & Norris, 1958). Apart from the well-established age-involution of nervous tissue and the appearance of 'age-pigment' in neurones of the central nervous system there are few indications of the cause of the failure of nervous co-ordination in old individuals. The excitation threshold for electrical stimulation of nerves is higher in old animals than in young ones. Also there is a decrease in the activity of choline acetylase and choline esterase (Frolkis, 1966), the key enzymes in transmission of the nerve impulse. It may well be that the decline in the number of neurones and motor end-plates with age also contributes to the loss of flexibility of nervous responses (Gutmann, Hanzlíková & Jakovbek, 1968). However, the most likely cause of the decline in vigour in relation to long-term environmental changes lies in the organization of the endocrine system. This possibility will now be considered in detail.

AGE-CHANGES IN ENDOCRINE ACTIVITY

Hormones are integrated into the pattern of development in two ways. They act as controllers, in that they release or trigger the developmental potential inherent in certain tissues. Actions of this type are sometimes irreversible, but usually the target tissue returns to its former state when the concentration of hormone falls to the initial level. Secondly, hormones invoke responses that offset undesirable changes in the external environment. From the latter point of view, the ageing of endocrine systems takes place in two stages. Initial development of endocrine responses make the animal more adaptable and independent of its surroundings. It is the later changes which decrease the capacity for adaptation to the environment that are usually discussed as ageing phenomena.

Several temporal aspects are of importance to the endocrinologist dealing with the ageing of endocrine systems. These are related to the source of the phenomenon that leads to the irreversible modification of the endocrine system. The age-change may be initiated from outside the organism. The chemical nature of the external stimulus is most clearly seen in the environmental changes encountered by euryhaline fish, where a spawning migration from fresh water to the sea brings about a profound change in the endocrine system (Chester Jones & Bellamy, 1964). However, even in this case the impulse to undertake the migration may be intrinsic.

An important aspect of developmental endocrinology centres on the differences in the biological role of animals at various stages of their life-span. The most obvious age-dependent modifications of the endocrine system that fall into this category are those linked with the attainment of sexual maturity. This applies particularly in the female, where there is a precise chemical control of the sequence of events in the reproductive cycle. In this context the corresponding alterations in the endocrine system are brought about by intrinsic mechanisms, although the various responses may be synchronized by external factors.

The above features of the ageing individual determine some of the changing patterns of homeostasis which may be regarded as increasing general biological efficiency. However, in order to discuss the ageing of endocrine systems, four aspects related to intrinsic ageing mechanisms will be considered only as they relate to a decline in vigour. Sexual maturation and metamorphosis clearly do not come within the scope of the discussion. The four aspects will be taken as follows: (1) changes in endocrine-controlled norms, (2) changes in the responsiveness of target tissues to hormones, (3) changes in the function of endocrine organs, (4) changes in the direction of the endocrine response.

CHANGES IN NORMS

At each stage of life the chemical composition of the body is maintained within narrowly defined limits, through the action of endocrine systems. However, as discussed above, new intracellular norms for enzymes and metabolites are established as the animal ages. Bearing in mind the fall in metabolic mass with age it is likely that a decreased rate of metabolism would tend to increase the concentration of most plasma metabolites which have a direct dietary origin. The fact that there is no general rise in the concentration of these substances with age suggests that adaptive changes in the endocrine system occur which offset this tendency, but we are largely ignorant of the details.

Every endocrine response, even though the end-point is the restoration or maintenance of a constant concentration of some chemical in the body fluids, is always associated with changes in the expenditure of energy. Thus, by its nature, a hormone brings about an alteration in metabolism and the temporary or permanent attainment of new norms. For example, secretion of a hormone may be accompanied by an increased rate of synthesis of the hormone and also, particularly in long-term hormone action, a stimulation of the formation of a component of the effector system in the target tissue. In this way there may be requirements for an increased supply of energy-yielding substances and precursors of macromolecules. Deteriorations of metabolism at this level are important in responses that result in protein synthesis as a necessary feature of the maintenance of a norm.

Where long-term hormone action increases the availability of metabolites, the extra metabolites are utilized with a minimum alteration in the composition of the intracellular environment. Under some conditions the organism achieves a constancy of internal composition by increasing the amount of enzyme available to deal with the increased supply of substrate. It is possible that the effect of the hormone on enzyme synthesis is mediated through a change in the concentration of specific metabolites. Thus, in principle, enzyme concentrations are adjusted, so altering the rate of substrate utilization, to regulate the concentration of intracellular substrate. Many hormones have been shown to produce an alteration in enzyme pattern that may well be of this homeostatic type. Taking this into account it is difficult to relate the observed changes in plasma metabolites to a variation in the endocrine organ, rather than to alterations in the enzyme activity of the target tissue. If there was a fall in the efficiency of either the appropriate endocrine organ or the enzymic process that previously maintained a particular metabolite at a constant level, the concentration in the body fluids would tend to change with age. However, in this complex control system, a change in one of the three potential variables (availability of metabolite, activity of endocrine gland and rate of enzyme synthesis) could well lead to alterations in the other two.

CHANGES IN RESPONSIVENESS

It is well known that all organs of the body undergo chemical and morphological changes with age. In discussing the ageing of tissues that are targets for hormones, only those features will be considered that have been experimentally linked with alterations in response to exogenous hormones.

The complicated nature of ageing at the endocrine level is illustrated by the control of renal function. Electrolyte control in the human kidney is not precise until 2–3 weeks after birth (McCance & Widdowson, 1957). This situation is thought to be connected with the slow development of the loop of Henle (Hubble, 1957). At the other end of the time-scale it is known that the kidney of old animals is not so versatile in its response to hormones as in early adult life. Experiments with the rat have shown that the anti-diuretic effect of exogenous vasopressin depends on the age of the animal. More hormone is required to produce a response in very young and old animals compared with animals of intermediate age. This will be considered in detail later.

With respect to other polypeptide hormones, and other targets, it has been shown that there is an age-dependent decrease in sensitivity to insulin (Shock, 1952; Silverstone, Brandfonbrener, Shock & Yiengst, 1957; Giarnieri & Lumia, 1961; Moore, 1968; Table 11), growth hormone (Moon, Koneff, Li & Simpson, 1956; Everitt, 1959; Root & Oski, 1969) and ACTH (Shaw, 1962). A loss of sensitivity is not general for all hormones. For example, the characteristics of lipid mobilization and deposition in humans treated with noradrenaline suggest that age has little effect on the endocrine response (Eisdorfer, Powell, Silverman & Bogdonoff, 1965; Table 8). However, previous exposure to stressful conditions reduces the effectiveness of catecholamines, particularly, with regard to lipolysis (Stuchlíková, Hruškova, Hrůza, Jelínková, Novák & Soukupová,

Table 8. *Effect of noradrenaline on resting and peak plasma free fatty acids* in humans†*

Mean Age (year)	Resting FFA (μ-equiv/l.)	Peak FFA (μ-equiv/l.)	Change in FFA (μ-equiv/l.)
35·0	690·7 ± 65·3	1308·5 ± 130·9	617·8 ± 126·2
68·8	544·1 ± 62·9	1106·3 ± 157·8	562·2 ± 138·2

* Means ± s.e.   † Eisdorfer *et al.* 1965.

1966). Similarly, adrenaline treatment produces the same rise in glucose and pyruvic and lactic acids in young and old rats (Hrůza & Jelinkova, 1965) and has the same effect on glycogenolysis in young and old humans (Stuchlíková *et al.* 1966).

Some hormones are more effective in old animals. Thus exogenous thyroid hormone may bring about a loss of body weight, which for a given dose is greater in old rats than in young ones (Bodansky & Duff, 1936). Also the effect of either thyroxine or thyrotrophic hormone on the basal oxygen consumption of rats is more marked in old animals (Belasco & Murlin, 1941). In apparent contradiction to these findings, thyroidectomy has more or less the same quantitative effect on the metabolism of old rats compared with young ones (Grad, 1969). Possibly in addition to differences in tissue sensitivity to thyroid hormones there are changes in the role of the thyroid gland with age.

It appears that there is a greater increase in ACTH output after adrenaline injection into old subjects than in young ones (Solomon & Shock, 1950). Also there has been a suggestion that old tissues are more sensitive to corticosteroids (Hennes, Wajchenberg, Fajans & Conn, 1957). Neither acute nor chronic administration of ACTH gives any evidence of an age impairment of either the adrenocortical response or the metabolic and haematological effects resulting from the increased adrenocortical activity (Solomon & Shock, 1950; Duncan, Solomon, Rosenberg, Nichols & Shock, 1952). On the other hand, in cattle where no age changes were detected in plasma cortisol and corticosterone, infusion of ACTH gave results which suggested that the adrenal cortex becomes much less responsive to ACTH with age (Riegle & Nellor, 1967).

During the early post-natal period there are large changes in the response of enzyme synthesis to glucocorticoids. The basic reaction in enzyme induction, the formation of RNA-polymerase, does not respond to the corticosteroids until 3 weeks after birth (Barnabei, Romano, Bitonto, Thomasi & Sereni, 1966). Also, the extent of enzyme induction after cortisol treatment depends on the age of the animal. Young rats show a large increase in glutamic-pyruvic transaminase but the percentage change in enzymic activity gradually diminishes with age (Harding, Rosen & Nichol, 1961). With tryptophan pyrrolase, another enzyme that is induced by cortisol, the kinetics of enzyme synthesis also vary with age (Correll, Turner & Haining, 1965).

The induction of glucose and fructose phosphatases by dexamethasone (a synthetic derivative of the adrenocorticoids) is also impaired as rats become old (Singhal, 1967a). The decrease in response is most marked between 1 and 9 months of age. Induction of phosphofructokinase in castrated rats using testosterone propionate is also age-dependent (Singhal, 1967 b). Very old animals have not been studied but there is a decline in response between 6 and 12 months which may well continue into old age. However, although it appears that in general the magnitude of enzyme induction is reduced with age, this does not take place progressively throughout the life-span and the available studies do not indicate that it is an important aspect of the senescent animal.

In none of the above studies which point to a variation in target response is it immediately clear that the differences in sensitivity resides in the target tissues. Although the alteration in an endocrine response may coincide with the development of new structures and function in the target tissues, it is difficult to delineate the fundamental process that produces a change in response. Once it has been established that the parameter being used to measure the response is not influenced by an age-change in some other tissue it is possible to use three experimental approaches to the general problem. The most fundamental method is concerned with a possible alteration in the effective concentration of the hormone. This involves an examination of the rate of hormone metabolism and the uptake and disposition of the hormone in the target cells throughout the life-span. A complementary method is to determine the nature of the hormone receptors on an age basis. The third approach requires an examination of the biochemistry of the cell in relation to possible age-variations in the enzymic potential of the target tissues to respond to the hormone.

Most studies on hormone uptake have involved the use of steroids where methods are available for the isolation and characterization of these compounds and their metabolites. However, a few workers have been concerned with age variations. In the human red cell, which is probably the simplest model system for the study of transport processes, there is a rise in the uptake of triiodothyronine with age, which reaches a plateau in early adult life (Soltz, Horonick & Chow, 1963). A study using radioactive adrenaline has shown that the old rat heart takes up more of a given dose of noradrenaline than young tissue. In this study, as in others concerned with changes in the uptake of hormones from the plasma, it is difficult to assess the contribution due to changes in the uptake mechanism as opposed to alterations in the proportion of total hormone available to the cell. This difficulty arises because the capacity of plasma proteins to bind the hormone and render it indiffusible may alter with age (Gala & Westphal, 1965). It has been found in this connection that the binding of [^{14}C]cholesterol to collagen fibres and aortic tissues from old animals decreases markedly with incubation in serum from young animals

(Hrůza, Babický & Hlaváčkova, 1967). With regard to cellular concentrations, two studies have been made on catecholamines. One of these (Gey, Burkard & Pletscher, 1965) suggests that noradrenaline in the myocardium decreases with age; the other shows that there are no significant age-changes in the endogenous catecholamines of adipose tissues (Stuchlíková, Jelínková, Hrůza & Hrušková, 1967). Results in this field appear to be inconsistent (Jelínková, Hrůza & Erdösová, 1967).

With reference to possible age variation in receptor structure, we are still largely ignorant of the chemical nature of the site of hormone action. For the steroids, it is known that cell proteins are important quantitatively, in the adsorption of hormones and metabolites (Bellamy, 1963) but no information is available on the age-related variations in protein structure and their influence on binding affinities. Physiochemical changes in collagen with age clearly affect the binding of ions and it may well be that similar proteins exist in target cells. It is thought that generally cell proteins reflect the process of ageing in their tertiary structure (Bjorksten, 1963). Such changes in configuration might affect the strength of bonds in steroid–protein complexes. But in skeletal muscle, which is a target tissue for steroid hormones and undergoes considerable changes in function with age, the age-related alterations in protein structure, as determined by enzymic methods, are not very marked (Kohn, 1963).

The response of target organs at the level of their ultrastructure has recently received attention. After thyroxine treatment swelling of liver mitochondria was more marked in old rats and there were indications of degeneration of the endoplasmic reticulum (Shamoto, 1968). Relatively high doses of thyroxine were used to produce these effects and it is possible that the age-differences reflect a failure of old tissues to resist the toxic effects of the hormone.

As to the biochemical potential of target tissues, as already indicated, it appears that the metabolism of some tissues is not affected by age, but this is not general. Muscle undergoes marked changes during the life-span. There is a decline in the work performance (Burke et al. 1953) which occurs at a time when there is a large fall in its relative cell mass. At the biochemical level there is a diminution in the capacity of aerobic energy production (Kiessling, 1962) and a rise in the activity of glycolytic enzymes (Stave, 1964). It has also been suggested that an observed drop of 80% in the magnesium-activated ATPase in the muscle of old rats is connected with the age-dependent loss of contractile efficiency (Rockstein & Brandt, 1962). All of these biochemical changes could well alter the response of muscle to hormones.

With regard to the renal response, it has already been pointed out that the ageing kidney becomes less sensitive to anti-diuretic hormone (Dicker & Nunn, 1958; Table 9). An age-dependent deterioration in renal function found in intact animals (Lewis & Alving, 1938) is also found in *in vitro* preparations of kidney (Beauchene, Fanestil & Barrows, 1965). The phenomenon coincides with a fall in the concentration of the ATPase thought to be a component of the active transport systems responsible for reabsorption from the glomerular filtrate. It is clear that the loss of hormone sensitivity is only one aspect of the ageing process in the kidney but it may well be linked with specific changes in the enzymes basic to ion transport. In summary, the generalization may be made that age-variations in metabolic pattern in target tissues are important factors leading to a decrease in the effectiveness of hormones.

Table 9. *Effect of age on the concentrating capacity of rat kidney**

Average body wt. (g)	Urine excretion (ml/100 g/24 h)	Osmotic pressure range (FP depression)	Anti-diuretic effect†	
			3 % NaCl	3·0 mμ vasopressin
250‡	0·62	−4·80 to −5·12	79·9±4·9	85·9±3·6
370§	0·97	−3·90 to −4·18	39·4±10·8	52·8±7·6

* Dicker & Nunn, 1958.
† Anti-diuretic effect (mean ± S.E.)

$$= \frac{\text{urine flow after injection} - \text{urine flow before injection}}{\text{urine flow before injection} \times 100}.$$

‡ Age about 100 days. § Age about 350 days.

CHANGES IN FUNCTION OF THE ENDOCRINE ORGANS

In the following discussion the function of an endocrine gland is defined relative to the rate of secretion of its products. Changes in secretion occur either through an ageing process intrinsic to the gland or through an alteration in the rate of metabolism or peripheral excretion of the hormone. In the latter two situations it is necessary to postulate a feedback mechanism that operates to maintain a fixed concentration of plasma hormone, i.e. a fall in the rate of hormone metabolism is counteracted by a decrease in the secretion rate. This appears to apply to the decline in the secretion of aldosterone and thyroxine in ageing humans (Verzár & Freydberg, 1956; Gregerman, Gaffney, Shock & Crowder, 1962; Tait, Rosemberg & Pincus, 1966; Table 10). Here the altered function of the endocrine organ may be compounded of several factors such as the quantity and concentration of metabolic enzymes, hepatic blood flow and the uptake of hormone by tissues other than the target organ.

Because of technical difficulties surrounding the methods of measurement, little is known of the variations in the concentration of circulating hormones throughout the life-span in most animals. From a biological point of view, a fall in the concentration of plasma growth hormone and the sex steroids would be expected with age, but little unequivocal work has been done on this aspect.

Table 10. *Thyroxine turnover in relation to age in humans**

Mean age (yr)	N	Plasma-bound iodine (μg%)	Distribution space (l.)	Half-life (days)	Fractional turnover rate	Degradation rate (μg/l./day)
21·1	12	6·90 ±1·10	12·34 ±3·46	6·61 ±0·87	0·107 ±0·015	88·65 ±24·88
53·7	18	6·06 ±1·00	12·01 ±3·15	7·94 ±1·22	0·089 ±0·014	63·08 ±14·45
65·1	13	6·83 ±1·07	11·65 ±2·39	9·10 ±1·35	0·078† ±0·012	58·84† ±17·34
74·4	15	6·54 ±0·60	10·79 ±1·61	9·17 ±1·06	0·076 ±0·008	53·63 ±9·30
83·2	13	6·24 ±0·60	9·34 ±1·05	9·25 ±0·89	0·076 ±0·008	43·89 ±6·66
90·5	2	5·70	9·05	8·19	0·086	42·38

* Gregerman *et al.* (means ± S.E.).   † $N = 12$, mean age 64·8.

A decreased concentration of plasma steroid hormones may be inferred from the age-related fall in the excretion of urinary 17-ketosteroids (Mason & Engstrom, 1950; Pincus, Romanoff & Carlo, 1954; Arvay & Takács, 1963). There is no change in the concentration of plasma-bound iodine with age (Rapport & Curtis, 1950), and from this it is inferred that the concentration of thyroid hormones is not affected in old animals. It appears that the constant steady-state level of these hormones is achieved despite a marked fall in the rate of turnover (Gregerman *et al.* 1962; Table 10). For the corticosteroids there is sound chemical evidence for changes in the concentration of circulating hormones. This applies mainly during the early phase of growth (Gala & Westphal, 1965). With regard to later-life there does not appear to be a marked difference in humans comparing the concentration of plasma corticosteroids in young adults with that in people over 65 years of age (Tyler, Eik-Nes, Sandberg & Florentin, 1955; Samuels, 1956). The rate of secretion and metabolism of corticosteroid hormones declines with age (Romanoff, Morris, Welch, Grace & Pincus, 1963; Grad, Kral, Payne & Berenson, 1967; Riegle, Przekop & Nellor, 1968), possibly related to a decrease in metabolic mass. A recent study emphasizes the small extent of the changes in the turnover rate of cortisol in humans and the lack of change in the plasma concentration (Serio, Piolanti, Capelli, Magistris, Ricci, Anzalone & Giusti, 1969).

However, plasma corticosterone in the mouse shows pronounced changes with age which are more marked in males compared with females (Grad & Khalid, 1968).

A complication in this work is that the concentration of total plasma hormone is not a guide to the concentration of the biologically active hormone since, for many hormones, a large proportion of the hormone interacts with the plasma proteins. The specific binding proteins in the plasma may vary independently of endocrine secretion, so that a variable proportion of the total hormone is in immediate diffusion equilibrium with the cells. In this connection, it is established that large changes in the plasma protein pattern occur during the life-span of a number of mammals (Rafsky, Newman & Krieger, 1949; Halliday & Kekwick, 1957; Das & Bhattacharya, 1961; Ringle & Dellenback, 1963) which may well be associated with difference in affinity for hormones.

However, the main problem in this field concerns the inadequate analytical methods that are available to measure small concentrations of hormones in peripheral plasma. An indirect approach is to study the capacity of preparations of endocrine organs to synthesize hormones *in vitro*. By this means it becomes easier to measure the larger amounts of steroids produced in the experiment. This method has been used to study age-differences in the rate of synthesis of androgens (Axelrod, 1965), corticosteroids (Kemény, Kemény & Vecsei, 1964) and growth hormone (Meites, Hopkins & Deuben, 1962).

A third approach to the problem of hormone availability may be made by measuring the amount of hormone stored in the endocrine gland. It is difficult for workers in this field to agree on a common interpretation of the changes in the quantities of stored hormones. For example, a decrease in the concentration of hormone in the endocrine gland may be regarded by some to indicate a fall in the rate of synthesis, whilst others would see the same phenomenon resulting from an increased rate of secretion with a concomitant high rate of synthesis.

With respect to the anterior pituitary, it appears that there are no age-related changes in the amount of stored growth hormone, ACTH and TSH (Blumenthal, 1954). In the case of growth hormone, it has been shown that a constant quantity is maintained in the gland throughout life without an impairment of the synthetic capacity of the tissue (Meites *et al.* 1962). The major factor that limits the amount of circulating growth hormone appears to be the concentration of releasing substance in the hypothalamus. Old rats have none of this releaser (Pecile, Müller, Falconi & Martini, 1965). In contrast to growth hormone, the amounts of pituitary gonadotrophins, FSH (follicle stimulating hormone) and LH (luteinizing hormone), in the pituitary increase considerably with age (Albert, Randall, Smith & Johnson, 1956).

Comparative work on the insulin content of the pancreas shows the dangers of generalizing from the situation with respect to stored hormones in a single species. In ageing cattle there is a decrease in the insulin concentration (Fisher & Scott, 1934), in rats an increase (Griffiths, 1941) and in humans no change, with age (Scott & Fisher, 1938). It is not known how these age differences are related to secretory activity.

Two factors that have bearing on changes in the magnitude of endocrine activity are connected with variation in the structure of endocrine cells and the amount of endocrine tissue relative to body volume. Like all other tissues in the body, endocrine glands show marked histological changes with age but none of the alterations are sufficiently unique to show that endocrine cells have a different pattern of ageing compared with other tissue types (Pokrajac, Rabadija, Vranic & Allegretti, 1959; Charipper, Pearlstein & Bourne, 1961; Bourne & Jayne, 1961; Steward & Brandes, 1961; Batali, Rogers & Blumenthal, 1961; Jayne, 1963). Most glands in higher vertebrates generally develop an increased amount of connective tissue with age. This appears as a thickening of the capsule, a rise in the density of collagen fibres and the replacement of secretory cells with connective tissue elements. In the endocrine cell, fragmentation of mitochondria and nuclear damage usually appear with age and there may be a decreased rate of mitosis. Ageing of endocrine glands in line with this pattern also occurs in lower vertebrates (Woodhead & Ellett, 1967).

In some glands specific histological changes are related to the characteristic structure or function of the tissue. For example, in the thyroid, age brings about a loss of follicles and a drop in the amount of stainable colloid. In the adrenal cortex there is an alteration in the lipid composition and pattern of zonation. Vasodilation also occurs, with an increasing frequency of haemorrhage. With regard to aldosterone secretion, a reduction in the granularity of the juxta-glomerular cells and a decreased width of the zona glomerulosa points to a morphological basis for the fall in the aldosterone secretion rate (Dunihue, 1965). However, as with other specialized tissues, in no case have histological changes in endocrine organs been linked directly with a specific endogenous variation in functional activity.

The proportion of endocrine tissue in the body varies with age. In the case of the thyroid there is a decrease in the weight of the gland relative to the body (Robertson, 1916). The change is most prominent during the early stages of postnatal life. For example, in the dog, during the first 8 weeks of life, there is an almost 50% fall in the relative weight of the thyroid (Haensly, Jermier & Getty, 1964). The relative weight of the adrenals and parathyroids also declines with age (Haensly & Getty, 1965; Baca & Chiodi, 1965; Raisz, O'Brien & Au, 1965). Complementary changes in the weight of the gonads also occur. These alterations in the relative weight of endocrine organs might have a bearing on the effective concentration of circulating hormone because of corresponding variation in the relative volume for distribution of the hormone, (Romanoff *et al.* 1963).

Little is known of age-related changes in the sensitivity of endocrine organs to the primary stimulus that results in hormone secretion. From this point of view the effectiveness of trophic hormones in influencing peripheral endocrine glands has been examined. After treatment with either ACTH or gonadotrophin, total urinary 17-ketosteroids was increased to a small extent in aged humans (Antonini, Porrow, Serio & Tinti, 1968). The problems are similar to those encountered in the assessment of the mechanisms responsible for changes in target-tissue sensitivity. It is established that homeostatic adjustments to the ingestion of water, sodium and potassium initially become more effective with age (Talbot & Richie, 1958), but more fundamental information is needed on the relationship between stimulus and response. For example, from the point of view of changes in the endocrine organs, it would be of interest to know the effect of age on both the secretory capacity and the time of response of the adrenal cortex to various concentrations of plasma sodium.

CHANGES IN THE DIRECTION OF THE ENDOCRINE RESPONSE

Most of the evidence that points to age-dependent changes in the direction of response has come from experiments on the action of exogenous corticosteroids and neurohypophysial hormones. The underlying principle is that age-dependent changes in the characteristics of the response to exogenous hormones are related to intrinsic changes in the homeostatic mechanisms due to new environmental demands (Krecek, Dlouha, Jelinek, Kreckova, Vacek, 1958). This applies particularly to the action of neurohypophysial hormones and adrenocorticosteroids on electrolyte excretion in rats. In the early stages of life vasopressin at a standard dose increases renal water loss in very young animals, but inhibits water excretion in old animals (Krecek *et al.* 1958). Similarly, cortisol increases the excretion of sodium and water in young rats but reduces water loss in older animals. The age difference here is most likely related to the fact that the younger, in contrast to the older, animals had not reached an age at which they were normally weaned. It is in this period that a difference is noted in the metabolic effects of adrenal hormones. For example, after adrenalectomy a 12 h period of starvation proved fatal to the majority of animals in a group of very young rats (20 days old) but old and young adult rats (60 days old up to 14–18 months) survived longer than 48 h (Vranic & Pokrajac, 1961). Again, this difference is probably linked with the endocrine control of energy stores and related to a change in the feeding behaviour after weaning. However, qualitative differences in hormone action are not confined to this early period of life. Aldosterone has little or no action on the excretion of potassium in 5-week-old weaned rats, but produces a marked inhibition of potassium output in rats 1 year old or more (Desaulles, 1958).

It is not known how these different effects of hormones are related to the development of homeostatic systems. Indeed, it may be that an explanation for some of the alterations in response is connected with variation in hormone sensitivity. For a number of hormones it is established that the effect of a small dose may be reversed with a high dose.

HOMEOSTASIS AND MORTALITY

With regard to the failure of homeostasis during ageing it has been suggested that, in general, there is a diminution of neural influences upon effector organs whilst the sensitivity to hormones increases (Frolkis, 1968). As already pointed out, there is much evidence in support of the failure of nervous co-ordination. The evidence for increased hormonal sensitivity comes mainly from experiments on the action of exogenous thyroxine. The data for other hormones are in favour of a decreased sensitivity (Table 11), although, as in the case of catecholamines, the degree of hormonal stimulation may depend upon the previous experience of the animal. The drop in sensitivity is also associated with a fall in the rate of hormone turnover.

Table 11. *Effect of insulin on glucose tolerance in humans**

Mean age (year)	N	Rate of disappearance of glucose from venous blood ($\%$/min)		
		—	Insulin†	Difference
31	12	$1 \cdot 68 \pm 0 \cdot 19$	$6 \cdot 39 \pm 0 \cdot 60$	$4 \cdot 12 \pm 0 \cdot 59$
49	11	$1 \cdot 44 \pm 0 \cdot 21$	$3 \cdot 61 \pm 0 \cdot 25$	$2 \cdot 17 \pm 0 \cdot 26$
78	12	$0 \cdot 98 \pm 0 \cdot 08$	$2 \cdot 49 \pm 0 \cdot 02$	$1 \cdot 52 \pm 0 \cdot 17$

* Silverstone *et al.* 1957.   † 5 units insulin/m² body surface.

There is a marked age-deterioration in many organs, particularly those with an endocrine function, and the cardiovascular system (Pickering, 1936). Despite this, the individual cells of liver, kidney and brain do not appear to undergo a drastic loss of functional capacity: it is probably true to say that for the latter two organs the deficiency in old age lies in the loss of cells. Despite these changes the ability to maintain day-to-day resting stability in an unchanging environment is largely unimpaired. Lack of adaptability in a disturbed environment is not due to the failure of any single component. The endocrine system probably fails to respond on several counts, bound up with the decreased capacity to secrete sufficient hormone and the failure of the hormone to evoke the correct degree of response in the target organ.

The major problem in the field of gerontology is to relate age changes in function to the increasing probability of death. Where a particular reflex action or rapid neuromuscular response is of importance to an organism in its search for food and shelter the reasons for the increased chances of mortality may be more or less defined. For example, the sensitivity of rat aorta to catecholamines is reduced in old animals. This is associated with a coarsening of the muscle fibres of the tunica media and an increased thickness and disarrangement of the elastic lamella. At a purely chemical level the old tissue appears to be more efficient in that it contains a higher concentration of protein and DNA. However, it is concluded that the structural changes alone are sufficient to account for the observed decrease in hormonal sensitivity through a limitation on the mechanical response (Tuttle, 1966). The gradual deterioration of effector organs containing muscle fibres and elastic tissue (Kirk & Chieffi, 1962) is probably an important factor in the poor adaptive responses of old organisms where the adaptation requires a rapid neurophysiological response at a much higher intensity than normal. However, animals still die over a narrow age-range when maintained under constant laboratory conditions where no sudden large demands are made upon the body.

It must be admitted that we are largely ignorant of the immediate cause of death in old age. This problem is distinct from the underlying molecular events that produce age-changes at a higher level of organisation. From this standpoint death is a consequence of events of a more complex order and is open to study without reference to the long-term progressive changes in function that lead up to it. It is true that the slow linear trend by which some physiological characters change with age is theoretically sufficient to account for the exponential increase in mortality observed at the population level (Sacher, 1951). But death is an individual process and involves events which take place over a small fraction of the life-span, possibly amounting to only a few hours. Death, in this context, is the result of an internal fluctuation which cannot be contained through homeostatic regulation; a small shift in metabolism which could be counteracted in youth becomes amplified to the point of preventing a vital function.

Thus death in old organisms may result from minor changes in the environment.

It has been a consistent finding that some, but by no means all, aspects of functional organization exhibit an increased variability with age (Tables 1, 3, 6–8). The increased variability may be due to the unmasking of genetic differences but there is also the possibility that homeostatic feedback systems are open to a greater degree of oscillation about the desired mean. Thus, the random chances of a fluctuation which exceeds the limits of homeostasis are increased with age. In this connection Simms (1942) pointed out that a large logarithmic increase in death-rate may result from a relatively small linear change in homeostasis which is coupled with the increased chance of random variability in function. He used haemorrhage as a measurable cause of death in healthy animals of various ages and was able to reproduce the known mortality curve for rats. On average, rats aged 50 days died after haemorrhage amounting to a 4·3 % loss of blood. Resistance to haemorrhage declined in a linear manner with age (amounting to −12 % over 724 days). There was also an increased variability in the value for the fatal loss of blood. Taking a loss of blood which was only 16 % lower than that necessary to kill the majority of young animals, the combined effect of the decreased resistance to death and the increased variability in response, resulted in a 16-fold increase in the probability of death occurring in old animals (Table 12).

Table 12. *Use of haemorrhage as a measurable cause of death in rats: loss of blood at the point of death in relation to age**

Mean age (days)	N	Mean body wt. (g)	Blood† (ml/100 g body wt.)	Log probability of death from haemorrhage
50	10	122	4·34 ± 0·23	−1·6
100	12	245	3·92 ± 0·19	−1·3
355	13	379	3·69 ± 0·11	−0·7
635	4	418	3·62 ± 0·38	−0·2
824	8	346	3·44 ± 0·26	−0·1

* Simms, 1942.
† Mean quantity of blood lost at time of death ± standard deviation.

Few experiments have been designed specifically to test the possibility that an increased degree of oscillation about the norm occurs in old animals. With regard to future work it is desirable to concentrate on the fluctuations in homeostatic control. In the past, too much emphasis has been placed on the maintenance of mean values which are regarded as constant norms. Data for many individual animals are required to establish these statistical norms and the short-term fluctuations within any one animal are largely ignored. In order to understand the increased tendency for homeostasis to fail during ageing we must first determine just how stable the 'constant' internal environment is at all ages, and investigate the reasons for changes in norms and 'overshoot phenomena' which in the old organism may ultimately result in death.

REFERENCES

ACKERMANN, P. G. & KHEIM, T. (1964). The effect of testosterone on plasma amino acid levels in elderly individuals. *J. Geront.* **19**, 207–14.

ALBERT, A., RANDALL, R. V., SMITH, R. A. & JOHNSON, C. E. (1956). *Hormones and Ageing*. New York and London: Academic Press.

ALBERTINI, A. (1952). What is ageing? *J. Geront.* **7**, 452–63.

ANTONINI, F. M., PORRO, A., SERIO, M. & TINTI, P. (1968). Gas chromatographic analysis of urinary 17-keto steroids response to gonadotrophin and ACTH in young and old persons. *Exp. Geront.* **3**, 181–92.

ALLEN, T. H., ANDERSON, E. C. & LANGERHAM, W. H. (1960). Total body potassium and gross body composition in relation to age. *J. Geront.* **15**, 348–57.

ARVAY, A. VON & TAKACS, I. (1963). Einfluss der sexual function auf die Ausscheidung der Nebennierenrindensteroide in laufe der Alterung von Ratten. *Gerontologia* **8**, 81–91.

AXELROD, L. R. (1965). Metabolic patterns of steroid biosynthesis in young and aged human testes. *Biochim. biophys. Acta* **97**, 551–6.

BACA, Z. V. & CHIODI, H. (1965). Developmental changes in the size and ascorbic acid content of the adrenals of white rats. *Endocrinology* **76**, 1208–12.

BARNABEI, O., ROMANO, B., BITONTO, G. DI, THOMASI, V. & SERENI, F. (1966). Factors influencing the glucocorticoid-induced increase of ribonucleic acid polymerase activity of rat liver nuclei. *Archs Biochem. Biophys.* **113**, 478–86.

BARROWS, C. H. & ROEDER, L. M. (1961). Effect of age on protein synthesis in rats. *J. Geront.* **16**, 321–5.

BARROWS, C. H., FALZONE, J. H. & SHOCK, N. W. (1960). Age differences in the succinoxidase activity of homogenates and mitochondria from the livers and kidneys of rats. *J. Geront.* **15**, 130–3.

BARROWS, C. H., ROEDER, L. M. & FALZONE, J. A. (1962). Effect of age on the activities of enzymes and the concentrations of nucleic acids in the tissues of female wild rats. *J. Geront.* **17**, 144–7.

BARROWS, C. H., ROEDER, L. M. & OLEWINE, D. A. (1962). Effect of age on renal compensatory hypertrophy following unilateral nephrectomy in the rat. *J. Geront.* **17**, 148–50.

BARROWS, C. H., YIENGST, M. J. & SHOCK, N. W. (1958). Senescence and the metabolism of various tissues of rats. *J. Geront.* **13**, 351–5.

BASHEY, R. I., TORII, S. & ANGRIST, A. (1967). Age related collagen and elastin content of human heart valves. *J. Geront.* **22**, 203–8.

BATALI, M., ROGERS, J. B. & BLUMENTHAL, H. T. (1961). Ageing processes in the guinea pig ovary. *J. Geront.* **16**, 230–8.

BEAUCHENE, R. E., FANESTIL, D. D. & BARROWS, C. H. (1965). The effect of age on active transport and sodium-potassium-activated ATPase activity in renal tissue of rats. *J. Geront.* **20**, 306–10.

BEAUCHENE, R. E., ROEDER, L. M. & BARROWS, C. H. (1967). The effect of age and of ethionine feeding on the ribonucleic acid and protein synthesis of rats. *J. Geront.* **22**, 318–24.

BELASCO, I. J. & MURLIN, J. R. (1941). The effect of thyroxin and thyrotropic hormone on the basal metabolism and thyroid tissue respiration of rats at various ages. *Endocrinology* **28**, 145–52.

BELLAMY, D. (1963). The adsorption of corticosteroids to particulate preparations of rat liver. *Biochem J.* **87**, 334–40.

BERG, B. N. & HARMISTON, C. R. (1955). Blood pressure and heart size in ageing rats. *J. Geront.* **10**, 416–19.

BIRREN, J. E. (1955). Age differences in startle reaction time of the rat to noise and electric shock. *J. Geront.* **10**, 437–40.

BIRREN, J. E. & BOTWINICK, J. (1955). Age differences in finger, jaw and foot reaction time to auditory stimuli. *J. Geront.* **10**, 429–32.

BIRREN, J. E. & KAY, H. (1958). Swimming speed of the albino rat. Age and sex differences. *J. Geront.* **13**, 374–7.

BJORKSTEN, J. (1963). Ageing, primary mechanism. *Gerontologia* **8**, 179–92.

BLUMENTHAL, H. T. (1954). Relation of age to the hormonal content of the human anterior hypophysis. *Archs Path.* **57**, 481–94.

BODANSKY, M. & DUFF, V. B. (1936). Age as a factor in the resistance of the albino rat to thyroxine with further observations on the creatine content of tissues in experimental hyperthyroidism. *Endocrinology* **20**, 541–5.

BOGART, B. I. (1967). The effect of ageing on the histochemistry of the rat submandibular gland. *J. Geront.* **22**, 372–5.

BOURNE, G. H. & JAYNE, E. P. (1961). The adrenal gland, pp. 303–24. In *Structural Aspects of Ageing* (ed. G. H. Bourne). London: Pitman.

BUCHER, N. L. R. & GLINOS, A. D. (1950). The effect of age on regeneration of rat liver. *Cancer Res.* **10**, 324–32.

BURKE, W. E., TUTTLE, W. W., THOMPSON, C. W., JANNEY, C. D. & WEBER, R. J. (1953). The relation of grip strength and grip strength endurance to age. *J. appl. Physiol.* **5**, 628–30.

CHANDLER, A. R. (1952). A note on the meaning of ageing. *J. Geront.* **7**, 437–8.

CHARIPPER, H. A., PEARLSTEIN, A. & BOURNE, G. H. (1961). Ageing changes in the thyroid and pituitary glands. In *Structural Aspects of Ageing*, pp. 261–76 (ed. G. H. Bourne). London: Pitman.

CHESTER JONES, I. & BELLAMY, D. (1964). Hormonal mechanisms in the homeostatic regulation of the vertebrate body with special reference to the adrenal cortex. *Soc. exp. Biol. Symp.* **18**, 195–236.

CHVAPIL, M. & ROTH, Z. (1964). Connective tissue changes in wild and domesticated rats. *J. Geront.* **19**, 414–18.

COLLINS, D. D. (1950). Mortality in traumatic abdominal injuries in the elderly. *J. Geront.* **5**, 241–4.

CONARD, R. A. (1960). An attempt to quantify some clinical criteria of ageing. *J. Geront.* **15**, 358–65.

CORRELL, W. W., TURNER, M. D. & HAINING, J. L. (1965). Changes in trytophan pyrrolase induction with age. *J. Geront.* **20**, 507–10.

DAS, B. C. (1964). Indices of blood biochemistry in relation to age, height and weight. *Gerontologia* **9**, 179–92.

DAS, B. C. & BHATTACHARYA, S. K. (1961). Changes in human serum protein fractions with age and weight. *Can. J. Biochem. Physiol.* **39**, 569–79.

DESAULLES, P. A. (1958). *Ciba Fdn Colloq. Ageing.* **4**, p. 180 (eds. G. E. W. Wolstenholme and M. O'Connor). London: Churchill.

DETWILER, T. C. & DRAPER, H. H. (1962). Physiological aspects of ageing. IV. Senescent changes in the metabolism and composition of nucleic acids of the liver and muscle of the rat. *J. Geront.* **17**, 138–43.

DICKER, S. E. & NUNN, J. (1958). Antidiuresis in adult and old rats. *J. Physiol.,* *Lond.* **141**, 332–6.

DUNCAN, L. E., SOLOMON, D. H., ROSENBERG, E. K., NICHOLS, M. P. & SHOCK. N. W. (1952). The metabolic and hematologic effects of the chronic administration of ACTH to young and old men. *J. Geront.* **7**, 351–74.

DUNIHUE, F. W. (1965). Reduced juxtaglomerular cell granularity, pituitary neurosecretory material and width of the zona glomerulosa in ageing rats. *Endocrinology* **77**, 948–51.

EISDORFER, C., POWELL, A. H., SILVERMAN, G. & BOGDONOFF, M. D. (1965). The characteristics of lipid mobilization and peripheral disposition in aged individuals. *J. Geront.* **20**, 511–14.

ENESCO, H. E. (1967). A cytophotometric analysis of DNA content of rat nuclei in ageing. *J. Geront.* **22**, 445–8.

ENZMANN, E. V. & PINCUS, G. (1934). The extinction of reflexes in spinal mice of different ages as an indicator of the decline of anaerobiosis. *J. gen. Physiol.* **18**, 163–9.

EVERITT, A. V. (1959). The effect of pituitary growth hormone on the ageing male rat. *J. Geront.* **14**, 415–24.

EVERITT, A. V. & WEBB, C. (1958). The blood picture of the ageing male rat. *J. Geront.* **13**, 255–60.

FALZONE, J. A., BARROWS, C. H. & SHOCK, N. W. (1959). Age and polyploidy of rat liver nucleii as measured by volume and DNA content. *J. Geront.* **14**, 2–8.

FEDORČÁKOVA, A. M., BACHLEDOVÁ, E., NIEDERLAND, T. R., & BÓZNER, A. (1968). Some organ changes in old and young rats after a pathogenic diet. *Exp. Geront.* **3**, 63–7.

FISHER, A. M. & SCOTT, D. A. (1934). The insulin content of the pancreas in cattle of various ages. *J. biol. Chem.* **106**, 305–10.

FLETCHER, M. J. & SANADI, D. R. (1961). Turnover of liver mitochondrial components in adult and senescent rats. *J. Geront.* **16**, 255–7.

FRIEDFELD, L. (1949). Heat reaction states in the aged. *Geriatrics* **4**, 211–16.

FROLKIS, V. V. (1966). Neuro-humoral regulations in the ageing organism. *J. Geront.* **21**, 161–7.

FROLKIS, V. V. (1968). Regulatory process in the mechanism of ageing. *Exp. Geront.* **3**, 113–23.

FLÜCKIGER, E. & VERZÁR, F. (1955). Lack of adaptation to low oxygen pressure in aged animals. *J. Geront.* **10**, 306–11.

GALA, R. R. & WESTPHAL, U. (1965). Corticosteroid-binding globulin in the rat: possible role in the initiation of lactation. *Endocrinology* **76**, 1079–88.

GERTLER, M. M. & OPPENHEIMER, B. S. (1953). Serum uric acid levels in men and women past the age of 65 years. *J. Geront.* **8**, 465–71.

GEY, K. F., BURKARD, W. P. & PLETSCHER, A. (1965). Variation of the norepinephrine metabolism of the rat heart with age. *Gerontologia* **11**, 1.

GIARNIERI, D. & LUMIA, V. (1961). Action of insulin on blood diphosphothiamine (DPT) in young and aged diabetic patients. *Clinica chim. Acta* **6**, 144–5.

GOLD, P. H., GEE, M. V. & STREHLER, B. L. (1968). Effect of age on oxidative phosphorylation in the rat. *J. Geront.* **23**, 509–12.

GRAD, B. (1969). The metabolic responsiveness of young and old female rats to thyroxine. *J. Geront.* **24**, 5–11.

GRAD, B. & KHALID, R. (1968). Circulating corticosterone levels of young and old, male and female C57B1/6J mice. *J. Geront.* **23**, 522–8.

GRAD, B. & KRAL, V. A. (1956). The effect of senescence on resistance to stress. 1. Response of young and old mice to cold. *J. Geront.* **12**, 172–81.

GRAD, B., KRAL, A. V., PAYNE, R. C. & BERENSON, J. (1967). Plasma and urinary corticoids in young and old persons. *J. Geront.* **22**, 66–71.

GREGERMANN, R. I., GAFFNEY, G. W., SHOCK, N. W. & CROWDER, S. E. (1962). Thyroxine turnover in euthyroid man with special reference to changes with age. *J. clin. Invest.* **41**, 2065–74.

GRIFFITHS, M. (1941). The influence of anterior pituitary extracts on the insulin content of the pancreas of the hypophysectomized rat. *J. Physiol., Lond.* **100**, 104–11.

GRISWOLD, R. L. & PACE, N. (1956). A comparison of intracellular nitrogen and metal ion distribution patterns in the livers of young and old rats. *J. Geront.* **12**, 150–2.

GUTMANN, E., HANZLÍKOVÁ, V. & JAKOVBEK, B. (1968). Changes in the neuro-muscular system during old age. *Exp. Geront.* **3**, 141–6.

HAENSLY, W. E. & GETTY, R. (1965). Age changes in the weight of the adrenal glands of the dog. *J. Geront.* **20**, 544–7.

HAENSLY, W. E., JERMIER, J. A. & GETTY, R. (1964). Age changes in the weight of the thyroid gland of the dog from birth to senescence. *J. Geront.* **19**, 54–6.

HALL, D. A., KEECH, M. K., REED, R., SAXL, H., TUNBRIDGE, R. E. & WOOD, M. J. (1955). Collagen and elastin in connective tissue. *J. Geront.* **10**, 388–400.

HALLIDAY, R. & KEKWICK, R. A. (1957). Electrophoretic analysis of the sera of young rats. *Proc. Roy. Soc. Lond.* **146**, 431–7.

HARDING, H. R., ROSEN, F. & NICHOL, C. A. (1961). Influence of age, adrenalec-tomy and corticosteroids on hepatic transaminase activity. *Am. J. Physiol.* **201**, 271–5.

HELLON, R. F. & LIND, A. R. (1958). The influence of age on peripheral vasodila-tion in a hot environment. *J. Physiol., Lond.* **141**, 262–72.

HENNES, A. R., WAJCHENBERG, B. L., FAJANS, S. S. & CONN, J. W. (1957). The effect of adrenal steroids on blood levels of pyruvic and α ketoglutaric acids in normal subjects. *Metabolism* **6**, 339.

HENRY, J. P., MEEHAN, J. P., STEPHENS, P. & SANTISTEBAN, G. A. (1965). Arterial pressure in CBA mice as related to age. *J. Geront.* **20**, 239–43.

HODGKINS, J. (1962). Influence of age on the speed of reaction and movement in females. *J. Geront.* **17**, 385–9.

HOLLANDER, J. & BARROWS, C. H. (1968). Enzymatic studies in senescent rodent brains. *J. Geront.* **23**, 174–9.

HORVATH, S. M. (1946). The influence of the ageing process on the distribution of certain components of the blood and the gastrocnemius muscle of the albino rat. *J. Geront.* **1**, 213–23.

HRŮZA, Z., BABICKÝ, A. & HLAVÁČKOVÁ, V. (1967). Binding of cholesterol to the rat aorta and collagen fibres during ageing and atherosclerosis *in vitro*. *Exp. Geront.* **2**, 101–7.

HRŮZA, Z. & JELÍNKOVÁ, M. (1965). Carbohydrate metabolism after epinephrine glucose and stress in young and old rats. *Exp. Geront.* **1**, 139–47.

HUBBLE, D. (1957). Some principles of homeostasis. *Lancet* ii, 301–5.

JAKUBCZAK, L. F. (1967). Age differences in the effects of terminal food deprivation (starvation) on activity, weight loss, and survival of rats. *J. Geront.* **22**, 421–6.

JAYNE, E. P. (1963). A histo-cytologic study of the adrenal cortex in mice as influ-enced by strain, sex and age. *J. Geront.* **18**, 227–34.

JELÍNKOVÁ, M., HRŮZA, Z. & ERDÖSOVÁ, R. (1967). The effect of the application of epinephrine on its level in the adipose tissue in rats of different age. *Exp. Geront.* **2**, 63–71.

KANABROCKI, E. L., FELS, I. G., DECKER, C. F. & KAPLAN, E. (1963). Total hexo-samine, sulphur and nitrogen levels in human aortae. *J. Geront.* **18**, 18–22.

KEMÉNY, V., KEMÉNY, A. & VECSEI, P. (1964). Adrenal function of new born and adult rats. *Acta physiol. hung.* **25**, 31–7.

KIESSLING, K. H. (1962). Respiration and oxidative phosphorylation of muscle mitochondria from rats of various ages. *Expl Cell. Res.* **28**, 145–50.

KIRK, J. E. (1959a). The adenylpyrophosphatase, inorganic pyrophosphatase and phosphomonoesterase activities of human arterial tissue in individuals of various ages. *J. Geront.* **14**, 181–8.

KIRK, J. E. (1959b). The 5-nucleotidase activity of human arterial tissue in individuals of various ages. *J. Geront.* **14**, 288–91.

KIRK, J. E. (1959c). The ribose-5-phosphate isomerase activity of arterial tissue in individuals of various ages. *J. Geront.* **14**, 447–9.

KIRK, J. E. (1960a). The leucine aminopeptidase activity of arterial tissue in individuals of various ages. *J. Geront.* **15**, 136–41.

KIRK, J. E. (1960b). The glyoxylase I activity of arterial tissue in individuals of various ages. *J. Geront.* **15**, 139–41.

KIRK, J. E. (1960c). The isocitric dehydrogenase and TPN-malic enzyme activities of arterial tissue in individuals of various ages. *J. Geront.* **15**, 262–6.

KIRK, J. E. (1961a). The aconitase activity of arterial tissue in individuals of various ages. *J. Geront.* **16**, 25–8.

KIRK, J. E. (1961b). The purine nucleoside phosphorylase activity of arterial tissue in individuals of various ages. *J. Geront.* **16**, 243–6.

KIRK, J. E. (1962a). The glycogen phosphorylase activity of arterial tissue in individuals of various ages. *J. Geront.* **17**, 154–57.

KIRK, J. E. (1962b). The cathepsin activity of arterial tissue in individuals of various ages. *J. Geront.* **17**, 158–62.

KIRK, J. E. (1962c). The diaphorase and cytochrome C reductase activities of arterial tissue in individuals of various ages. *J. Geront.* **17**, 276–80.

KIRK, J. E. (1962d). Variation with age in the creatine phosphokinase activity of human aortic tissue. *J. Geront.* **17**, 369–72.

KIRK, J. E. (1965). The glutamic dehydrogenase and glutathione reductase activities of arterial tissue in individuals of various ages. *J. Geront.* **20**, 357–62.

KIRK, J. E. (1966). The phosphoglucomutase, phosphoglyceric acid mutase and phosphomannose isomerase activities of arterial tissue of various ages. *J. Geront.* **21**, 420–5.

KIRK, J. E. & CHIEFFI, M. (1962). Variation with age in elasticity of skin and subcutaneous tissue in human individuals. *J. Geront.* **17**, 373–80.

KIRK, J. E. & DYRBYE, M. (1956a). The phenolsulfatase activity of aortic and pulmonary artery tissue in individuals of various ages. *J. Geront.* **11**, 129–33.

KIRK, J. E. & DYRBYE, M. (1956b). Hexosamine and acid-hydrolyzable sulphate concentrations of the aorta and pulmonary artery in individuals of various ages. *J. Geront.* **11**, 273–81.

KIRK, J. E., & RITZ, E. (1967). The glyceraldehyde-3-phosphate and α glycerophosphate dehydrogenase activities of arterial tissue in individuals of various ages. *J. Geront.* **22**, 427–38.

KIRK, J. E. & SØRENSEN, L. B. (1956). The aldolase activity of aortic and pulmonary artery tissue in individuals of various ages. *J. Geront.* **11**, 373–8.

KIRK, J. E., WANG, M. S. & BRANDSTRUP, N. (1959). The glucose-6-phosphate and 6-phospho-gluconate dehydrogenase activities of arterial tissue in individuals of various ages. *J. Geront.* **14**, 25–31.

KOHN, R. R. (1963). Age-related variation in susceptibility of protein in subcellular fractions of human muscle to digestion by pepsin. *J. Geront.* **18**, 14–17.

KRAG, C. L. & KOUNTZ, W. B. (1950). Stability of body function in the aged. 1. Effect of exposure of the body to cold. *J. Geront.* **5**, 227–35.

KRAG, C. L. & KOUNTZ, W. B. (1952). Stability of body function in the aged. 2. Effect of exposure of the body to heat. *J. Geront.* **7**, 61–70.

KRECEK, J., DLOUHA, H., JELINEK, J., KRECKOVA, J. & VACEK, Z. (1958). *Ciba Fdn Colloq. Ageing.* **4**, 165 (eds. G. E. W. Wolstenholme and M. O'Connor). London: Churchill.

KURNICK, N. B. & KERNEN, R. L. (1962). The effect of ageing on the desoxyribo-nuclease system, body and organ weight and cellular content. *J. Geront.* **17**, 245–53.

LANSING, A. I., ROSENTHAL, T. B. & ALEX, M. (1950). Significance of medial age changes in the human pulmonary artery. *J. Geront.* **5**, 211–15.

LEWIS, W. H. & ALVING, A. S. (1938). Changes with age in the blood pressures in adult men. *Am. J. Physiol.* **123**, 500–5.

LOWRY, O. H., HASTINGS, A. B., McCAY, C. M. & BROWN, A. N. (1946). Histo-chemical changes associated with ageing. IV. Liver, brain and kidney in the rat. *J. Geront.* **1**, 345–57.

MACNIDER, W. DE B. (1917). A consideration of the relative toxicity of uranium nitrate for animals of different ages. *J. exp. Med.* **26**, 1–17.

MACNIDER, W. DE B. (1943). Stability of acid-base equilibrium of blood in animals falling in different age periods. *Proc. Soc. exp. Biol. Med.* **53**, 1–8.

MACNIDER, W. DE B. (1946). The factor of age in determining the toxicity of certain poisons. *J. Geront.* **1**, 189–95.

MAGLADERY, J. W., TEASDALL, R. D. & NORRIS, A. H. (1958). Effect of ageing on plantar flexor and superficial abdominal reflexes in man—a clinical and electro-myographic study. *J. Geront.* **13**, 282–8.

MASON, H. L. & ENGSTROM, W. W. (1950). The 17-ketosteroids: their origin determination and significance. *Physiol. Rev.* **30**, 321–74.

McCANCE, R. A. & WIDDOWSON, E. M. (1957). The physiology of the new born animal. *Lancet* ii, 585–8.

McKAY, L. L., McKAY, E. M. & ADDIS, T. (1924). The effect of various factors on the degree of compensatory hypertrophy after unilateral nephrectomy. *J. Clin. Invest.* **1**, 576–7.

McFARLAND, R. A. & FISHER, M. B. (1955). Alterations in dark adaptation as a function of age. *J. Geront.* **10**, 424–8.

MEITES, J., HOPKINS, T. F. & DEUBEN, R. (1962). Growth hormone production by rat pituitary in vitro. *Fedn Proc. Fedn Am. Socs. exp. Biol.* **21**, 196.

MILLER, J. H., McDONALD, R. K. & SHOCK, N. W. (1951). The renal extraction of p-aminohippurate in the aged individual. *J. Geront.* **6**, 213–16.

MILLER, J. H., McDONALD, R. K. & SHOCK, N. W. (1952). Age changes in the maximal rate of renal tubular reabsorption of glucose. *J. Geront.* **7**, 196–200.

MOON, H. D., KONEFF, A. A., LI, C. H. & SIMPSON, M. E. (1956). Pheochromo-cytomas of adrenals in male rats chronically injected with pituitary growth hormone. *Proc. Soc. exp. Biol. Med.* **93**, 74–7.

MOORE, R. O. (1968). Effect of age of rats on the response of adipose tissue to insulin and the multiple forms of hexokinase. *J. Geront.* **23**, 45–9.

MYERS, V. C. & LANG, W. C. (1946). Some chemical changes in the human thora-cic aorta accompanying the ageing process. *J. Geront.* **1**, 441–4.

NORRIS, A. H., SHOCK, N. W. & YIENGST, M. J. (1955). Age differences in ventilatory and gas exchange responses to graded exercise in males. *J. Geront.* **10**, 145–55.

PECILE, A., MULLER, E., FLACONI, G. & MARTINI, L. (1965). Growth hormone-releasing activity of hypothalamic extracts at different ages. *Endocrinology* **77**, 241–6.

PICKERING, G. W. (1936). The peripheral resistance in persistent hypertension. *Clin. Sci.* **2**, 209–35.

PIERSON, W. R. & MONTOYE, H. J. (1958). Movement time, reaction time and age. *J. Geront.* **13**, 418–21.

PINCUS, G., ROMANOFF, L. P. & CARLO, J. (1954). Excretion of urinary steroids by men and women of various ages. *J. Geront.* **9**, 113–17.

POKRAJAC, N., RABADIJA, L., VRANIC, M. & ALLEGRETTI, N. (1959). Das insel-system der Ratte in verschiedenen Lebensaltern. *Naturwissenschaften* **46**, 338–9.

PRAETORIUS, E. (1951). Plasma uric acid in aged and young persons. *J. Geront.* **6**, 135–7.

RAFSKY, H. A., NEWMAN, B. & HORONICK, A. (1952). Age differences in respiration of guinea pig tissues. *J. Geront.* **7**, 38–40.

RAFSKY, H. A., NEWMAN, B. & KRIEGER, C. I. (1949). Electrophoretic analysis of the plasma proteins in the aged. *Am. J. Med. Sci.* **217**, 206–10.

RAISZ, L. G., O'BRIEN, J. E. & AU, W. Y. W. (1965). Parathyroid size and uptake of alpha-aminoisobutyric acid in intact and hypophysectomized rats. *Proc. Soc. Exp. Biol. Med.* **119**, 1048–53.

RAPPORT, R. L. & CURTIS, G. M. (1950). The clinical significance of the blood iodine: a review. *J. Clin. Endocr. Metab.* **10**, 735–90.

REITER, R. J., McCREIGHT, C. E. & SULKIN, N. M. (1964). Age differences in cellular proliferation in rat kidneys. *J. Geront.* **19**, 485–9.

RIEGLE, G. D. & NELLOR, J. E. (1966). Changes in blood cellular and protein components during ageing. *J. Geront.* **21**, 435–8.

RIEGLE, G. D. & NELLOR, J. E. (1967). Changes in adrenocortical function during ageing in cattle. *J. Geront.* **22**, 83–7.

RIEGLE, G. D., PRZEKOP, F. & NELLOR, J. E. (1968). Changes in adrenocortical responsiveness to ACTH infusion in ageing goats. *J. Geront.* **23**, 187–90.

RING, G. C., DUPUCH, G. H. & EMERIC, D. (1964). Ageing and the whole body metabolism. *J. Geront.* **19**, 215–19.

RINGLE, D. A. & DELLENBACK, R. J. (1963). Age changes in plasma proteins of the Mongolian gerbil. *Am. J. Physiol.* **204**, 275–8.

RITZ, E. & KIRK, J. E. (1967). The phosphofructokinase and sorbitol dehydrogenase activities of arterial tissue in individuals of various ages. *J. Geront.* **22**, 433–8.

ROBERTSON, T. B. (1916). Experimental studies on growth. II. The normal growth of the white mouse. *J. biol. Chem.* **24**, 363–83.

ROCKSTEIN, M. & BRANDT, K. (1962). Muscle enzyme activity and changes in weight in ageing white rats. *Nature, Lond.* **196**, 142–3.

ROMANOFF, L. P., MORRIS, C. W., WELCH, P., GRACE, M. P. & PINCUS, G. (1963). Metabolism of progesterone-4-C^{14} in young and elderly men. *J. clin. Endocr. Metab.* **23**, 286.

ROGERS, J. B. (1951). The ageing process in the guinea pig. *J. Geront.* **6**, 13–16.

ROMAN, L. & STRONG, L. C. (1961). Age, gestation, mortality and litter size in mice. *J. Geront.* **17**, 37–9.

ROOT, A. W. & OSKI, F. A. (1969). Effects of human growth hormone in elderly males. *J. Geront.* **24**, 97–104.

RUGER, H. A. & STOESSIGER, B. (1927). On the growth curves of certain characters in man (Males). *Ann. Eugen.* **2**, 76–110.

SACHER, G. A. (1951). On the statistical nature of mortality with especial reference to chronic radiation mortality. *Radiology* **67**, 250–8.

SAMSON, F. E., BALFOUR, W. M. & DAHL, N. A. (1958). The effect of age and temperature on the cerebral energy requirement in the rat. *J. Geront.* **13**, 248–51.

SAMUELS, L. T. (1956). *Hormones and the Ageing Process.* New York and London: Academic Press.

SCHAUB, M. C. (1963). Qualitative and quantitative changes of collagen in paren-chymatous organs of the rat during ageing. *Gerontologia* **8**, 114–22.

SCHAUS, R. & KIRK, J. E. (1956). The riboflavin concentration of brain, heart and skeletal muscle in individuals of various ages. *J. Geront.* **11**, 147–50.

SCHAUS, R., KIRK, J. E. & LAURSEN, T. J. S. (1955). The riboflavin content of human aortic tissue. *J. Geront.* **10**, 170–7.

SCHMUKLER, M. & BARROWS, C. H. (1967). Effect of age on dehydrogenase hetero-geneity in the rat. *J. Geront.* **22**, 1–7.

SCHNEEBERG, N. G. & FINESTONE, I. (1952). The effect of age on the intravenous glucose tolerance test. *J. Geront.* **7**, 54–60.

SCOTT, D. A. & FISHER, A. M. (1938). The insulin and the zinc content of normal and diabetic pancreas. *J. clin. Invest.* **17**, 725–8.

SERIO, M., PIOLANTI, P., CAPELLI, G., MAGISTRIS, L. DE, RICCI, F., ANZALONE, M. & GIUSTI, G. (1969). The miscible pool and turnover rate of cortisol in the aged and variations in relation to time of day. *Exp. Geront.* **4**, 95–101.

SHAMOTO, M. (1968). Age differences in the ultrastructure of hepatic cells of thyroxine-treated rats. *J. Geront.* **23**, 1–8.

SHATTUCK, G. C. & HILFERTY, A. (1932). Sunstroke and allied conditions in the United States. *Am. J. Trop. Med.* **12**, 223–45.

SHAW, K. E. & NICHOLS, R. E. (1962). The influence of age upon the circulating 17-hydroxycorticosteroids of cattle subjected to blood sampling and exo-genous adrenocorticotrophic hormone and hydrocortisone. *Am. J. vet. Res.* **23**, 1217–18.

SHOCK, N. W. (1952). *Cowdry's Problems of Ageing*, pp. 415–46. Baltimore: Williams Wilkins.

SHOCK, N. W. & YIENGST, M. J. (1950). Age changes in the acid-base equilibrium of the blood of males. *J. Geront.* **5**, 1–4.

SHOCK, N. W. & YIENGST, M. J. (1955). Age changes in basal respiratory measure-ments and metabolism in males. *J. Geront.* **10**, 31–40.

SHOCK, N. W., WATKIN, D. M., YIENGST, M. J., NORRIS, A. H., GAFFNEY, G. W., GREGERMAN, R. I. & FALZONE, J. A. (1963). Age differences in the water con-tent of the body as related to basal oxygen consumption in males. *J. Geront.* **18**, 1–8.

SHUKLA, S. P. & KANUNGO, M. S. (1969). Effect of age on the activity of arginase of the liver and kidney cortex of rat. *Exp. Geront.* **4**, 57–60.

SILVERSTONE, F. A., BRANDFONBRENER, M., SHOCK, N. W. & YIENGST, M. J. (1957). Age differences in the intravenous glucose tolerance tests and the response to insulin. *J. clin. Invest.* **36**, 504–14.

SIMMS, H. S. (1942). The use of a measurable cause of death, (haemorrhage) for the evaluation of ageing. *J. gen. Physiol.* **26**, 169–78.

SINGHAL, R. L. (1967a). Effect of age on the induction of glucose-6-phosphatase and fructose-1-6 diphosphatase in rat liver. *J. Geront.* **22**, 77–82.

SINGHAL, R. L. (1967b). Effect of age on phosphofructokinase induction in rat prostate and seminal vesicle. *J. Geront.* **22**, 343–7.

SOBEL, H. & HEWLETT, M. J. (1967). Effect of age on hyaluronic acid in hearts of dogs. *J. Geront.* **22**, 196–8.

SOBEL, H., HRUBANT, H. E. & HEWLETT, M. J. (1968). Changes in the body composition of C57BL/6-aa mice with age. *J. Geront.* **23**, 387–9.

SODERWALL, A. L., KENT, H. A., TURBYFILL, C. L. & BRITENBAKER, A. L. (1960). Variation in gestation length and litter size of the golden hamster, *Mesocricetus auratus*. *J. Geront.* **15**, 246–8.

SOLOMON, D. H. & SHOCK, N. W. (1950). Studies of adrenal cortical and anterior pituitary function in elderly men. *J. Geront.* **5**, 302–13.

SOLTZ, W. B., HORONICK, A. & CHOW, B. F. (1963). Age wise difference in the uptake of radioactivive triiodothyronine (T3) by red blood cells. *J. Geront.* **18**, 151–4.

STAVE, V. (1964). Age-dependent changes of metabolism. I. Studies of enzyme patterns of rabbit organs. *Biol. Neonat.* **6**, 128–47.

STEWARD, V. W. & BRANDES, D. (1961). *Structural Aspects of Ageing*, pp. 398–414. London: Pitman.

STREICHER, E. (1958). Age changes in the calcium content of rat brain. *J. Geront.* **13**, 356–8.

STREICHER, E. & GARBUS, J. (1955). The effect of age and sex on the duration of hexobarbital anesthesia in rats. *J. Geront.* **10**, 441–4.

STUCHLÍKOVÁ, J. HRUŚKOVÁ, Z., HRŮZA, Z., JELÍNKOVÁ, M., NOVÁK, P. & SOUKU- POVÁ, K. (1966). The effect of adrenaline on lipolysis and glycogenolysis in relation to age and stress. *Exp. Geront.* **2**, 15–21.

STUCHLÍKOVÁ, E., JELÍNKOVÁ, M. HRŮZA, Z. & HRUŚKOVÁ, J. (1967). The level of endogenous catechol amines in adipose tissue of old and obese persons. *Exp. Geront.* **2**, 57–62.

SULKIN, N. M. & SULKIN, D. F. (1967). Age differences in response to chronic hypoxia on the fine structure of cardiac muscle and autonomic ganglion cells. *J. Geront.* **22**, 485–501.

SWANSON, P., LEVERTON, R., GRAM, M. R., ROBERTS, H. & PESEK, I. (1955). Blood values of women: cholesterol. *J. Geront.* **10**, 41–7.

TAIT, J. F., ROSEMBERG, E. & PINCUS, G. (1966). Aldosterone secretion and clearance rates. *Abstracts 7th Int. Cong. Geront.* p. 64. Vienna: International Association of Gerontology.

TALBOT, N. B. & RICHIE, R. (1958). The effect of age on the body's tolerance for fasting, thirsting and for overloading with water and certain electrolytes. In *Water and Electrolyte Metabolism in Relation to Age and Sex.* (eds. G. E. W. Wolstenholme and M. O'Connor). *Ciba Fdn Coll. Ageing*, **4**, 139. London: Churchill.

TAUCHI, H. & SATO, T. (1962). Some micromeasuring studies of hepatic cells in senility. *J. Geront.* **17**, 254–9.

THUNG, P. J. (1966). Note on the thymic epithelium in senile mice. *Exp. Geront.* **1**, 337–40.

TUTTLE, R. S. (1966). Age changes in the sensitivity of rat aortic strips to nore- pinephrine and associated chemical and structural alterations. *J. Geront.* **21**, 510–16.

TUTTLE, W. W., HORVATH, S. M., PRESSON, L. F. & DAUM, K. (1952). Specific dynamic action of protein in men past 60 years of age. *J. appl. Physiol.* **5**, 631–4.

TYLER, F. H., EIK-NES, K., SANDBERG, A. A., FLORENTIN, A. A. & SAMUELS, L. T. (1955). Adrenocortical capacity and metabolism of cortisol in elderly patients. *J. Am. Geriat. Soc.* **3**, 79–84.

VERZÁR, F. & FREYDBERG, V. (1956). Changes of thyroid activity in the rat in old age. *J. Geront.* **11**, 53–7.

VRANIC, M. & POKRAJAC, N. (1961). The effect of age and fasting on blood sugar level in normal and adrenalectomized rats. *J. Geront.* **16**, 110–13.

WANG, I. & KIRK, J. E. (1959). The enolase activity of arterial tissue in individuals of various ages. *J. Geront.* **14**, 444–6.

WOODHEAD, A. D. & ELLETT, S. (1967). Endocrine aspects of ageing in the guppy *Lebistes reticulatus* (Peters). II. The interrenal gland. *Exp. Geront.* **2**, 159–71.

WULFF, V. J., PIEKIELNIAK, M. & WAYNER, M. J. (1963). The ribonucleic acid content of tissues of rats of different ages. *J. Geront.* **18**, 322–5.

WULFF, V. J., QUASTLER, H. & SHERMAN, F. G. (1964). The incorporation of H³ cytidine into some viscera and skeletal muscle of young and old mice. *J. Geront.* **19**, 294–300.

YU, S. Y. & BLUMENTHAL, H. T. (1963). The calcification of elastic fibres. I. Biochemical studies. *J. Geront.* **18**, 119–26.

YIENGST, M. J., BARROWS, C. H. & SHOCK, N. W. (1959). Age changes in the chemical composition of muscle and liver in the rat. *J. Geront.* **14**, 400–4.

ZORZOLI, A. & LI, J. B. (1967). Gluconeogenesis in mouse kidney cortex. Effect of age and fasting on glucose production and enzyme activities. *J. Geront.* **22**, 151–7.

DISCUSSION

OLIVEREAU: What do you think of hibernation as a possible way to slow down the ageing process in some mammals?

BELLAMY: Lowering the body temperature obviously retards the rate of deterioration of extracellular substances such as collagen. There is evidence from hibernators that the prevention of hibernation shortens life, but this does not mean that the ageing process has been accelerated.

QUAY: I wish to very briefly suggest an alternative and possibly more advantageous strategy in such research concerning mechanisms in ageing. Over several decades we have seen increasing numbers of such statistical analyses of quantitative decreases in functional levels of various systems, but we are no closer to understanding mechanisms of ageing. At this juncture it would seem to be more promising to test the 'error hypothesis' of ageing—that is, to test the possibility that ageing consists essentially of increasing molecular errors in the cellular synthesis of specific proteins somewhere in the sequence from genetic material (DNA) through nucleo-protein to cellular protein. Such errors may be manifest in the specificity of the protein product in qualitative terms—as in substrate specificity of an enzyme protein or in other characteristics.

BELLAMY: I agree that there are other ways of studying the process of ageing apart from taking a physiological approach. Events at a high level of organization obviously have their origin in fundamental changes at a biochemical level.

MADERSON: Isn't the rat a poor subject for experimental investigation of ageing in view of the unusual pattern of growth in this form by comparison with other mammals, i.e. the rat continues to grow for longer into its life span than is usual.

BELLAMY: I do not take the view that ageing, in the sense that I defined it at the beginning of this talk, only starts after growth has stopped. It is possible that experiments with the growing rat may be weighted against the detection of early degenerative changes in some organs. On the other hand there is evidence that physiological ageing occupies about 60% of the life-span.

MADERSON: With the available evidence which suggests that fish do not age in the same way that mammals do, do you feel that this reflects different degrees of functional stress associated with the environment, i.e. aquatic versus terrestrial, or from the grade of organization, i.e. ectotherm versus endotherm?

BELLAMY: Fish appear to have a finite life-span as for higher vertebrates. There is little evidence that the process of ageing in fish is fundamentally different from that in higher vertebrates. However, there are differences in the mode of development. I would anticipate that the environmental temperature would be the main factor influencing longevity of individuals of a particular species.

CONTROL AND FUNCTION OF SKIN GLANDS IN MAMMALS

By J. S. STRAUSS and F. J. EBLING

INTRODUCTION

In a symposium on hormones and the environment there are logical reasons to include a discussion of the hormonal control and function of mammalian skin glands. First, the sebaceous glands and some sweat glands are under the control of steroid and polypeptide hormones. Therefore, the glandular activity may vary with changes in the biotic environment, such as may occur during the mating season. Secondly, the glands of the skin are responsible for the secretion of odoriferous substances which probably act as pheromones and are of importance in sexual attraction. Some of the odoriferous materials also are undoubtedly of territorial importance. Thus, the glands of the skin, in addition to their dependence on the internal biotic environment, enter into biotic interactions with the external environment.

NATURE OF THE CUTANEOUS GLANDS

There are three different mechanisms of formation of the secretory products of the cutaneous glands of mammals, and the classification of the glands is based upon the respective secretory pattern. In holocrine glands the entire mature cell is cast off into the excretory stream. This type of secretion is typical of the sebaceous glands, whose product, sebum, is a viscous or semi-solid product. In the mature sebaceous cell the cytoplasm is packed with lipid droplets which compress the remaining cellular components into thin strands between the lipid. Therefore, while a small amount of cellular debris is present in sebum, the material is almost exclusively composed of lipid. In the apocrine sweat glands, the secretory product appears to be formed by the pinching off of only the luminal tip of the secretory cell. Apocrine sweat is a milky fluid that contains a variable amount of lipid. The third type of cutaneous gland is the eccrine sweat gland. Its product is primarily an aqueous salt solution, and in contrast to the other two types of cutaneous glands, secretion is not accompanied by any visible loss of integrity of the cells.

The eccrine sweat glands have reached their ultimate phylogenetic development in man, where they are found in a density averaging 143–

339/cm² (Kuno, 1956). They are found all over the surface except for the lips, the glans penis, the inner surface of the prepuce, the clitoris and the labia minora. They are of extreme importance in thermoregulation in man, and their blockage may lead to symptoms of heat intolerance. Varying degrees of development of the eccrine sweat glands may occur in some of the higher simian primates, and the chimpanzee and gorilla have more eccrine glands than apocrine glands (Montagna, 1962 a). In other mammals eccrine sweat glands are usually restricted to the volar surface of the paws or digits, or to glaborous skin surfaces that are characterized by a highly differentiated friction surface, such as the knuckle pads of the gorilla or the prehensile tail of some South American monkeys (Montagna, 1962 a). Eccrine sweat glands open directly on to the skin surface and are independent of the pilosebaceous apparatus. Despite reports from one laboratory that eccrine sweat-gland hypertrophy occurs from local inunction of testosterone (Papa & Kligman, 1965; Papa, 1967), there is no other evidence that eccrine sweat gland development or secretion is under hormonal control. These glands are stimulated by neurohumoral stimuli, which in the human being are principally, if not solely, cholinergic in type. Since the topic of this symposium is the hormonal aspects of the biotic environment, the eccrine sweat glands will not be considered any further.

The apocrine glands are tubular structures. They are composed of a coiled secretory portion and a duct which opens into the upper portion of the hair follicle. In man, the apocrine sweat glands are localized in a few circumscribed areas such as the axilla, the anogenital region, the mammary areola, the ear canal (ceruminous glands), and the eyelids. A few glands may be found scattered over the body surface. In contrast to the eccrine sweat glands, apocrine sweat glands have a much wider distribution over the body surface of lower mammals. Simian primates have both apocrine and eccrine sweat glands over the general body surface, whereas prosimian primates have primarily apocrine glands (Schiefferdecker, 1922). The sweat glands of the horse have been studied by Hurley & Shelley (1960). While the glands have some histologic and pharmacologic features of eccrine glands, they conclude that they are most likely apocrine glands. Other examples of mammals with apocrine glands over the entire body surface are sheep and dogs. Apocrine glands may also be grouped in special areas. Such is the case in rabbits, in which clusters of apocrine glands are found in the submandibular, inguinal and anal areas (Mykytowycz, 1965, 1966a, b; Wales & Ebling, unpublished). While the apocrine glands, like eccrine sweat glands, are stimulated by neurohumoral agents, the fragmentary evidence that is available indicates that the development of the glands is under the control of the steroidal sex hormones. The apocrine glands also appear to be of great importance in territorialism and in sexual attraction. Therefore, the apocrine sweat glands will be discussed in greater detail later in this review.

The regional distribution and specialization of the sebaceous glands is probably more complex than that of the apocrine glands. In man, sebaceous glands are found over the entire body except for the palms, soles and the dorsal surface of the feet. In the adult human being, the glands are best developed on the face, scalp and, to a lesser extent, the midline of the chest and back. The glands in these areas are complex multiacinar structures. Over the rest of the body the sebaceous glands are much smaller and less numerous. Specialized areas of sebaceous-gland development in man include Tyson's glands of the mucocutaneous surface of the penis, the glands of Zeis, which open into the follicles of the eyelashes, and the giant, elongated Meibomian glands of the tarsal plate of the eyelids. Sebaceous glands which open directly to the surface are found in the mucous membranes. For a discussion of the glands of the mucous membrane and the ectopic glands of the human, the reader is referred to the reviews of Miles (1963) and Hyman & Guiducci (1963).

The comparative aspects of the sebaceous glands of mammals have been reviewed by Schaffer (1940) and by Montagna (1963). The number, size, and configuration of the sebaceous glands of the general body surface vary greatly. In general, in animals the largest sebaceous glands are found in the skin of the muzzle, the external auditory meatus and the anogenital areas. The sebaceous glands of the general body surface in many species, such as the rat, are paired uni- or bi-lobed structures that lie along side the hair canal and empty into the canal. In addition, specialized clusters of sebaceous glands occur at various sites. Because the glands tend to be large in the clusters, they have often been the subject of special study. The preputial glands, found in rats and other rodents, are representative of such specialized structures. They have been subjected to more study than any other specialized sebaceous structure. In the rat, the preputial glands are two elongated organs which lie in the loose connective tissue of the prepuce and open on to the inner surface of the prepuce. Because of their location in the loose connective tissue, they are easily dissected out. In the rat they measure up to 15 mm in length and weigh approximately 10–15 mg/100 g of body weight. The glands are tubulo-alveolar in type with multiple lateral ducts which open into a large central collecting duct.

Another specialized sebaceous gland is the ventral gland pad. This is a grossly visible linear midline streak found in the gerbil (Glenn & Gray, 1965; Thiessen, Friend & Lindzey, 1968) and other small mammals (Eadie, 1938; Richmond & Roslund, 1952; Quay & Tomich, 1963; Dryden

& Conaway, 1967). The brachial glands of the male lemur and the costo-vertebral glands of the hamster are also localized, grossly visible sebaceous gland clusters. The former are almond-shaped, palpable aggregates of sebaceous glands which open to the surface at the junction of the clavicle and the scapula (Montagna, 1962b; Montagna & Yun, 1962). The costo-vertebral gland of the hamster is a 5–7 mm diameter area containing large closely packed sebaceous glands associated with coarse hairs. The costo-vertebral gland is particularly well developed in the male hamster since it is an androgen-stimulated secondary sex characteristic (Kuppermann, 1944; Montagna & Hamilton, 1949; Hamilton & Montagna, 1950). Other examples of specialized sebaceous glands are the supracaudal gland of the guinea-pig (Martan, 1962; Martan & Price, 1967), the dermal side glands of shrews (Eadie, 1938; Dryden & Conaway, 1967), the glands of the oral angle of rodents (Quay, 1965), and the posterolateral glands of the micro-tine rodents (Quay, 1968).

The responses of the various sebaceous glands to steroidal and poly-peptide hormones are discussed together since in most instances the glands show a similar reaction pattern. However, there are instances in which the specialized glands do not show the same pattern of response as the sebaceous glands of the body surface. These will be so indicated.

THE ROLE OF STEROID AND PITUITARY HORMONES IN THE CONTROL OF SEBACEOUS GLANDS

There is no question but that more information on hormonal control is available for the sebaceous glands than for the apocrine glands. The experimental models that have been most extensively studied have been the preputial gland of the rat, the sebaceous glands of the body of the rat, and the sebaceous glands of the human face. Most of the earlier measurements in the rat skin were based upon one or another histologic parameter of gland size. Gland size, *per se*, may not be an accurate estima-tion of glandular activity since it is dependent not only upon the rate of production but on the turnover time of the cells as well as on cell size (Ebling, 1963). However, as Ebling has shown, size still is an indicator of glandular function. In man the sebaceous glands are less uniform than in the rat, and histologic assessment of glandular development poses much greater problems. Furthermore, especially when studying sebaceous gland changes on an area such as the face, the number and size of biopsies that can be obtained is strictly limited. The amount of sebum that is secreted is related to sebaceous gland size (Miescher & Schönberg, 1944). Therefore, Strauss & Pochi (1961) have used measurements of sebum production as indicators of the response of the sebaceous glands to hormonal stimuli. Recently, sebum secretion also has been determined in rats either by dipping the whole animal in a lipid solvent (Nikkari & Valavaara, 1969) or by extracting the fat from a known amount of hair clipped from the animal at a known interval after shampooing the animal (Ebling & Skinner, 1967).

Androgens

There is general agreement that androgens increase the size and functional capacity of the sebaceous glands in all species. Castration results in a decrease in gland volume in the rat (Ebling, 1957 a, 1963). Sebum production is low in castrated men (Pochi, Strauss & Mescon, 1962) and drops after orchiectomy in elderly men (Pochi & Strauss, 1963). Atrophy of the sebaceous glands of the side glands of the male shrew (Dryden & Conaway, 1967), of the costovertebral spot of the hamster (Kuppermann, 1944), of the ventral gland of the gerbil (Mitchell, 1965; Thiessen et al. 1968), and of the supracaudal gland of the guinea-pig (Martan, 1962; Martan & Price, 1967) have also been observed after castration. Furthermore, sebum production is higher in adult men than in women (Pochi & Strauss, 1965 a), the costovertebral spot is larger in male guinea-pigs (Kuppermann, 1944; Montagna & Hamilton, 1949; Hamilton & Montagna, 1950), the supracaudal gland is larger in the male guinea-pig (Martan, 1962; Martan & Price, 1967) and the ventral gland of the gerbil is larger in the male (Glenn & Gray, 1965).

Testosterone administration has resulted in enlargement of the sebaceous glands of the pubic region in boys (Rony & Zakon, 1943) and of the sebaceous glands of the face in prepuberal children (Strauss, Kligman & Pochi, 1962; Strauss & Pochi, 1963). When testosterone is applied topically, local stimulation of the sebaceous glands can be produced without altering the size of the sebaceous glands at distant sites (Strauss, Kligman & Pochi, 1962; Strauss & Pochi, 1963). Enlargement of the sebaceous glands has followed the administration of androgen in rats (Ebling, 1948, 1957 a, b, 1963; Haskin, Lasher & Rothman, 1953), rabbits (Montagna & Kenyon, 1949), hamsters (Kuppermann, 1944; Montagna & Hamilton, 1949; Hamilton & Montagna, 1950) and mice (Lapière, 1953). Ebling (1957 a, b, 1963) has shown that this increase in size is due to an increase in both the mitotic rate and the cell size. At the same time, testosterone increases the rate of cell turnover, a change which would tend to decrease gland size (Ebling, 1963). Obviously the latter effect is less than the former effect.

The specialized sebaceous glands show a similar response to androgen. The preputial gland has been extensively studied by Korenchevsky and his co-workers and has been found to increase in size when various androgens are given (Korenchevsky, Dennison & Simpson, 1935; Korenchevsky & Dennison, 1936 a, b; Korenchevsky, Dennison & Eldridge, 1937 a, b; Korenchevsky, Dennison & Hall, 1937; Korenchevsky, Hall & Burbank, 1939; Korenchevsky, Hall & Ross, 1939). Others (Seyle, 1940; Glenn, Richardson, Bowman & Lyster, 1960) have demonstrated a similar androgen response in the preputial gland. Salmon (1938) has shown that the effect of testosterone is rapid in that preputial gland stimulation occurs within 48 h. Because of its response to androgens, Glenn and his co-workers (Glenn, Richardson & Bowman, 1959 a, b; Glenn et al. 1960) have used preputial gland weight increase as a bioassay for androgenic activity. Other specialized sebaceous glands showing an increase in size due to testosterone include the guinea-pig costovertebral spot (Kuppermann, 1944; Montagna & Hamilton, 1949; Hamilton & Montagna, 1950), the supracaudal gland of the male guinea-pig (Martan, 1962; Martan & Price, 1967), the side glands of the shrews (Dryden & Conaway, 1967), and the ventral gland of the gerbil (Glenn & Gray, 1965; Mitchell, 1967; Thiessen et al. 1968). The stimulation of these specialized sebaceous glands by testosterone may be greater than that observed in the sebaceous glands of the rest of the body. This has been shown for the costovertebral spot in the hamster (Hamilton & Montagna, 1950) and for the supracaudal gland of the guinea-pig (Martan & Price, 1967).

Sebum production is increased in man by testosterone (Strauss et al. 1962; Strauss & Pochi, 1963). The rise is detectable within 2 weeks (Fig. 1), and is produced by as little as 5 mg of methyl testosterone given orally. Because of the extreme sensitivity of the glands to androgens, the sebaceous glands of adult men appear to be maximally stimulated, and testosterone administration does not produce an increase in sebum production in this group. However, sebum production can be increased by testosterone in prepuberal children, castrated men and in post-menopausal women. Testosterone also increases sebum production in rats (Ebling & Skinner, 1967; Ebling, 1967).

The above changes in sebum production reflect an increase in gland size. There is recent evidence, albeit fragmentary, that androgens also may affect the composition of sebum. Nikkari (1965) has found a decrease in the proportion of longer-chain aliphatic alcohols, a decrease in the proportion of the mono-unsaturated alcohols and fatty acids, and an increase in the proportion of branched-chain anteiso alcohols and fatty acids in young male rats treated with androgens. Wilde & Ebling (1969) have reported that the treatment of castrated rats with androgen increases the palmitate/sterate ratio (16:0/18:0). They also have found a decrease in the sterate/oleate ratio (18:0/18:1), an effect that is opposite to that found by Nikkari (1965) for the unsaturated fatty acids. The reason for this difference is unknown. Nikkari (1965) used intact rats while Wilde & Ebling (1969) used castrated rats, but it is not known whether this is responsible for the different results.

Since sebaceous glands are stimulated by androgens, anti-androgens should suppress their activity. Among dermatologists there is a keen desire

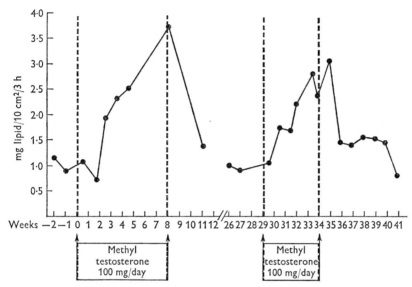

Fig. 1. Sebum production in a 12-year-old prepuberal boy given two courses of methyl testosterone orally (100 mg/day). Sebaceous secretion promptly rose with androgenic stimulation and quickly declined to the original level when the drug was stopped. Reproduced from Strauss & Pochi (1961).

for a locally acting anti-androgen, for such a compound might have therapeutic value in acne, a disease in which sebum plays a pathogenic role. Cyproterone acetate, a potent progestational agent, possesses anti-androgenic activity for mouse sebaceous glands in that it antagonizes the stimulatory effect of testosterone on these glands (Neumann & Elger, 1966). Another anti-androgen, 17α-methyl-B-nortestosterone, prevents the increase in sebum production produced by exogenous testosterone in spayed female rats (Ebling, 1967). The 6-cyclopropyl substituent, 6α,6β-ethylene 17α-methyl-B-nortestosterone produces a similar decrease in sebum production (Saunders & Ebling, 1969). The inhibition of sebum production is accompanied by a proportional decrease in cell division which has been interpreted as indicating that this compound exerts a direct peripheral action on the sebaceous glands (Fig. 2). As will be subsequently discussed, oestrogens do not decrease cell division. Therefore, Ebling (1967) has concluded that oestrogens and anti-androgens inhibit sebum secretion through different mechanisms.

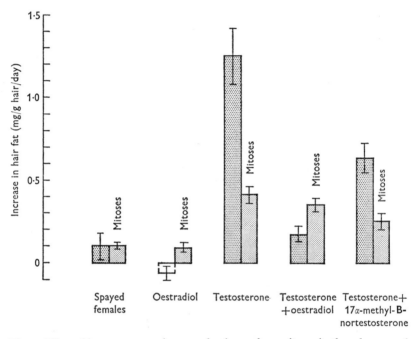

Fig. 2. Effect of hormones on sebum production and on mitoses in the sebaceous glands of female rats. Sebum production is expressed as increase in hair fat/g hair/day measured over an 8-day period after washing. Testosterone increased sebum production and oestrogen decreased it both in spayed and testosterone treated-animals. The increase due to testosterone was due to an increase in cell division. The decrease produced by oestrogen did not involve a proportional decrease in mitoses, indicating an inhibition of intracellular synthesis rather than an inhibition of cell division. In contrast, 17α-methyl-B-nortestosterone decreased sebum production by producing a proportional reduction in cell division. Date from Ebling (1967).

Most of the available data on the effect of anti-androgens in man has been obtained in subjects treated with 17α-methyl-B-nortestosterone. When administered either systemically or locally, this compound decreases sebum production (Zarate, Mahesh & Greenblatt, 1966; Strauss, Pochi, Sarda & Wotiz, 1969). Whether this compound has a direct local effect in man has not been determined. There is some evidence that it may have a local effect, although topical application of the compound may induce a decrease in plasma testosterone (Strauss et al. 1969). This decrease in plasma testosterone is not correlated with the degree of sebum suppression.

Progestogens

In contrast to the uniformity of opinion as to the effect of androgens on the sebaceous glands, a consensus does not exist in regard to the effect of progestogens on the sebaceous glands. While Ebling (1948) has not produced any increase in the sebaceous gland size in immature female rats with progesterone, Haskin *et al.* (1953), Lasher, Lorincz & Rothman (1954) and Lorincz (1963) have demonstrated sebaceous-gland enlargement in castrated rats treated with 1 mg/day or more of progesterone. Ebling (1961, 1963) has shown that the difference in the results is dose-dependent, in that while low dosages of progesterone (0·1–0·2 mg/day) are unable to produce sebaceous gland enlargement, large amounts (10 mg/day) do produce slight glandular enlargement, a finding confirmed by de Groot, Lely & Kooij (1965). However, sebum production does not increase in rats given 10 mg/day of progesterone (Ebling, Ebling & Skinner, 1969 *b*).

The specialized sebaceous structures, which as already indicated may be more easily stimulated by testosterone, appear for the most part to be progesterone-sensitive. Progesterone induces preputial gland enlargement (Lorincz & Lancaster, 1957; Lorincz, 1963; Ebling, 1963). On the other hand, Glenn *et al.* (1959 *b*) have not demonstrated any change in the preputial glands of rats given 8 mg/day of progesterone or the synthetic progesterone, 17-α-hydroxyprogesterone acetate. Progesterone administration increases the size of the gerbil ventral gland (Glenn & Gray, 1965) and the side gland of the shrew (Dryden & Conaway, 1967).

In man, only Smith (1959) has presented experimental data that a progestogen can increase sebaceous-gland secretion. In contrast, Jarrett (1959) has not detected any increase in sebaceous-gland activity with 25 mg/day of progesterone, and Strauss & Kligman (1961 *a*) have not detected any change in sebum production in prepuberal, post-puberal or postmenopausal subjects given 50 mg/day of progesterone. Synthetic progestogens such as 17α-hydroxyprogesterone-17-*n*-caproate, norethynodrel and medroxyprogesterone acetate do not alter sebum secretion (Strauss & Kligman, 1961 *a*; Strauss & Pochi, 1963). However, norethindrone (17α-ethynyl-19-nortestosterone) increases sebaceous-gland size and sebum production (Strauss & Kligman, 1961 *a*, *b*). The increase in sebaceous-gland activity has been accompanied by other signs of masculinization such as penile enlargement and pubic hair growth. Therefore, the sebaceous gland stimulation is obviously an androgenic side-effect. It should be noted that this compound has been incriminated as a cause of pseudohermaphrodism in female infants born of mothers receiving the agent during pregnancy (Grumbach, Ducharme & Moloshok, 1959; Wilkins,

1960). The effect of this compound on the sebaceous glands is dose-related and has not been seen with dosages of 5 mg/day or less (Pochi & Strauss, 1965 b)

Oestrogens

Oestrogens, like progesterone, have produced variable changes in the sebaceous glands. The side glands of the shrew (Dryden & Conaway, 1967) and the ventral gland of the gerbil (Glenn & Gray, 1965) have been stimulated by oestrogens. In the studies of Korenchevsky and his associates (Korenchevsky & Dennison, 1936 a, b; Korenchevsky, Hall & Ross, 1939; Korenchevsky, Hall & Burbank, 1939) oestrogens alone have not changed the weight of the preputial glands, but they have inhibited the increase in gland weight induced by androgens, particularly in spayed female rats. However, even this inhibition of concomitantly administered androgens is variable and inconsistent. Glenn et al. (1959 a, b) have noted a slight increase in rat preputial gland weight with oestradiol in varying dosages up to 80 μg/kg.

These studies once again illustrate the unique position of the specialized sebaceous-gland organs, for it has been clearly shown that oestrogens produce atrophy of the sebaceous glands of the skin of the body of rats (Ebling, 1948, 1951, 1963), and also decrease sebum production in the rat (Ebling & Skinner, 1967; Ebling, 1967). Suppression of the sebaceous glands has been observed with oestrogens of varying degrees of activity (Bullough & Laurence, 1960; Ebling, 1964). The effect may be biphasic, for Ebling (1951, 1963) has shown that after oestrogen pellet implantation in immature rats there is an increase in sebaceous-gland size before the glands decrease in size. Such a finding may explain the reported increase in mitoses in the sebaceous glands of the mouse at pro-oestrus (Bullough, 1946) and the increase in the size of rat sebaceous glands in pro-oestrus (Ebling, 1954, 1963).

Adequate dosages of oestrogen (Fig. 3) have also decreased the size and functional capacity of the sebaceous glands of the human being (Jarrett, 1959; Strauss et al. 1962; Strauss & Pochi, 1963). In the human female, sebum production is significantly inhibited by approximately 0·1 mg/day of ethynyl oestradiol or its equivalent when administered for 3 out of 4 weeks each month (Strauss & Pochi, 1964).

Ebling (1955) has shown that oestrogens produce as much suppression in hypophysectomized rats as in intact rats. He has also shown that oestrogens will decrease sebaceous activity in adrenalectomized-castrated male rats in which there is no obvious source of androgen (Ebling, unpublished). Therefore, it would appear that oestrogens act peripherally and independently of mediation through the pituitary gland or gonads. Nevertheless, in rats given testosterone and oestrogen simultaneously, the sebaceous glands are smaller (Ebling, 1957) and sebum production is not as great (Ebling & Skinner, 1967; Ebling, 1967) as when only testosterone is administered. Ebling's (1967) studies indicate that this inhibition is not a direct antagonism of the androgen on cell division, but rather is due to the inhibition of intracellular synthesis. There is no proportional drop in cell division in oestrogen-treated animals as has been seen with anti-androgens. Thus, by combining measurements of sebum production and cell mitosis, he has been able to differentiate between the site of action of oestrogens and anti-androgens on the sebaceous glands.

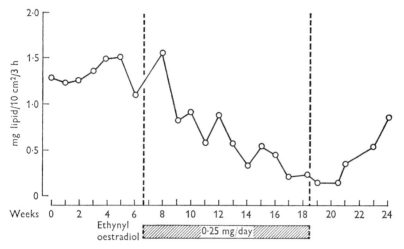

Fig. 3. Reduction in sebum production in an adult male given 0·25 mg of ethynyl oestradiol orally daily. Reproduced from Strauss & Pochi (1963).

In man, the same conclusions have not been reached. Strauss *et al.* (1962) and Strauss & Pochi (1963) have not been able to demonstrate a peripheral effect of oestrogen. They have applied ethynyl oestradiol creams of varying concentrations to one side of the forehead of adult men and have made simultaneous measurements of sebum production from both the treated and untreated sides of the forehead. Whenever suppression of sebum production has occurred on the treated side, it has also occurred on the contralateral control side (Fig. 4). Furthermore, the degree of suppression always has been equal. Thus, they have been unable to separate a systemic effect from a local one and for this reason have proposed that oestrogens influence sebaceous-gland function through a central inhibitory pathway. In support of this contention is the additional finding that when radio-

active thymidine is injected intracutaneously during oestrogen treatment
there is a decreased labelling index for both the germinative layer and the
whole gland (Sweeney, Szarnicki, Strauss & Pochi, 1969).

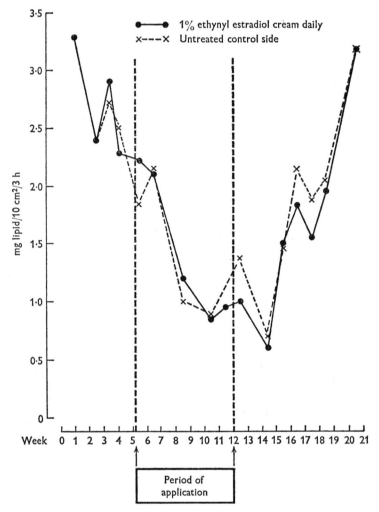

Fig. 4. Decreased sebum output after daily inunction of approximately o·1 g of 1%
ethynyl estradiol ointment to one side of the forehead of an adult male. Sebum production
was assayed simultaneously on both the treated and untreated sides. Sebaceous secretion
decreased sharply, but equally on both sides. Reproduced from Strauss, Kligman & Pochi
(1962).

Strauss & Pochi (1963) also have shown that there is no peripheral
antagonism of androgen by oestrogen for the human sebaceous gland. If
an oestrogen is administered until sebum production is decreased and then

an androgen is given concomitantly with the oestrogen, sebum production increases (Fig. 5). The lack of such peripheral antagonism has served
as a basis for a bioassay of androgenicity in the human (Strauss & Pochi,
1963; Pochi & Strauss, 1969). Some of the data on progestogens that are
reported earlier in this paper have been derived from this assay.

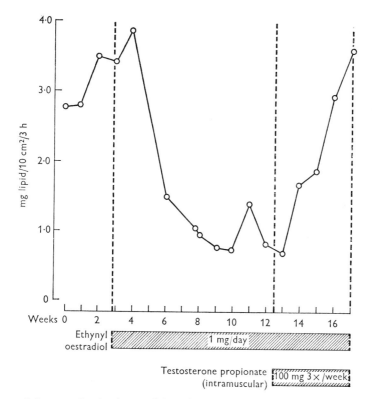

Fig. 5. Sebum production in an adult male given ethynyl estradiol (1 mg/day orally) to
suppress sebum production, and then testosterone propionate (100 mg intramuscularly
3 times/week) while the oestrogen was continued. Sebum production promptly rose when
testosterone propionate was given.

Adrenal hormones

In discussing the role of the adrenal in the control of sebaceous-gland
function, it is necessary to consider both glucocorticoids and adrenal
androgens. Therefore, careful interpretation is necessary of the results of
studies based upon suppression or stimulation of the adrenal. Pochi &
Strauss (1967) have shown that while adrenal suppression with 10–20 mg
of prednisone is without effect in intact men, the same dosage of prednisone
produces a 20% decrease in sebum production in women and a 40% decrease in sebum production in castrated males. This decrease is probably

not a direct effect, but is most likely due to adrenal androgen suppression since 17-ketosteroid excretion decreases in the castrate group. Furthermore, Pochi, Strauss & Mescon (1963), using the assay for androgens described in the previous section, have been unable to demonstrate a direct stimulatory effect of prednisone on the sebaceous glands. They have, however, demonstrated that adrenal androgens (Δ^4-androstenedione and dehydroepiandrosterone) can increase sebum production in the androgen assay (Pochi & Strauss, 1969).

At present, data derived from animal experiments is fragmentary and contradictory. Cortisone has been reported to decrease sebaceous gland size in the rat (Baker & Whitaker, 1948; Castor & Baker, 1950) and ACTH has caused glandular hyperplasia, even in hypophysectomized and gonadectomized rats (Haskin et al. 1953; de Graaf & Kooy, 1955). In contrast, hydrocortisone in a dosage of 8 mg/kg/day has not changed preputial glands (Glenn, Richardson & Bowman, 1959a, 1960). The gerbil ventral gland is also unchanged by hydrocortisone administration (Glenn & Gray, 1965). Some investigators have reported no change in the preputial gland after adrenalectomy (Montagna & Noback, 1946; Huggins, Parsons & Jensen, 1955; Noble & Collip, 1941), but Hess, Hall, Hall & Finerty (1952) have reported hypertrophy of the preputial gland.

The pituitary control of sebaceous glands

Experimental data on the pituitary control of the sebaceous gland is restricted to studies of the sebaceous glands of the body and the preputial glands of the rat. Even in castrated animals, hypophysectomy appears to cause atrophy of the sebaceous glands (Lasher et al. 1954) and of the preputial gland (Noble & Collip, 1941; Huggins et al. 1955; Hess, Rennels & Finerty, 1953). Sebum production also decreases after hypophysectomy (Ebling, Ebling & Skinner, 1969a). Ebling has been unable to demonstrate a significant increase in gland size (Ebling, 1957a) or sebum production (Ebling et al. 1969a) when testosterone is administered to hypophysectomized rats (Fig. 6). However, Lasher et al. (1954) have shown that there is a minimal increase in gland size after the injection of testosterone in hypophysectomized animals, and Nikkari & Valavaara (1969) have demonstrated a slight increase in sebum production when testosterone is administered to hypophysectomized rats. Be that as it may, it is necessary to emphasize that the testosterone response is minimal and there obviously is a pituitary factor that supports sebaceous gland activity.

The questions to be answered are what is the nature of the pituitary factor, and does it act directly or is it a permissive factor which allows the gland to respond to other hormones such as testosterone? Growth hormone, prolactin, and ACTH have all been reported to cause an increase in preputial gland size in intact rats, an effect which is increased when the three hormones are given simultaneously (Bates, Milkovic & Garrison, 1964). Jacot & Seyle (1951) have reported that ACTH has a direct effect on the preputial glands in ovariectomized-adrenalectomized animals, and Hess *et al.* (1953) have reported that ACTH increases preputial gland weight in hypophysectomized rats. An extra-adrenal stimulation has been confirmed by Yip & Freinkel (1964), who have shown both a weight increase and stimulation of [¹⁴C]glucose uptake in the rat preputial gland

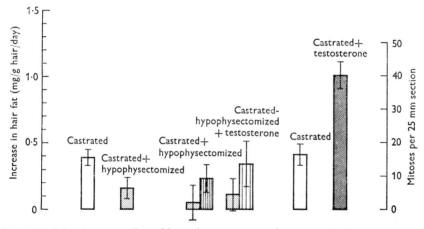

Fig. 6. Left-hand group: effect of hypophysectomy on sebum secretion in castrated rats (12 litter mate pairs). Means ± s.e. are shown. Centre group: non-effect of testosterone on sebum secretion (left-hand columns) and mitoses in the sebaceous glands (right-hand columns) in hypophysectomized-castrated rats (nine litter-mate pairs). Right-hand group: effect of testosterone on sebum secretion in castrated rats (ten litter-mate pairs). Data from Ebling & Skinner (1967). and Ebling *et al.* (1969a).

after ACTH stimulation of castrated-adrenalectomized rats. In the absence of the adrenals, this must be a direct rather than an indirect effect.

The possibility of there being a pituitary gland tropic factor has been raised by Lasher *et al.* (1954) and Lorincz (1963). These investigators have found that while sebaceous gland stimulation by both progesterone and testosterone is minimal in hypophysectomized animals treated with ACTH, growth hormone, FSH, prolactin, pituitrin, pitressin and chorionic gonadotrophin, the sebaceous-gland response to progesterone and testosterone is restored by mare serum gonadotrophin or crude pituitary gonadotrophin. They have shown that the 'sebotropic' factor varies independently of gonadotrophic activity.

Much of the subsequent work on the isolation of the sebotropic factor has involved studies utilizing the response of the preputial gland to progesterone (Lorincz & Lancaster, 1957; Lorincz, 1963; Woodbury, Lorincz & Ortega, 1965 a, b). This may not be a satisfactory assay system since the preputial gland will respond maximally to testosterone in the hypophysectomized rat (Woodbury et al. 1965 b; Ebling et al. 1969 b). Moreover, the sebaceous glands show little or no response to progesterone, whereas the preputial glands do show such an effect, even in the intact rat. Therefore, the 'sebotropic' factor for the preputial glands may not be identical with the pituitary potentiator for the sebaceous glands of the skin.

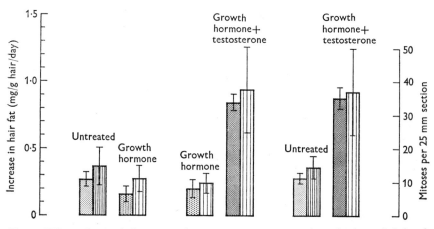

Fig. 7. Effect of growth hormone plus testosterone on secretion of sebum (left-hand columns) and mitosis in the sebaceous glands (right-hand columns) in hypophysectomized-castrated rats. Comparisons between litter-mates; 9, 13 and 12 pairs respectively. Means ±S.E. are shown. Data from Ebling et al. (1969a).

Such a conclusion is strengthened by recent studies by Ebling et al. (1969a) in which it has proved to be impossible to restore the stimulatory activity of testosterone in hypophysectomized-castrated rats treated with a pituitary 'sebotropic factor' prepared according to the method of Woodbury et al. (1965b). While Ebling et al. (1969a) have acknowledged that the pretreatment of the pituitary material may have been different, they have been able to restore the responsiveness of the glands to testosterone with both a porcine growth hormone (Fig. 7) and prolactin. Neither of these hormone preparations have possessed any adrenocortico-trophic activity and the prolactin has been found to be free of any somato-trophic activity. Therefore, they have concluded that the 'sebaceous activity factor' is free of any association with either adrenocorticotrophic or somatotrophic action. Ebling et al. (1969b) have found that their somato-trophin preparation is inactive as a potentiator of the stimulation of the preputial gland by progesterone in hypophysectomized-castrated rats. On this basis they have proposed that the factor that permits the response of the sebaceous glands to androgens is distinct from the one that facilitates the response of the preputial gland to progesterone. If this theory is confirmed, it will once again illustrate the difficulty inherent in making generalizations about all sebaceous glands based upon studies of the preputial glands.

THE HORMONAL CONTROL OF APOCRINE SWEAT GLANDS

Information on the hormonal control of the apocrine glands is much less complete. The fact that these glands develop at puberty in man makes it attractive to consider that they are stimulated by steroidal hormones. There is, however, essentially no direct evidence to support this theory. Shelley & Cahn (1955) have not detected any histologic changes in the apocrine sweat glands of a group of 20 to 40-year-old-men after the daily application of wet compresses for 8 weeks with ethynyl oestradiol, testosterone and progesterone. These investigators also have given these hormones singularly and in combination, as well as growth hormone, prolactin, and gonadotrophins systemically. No histologic changes were noted in the apocrine glands, but this study is not particularly impressive since the hormones have been administered for only 3 weeks and the dosages were often small. Shelley & Hurley (1957) have implanted pellets of testosterone and oestradiol in the axillae of adult males for as long as 5 months. Although gynecomastia has been produced in some of the subjects, no histologic changes have been found in the apocrine sweat glands.

The submandibular, inguinal and anal glands of the male wild rabbit, *Oryctolagus cuniculus*, are larger than in the female (Mykytowycz, 1962, 1965, 1966 a, b, 1968). Early castration of the male rabbit inhibits apocrine gland development (Coujard, 1947; Mykytowycz, 1962, 1965, 1966 a, b, 1968), but apocrine gland development in the female rabbit is not inhibited by castration (Mykytowycz, 1962, 1965, 1966 a, b, 1968). An increase in apocrine gland size is induced by testosterone in the castrate male rabbit (Mykytowycz, 1962, 1968). More extensive studies of the hormonal response of the apocrine glands of the rabbit have just been completed (Wales & Ebling, unpublished). These investigators have confirmed the greater enlargement in the male rabbit, the inhibition of apocrine growth in castrated rabbits, and the stimulation of the apocrine glands in the castrate rabbit by testosterone (Fig. 8). In addition, they have found that oestrogen causes a decrease in the size of the apocrine sweat glands in male rabbits, and oestrogen also antagonizes the effect of testosterone in the castrated animal. Small amounts of progesterone do not stimulate the apocrine sweat glands. Thus, it can be seen that the apocrine gland shows the same pattern of response to hormonal agents as observed in the sebaceous gland.

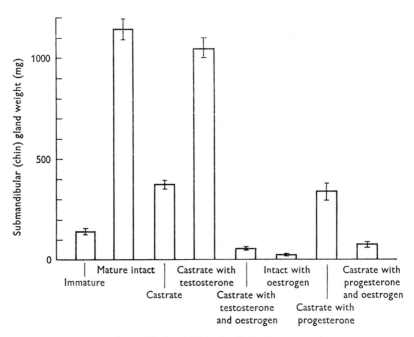

Fig. 8. Weight of submandibular (chin) glands in immature, intact mature, castrated mature, and hormonally treated rabbits. Means ± s.e. for eight animals in each group (Wales & Ebling, unpublished).

Regional differences have been demonstrated by the latter investigators. The submandibular glands show the greatest response to testosterone and the greatest suppression by oestrogen. The submandibular glands also have the greatest change in total polysaccharide, Schiff-positive staining, and extractable lipids accompanying the administration of androgen or oestrogen.

FUNCTION OF THE GLANDS OF THE SKIN

There is little evidence that either the apocrine sweat glands or sebaceous glands serve any vital function in man. Apocrine sweat does have an odour after bacterial decomposition, but this is not a desirous trait and a multimillion dollar industry is devoted to its eradication. Sebum has weak antifungal and antibacterial qualities, but as pointed out by Kligman (1963), such activity is so weak as to be unimportant. Kligman has indicated that sebum does not have any true function in the human being, and that the skin will maintain its functional integrity without compromise in the child in whom sebaceous secretion is minimal.

In animals, sebum probably serves to coat the hair and thus protect it against wetting. Furthermore, in animals, both apocrine and sebaceous secretions may be odoriferous. The examples of cutaneous glands which have an odoriferous secretion are too numerous to fully detail. Examples are the supracaudal sebaceous gland of the guinea-pig (Martan & Price, 1967) and the mid-dorsal scent gland of the New World swine (Peccaries), which is an apocrine gland (Werner, Dalquest & Roberts, 1952). A species in which both apocrine sweat glands and sebaceous glands are important and, in fact, complementary is the shrew. Eadie (1938) has suggested that the specialized sebaceous glands in the side glands of the shrew act as a sexual attractant since they enlarge at the time of testicular growth and in the breeding period in the male. He also has concluded that the specialized ventral gland, also a sebaceous gland, serves a territorial function emitting a pungent odoriferous material. Dryden & Conaway (1967) have concluded that the apocrine glands of the post-auricular area secrete the musk, in that musk is still formed after the removal of the side glands. However, the product of the side glands functions to trap the odoriferous material so that its release is prolonged. Thus, it contains the 'perfume carrier' and therefore is also important in territorial functions.

Whereas in our present culture human body odour is regarded as an undesirable trait, there can be little doubt that in most other mammals scent plays an important protective territorial and sexual function. The reader is referred to the excellent review of such activities by Wynne-Edwards (1962). As he points out, urine, faeces and the product of the cutaneous glands all may be important olfactory signals. The unique physical position of some of the cutaneous glands makes them most suitable for their role. For example, the ventral gland of the gerbil can easily be pressed against the ground, and Thiessen et al. (1968) have observed abdominal skimming action in the gerbil with the deposition of some of the oily secretion on the surface that is touched. Mitchell (1967) has also proposed that the product of the ventral gland provides an olfactory signal which he believes to be sexual.

Probably the most extensive studies on the role of the secretion of skin glands in providing olfactory signals has been done in investigations of the anal, inguinal and submandibular apocrine glands of rabbits by Mykytowycz (1962, 1965, 1966a, b, 1968) and Mykytowycz & Dudzinski (1966).

The anal glands secrete their product with the faeces, but it has been shown conclusively that the product of the glands is not essential as a faecal lubricant (Takagi & Tagawa, 1961). Furthermore, as pointed out by Mykytowycz (1966a), sexual dimorphism of these glands would be unlikely if they were only of importance as a faecal lubricant. The anal glands, as well as the inguinal and submandibular glands, are larger in male than in female rabbits and are particularly large in the most dominant bucks. When the rabbit chins himself, the product of the submandibular glands is deposited on the object that is contacted. Chinning is ten times more common in the male rabbits, but it can be increased in the female rabbit by testosterone administration. There is a correlation of apocrine gland size with territorial properties (Mykytowycz 1962, 1965, 1968; Mykytowycz & Dudzinski, 1966). Thus the swamp rabbit (*Sylvilagus aquaticus*), which is strongly territorial, has larger chin glands than the wide-ranging cottontail rabbit (*Sylvilagus floridanus*) (Mykytowycz, 1968). The hare (*Lepus europaeus*), which is less territorial than the Australian wild rabbit (*Oryctolagus cuniculus*), has smaller apocrine glands of the submandibular and anal areas (Mykytowycz, 1962, 1965, 1966a). However, the inguinal glands are larger in the male hare than in the male wild rabbit (Mykytowycz, 1966b). This he interprets as indicating that the inguinal glands are important as a sexual attractant, whereas the anal and submandibular glands are territorial.

The difference between the anal and submandibular apocrine glands on the one hand and the inguinal apocrine glands on the other has been confirmed by studies of the composition of the odoriferous material of these glands (Wales & Ebling, unpublished). The inguinal-gland secretory product has a molecular weight of 400 and a steroidal pattern on mass spectrometry (Fig. 9). In contrast, the odoriferous component of both the anal and chin glands is non-steroidal. They have also concluded that the anal and chin gland secretion probably serves a territorial function, whereas the inguinal gland product acts as a sexual attractant.

There is increasing evidence that the skin can convert steroidal hormones to new metabolites. Therefore, if the skin does excrete a pheromone that is steroidal, it could be synthesized in the skin. Evidence for such steroid conversions are particularly well developed for androgens. It is beyond the scope of this review to disucss this field in detail. However, a few pertinent points must be mentioned. Hydroxysteroid dehydrogenases, such as the 3β-dehydrogenase, which is responsible for the conversion of pregnenolone to progesterone and dehydroepiandrosterone to Δ^4-androstenedione, are found in the skin (Baillie, Calman & Milne, 1965; Baillie, Thomson & Milne, 1966). This dehydrogenase activity is present in the sebaceous glands of the face, neck, back, chest, and epigastric regions. Specific androgenic steroids have been identified on the skin surface (Dubovyi, 1967; Oertel & Treiber, 1969), and it has been shown

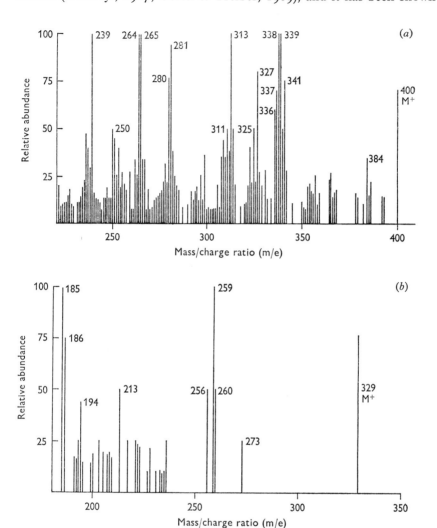

Fig. 9. Mass spectra of odoriferous secretion from apocrine glands of rabbit. (a) Inguinal glands; (b) Anal glands. The chin glands give an identical spectrum to that of the anal glands. (Wales, unpublished)

that the skin can metabolize dehydroepiandrosterone (Gallegos & Berliner, 1967; Faredin, Fazekas, Kókai, Tóth & Julesz, 1967) and testosterone (Rongone, 1966; Gomez & Hsia, 1968). It is interesting and probably significant that the major metabolite that Gomez & Hsia (1968) have identified when the skin is incubated with testosterone is the 5α-reductase product of testosterone, namely dihydrotestosterone. Dihydrotestosterone has been shown by Bruchovsky & Wilson (1968 *a*, *b*) to be most probably the active tissue androgen. Recently, dihydrotestosterone has been identified in human skin (Wilson & Walker, 1969). Thus, we may close this discussion of the relationship of the skin glands to the biotic environment with the statement that the skin may, indeed, not only be influenced by it, but also contribute directly to it.

Supported in part by U.S.P.H.S. grant AM 07084, National Institute of Arthritis and Metabolic Desease, National Institutes of Health.

REFERENCES

BAILLIE, A. H., CALMAN, K. C. & MILNE, J. A. (1965). Histochemical distribution of hydroxysteroid dehydrogenases in human skin. *Br. J. Derm.* **77**, 610–16.

BAILLIE, A. H., THOMSON, J. & MILNE, J. S. (1966). The distribution of hydroxysteroid dehydrogenase in human sebaceous glands. *Br. J. Derm.* **78**, 451–7.

BAKER, B. L. & WHITAKER, W. L. (1948). Growth inhibition in the skin following direct application of adrenal cortical preparations. *Anat. Rec.* **102**, 333–47.

BATES, R. W., MILKOVIC, S. & GARRISON, N. M. (1964). Effects of prolactin, growth hormone and ACTH, alone and in combination, upon organ weights and adrenal function in normal rats. *Endocrinology* **74**, 714–23.

BRUCHOVSKY, N. & WILSON, J. D. (1968*a*). The conversion of testosterone to 5α-androstan-17β-ol-3-one by rat prostate *in vivo* and *in vitro*. *J. biol. Chem.* **243**, 2012–21.

BRUCHOVSKY, N. & WILSON, J. D. (1968*b*). The intranuclear binding of testosterone and 5α-androstan-17β-ol-3-one rat prostate. *J. biol. Chem.* **243**, 5953–60.

BULLOUGH, W. (1946). Mitotic activity in the adult female mouse, *Mus musculus* L. A study of its relation to the oestrous cycle in normal and abnormal conditions. *Phil. Trans. R. Soc.* B **231**, 453–516.

BULLOUGH, W. S. & LAURENCE, E. B. (1960). Experimental sebaceous gland suppression in the adult male mouse. *J. invest. Derm.* **35**, 37–42.

CASTOR, C. W. & BAKER, B. L. (1950). The local action of adrenocortical steroids on epidermis and connective tissue of skin. *Endocrinology* **47**, 234–41.

COUJARD, R. (1947). Etudes des glandes odorantes du lapin et de leur influencement par les hormones sexuelles. *Revue can. Biol.* **6**, 3–15.

DRYDEN, G. L. & CONAWAY, C. H. (1967). The origin and hormonal control of scent production in *Suncus murinus*. *J. Mammol.* **48**, 420–8.

DUBOVYI, M. I. (1967). Androgenic steroids in sebum. *Vest. Derm. Vener.* **41**, 14–21.

EADIE, W. R. (1938). The dermal glands of shrews. *J. Mammol.* **19**, 171–4.

EBLING, F. J. (1948). Sebaceous glands. 1. The effect of sex hormones on the sebaceous glands of the female albino rat. *J. Endocr.* **5**, 297–302.

EBLING, F. J. (1951). Sebaceous glands. 2. Changes in the sebaceous glands following the implantation of oestradiol benzoate in the female albino rat. *J. Endocr.* **7**, 288–98.

EBLING, F. J. (1954). Changes in the sebaceous glands and epidermis during the oestrous cycle of the albino rat. *J. Endocr.* **10**, 147–54.

EBLING, F. J. (1955). Endocrine factors affecting cell replacement and cell loss in the epidermis and sebaceous glands of the female albino rat. *J. Endocr.* **12**, 38–49.

EBLING, F. J. (1957 a). The action of testosterone on the sebaceous glands and epidermis in castrated and hypophysectomized male rats. *J. Endocr.* **15**, 297–306.

EBLING, F. J. (1957 b). The action of testosterone and oestradiol on the sebaceous glands and epidermis of the rat. *J. Embryol. exp. Morph.* **5**, 74–82.

EBLING, F. J. (1961). Failure of progesterone to enlarge sebaceous glands in the female rat. *Br. J. Derm.* **73**, 65–68.

EBLING, F. J. (1963). Hormonal control of sebaceous glands in experimental animals. In *Advances in Biology of Skin*, vol. IV. chap. XIII, pp. 200–19 (eds. W. Montagna, R. A. Ellis and A. F. Silver). Oxford: Pergamon Press.

EBLING, F. J. (1964). The action of steroids on the skin. In *Hormonal Steroids. Biochemistry, Pharmacology, and Therapeutics. Proc. First Int. Congr. Hormonal Steroids*, vol. 1, pp. 537–51 (eds. L. Martini and A. Pecile). New York and London: Academic Press.

EBLING, F. J. (1967). The action of an anti-androgenic steroid, 17α-methyl-B-nortestosterone, on sebum secretion in rats treated with testosterone. *J. Endocr.* **38**, 181–5.

EBLING, F. J., EBLING, E. & SKINNER, J. (1969 a). The influence of pituitary hormones on the response of the sebaceous glands of the male rat to testosterone. *J. Endocr.* **45**, 245–56.

EBLING, F. J., EBLING, E. & SKINNER, J. (1969 b). The influence of the pituitary on the response of the sebaceous and preputial glands of the rat to progesterone. *J. Endocr.* **45**, 257–63.

EBLING, F. J. & SKINNER, J. (1967). The measurement of sebum production in rats treated with testosterone and oestradiol. *Br. J. Derm.* **79**, 386–92.

FAREDIN, I., FAZEKAS, A. G., KÓKAI, K., TÓTH, I. & JULESZ, M. (1967). The *in vitro* metabolism of 4-¹⁴C-dehydroepiandrosterone by human male pubic skin. *Eur. J. Steroids*, **2**, 223–42.

GALLEGOS, A. J. & BERLINER, D. L. (1967). Transformation and conjugation of dehydroepiandrosterone by human skin. *J. clin. Endocr. Metab.* **27**, 1214–18.

GLENN, E. M. & GRAY, J. (1965). Effect of various hormones on the growth and histology of the gerbil (*Meriones unguiculatus*) abdominal sebaceous gland pad. *Endocrinology* **76**, 1115–23.

GLENN, E. M., RICHARDSON, S. L. & BOWMAN, B. J. (1959 a). A method of assay of antitumor activity using a rat mammary fibroadenoma. *Endocrinology* **64**, 379–89.

GLENN, E. M., RICHARDSON, S. L. & BOWMAN, B. J. (1959 b). Biologic activity of 6-alpha-methyl compounds corresponding to progesterone, 17-alpha-hydroxy-progesterone acetate and compound S. *Metabolism* **8**, 265–85.

GLENN, E. M., RICHARDSON, S. L., BOWMAN, G. J. & LYSTER, S. C. (1960). Steroids and experimental mammary cancer. In *Biological Activities of Steroids in Relation to Cancer*, pp. 257–305 (eds. G. Pincus and E. P. Vollmer). New York and London: Academic Press.

GOMEZ, E. C. & HSIA, S. L. (1968). *In vitro* metabolism of testosterone-4-¹⁴C and Δ⁴-androstene-3,17-dione-4-¹⁴C in human skin. *Biochemistry* **7**, 24–32.

DE GRAAF, H. J. & KOOY, R. (1955). The effect of ACTH on the sebaceous glands of the rat. *Acta physiol. pharmac. néerl.* **4**, 201–6.

DE GROOT, C. A., LELY, M. A. v.D. & KOOY, R. (1965). The effect of progesterone on the sebaceous glands of the rat. *Br. J. Derm.* **77**, 617–21.

364   J. S. STRAUSS AND F. J. EBLING

GRUMBACH, M. M., DUCHARME, J. R. & MOLOSHOK, R. E. (1959). On the fetal
    masculinizing action of certain oral progestins. *J. clin. Endocr. Metab.* **19**,
    1369–80.
HAMILTON, J. B. & MONTAGNA, W. (1950). The sebaceous glands of the hamster.
    I. Morphological effects of androgens on integumentary structures. *Am. J.
    Anat.* **50**, 191–233.
HASKIN, D., LASHER, N. & ROTHMAN, S. (1953). Some effects of ACTH, cortisone
    progesterone and testosterone on sebaceous glands in the white rat. *J. invest.
    Derm.* **20**, 207–12.
HESS, M., HALL, O., HALL, C. E. & FINERTY, J. C. (1952). Endocrine factors
    affecting weight and ascorbic acid content of rat preputial glands. *Proc. Soc.
    exp. Biol. Med.* **79**, 290–2.
HESS, M., RENNELS, E. G. & FINERTY, J. C. (1953). Response of preputial and
    adrenal glands of hypophysectomized rats to ACTH. *Endocrinology* **52**,
    223–7.
HUGGINS, C., PARSONS, F. M. & JENSEN, E. V. (1955). Promotion of growth of
    preputial glands by steroids and the pituitary growth hormone. *Endocrinology*
    57, 25–32.
HURLEY, H. J. & SHELLEY, W. B. (1960). *The Human Apocrine Sweat Gland in
    Health and Disease*. Springfield, Illinois: Charles C Thomas.
HYMAN, A. F. & GUIDUCCI, A. A. (1963). Ectopic sebaceous glands. In *Advances
    in Biology of Skin*, vol. IV, chap. V, pp. 78–93 (eds. W. Montagna, R. A. Ellis
    and A. F. Silver). Oxford: Pergamon Press.
JACOT, B. & SEYLE, H. (1951). A non adrenal-mediated action of ACTH on the
    preputial glands. *Proc. Soc. exp. Biol. Med.* **78**, 46–8.
JARRETT, A. (1959). The effects of progesterone and testosterone on surface sebum
    and acne vulgaris. *Br. J. Derm.* **71**, 102–16.
KLIGMAN, A. M. (1963). The uses of sebum? In *Advances in Biology of Skin*, vol.
    IV, chap. VII, pp. 110–24 (eds. W. Montagna, R. A. Ellis and A. F. Silver).
    Oxford: Pergamon Press.
KORENCHEVSKY, V. & DENNISON, M. (1936*a*). The histology of the sex organs of
    ovariectomized rats treated with male or female sex hormone alone or with
    both simultaneously. *J. Path. Bact.* **42**, 91–104.
KORENCHEVSKY, V. & DENNISON, M. (1936*b*). The histological changes in the sex
    organs of spayed rats induced by testosterone and oestrone. *J. Path. Bact.* **43**,
    345–56.
KORENCHEVSKY, V., DENNISON, M. & ELDRIDGE, M. (1937*a*). The effects of
    Δ^4-androstenedione and Δ^5-androstenediol on castrated and ovariectomized
    rats. *Biochem. J.* **31**, 467–74.
KORENCHEVSKY, V., DENNISON, M. & ELDRIDGE, M. (1937*b*). The prolonged treat-
    ment of castrated and ovariectomized rats with testosterone propionate.
    Biochem. J. **31**, 475–85.
KORENCHEVSKY, V., DENNISON, M. & HALL, K. (1937). The action of testosterone
    propionate on normal adult female rats. *Biochem. J.* **31**, 780–5.
KORENCHEVSKY, V., DENNISON, M. & SIMPSON, S. L. (1935). The prolonged treat-
    ment of male and female rats with androsterone and its derivatives, alone or
    together with oestrone. *Biochem. J.* **29**, 2534–52.
KORENCHEVSKY, V., HALL, K. & BURBANK, R. (1939). The manifold effects of
    prolonged administration of sex hormones to female rats. *Biochem. J.* **33**,
    373–80.
KORENCHEVSKY, V., HALL, K. & ROSS, M. A. (1939). Prolonged administration of
    sex hormones to castrated rats. *Biochem. J.* **33**, 213–22.
KUNO, Y. (1956). *Human Perspiration*. Springfield, Illinois: Charles C Thomas.

KUPPERMANN, H. S. (1944). Hormone control of a dimorphic pigmentation area in the golden hamster (*Cricetus auratus*). *Anat. Rec.* **88**, 442.

LAPIÈRE, C. (1953). Modications des glandes sébacées par des hormones sexuelles appliquées localement sur la peau souris. *C. r. Séanc. Soc. Biol.* **147**, 1302–6.

LASHER, N., LORINCZ, A. L. & ROTHMAN, S. (1954). Hormonal effects on sebaceous glands in the white rat. II. The effect of the pituitary-adrenal axis. *J. invest. Derm.* **22**, 25–31.

LORINCZ, A. L. (1963). The effects of progesterone and a pituitary preparation with sebotropic activity on sebaceous glands. In *Advances in Biology of Skin*, vol. IV, chap. XII, pp. 188–99 (eds. W. Montagna, R. A. Ellis and A. F. Silver). Oxford: Pergamon Press.

LORINCZ, A. L. & LANCASTER, G. (1957). Anterior pituitary preparation with tropic activity for sebaceous, preputial and Harderian glands. *Science, N.Y.* **126**, 124–5.

MARTAN, J. (1962). Effect of castration and androgen replacement on the supracaudal gland of the male guinea pig. *J. Morph.* **110**, 285–93.

MARTAN, J. & PRICE, D. (1967). Comparative responsiveness of supracaudal and other sebaceous glands in male and female guinea pigs to hormones. *J. Morph.* **121**, 209–21.

MIESCHER, G. & SCHÖNBERG, A. (1944). Untersuchungen über die Funktion der Talgdrüsen. *Bull. schweiz. Akad. med. Wiss.* **1**, 101–14.

MILES, A. E. W. (1963). Sebaceous glands in oral and lip mucosa. In *Advances in Biology of Skin*, vol. IV, chap. IV, pp. 46–77 (eds. W. Montagna, R. A. Ellis and A. F. Silver). Oxford: Pergamon Press.

MITCHELL, O. G. (1965). Effect of castration and transplantation on ventral gland of the gerbil. *Proc. Soc. exp. Biol. Med.* **119**, 953–55.

MITCHELL, O. G. (1967). The supposed role of the gerbil ventral gland in reproduction. *J. Mammol.* **48**, 142.

MONTAGNA, W. (1962a). *The Structure and Function of Skin*, 2nd ed. New York and London: Academic Press.

MONTAGNA, W. (1962b). The skin of lemurs. *Ann. N.Y. Acad. Sci.* **102**, 190–209.

MONTAGNA, W. (1963). Comparative aspects of sebaceous glands. In *Advances in Biology of Skin*, vol. IV, chap. III, pp. 32–45 (eds. W. Montagna, R. A. Ellis and A. F. Silver). Oxford: Pergamon Press.

MONTAGNA, W. & HAMILTON, J. B. (1949). The sebaceous glands of the hamster. II. Some cytochemical studies in normal and experimental animals. *Am. J. Anat.* **84**, 365–95.

MONTAGNA, W. & KENYON, P. (1949). Growth potentials and mitotic division in the sebaceous glands of the rabbit. *Anat. Rec.* **103**, 365–79.

MONTAGNA, W. & NOBACK, C. R. (1946). The histochemistry of the preputial gland of the rat. *Anat. Rec.* **96**, 111–27.

MONTAGNA, W. & YUN, J. S. (1962). The skin of primates. X. The skin of the ring-tailed lemur (*Lemur catta*). *Am. J. phys. Anthrop.* **20**, 95–117.

MYKYTOWYCZ, R. (1962). Territorial function of chin gland secretion in the rabbit, *Oryctolagus cuniculus* (L.). *Nature, Lond.* **193**, 799.

MYKYTOWYCZ, R. (1965). Further observations on the territorial function and histology of the submandibular cutaneous (chin) glands in the rabbit, *Oryctolagus cuniculus* (L.) *Anim. Behav.* **13**, 400–12.

MYKYTOWYCZ, R. (1966a). Observations on odoriferous and other glands in the Australian wild rabbit, *Oryctolagus cuniculus* (L.), and the hare, *Lepus europaeus* P. I. The anal gland. *C.S.I.R.O. Wildl. Res.* **11**, 11–29.

MYKYTOWYCZ, R. (1966b). Observations on odoriferous and other glands in the Australian wild rabbit, *Oryctolagus cuniculus* (L.), and the hare, *Lepus europaeus* P. II. The inguinal glands. *C.S.I.R.O. Wildl. Res.* **11**, 49–64.

MYKYTOWYCZ, R. (1968). Territorial marking by rabbits. *Scient. Am.* **218** (May), 116–26.

MYKYTOWYCZ, R. & DUDZINSKI, M. L. (1966). A study of the weight of odoriferous and other glands in relation to social status and degree of sexual activity in the wild rabbit, *Oryctolagus cuniculus* (L.). *C.S.I.R.O. Wildl. Res.* **11**, 31–47.

NEUMANN, F. & ELGER, W. (1966). The effect of a new antiandrogenic steroid, 6-chloro-17-hydroxy-1α,2α-methylenepregna-4,6-diene-3,20-dione acetate (cyproterone acetate) on the sebaceous glands of mice. *J. invest. Derm.* **46**, 561–72.

NIKKARI, T. (1965). Composition and secretion of the skin surface lipids of the rat; effects of dietary lipids and hormones. *Scand. J. clin. Lab. Invest.* **17** (Suppl. 85), 1–140.

NIKKARI, T. & VALAVAARA, M. (1969). The production of sebum in young rats: Effects of age, sex, hypophysectomy and treatment with somatotrophic hormone and sex hormones. *J. Endocr.* **43**, 113–18.

NOBLE, R. L. & COLLIP, J. B. (1941). A possible direct control of the preputial glands of the female rat by the pituitary gland and indirect effects produced through the adrenals and gonads by augmented pituitary extracts. *Endocrinology* **29**, 943–51.

OERTEL, G. W. & TREIBER, L. (1969). Metabolism and excretion of C_{19}- and C_{18}-steroids by human skin. *Eur. J. Biochem.* **7**, 234–8.

PAPA, C. M. (1967). Effect of topical hormones on aging skin. *J. Soc. cosmet. Chem.* **18**, 549–62.

PAPA, C. M. & KLIGMAN, A. M. (1965). The effect of topical steroids on the aged human axilla. In *Advances in Biology of Skin*, vol. VI, chap. XI, pp. 177–198 (ed. W. Montagna). Oxford: Pergamon Press.

POCHI, P. E. & STRAUSS, J. S. (1963). Sebaceous gland function before and after bilateral orchiectomy. *Archs Derm.* **88**, 729–31.

POCHI, P. E. & STRAUSS, J. S. (1965a). The effect of aging on the activity of the sebaceous gland in man. In *Advances in Biology of Skin*, vol. VI, chap. VII, pp. 121–7 (ed. W. Montagna). Oxford: Pergamon Press.

POCHI, P. E. & STRAUSS, J. S. (1965b). Lack of androgen effect on human sebaceous glands with low-dosage norethindrone. *Am. J. Obstet. Gynec.* **93**, 1002–4.

POCHI, P. E. & STRAUSS, J. S. (1967). Effect of prednisone on sebaceous gland secretion. *J. invest. Derm.* **49**, 456–9.

POCHI, P. E. & STRAUSS, J. S. (1969). Sebaceous gland response in man to the administration of testosterone, Δ^4-androstenedione, and dehydroisoandrosterone. *J. invest. Derm.* **52**, 32–6.

POCHI, P. E., STRAUSS, J. S. & MESCON, H. (1962). Sebum secretion and urinary fractional 17-ketosteroid and total 17-hydroxycorticoid excretion in male castrates. *J. invest. Derm.* **39**, 475–83.

POCHI, P. E., STRAUSS, J. S. & MESCON, H. (1963). The role of adrenocortical steroids in the control of human sebaceous gland activity. *J. invest. Derm.* **41**, 391–9.

QUAY, W. B. (1965). Comparative survey of the sebaceous and sudoriferous glands of the oral lips and angle in rodents. *J. Mammol.* **46**, 23–37.

QUAY, W. B. (1968). The specialized posterolateral sebaceous glandular regions in microtine rodents. *J. Mammol.* **49**, 427–45.

QUAY, W. B. & TOMICH, P. Q. (1963). A specialized midventral sebaceous glandular area in *Rattus exulans*. *J. Mammol.* **44**, 537–42.

RICHMOND, N. D. & ROSLUND, H. R. (1952). A mid-ventral dermal gland in *Peromyscus maniculatus*. *J. Mammal.* **33**, 103–4.

RONGONE, E. L. (1966). Testosterone metabolism by human male mammary skin. *Steroids* **7**, 489–504.

RONY, H. R. & ZAKON, S. J. (1943). Effect of androgen on the sebaceous glands of human skin. *Archs Derm. Syph.* **48**, 601–4.

SALMON, U. J. (1938). The effect of testosterone propionate on the genital tract of the immature female rat. *Endocrinology* **23**, 779–83.

SAUNDERS, H. L. & EBLING, F. J. (1969). The antiandrogenic and sebaceous gland inhibitory activity of 6α,6β-ethylene 17α-methyl-B-nortestosterone. *J. invest. Derm.* **52**, 163–8.

SCHAFFER, J. (1940). *Die Hautdrüsenorgane der Säugetiere, mit Gesonderer Berücksichtigung ihres histologischen Aufbaues und Bemerkungen über die Proktodäaldrüsen.* Berlin: Urban and Schwarzenburg.

SCHIEFFERDECKER, P. (1922). Die Hautdrüsen des Menschen und der Saugetiere, ihre Bedeutung, sowie die Muscularis sexualis. *Zoologica, Stuttg.* **72**, 1–154.

SEYLE, H. (1940). Interactions between various steroid hormones. *Can. med. Ass. J.* **42**, 113–16.

SHELLEY, W. B. & CAHN, M. M. (1955). Experimental studies on the effect of hormones on the human skin with reference to the axillary apocrine sweat gland. *J. invest. Derm.* **25**, 127–31.

SHELLEY, W. B. & HURLEY, H. J. (1957). An experimental study of the effects of subcutaneous implantation of androgens and estrogens on human skin. *J. invest. Derm.* **28**, 155–8.

SMITH, J. G., JR (1959). The aged human sebaceous gland. *Archs Derm. Syph.* **80**, 663–71.

STRAUSS, J. S. & KLIGMAN, A. M. (1961 a). The effect of progesterone and progesterone-like compounds on the human sebaceous gland. *J. invest. Derm.* **36**, 309–19.

STRAUSS, J. S. & KLIGMAN, A. M. (1961 b). Androgenic effects of a progestational compound, 17α-ethynyl-19-nortestosterone (Norlutin), on the human sebaceous gland. *J. clin. Endocr. Metab.* **21**, 215–19.

STRAUSS, J. S., KLIGMAN, A. M. & POCHI, P. E. (1962). The effect of androgens and estrogens on human sebaceous glands. *J. invest. Derm.* **39**, 139–55.

STRAUSS, J. S. & POCHI, P. E. (1961). The quantitive gravimetric determination of sebum production. *J. invest. Derm.* **36**, 293–8.

STRAUSS, J. S. & POCHI, P. E. (1963). The human sebaceous gland: its regulation by steroidal hormones and its use as an end organ for assaying androgenicity *in vivo*. *Recent Prog. Horm. Res.* **19**, 385–444.

STRAUSS, J. S. & POCHI, P. E. (1964). Effect of cyclic progestin-estrogen therapy on sebum and acne in women. *J. Am. med. Ass.* **190**, 815–19.

STRAUSS, J. S., POCHI, P. E., SARDA, I. R. & WOTIZ, H. H. (1969). Effect of oral and topical 17α-methyl-B-nortestosterone on sebum production and plasma testosterone. *J. invest. Derm.* **52**, 95–9.

SWEENEY, T. M., SZARNICKI, R. J., STRAUSS, J. S. & POCHI, P. E. (1969). The effect of estrogen and androgen on the sebaceous gland turnover time. *J. invest. Derm.* **53**, 8–10.

TAKAGI, S. & TAGAWA, M. (1961). Dry and soft feces of the rabbit and their relation to the anal secretion and to the cecal contents. *Zool. Mag., Tokyo* **70**, 248–52.

THIESSEN, D. D., FRIEND, H. C. & LINDZEY, G. (1968). Androgen control of territorial marking in the mongolian gerbil. *Science, N.Y.* **160**, 432–4.

WERNER, H. J., DALQUEST, W. W. & ROBERTS, J. H. (1952). Histology of the sweat gland of the peccaries. *Anat. Rec.* **113**, 71–80.

WILDE, P. F. & EBLING, F. J. (1969). Preliminary observations on the composition of skin surface fat from rats treated with testosterone and estradiol. *J. invest. Derm.* **52**, 362–5.

WILKINS, L. (1960). Masculinization of female fetus due to use of orally given progestins. *J. Am. med. Ass.* **172**, 1028–32.

WILSON, J. D. & WALKER, J. D. (1969). The conversion of testosterone to 5α-androstan-17β-ol-3-one (dihydrostestosterone) by skin slices of man. *J. clin. Invest.* **48**, 371–9.

WOODBURY, L. P., LORINCZ, A. L. & ORTEGA, P. (1965a). Studies on pituitary sebotropic activity. I. A new sensitive assay method for sebotropic activity based on beta-glucuronidase content of preputial glands. *J. invest. Derm.* **45**, 362–3.

WOODBURY, L. P., LORINCZ, A. L. & ORTEGA, P (1965b). Studies on pituitary sebotropic activity. II. Further purification of a pituitary preparation with sebotropic activity. *J. invest. Derm.* **45**, 364–7.

WYNNE-EDWARDS, V. C. (1962). *Animal Dispersion in Relation to Social Behavior.* Edinburgh and London: Oliver and Boyd, Ltd.

YIP, S. Y. & FREINKEL, R. K. (1964). The direct effect of ACTH on the rat pre-putial gland. *J. invest. Derm.* **43**, 389–93.

ZARATE, A., MAHESH, V. B. & GREENBLATT, R. B. (1966). Effect of an antiandrogen, 17α-methyl-B-nortestosterone, on acne and hirsutism. *J. clin. Endocr. Metab.* **26**, 1394–8.

DISCUSSION

HELLER: I was rather intrigued by Dr Strauss's reference to the endeavours of the chemical industry to suppress body odour, but I would like to ask him quite seriously whether we are not underrating the role of these glands and of body smell by a sort of polite convention. Would he not agree that human beings do smell, both woman and man, and that this has some significance in sexual life. In fact, if as a medical person one goes into disturbed sexual relations one finds that body smell is of considerable importance in sex life in human beings, if perhaps only in the negative rather than in the positive way. In other words, a man or a women will not accept a partner who, however much anti-body-odour agents they use, does not smell attractively. I would be interested to hear his views on that.

STRAUSS: Body odour, as pointed out by Professor Heller, is obviously important in a negative sense. In other words there is no evidence that it acts as an attractant, but its presence may make an individual repulsive to others. There are two phases to apocrine-gland function in man. First, the apocrine gland forms its product, which is stored in the tubules of the gland. Secondly, as a result of adrenergic stimulation of the myoepithelium of the gland, which obviously might occur as a result of sexual excitement, the gland discharges its contents to the skin surface. There is an additional factor that must be taken into consideration and that is the fact that freshly secreted apocrine sweat is odourless. It attains its odour as a result of changes in the composition of the secretory product by bacterial organisms in the upper portions of the follicle or on the skin surface. Eccrine sweating and the amount of hair in the axilla also influence the rate and extent of odour dispersement. Other than in helping with the dispersion of apocrine sweat, eccrine sweat is not known to be important as a source of odour. Sebaceous material may be a source of odour. This is particularly true of the discharge from cysts, etc. Such a discharge may have a distinct odour characteristic of short-chain fatty acids, which is not to be unexpected since approximately two-thirds of sebum is triglyceride.

MADERSON: I have been doing a considerable amount of work on epidermal glands in lizards, and contrary to the information contained in all the text-books I have not yet come across a lizard which does not have holocrine secretory tissue somewhere in the epidermal system. It is patently obvious that there is a great probability of multiple parallel or convergent evolution in these forms, and therefore I was not aware of the incredible size and diversity that there are in different mammalian groups. The probability, therefore, that these systems arose in the course of multiple evolutionary pathways would make it predictable that there would be great diversity in the subtleties of the activation of the control mechanisms and in the precise chemical nature of the excreted materials. The second point I would like to make is the fact that in all lizards so far looked at there seems to be these specializations; it is also true that many snakes have these structures, but they are not usually on the general epidermal surface, but in the cloacal region. Chelonians and crocodiles have such specializations and we have a picture which enhances Montagna's original statement that skin is just a great holocrine system and we should face the possibility that cutaneous secretions are of exceptional importance in all amniotes, although we cannot say very much about amphibians because we are dealing with such specialized animals. Probably a number of people in the audience are aware of the extra-sensory structure in many reptiles and the majority of mammals called Jacobson's organ. Jacobson's organ is a median sensory structure in the nasal organ which is incredibly well elaborated in lizard families, and although it has been shown that this is a specialized olfactory structure, nobody has pointed out exactly what chemical stimuli it is supposed to pick up. It would be valuable, since Jacobson's organ is found in various degrees of elaboration in many mammalian orders as it is in lizard families, to establish a correlation between the degree of development of the organ and other epidermal glands. Another point I should like to make is the frequency with which these structures are found in and around the anal cloacal region. This again, as a result probably of multiple parallel evolution in lizards, is always found in the

cloacal region itself, i.e. the glandular derivatives come from the procto-
daeal epithelium or else from the general cutaneous area in the inguinal
interfenoral or pericloacal region; and finally the male and female situation
shows great intersexual differences in mammals. This is also paralleled in
many lizards and there is a spectrum of genera in which glands are equally
well developed in both sexes all the way down the line to forms in which
only the males have them, and this raises the question that there are sometimes issues for territorial identification alone, sometimes issues for species
recognition alone and sometimes it may be very important in sexual
behaviour.

STRAUSS: Most of what you have said is self-explanatory and requires no
answer. I would like to point out, however, that it is known that there are
marked differences in sebum composition throughout the animal kingdom. Secondly, I would like to emphasize that we do definitely find
differences in sebum production in men and women. Mean sebum production is higher in adult men than in women, although the values for sebum
production may show considerable overlap when individual results are
examined. This difference is exaggerated later in life, since sebum production decreases in women after menopause, but shows very little change in
men until after the age of 70.

BRIGGS: I would like to ask one or two questions to try and cover one or
two of the apparent discrepancies between the effects of hormones upon
the human and Professor Ebling's data on the animal. I believe you used
ethynyl oestradiol in your human studies and doses of up to 1,000 μg a
day, which is an extremely high dose of this compound. I would think
that at this dosage you would get virtual suppression of pituitary activity,
including suppression of growth hormone, prolactin and so on, which,
according to the animal data, might also be of influence on sebum production. Professor Ebling used oestradiol, which is only weakly antigonado-
trophic. It is probably this difference on endogenous hormone production
that is of importance. Secondly, this reported suppressive effect of high
doses of progesterone. Do you consider that this is perhaps due to androgenic metabolites, and is it possible to suppress the effects of high doses
of progesterone with anti-androgens?

EBLING: May I comment on these points? In respect of the first one I
would not wish to deny the possibility that some of the effects of oestradiol
could be due to suppression of endogenous androgen production or some
other central effect. What I do maintain is that there is ample evidence
that oestrogens also have a peripheral action. With regard to the second
point, I would like to say that we have recently been using a dose of
progesterone of 10 mg a day in the rat, and we have been quite unable to

demonstrate any changes in sebum secretion as measured by hair fat levels. Finally, I suppose it is not impossible that the skin (or some other organ) might metabolize these very large doses of progesterone to androgens, but we have not any evidence on this problem—except that the sebaceous glands do not appear to be stimulated!

STRAUSS: The only instance in which a progestogen has caused an increase in glandular size or sebum production has occurred with the administration of 10–20 mg of norethindrone daily. These dosages are much larger than those used in oral contraceptives containing this compound and in our studies we saw other androgenic side-effects. In regards to your question on the use of a high dose of a potent oestrogen (ethynyl oestradiol), I only presented this data because of time limitations. We have used various oestrogens (diethylstilboestrol, conjugated equineoestrogens), and seen the same results. From our studies on the therapeutic use of oestrogens in acne, it would appear that 0·075–0·1 mg of ethynyl oestradiol or its 3-methyl ether is the dosage of oestrogen necessary to decrease sebum production in women given the drug in a cyclic fashion as for oral contraception.

LEATHEM: Since we are considering the influence of steroids on glands I would like to ask what effect pregnancy has on these glands.

STRAUSS: Our studies of pregnancy are very incomplete, and no pattern has been found. Sebum production increases in some women and decreases in others during pregnancy.

MARTINI: I would like to know whether the progestational agent norethynodrel, which is metabolized through oestrogenic pathways, has some effect on sebaceous glands. It would also be interesting to know whether growth hormone has any effect on the enzymes which convert testosterone into dihydrotestosterone, which apparently is the active form of androgenic hormones.

STRAUSS: Norethynodrel has not had any effect on sebum production in our studies. This, of course, is not so for norethindrone, as I have previously pointed out. I have no data on growth hormone, but perhaps Professor Ebling can answer that question.

EBLING: It is a very good question, but we haven't any data either.

PHEROMONE–ENDOCRINE INTERACTIONS IN INSECTS

By R. H. BARTH Jr

INTRODUCTION

Pheromones represent a small but often highly significant portion of the biotic environment of many if not most animals. Pheromones are environmental factors which serve a signal function in individuals able to perceive them. As originally defined by Karlson & Butenandt (1959), pheromones are chemical substances produced by exocrine glands which modify the behaviour and/or development of conspecific individuals. As Wilson (1968) has pointed out, pheromones were probably the first signals to be utilized in the evolution of animal communication. Although chemical communication systems have been revealed in most animals in which a diligent search has been undertaken, research efforts have focused chiefly on insects and it is from this group that the majority of known pheromones has been described (for recent reviews of chemical communication in insects see Schneider, 1966; Butler, 1967a; Regnier & Law, 1968).

Pheromonal and hormonal communication systems have doubtless been closely intertwined throughout metazoan evolution. Haldane (1955) has argued that communication among unicellular organisms must have preceded the formation of metazoans, and almost certainly this communication was chemical. Within the metazoan body chemical communication between cells is frequently carried out by means of hormones. It is therefore not surprising that there is often a close relationship in higher animals between the endocrine system and inter-individual communication mediated by pheromones. In higher animals such as insects, pheromonal communication (broadly construed to include all aspects of the process) may be influenced by the endocrine system of the participants at two separate points: (1) in the animal producing the signal, hormones may regulate the synthesis and/or release of the pheromone, (2) in the receiver of the signal, the endocrine system may be involved in various ways in the mediation of the response to the pheromone. In contrast to 'releaser' pheromones which have an immediate behavioural effect on the recipient (a matter of nervous mediation via classical receptor-conductor-effector pathways), pheromones which affect the endocrine system typically have a delayed effect on the recipient. Such pheromones have been termed 'primer' pheromones by Wilson & Bossert (1963), as they act to set the stage for subsequent behavioural or developmental events.

As noted above, the term 'pheromone' is restricted to chemical signals employed in intraspecific communication. The scope of this review will be extended somewhat beyond this narrow definition to include some interspecific interactions involving chemical communication in which the signals resemble primer pheromones in influencing the endocrine systems of recipients, and some cases in which hormones themselves serve as agents of chemical communication between individual organisms of different species. The distinction between intra- and interspecific communication upon which the definition of the term 'pheromone' was originally based would seem on physiological grounds to be an artificial one, as many if not all of the same phenomena are exhibited in both types of communciation—an observation also made by Wilson (1968). For this reason it would seem inadvisable to limit our discussion to intraspecific interactions.

Rather than attempting to be comprehensive, this review will concentrate on a number of specific examples selected to illustrate the variety of pheromone-endocrine interactions which have been revealed in recent years. To achieve this aim, the following examples will be discussed: (1) the endocrine control of mating behaviour in insects through control of the synthesis and/or release of sex pheromones; (2) the synchronization of reproduction and other developmental events in populations of certain insects as achieved by means of primer pheromones and plant-derived chemical factors having primer effects; (3) caste determination in social insects as an example of endocrine mediated developmental effects of primer pheromones; (4) interspecific interactions involving insects and other animals (chiefly prey–predator and host–parasite relationships) in which a compound which is a hormone for one species acts as a chemical signal with a releaser or primer effect on another species; (5) compounds produced by plants which mimic the actions of insect developmental hormones.

In restricting our discussion to insects we should not lose sight of the fact that analogous pheromone–endocrine interactions are also being revealed in vertebrates particularly mammals. The most familiar examples are those drawn from the reproductive physiology of rodents and include the Lee–Boot effect (suppression of oestrus in laboratory mice by all-female grouping—van der Lee & Boot, 1955, 1956), the Whitten effect (synchronization by acceleration of the oestrous cycle of grouped female mice by exposure to male urine—Whitten, 1957) and the Bruce effect (the failure of implantation in impregnated females exposed to the urine of alien males—Bruce, 1959). For recent discussions of the significance of pheromones in the life of mammals and other vertebrates, the reader is referred to the reviews of Whitten (1966), Bronson (1968), and Gleason & Reynierse (1969).

SEX PHEROMONES AND MATING BEHAVIOUR

Sex attractants or sex pheromones play an important role in the mating behaviour of many insects. New reports of the existence of sex pheromones in different insect species appear in the literature at an ever increasing rate. As Jacobson (1965) has summarized in an extensive review, sex pheromones may be produced by either sex of a species, and in some cases, at least, each sex may produce a pheromone with a distinctive role in courtship behaviour. Information currently available indicates that the vast majority of sex pheromones possess releaser functions alone (the queen substance of honeybees is a notable exception—see below). Thus they stimulate the chemoreceptors of the recipient insect and produce an essentially immediate behavioural effect via neural transmission of the information contained in the signal. Depending upon the nature of the behavioural response elicited, sex pheromones may function as attractants, excitants, identifiers (of both the species and sex of the producers), aphrodisiacs, or releasers of specific courtship patterns. In some cases, a sex pheromone may appear to have several of these effects. Whether these multiple behavioural effects result from an increasing concentration of a single substance activating increasing numbers or different types of chemoreceptors in the recipient (a differential threshold phenomenon) or whether they result from the production of medleys of substances by the producer is not clear, given the present rudimentary state of knowledge of pheromone chemistry. The efficacy of employing medleys of substances to increase signal diversity for short-range communication has been pointed out by Wilson (1968). At greater distances, however, the signal specificity of medleys is lost; therefore it may be predicted that single substances will be employed in long-range communication. The available evidence tends to bear out the predicted relationship between the utilization of medleys or single substances and the distance over which communication must be effected (Wilson, 1968; Regnier & Law, 1968).

Although sex pheromones typically do not have primer effects on recipients, the endocrine system of the producer insect is sometimes involved in control of their synthesis and/or release. Through control of these processes the endocrine system regulates mating behaviour so that mating occurs at appropriate stages in the reproductive cycle. Such a control system operates in females of the ovoviviparous cockroaches—cockroaches in which the ootheca is oviposited into a brood sac and incubated within the female until hatching. In these cockroaches the female's reproductive cycle consists of: a pre-oviposition period during which time the first set of oocytes is matured and females are sexually receptive; ovulation and

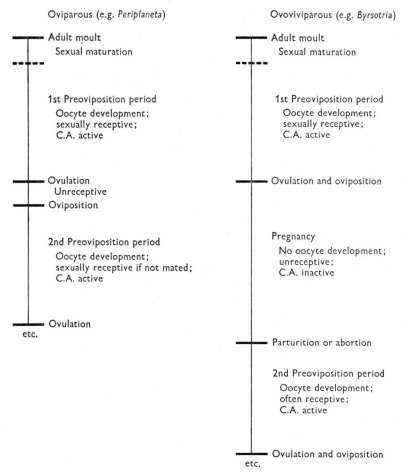

Fig. 1. A comparison of the female reproductive cycles of oviparous and ovoviviparous cockroaches.

oviposition; pregnancy, during which maturation of additional oocytes does not occur, mating is mechanically impossible, and females are unreceptive; and parturition (Fig. 1). Following parturition a second preoviposition period commences during which a second set of oocytes matures and mating may again occur (Barth, 1968). A number of these cycles may occur during the life of the female. In these insects the reproductive cycle is a result of cycles of activity in the corpora allata. These endocrine organs are active during the pre-oviposition period and inhibited during pregnancy (Engelmann, 1965, 1968; Barth, 1968).

The first evidence suggestive of a role for the endocrine system in governing the mating behaviour of any insect appeared in the report of Engelmann (1960), who showed that a substantial percentage of females of the ovoviviparous cockroach, *Leucophaea maderae*, failed to mate during the first pre-oviposition period if deprived of their corpora allata 1 day after the adult moult. Subsequently it was shown by Barth (1961, 1962) that allatectomized females of another ovoviviparous cockroach, *Byrsotria fumigata*, failed to mate owing to the absence of a sex pheromone normally produced by virgin females during the pre-oviposition period. This phero-mone plays an essential role in mating behaviour in attracting males, permitting sex recognition, and releasing the preliminary courting beha-viour of males (Barth, 1964). Normal mating behaviour was observed in allatectomized females artificially coated with sex pheromone (Roth & Barth, 1964), suggesting that stimulation of pheromone production is indeed the means by which the corpora allata regulate mating behaviour. In allatectomized females the ability to produce the pheromone is restored by implantation of active corpora allata (Barth, 1962), injection of farnesyl methyl ether—a compound which mimics the action of the corpus allatum hormone (Emmerich & Barth, 1968)—or injection of synthetic juvenile hormone (corpus allatum hormone) (a gift of H. Röller) (Barth, un-published observations). As would be expected from the cyclic activity of the corpora allata in females, the sex pheromone is produced during the pre-oviposition period when the corpora allata are active, is absent during pregnancy when the corpora allata are inhibited, and reappears after parturition when the corpora allata are reactivated (Barth, 1962). In this way Nature ensures that females will mate only at appropriate times during the reproductive cycle, i.e. when maturing oocytes which will soon be ready for fertilization are present. The act of mating itself also inhibits further production of the sex pheromone at least until after the succeeding pregnancy (Barth, 1968). The nature of this inhibition is uncertain but evidence from recent experiments suggests that it may be merely a matter of the increased juvenile hormone titre resulting from mating. The act of mating is known to stimulate the activity of the corpora allata in females of a number of species of cockroaches (Barth, 1968). Injection of the larger and/or repeated doses of juvenile hormone (2–4 μg) necessary for complete oocyte maturation is markedly less effective in stimulating pheromone production than injection of single small doses (1 μg) (Barth, unpublished observations).

It should be pointed out that the corpora allata may influence the occurrence of mating behaviour in ways other than through control of pheromone production. In an attempt to explain the effect of allatectomy on mating behaviour in *Leucophaea*, Engelmann (1960) suggested that the corpora allata govern the level of sexual receptivity in females and that consequently females remain unreceptive in the absence of these organs. Subsequent investigation has shown that at least in *Leucophaea* the corpora allata do influence sexual receptivity (Englemann & Barth, 1968), though in most cockroaches the primary control centre for sexual receptivity appears to reside in the neurosecretory cells of the brain (Roth & Barth, 1964; Barth, 1968).

The question as to whether the juvenile hormone control synthesis or merely release of the sex pheromone remains unresolved although the available evidence suggests that it controls synthesis of the pheromone. Recent experiments indicate that in *Byrsotria* the pheromone is a product of the female's reproductive tract; however, it has not been possible to locate any storage reservoir from which release could be triggered by the hormone (Barth, unpublished results). In other cockroaches (e.g. *Periplaneta americana* and *Leucophaea*) the juvenile hormone stimulates synthesis of certain substances of importance in reproduction, namely colleterial gland materials which are employed in oothecal construction (Willis & Brunet, 1966) and vitellogenins which are taken up by the oocytes (Bell, 1969; Engelmann, 1969). In *Byrsotria*, sex pheromone production, vitellogenin synthesis, and synthesis of colleterial gland materials, all of which require active corpora allata, commence at approximately the same time during the first preoviposition period, suggesting that the appearance of juvenile hormone in the blood at this time stimulates the synthesis of all of these materials (Barth & Bell, unpublished observations).

In another ovoviviparous cockroach, *Pycnoscelus indicus*, the corpora allata control the production of female sex pheromone as they do in *Byrsotria*. However, in a closely related parthenogenetic species, *P. surinamensis*, corpus allatum control of sex pheromone production has been lost. These parthenogenetic females produce a sex pheromone (which stimulates courtship behaviour in males of *P. indicus*) throughout their reproductive cycle regardless of the state of or even in the absence of their corpora allata (Barth, 1965, 1968). In this case, the male of course no longer has any role in the perpetuation of the species so there can no longer be any selective advantage in being able to signal to the male the female's readiness to mate; hence, the endocrine control over the underlying communication system has been lost. These findings led Barth to propose an hypothesis relating the existence of endocrine control of mating behaviour in an insect to the type of reproductive cycle exhibited (Barth, 1965). Briefly, the hypothesis stated that endocrine control of mating behaviour by whatever means would occur only in those insects which are long-lived as adults and which have repeated reproductive cycles containing periods during which mating is not possible. In insects which are short-lived as adults, which lay eggs and die within a few days, the female must attract a mate within a brief time-span. Here it would seem advantageous to have a simpler arrangement in which the communication system for mating is genetically built in as a part of the adult developmental process so that it appears automatically as soon as the adult stage is reached. For example, in moths having short non-feeding adult stages, one would not expect to find an endocrine control mechanism influencing the production of sex pheromone. The failure of allatectomy to affect pheromone production in *Galleria mellonella* (Röller, Piepho & Holz, 1963) and *Antheraea pernyi* (Barth, 1965), two species of moths exhibiting this type of life-cycle, tends to support this hypothesis.

In relation to this hypothesis, an interesting case has recently come to light concerning the control of mating behaviour in the polyphemus moth (*Antheraea polyphemus*). This moth will mate only in the immediate vicinity of oak leaves—a favourite food plant of the larvae (Riddiford & Williams, 1967*a*). Thus some chemical emanation from oak leaves is necessary for mating to occur. Riddiford and Williams demonstrated that the volatile material from oak leaves acts only on the female and is perceived by her antennae. Reception of the oak-leaf emanation stimulates release of the female's sex pheromone; the pheromone in turn attracts males and releases their sexual behaviour. Subsequently the oak-leaf emanation was shown to be trans-2-hexenal (Riddiford, 1967). Here we have an example of interspecific chemical signalling regulating mating behaviour so as to ensure that it occurs on or near the food plant of the larvae so that the female can oviposit with a minimum of delay. The obvious adaptive value of such an arrangement leads to the prediction that it may prove to be a widespread phenomenon in phytophagous insects. The polyphemus moth case is interesting from our point of view in that there does seem to be some sort of neuroendocrine relay involved between activation of the sensory receptors on the antennae and the response of the animal. Riddiford & Williams (1967*b*) found that the prior removal of the corpora cardiaca in females prevents the calling behaviour normally seen in response to trans-2-hexenal from oak leaves. Calling refers to the abdominal extension necessary to expose the openings of the pheromone containing glands located on the intersegmental membranes near the posterior tip of the abdomen. This behaviour is essential for release of the pheromone. The fact that the corpora cardiaca appear to promote release of stored pheromone rather than *de novo* synthesis would be in line with the well-known excitatory effects of corpus cardiacum extracts on various kinds of nervous and muscular activity (Özbas & Hodgson, 1958; Milburn, Weiant, & Roeder, 1960). According to Riddiford & Williams (1967*b*) removal of the corpora cardiaca from males also prevents the appearance of attraction and courtship behaviour in response to the female sex pheromone. In recent experiments, Riddiford has shown that the corpora cardiaca are necessary for calling behaviour in another saturniid, *Hyalophora cecropia*, (though no effect of the food plant has been implicated in this case) but not in a third species, *A. pernyi* (Riddiford, personal communication).

MATURATION AND SYNCHRONIZATION PHEROMONES

The first true insect primer pheromone to be discussed is the maturation pheromone of the desert locust *Schistocerca gregaria*. The biology of this pheromone was detailed by Loher (1960) in an elegant series of experiments. The maturation pheromone is produced by mature gregarious-phase males and aids in the synchronization of the reproductive states of other members of the population by accelerating the maturation process of young adult locusts. Four aspects of maturation in adult males were studied, namely a colour change from brownish pink to bright yellow, production of the maturation pheromone itself, the appearance of sexual behaviour, and the development of the accessory glands. All of these processes are under the control of the corpora allata and do not occur in the absence of these organs. Loher was thus the first to document a case of hormonal control of pheromone production in an insect. The pheromone is a volatile lipophilic substance produced by generally distributed epidermal cells. It is perceived by receptors on the antennae of recipients, but is much more effective in promoting maturation if recipients are allowed direct contact with producers. Perception of the maturation pheromone elicits a vibration response in young locusts in which antennae, palpi and hind femora participate. The maturation pheromone therefore has a releaser as well as a primer effect, though the significance of the vibration behaviour is uncertain. The primer effect of the pheromone is of course mediated through the activity of the corpora allata as all four criteria of maturation—colour change, production of the maturation pheromone, appearance of sexual behaviour and accessory gland development—are under corpus allatum control. Allatectomized young adults failed to mature in the presence of a plentiful supply of maturation pheromone. According to Norris (1954) mature females also have an accelerating effect upon the maturation of young locusts, though it is much less pronounced than the corresponding effect of males. Loher (1960) was not able to demonstrate production of a maturation pheromone by females.

In direct contrast to the accelerating effect of the maturation pheromone on young locusts is the retarding effect of very young locusts. Norris (1954, 1964a) reports that very young adult locusts within a week of emergence retard the maturation of young males. Norris & Pener (1965) were able to prolong the retarding effect of very young locusts by allatectomy. The results of their experiments indicate that the presence of allatectomized males or females retards the maturation of males while normal females are neutral in effect. Allatectomy therefore, in addition to preventing maturation, seems to prolong the period during which an inhibitory influence is exerted on the maturation of others. It would appear that young adults with inactive corpora allata produce an inhibitory pheromone though the existence of such a substance remains to be demonstrated. In achieving synchronization, the maturation pheromone would work together with the hypothetical inhibitory pheromone, accelerating maturation of the laggards when the majority of the population are approaching sexual maturity, retarding potentially precocious individuals when most are still juvenile (Butler, 1967a).

Environmental factors of a chemical nature which influence maturation have been revealed by the combined field and laboratory investigations of Carlisle and associates. Under field conditions local populations of desert locusts often mature simultaneously, but the time-interval between adult emergence and sexual maturity (egg laying) may be as short as 3 weeks or as long as 9 months, depending upon local conditions (Carlisle, Ellis & Betts, 1965). Egg laying can be correlated with the onset of the rainy season, but maturation commences before the rains so some other environmental factor must trigger the onset of maturation. Obvious factors such as changes in temperature, humidity or photoperiod do not play an important role (Norris, 1964b). Field studies in the Somali Peninsula indicated that bud burst in various aromatic desert shrubs (primarily of the family *Burseraceae* among the members of which are species producing frankincense and myrrh) precedes the autumn rains by a week or more and the appearance of annual vegetation by at least 3 weeks (Carlisle *et al.* 1965). At the time the rains started the locusts were beginning to copulate. A week later they were laying eggs even though as yet no annual vegetation had appeared. However, by the time the eggs hatched 2 weeks later, food in the form of annual vegetation was plentiful for the young first instar hoppers. In order for this sequence of events to have occurred, maturation in the adult locusts must have been initiated about the same time as the occurrence of bud burst in the desert shrubs. Carlisle and his associates hypothesized that some constituent of the desert shrubs serves as the signal for the onset of maturation even though the young green leaves of the shrubs could have offered no more than a small supplement to the diet (Carlisle *et al.* 1965). In experiments designed to test the ability of the shrubs to influence maturation, pieces of frankincense and myrrh, resinous products of the budding shrubs, were placed in cages of young adult locusts and it was found that myrrh was particularly effective in accelerating the colour change associated with maturation and also stimulated vitellogenesis in females. Next, the major constituents of myrrh—the terpenoids eugenol, δ-pinene, β-pinene and limonene—were tested singly by topical application to individual locusts. All of the materials stimulated maturation to varying degrees as compared with the controls. On the basis of these results the authors suggest that it is the terpenoids of the aromatic shrubs, present in highest concentration at bud burst, which provide 'the environmental cue by which desert locusts gear their breeding seasons to the rains' (Carlisle, Ellis & Betts, 1965). Thus a chemical signal from vegetation in the environment of the locusts appears to have a primer effect on their maturation process. Whether the shrub terpenoids are acting as true primers in activating the locusts' endocrine system (primarily the corpora allata) which in turn stimulates maturation, or whether they are merely mimicking the action of the juvenile hormone as many terpenoids are known to do (Staal, 1967), is uncertain.

Carlisle and his associates have also studied the effect of feeding on senescent vegetation on sexual maturation in desert locusts (Ellis, Carlisle & Osborne, 1965). A diet of senescent vegetation, they find, has a substantial delaying effect on the attainment of maturation. They attribute this effect to the absence of sufficient quantities of the plant hormone gibberellin A_3 in such vegetation and demonstrate that by adding gibberellin to the diet of locusts fed on senescent vegetation, maturation may be effected as quickly as in locusts fed on fresh vegetation. The often prolonged period of sexual immaturity amounting to a sort of reproductive diapause which persists in the field during the dry season is thus attributed to the restricted diet of dry senescent vegetation available to the locusts at that season.

These several studies on sexual maturation in desert locusts indicate that the control of reproductive synchrony, a matter of obvious selective value to the species, is a highly complex matter involving endocrine responses to both intra- and interspecific chemical signals. The rate at which maturation proceeds depends upon the over-all balance of maturation accelerating and maturation retarding factors. Whether maturation is similarly governed in other gregarious migratory locusts is uncertain.

The finding that synchronous maturation occurs in *Locusta migratoria* (Norris, 1954) suggests that similar control systems are likely to be revealed in other species.

Another rather similar phenomenon which may ultimately turn out to result from the effects of primer pheromones is the acceleration of development which occurs in crowded *v.* isolated nymphal insects such as cockroaches (Willis, Riser & Roth, 1958) and crickets (Chauvin, 1958). In isolated nymphs of a number of species of cockroaches a grey viscous secretion accumulates on the terminal abdominal segments (Roth & Stahl, 1956). These authors, noting the absence of this material in crowded nymphs, suggest that it is eaten or rubbed off by the other insects and that it may serve some nutritional role as a dietary supplement, thus acting to accelerate development. Recently Ishii & Kuwahara (1967, 1968) have extracted an aggregation promoting substance from young nymphs of *Blattella germanica*. The active principle was found in faeces and ether washings of the body surface, particularly the surface of the abdominal tip. Additional investigations led the authors to conclude that the aggregation pheromone is produced by the glandular rectal pad cells. Whether this substance is the same as or is a component of the greyish viscous secretion described by Roth and Stahl is uncertain, as is also the question whether the substance or the secretion has any primer effects on developmental rates. In crickets, McFarlane (1966, 1968) has reported that the methyl esters of two fatty acids, Me linolenate and Me laurate, stimulate the growth rate of nymphs reared singly but not of those reared in groups when adsorbed on to filter papers placed in the rearing containers. According to McFarlane, the effect is not a dietary one and he suggests that it may resemble that of a primer pheromone such as is presumed to be responsible for the developmental acceleration observed in grouped animals. However, it remains unknown whether nymphs actually produce substances having primer effects on development, let alone whether fatty-acid esters are involved.

CASTE DETERMINATION IN SOCIAL INSECTS

If one were to make an *a priori* prediction as to where one would find the most complex and highly evolved interactions between pheromones and endocrine systems, the social insects would be the obvious choice. Although evidence of such interactions in social insects is surprisingly meagre, that which is available suggests that primer pheromones are of particular importance in caste determination both in the social Hymenoptera and the Isoptera.

Best known of all the primer pheromones involved in caste determination, thanks to the extensive and detailed studies of Butler and his associates (see Butler, 1964, and Butler, 1967a, for reviews), is the queen
substance of honeybees (*Apis mellifera*). This compound has a number of
important roles in the life of the honeybee colony, though its role in caste
determination is an indirect one. The queen substance has both releaser
and primer effects. In virgin queens it acts as a sex pheromone during
the nuptial flight, attracting drones from a distance and releasing their
courtship behaviour (Gary, 1962; Butler, 1967b). In most cases mated
female insects cease to produce their sex pheromone after mating. In
honeybees, however, the mated queen continues to produce the queen
substance after her return to the hive. This substance has two important
inhibitory effects on the members of the honeybee colony, namely it
prevents maturation of worker ovaries and it suppresses queen-rearing
behaviour on the part of the workers.

If the queen is removed from a colony, the workers become increasingly
restless and within a few hours begin to modify one or more worker brood
cells into the larger queen cells within which new queens will be reared. A
few days after removal of the queen, ovarian development begins in some
worker bees. These behavioural and developmental effects on workers
were shown by Butler to be the result of the removal of a pheromone
present in the queen (Butler, 1954). This inhibitory queen substance is
present on all parts of the queen's body; workers obtain it by licking the
queen's body and share it with other workers in regurgitated food (Butler,
1956). In this way all workers are kept informed of the queen's presence
in the hive. Queen substance is produced in the queen's mandibular
glands (Butler & Simpson, 1958). The chief component of the mandibular
gland secretion was characterized as 9-oxodec-trans-2-enoic acid (= trans-
9-keto-2-decenoic acid) by Butler, Callow & Johnston (1961) and also by
Barbier & Lederer (1960). Tests employing synthetic 9-oxodecenoic acid
indicate that this compound is largely responsible for the inhibition of
queen rearing (Butler, Callow & Johnston, 1961), though an unidentified
volatile 'inhibitory scent' appears to act synergically with the acid to
produce complete inhibition (Butler, 1961). In order for a normal colony
to produce new queens during the annual breeding season the inhibition
to queen rearing owing to the presence of queen substance must somehow
be negated. Butler (1960) demonstrated that not only is the number of
workers particularly large at the time of swarming but also that the amount
of queen substance present in queens from colonies engaging in normal
reproduction is substantially less than the amount present in queens from
non-reproducing colonies. Queen rearing by colonies in the pre-swarming

stage would thus appear to result from the fact that the amount of queen substance reaching the workers is insufficient to inhibit queen-rearing behaviour.

The 9-keto-2-decenoic acid thus has an indirect effect on caste determination by suppressing queen-rearing behaviour. Queen-worker caste determination occurs early in the life of the female larvae and depends upon the diet which the larvae are fed. Larvae destined to become workers are fed royal jelly, a product of the pharyngeal glands of young workers, for about $2\frac{1}{2}$ days, then they receive honey and pollen, while larvae residing in the specially constructed queen cells are fed royal jelly exclusively during larval life (Michener, 1961). Royal jelly is chemically highly complex, and exactly what features of this material promote the development of queens remain obscure. The effect appears to be more than a general nutritional one; however, the evidence that a specific queen-promoting material is present is royal jelly in inconclusive at present (see Weaver, 1966).

The second inhibitory effect of queen substance is the prevention of ovarian maturation in workers. In the absence of queen substance the atrophied ovaries of many workers begin to develop, some reaching the stage of actually laying eggs, though of course being unfertilized such eggs give rise only to drones. Butler and his associates have found that at least two components of the queen's mandibular gland secretion are involved in the normal inhibition of worker ovaries, the 9-oxodecenoic acid and an unidentified scented substance (Butler *et al.* 1961; Butler & Fairey, 1963). The 9-oxodecenoic acid alone is a fairly effective inhibitor of ovarian development in workers; more complete inhibition is obtained with the addition of the second substance. The two substances together are, however, less effective than a live queen, so possibly a third factor is involved (Butler & Fairey, 1963). The primer effect of the queen substance is presumably mediated through the corpora allata of the recipient workers. Lüscher & Walker (1963) found that the corpora allata of workers increase in size in the days following removal of the queen from the colony and hypothesized that queen substance suppresses ovarian development by inhibiting secretion of the corpus allatum hormone, though it should be pointed out that increasing size is not always a reliable criterion of corpus allatum activity (Engelmann, 1968; Lea & Thomsen, 1969; see also Mordue *et al.* this symposium, p. 111).

Suppression of the development of worker ovaries by means of inhibitory pheromones probably occurs in many social Hymenoptera. In a number of species of bees, wasps and ants it is known that the presence of a queen in the colony will suppress the development of worker ovaries.

Removal of the queen, by contrast, results in increased ovarian development in workers of many of these species (see Butler, 1967a, for review). Further, extracts of queens from some but not all the social insect species tested are able to inhibit the ovaries of worker honeybees (Butler, 1965). Recently Butler, Calam & Callow (1967) have shown that queens of two additional species of *Apis*, *A. cerana* and *A. flora*, produce 9-oxodecenoic acid and employ it in ways similar in most respects to its employment by *A. mellifera* queens.

How 9-keto-2-decenoic acid effects the inhibition of queen-rearing behaviour and ovarian development in workers is not entirely clear. The odour of this substance alone is sufficient to inhibit both of these phenomena to some extent though the compound is much more effective if workers are allowed contact with and ingestion of material containing it (Butler, 1967a). Any effect produced as a result of olfactory perception of the pheromone must be mediated at least in part by the nervous system. Information reaching the brain from the chemoreceptors could then be relayed to the endocrine system; thus the target organs could receive information via either endocrine or nervous channels or both. Contact with or ingestion of the pheromone affords a further possibility. Chemoreceptors could of course still be involved in reception of the pheromone but the pheromone might also be absorbed directly through the cuticle or the wall of the gut and circulate in the blood affecting target organs directly like a hormone, thereby by-passing the nervous system entirely. The two separate effects of 9-keto-2-decenoic acid on worker bees appear to be mediated in different ways, for Butler & Fairey (1963) have found that when the compound is injected into the body cavity of workers, thus by-passing the external chemoreceptors, it inhibits ovarian development but not queen-rearing behaviour. Thus for the inhibition of queen rearing, the pheromone must be perceived by sensory receptors and the information transmitted via nervous channels to the brain; however, for inhibition of ovarian development, the pheromone may be transported by the blood, functioning essentially as a hormone. Whether the pheromone actually functions in the latter way under natural conditions is uncertain. If it does, it would be interesting to know whether it can act directly on the ovary or whether it acts via the corpora allata as suggested by Lüscher & Walker (1963). The problem of the method of uptake of primer pheromones and the means by which their effects are mediated in recipients has also been discussed by Loher (1960) with reference to the maturation pheromone of locusts and by Lüscher in connection with caste-determining pheromones in termites (see below).

In order to account for the manifestation of queenless behaviour in workers occurring shortly after removal of the queen, one must suppose that ingested queen substance is rapidly metabolized. Johnston, Law & Weaver (1965) have shown that this is indeed the case for 9-keto-2-decenoic acid. Using a radioactive form of the compound, they found that 95 % of the pheromone fed to workers was converted to inactive substances within 72 h. These investigators have also proposed the operation of a pheromone cycle in which the inactive molecules would be passed back to the queen as part of the regurgitated material with which workers feed the queen. The queen could then convert them into the active form by simple enzymic reactions, thus saving herself the enormous energy expenditure required for synthesis of the complete molecule. In connection with this novel hypothesis it is interesting to note that the queen at any given moment contains only 100 μg of the ketodecenoic acid, yet she must provide approximately 0·1 μg of this compound per worker per day in order to suppress queen rearing in her colony (Butler & Paton, 1962). Obviously the pheromone must have a high turnover rate in the colony.

Information concerning the possible role of primer pheromones in caste determination in ants is virtually non-existent. Examples of partial or complete inhibition of ovarian development in workers in the presence of the queen have been found in some ant species (e.g. *Formica rufa* (Bier, 1956) and *Leptothorax tuberum* (Bier, 1954) but not in others (e.g. *Myrmica rubra*, a species in which certain workers normally lay eggs which are apparently required for feeding the brood (Brian, 1953)). Schneirla & Rosenblatt (1961) conclude that one or more pheromones from the queen inhibit queen rearing in army ants of the genus *Eciton*. According to Bier (1958) a queen-produced material that inhibits queen rearing is distributed among the workers in colonies of *Formica rufa*. Brian (1963) and Brian & Hibble (1963) have demonstrated that a material which acts similarly is produced in the head of queens of *Myrmica rubra*. In *Pheidole morrisi* a parallel inhibitory effect occurs in that soldiers inhibit to some extent the production of additional soldiers (Gregg, 1942). In a review article on the biology of ants Wilson (1963) concluded that 'perhaps the most likely approach to the problem of caste control would be a direct test of the inhibitory pheromone hypothesis'. To my knowledge, no such test has been undertaken since that time and it would still seem to be the best available approach. However, the solution to the problem of caste determination in ants is certainly not likely to be simple nor is there any reason to believe that the control mechanisms will be the same or even similar in all species.

From this review of the role of pheromones in caste determination in and reproduction of colonies of social Hymenoptera it is clear that the evolutionarily more advanced groups have evolved pheromonal control of reproduction by the queen. An analogous development has occurred in an entirely unrelated group of social insects, the Isoptera (see below). The evolution of pheromonal control has been discussed by Wilson (1965). As he points out, control mechanisms are less refined in the more primitive groups of Hymenoptera, where caste systems are less distinct. For example, in some wasps (*Polistes*) and bumblebees (*Bombus*) the queen resembles the workers closely and dominates them by threat behaviour in addition to having a distinctive recognition odour. In some primitive ants reproductive dominance is achieved by certain females who remove the eggs of rivals from the brood cells and substitute their own. Wilson concludes that during the evolution of the control of reproduction by a queen caste in Hymenoptera, a transition has occurred from direct intervention among rival females to indirect control by pheromones acting to suppress reproduction by workers and to promote behaviour in workers appropriate to the needs of the colony.

In termites, caste determination is controlled by a complex series of pheromone-endocrine interactions. The important role of pheromones in caste determination was appreciated by the early students of termite biology long before pheromones were generally recognized as a functionally discrete class of substances (Light, 1943). Light (1944) demonstrated in *Zootermopsis* that the primary reproductives produce substances which inhibit the development of supplementary reproductives. In the absence of the queen some of the larvae develop into supplementary reproductives; if the orphaned larvae are fed extracts of reproductives, fewer develop into reproductives than in orphaned colonies not receiving the extract.

An extensive series of studies undertaken by Lüscher and his colleagues have given us a remarkably complete account of the pheromone–endocrine interactions involved in caste determination in a single species of termite, *Kalotermes flavicollis*, though a number of details are still missing. The account of these interactions given here is based largely on Lüscher's reviews of this work (Lüscher, 1960, 1961, 1962). Unlike those of social Hymenoptera, termite colonies are bisexual and in most termites both sexes are represented in roughly equal numbers in all castes. Termite societies are headed by a pair of primary reproductives, the king and queen, who founded the colony as winged adults. If these die they are replaced by supplementary reproductives which develop directly from larvae or nymphs (nymphs are distinguished from larvae by the presence of wing pads). Most members of the colony are sterile workers or soldiers. *Kalotermes* has three adult castes: winged forms which become primary reproductives, supplementary reproductives and soldiers. There is no fixed worker caste in this species, the functions of workers being carried out by full-grown larvae termed 'pseudergates.' Of particular interest to us are the developmental possibilities open to the pseudergate. Depending upon conditions in the colony, pseudergates may moult to a pseudergate

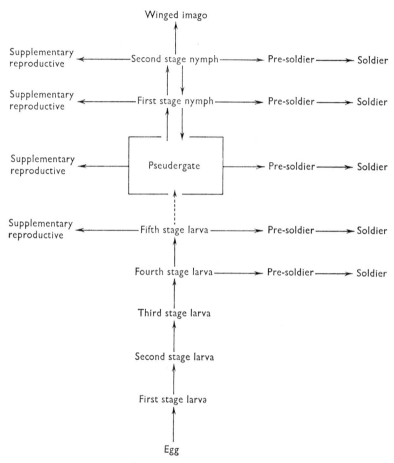

Fig. 2. The course of development in the termite *Kalotermes flavicollis*. Note the developmental possibilities open to the pseudergate. (After Lüscher, 1961).

(a stationary moult), a supplementary reproductive, a presoldier, or a first-stage nymph. The supplementary reproductive moults no more; presoldiers moult to become soldiers; first-stage numphs may moult to become second-stage nymphs which in turn moult to become winged adults, or nymphs may undergo regressive moults to an earlier stage, or moult to supplementary reproductives or soldiers (see Fig. 2).

The role of pheromones in caste determination has been best analyzed in connection with the formation of supplementary reproductives (see Fig. 3). As long as a pair of reproductives, either primary or supplementary, are present, no supplementaries are produced. This suppression is

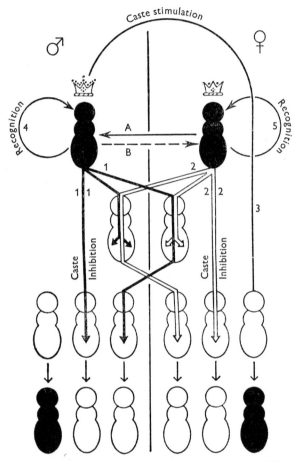

Fig. 3. Pheromonal control of supplementary reproductive caste differentiation in the termite *Kalotermes flavicollis*. 'Crowned' figures in top row represent the functional pair of reproductives. Other figures represent pseudergates. Solid-coloured pseudergates differentiate into reproductives. King and queen produce substances (1 and 2) which inhibit differentiation of pseudergates into their own royal castes. These inhibitory pheromones are passed from reproductive to pseudergate and from pseudergate to pseudergate. Female pseudergates absorb the female inhibitory pheromone and transmit the male inhibitory pheromone to other pseudergates and vice versa. Pheromone 3 from the king stimulates differentiation of female supplementary reproductives. When supernumerary male reproductives are present, they recognize each other through pheromone 4 and fight; supernumerary female reproductives recognize each other through pheromone 5 and fight. The presence of a reproductive of a given sex stimulates the production of the inhibitory pheromone by the opposite sex. The nature of these stimuli, labelled *A* and *B*, is unknown. For further explanation see text. (After Wilson, 1965).

due to inhibitory pheromones produced by both king and queen. The inhibitory pheromone produced by the king inhibits development of male supplementary reproductives; that produced by the queen inhibits development of female supplementary reproductives. Both must be present in the colony if supplementary reproductive formation is to be completely inhibited, as production of the inhibitory pheromone by the king is stimulated by the presence of the queen and vice versa. These pheromones are given off with the excrement through the anus of the reproductives and are distributed throughout the colony via the anal–oral trophallaxis so characteristic of termites. To demonstrate that the inhibitory pheromones could be passed from pseudergate to pseudergate, Lüscher separated two colonies with a screen and placed a pseudergate in the screen with its head end in a colony with reproductives and its abdomen in an orphaned colony. Production of supplementary reproductives was suppressed in the orphaned colony so long as the pseudergate was present, indicating transmission of the pheromone through the pseudergate. Curiously, the nature of the suppression depends upon the sex of the pseudergate in the screen; male pseudergates are most effective at suppressing female supplementary reproductive differentiation and vice versa. This finding led Lüscher to hypothesize that the male inhibitor is absorbed by male pseudergates and utilized in the suppression of reproductive differentiation while the female inhibitor is excreted and thus passed on to other individuals. The reverse would then be true of the female inhibitor. In addition to the inhibitory pheromone, male reproductives in the absence of female reproductives produce a substance which stimulates the development of female reproductives. However, the reverse relationship does not occur. These three pheromones qualify as primer pheromones, influencing developmental events in recipients. As mentioned above in our earlier discussion of orally active primer pheromones, such substances could act through chemoreceptors or they could enter the haemolymph directly from the gut and thus function essentially as hormones. How these pheromones influence the endocrine system of pseudergates is not yet clear (see below). Only a small proportion of the pseudergates develop into supplementary reproductives when the inhibitory pheromones are removed. The reason for this, Lüscher discovered, is that only those pseudergates that have recently moulted when the pheromones are withdrawn are fully competent to differentiate. Competence is gradually lost during the intermoult period.

In addition to these three primer pheromones, two releaser pheromones are produced by the reproductives; one by the male and one by the female. These are recognition pheromones; when supernumerary male reproductives are present they recognize each other and fight, similarly supernumerary female reproductives recognize each other by means of the recognition pheromone and fight. When one reproductive in a fight is injured it is abandoned by the other and becomes a victim of cannibalism by larvae and nymphs (Lüscher, 1964a). In this way the number of supplementary reproductives in the colony is reduced to a single pair. These recognition pheromones are thought to be products of dermal glands scattered over the surface of the reproductive and to be perceived by contact chemoreception. A single screen barrier between two colonies prevents intercolony trophallaxis, hence supplementary reproductives are produced by the one orphaned colony. However, antennal contact with members of the adjoining colony enables workers in the orphaned colony to perceive the presence of the royal pair and the supplementary reproductives are killed. A double screen prevents antennal contact between colonies and allows survival of the supplementary reproductives; hence the reproductive recognition pheromones do not seem to be volatile.

None of these pheromones involved in the control of supplementary reproductive differentiation have been chemically identified. In fact, with the exception of the male pheromone which stimulates development of female reproductives, active extracts of these pheromones have yet to be prepared (Lüscher, 1964b).

The role of the endocrine system in termite caste determination is best understood in the case of soldier differentiation. The social control of soldier differentiation appears to be similar to that of supplementary reproductive differentiation. The presence of soldiers inhibits soldier differentiation; if an excess of soldiers is produced some will be destroyed; soldier differentiation is probably controlled by pheromones but there is as yet no proof of this. Active corpora allata are clearly involved in soldier differentiation. Implantation into pseudergates of corpora allata from primary or supplementary reproductives or soldiers is followed by differentiation into pre-soldiers at the next moult (Lüscher & Springhetti, 1960). Injection of synthetic juvenile hormone has a similar effect (Lüscher, personal communication). However, the implantation of corpora allata from pseudergates into pseudergates stimulated some pseudergates to differentiate into supplementary reproductives. This result along with others prompted Lüscher to suggest that the corpora allata produce more than one hormone and that perhaps each corpus allatum hormone promotes the differentiation of a different caste (Lüscher, 1963). Considering insects as a whole, this hypothesis has become less and less tenable since it was proposed, though the possibility still exists, that termites really are different from other insects in this regard. Lüscher has also suggested that the balance between the relative titres of three or more hormones from the corpora allata, prothoracic glands, and brain neurosecretory cells, along with fluctuations in the output of these hormones during the intermoult period, are probably important in determining the course of caste differentiation. Specifically he suggests on the basis of some histochemical observations that a massive release of brain hormone early in the intermoult period of the pseudergate is associated with supplementary reproductive differentiation (Lüscher, 1960). According to this hypothesis, removal of the inhibitory pheromones from the royal pair triggers a massive release of brain hormone in pseudergates competent to respond.

An interesting recent finding is that synthetic juvenile hormone stimulates soldier differentiation when fed to pseudergates as well as when injected (Lüscher, personal communication). Hence it would appear that in this case the juvenile hormone can potentially act as a pheromone. Whether it actually does so in the termine colony is not certain, but the fact that compounds with juvenile hormone activity have been isolated from the faeces of *Tenebrio* (Schmialek, 1961) lends credence to the idea.

It is clear from our discussion that hormonal interactions controlling caste determination in social insects are highly complex. Compounding the situation are the various environmental influences, pheromonal and otherwise, impinging on the control mechanism. Given the complexity of the situation together with the technical difficulties inherent in experimentation on social insects, it is not surprising that a complete understanding of caste determination continues to elude us.

INTERSPECIFIC INTERACTIONS

(a) Insects and other animals

Brief mention should be made of certain interspecific interactions involving insects and other animals in which a compound which is a hormone for one species acts as a chemical signal with a releaser or primer effect on another species. Interactions of this type tend to be found in rather specialized prey–predator and host–parasite relationships.

A remarkable case of this sort concerns the pygidial glands of the water beetle, *Dytiscus marginalis*, which have been shown to contain 11-deoxycorticosterone (Schildknecht, Siewerdt & Maschwitz, 1966). This material appears to be employed by the beetle as a defensive secretion against fish predators. According to Schildknecht *et al.* the beetle liberates the steroid in substantial amounts when seized by a fish, and the steroid, identical to a normal product of the vertebrate adrenal cortex, so disturbs the ionic balance of the body fluids of the fish that it suffers a partial paralysis, allowing the beetle to crawl out of its gullet to freedom.

The first well-documented case of a parasite, or in this case a symbiont, employing host hormones as a chemical signal is that of the hypermastigid flagellates living in the hind gut of the wood-eating cockroach, *Crypto-cercus*. Secretion of the moulting hormone ecdysone by the host at time of the moult initiates the sexual cycle in the protozoan symbionts. The sexual cycle of the protozoans can be induced artificially in adult cock-roaches by injection of ecdysone, a hormone normally not present in the adult (Cleveland, Burke & Karlson, 1960).

The rabbit flea (*Spilopsyllus cuniculi*) employs a number of the host's hormones in the regulation of its own sexual cycle in what is obviously a highly evolved relationship (Rothschild & Ford, 1966). The host hor-mones have in this case become chemical signals with important primer effects on the parasite. Female rabbit fleas undergo ovarian maturation only on female hosts during the last 10 days of pregnancy. At parturition they and the male fleas transfer to the new-born rabbits, where they undergo copulation and egg-laying. After about 10 days the fleas return to the lactating doe; any mature eggs remaining in the ovaries of the female fleas after this transfer are resorbed and the fleas remain reproductively quiescent until the doe again becomes pregnant (Rothschild & Ford, 1964a). In experiments involving the feeding of fleas on castrated bucks and ovariectomized does injected with the appropriate hormones, it was shown that under natural conditions the presence of high levels of corti-costeroids in the blood of the doe near the end of the pregnancy promotes ovarian maturation in the fleas while oestrogens and thyroxin appear to play a subsidiary role (Rothschild & Ford, 1964b). The rapid ovarian regression which occurs when the fleas return to the lactating doe has been shown to be due to the luteinizing hormone and progesterone circulating in the host blood at this time (Rothschild & Ford, 1966). Each of these hormones can effect ovarian regression independently. In Nature the fleas obtain the host hormones from the blood upon which they feed; however, spraying the fleas directly with the appropriate hormone induces either ovarian maturation or regression, indicating that the hormones act directly on the fleas and not indirectly through the host (Rothschild & Ford, 1964b, 1966). An additional host hormone has been shown to be important in stimulating copulatory behaviour in rabbit fleas. Copulation occurs only on newborn rabbits and Rothschild & Ford demonstrated that somatotrophin was the chief host hormone responsible. Females will respond rapidly to the somatotrophin only if they have previously experi-enced corticosteroid stimulated maturation; otherwise substantial quan-tities of the 'copulation factors' must be provided over a 10-day period. Males, on the other hand, no matter what their previous history, will respond rapidly to receptive females. How the rabbit hormones act on the flea is not certain; Rothschild & Ford (1966) suggest that they influence the release of the fleas' own hormones. Presumably the corpus allatum hormone is of primary importance in mediating the fleas' response to the host's hormones.

(b) Insects and plants

We have already discussed two examples of interactions between insects and plants; one involving the control of sex-pheromone release in female polyphemus moths, the other involving the control of sexual maturation in the desert locust. In this section brief mention will be made of some plant products which mimic the action of insect developmental hormones.

Plant substances known to mimic the juvenile hormone of insects are at the present time relatively limited in number. Farnesol, the first juvenile hormone mimic to be identified (Schmialek, 1961), is a component of

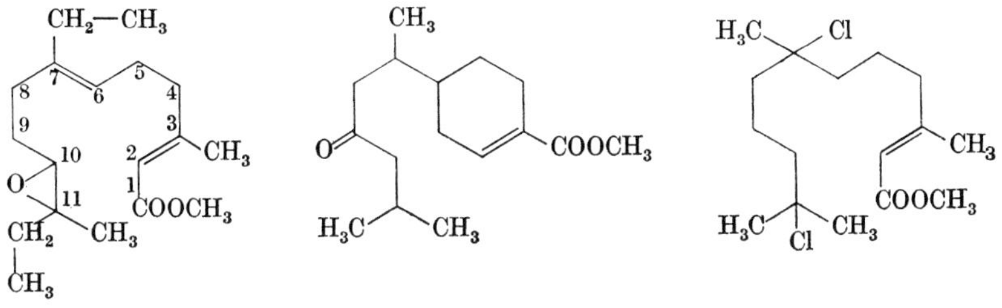

(a)   (b)   (c)

Fig. 4. Chemical structures of the juvenile hormone and some juvenile hormone analogues. (a) The juvenile hormone (Röller, Dahm, Sweely & Trost, 1967). (b) The paper factor from balsam fir (Bowers, Fales, Thompson & Uebel, 1966). (c) A dihydrochloride of methyl farnesoate (Romanuk, Slama & Sorm, 1967). (After Williams, 1967).

certain plant-derived essential oils (Staal, 1967). Most curious of the plant-derived juvenile hormone mimetic substances is the so-called paper factor first discovered by Slama & Williams (1965). This material, found in paper of American but not European origin, is highly active in the disruption of metamorphosis in bugs of the family Pyrrhocoridae but appears to be inactive in other insects (Slama & Williams, 1966). Slama & Williams (1966) demonstrated that the juvenile hormone activity originated in the balsam fir (*Abies balsamea*), used extensively in the manufacture of paper in North America. The compound chiefly responsible for this juvenile hormone activity was isolated from balsam fir by Bowers, Fales, Thompson & Uebel (1966) and was identified as todomatuic acid methyl-ester (juvabione). Fig. 4 compares the structure of this compound with that of the juvenile hormone identified by Röller, Dahm, Sweely & Trost

(1967). Although juvabione and the related dehydrojuvabione are the only compounds with juvenile hormone activity to have been first isolated and identified from plants, W. S. Bowers has obtained extracts with juvenile hormone activity in the *Tenebrio* test from 6 of 52 plant species chosen at random (cited by Williams & Robbins, 1968).

By contrast, the number of known plant substances with moulting hormone activity is increasing at a rapid rate (see reviews by Staal, 1967,

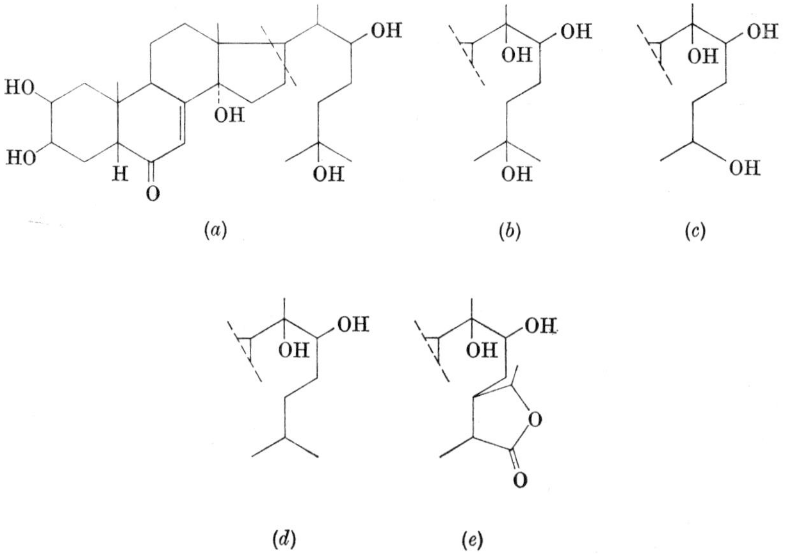

Fig. 5. Chemical structures of some moulting hormones occurring in insects and plants. (*a*) Ecdysone (Karlson, Hoffmeister, Hummel, Hocks & Spiteller, 1965). (*b*) Ecdysterone = crustecdysone (Hoffmeister, 1966). (*c*) Inokosterone (Takemoto, Ogawa, & Nishimoto 1967). (*d*) Ponasterone A (Nakanishi, Koreeda, Sasaki, Chang & Hsu, 1966). (*e*) Cyasterone (Takemoto, Hikino, Nomoto & Hikino, 1967). (After Staal, 1967.)

and Williams & Robbins, 1968). Already the structures of more than 15 phyto-ecdysones have been determined (see Fig. 5 for a comparison of the structures of several ecdysones from insects and plants). Phyto-ecdysones are widely distributed in the plant kingdom, having been found in at least some members of all major groups from ferns and conifers to the most advanced dicotyledons.

The role of these compounds in the physiology of plants is obscure and there has naturally been much speculation among insect biologists that these compounds have evolved as a means of defence against the attacks of phytophagous insects. Although a number of these hormone mimics do cause developmental anomalies when injected into certain insects and may even kill the insect if administered in suitable amounts at an appropriate stage of the life-cycle, it remains to be determined whether these compounds can produce these effects when introduced into the insect in a more natural way, namely in the diet. Carlisle & Ellis (1968) injected ecdysone derivatives from bracken fern (*Pteridium aquilinum*) into desert locusts and obtained acceleration of the moult; however, they obtained no effect on moulting, growth or development when the same material was fed in the diet. In locusts fed on bracken substantial amounts of ecdysone were found in the faeces, suggesting that the ingested ecdysone is not absorbed by the gut. However, in preliminary studies, Staal (1967) reports some effect on lepidopterous larvae if massive doses of ecdysone analogues are fed in the diet. Kaplanis & Robbins (cited by Williams & Robbins, 1968) found no effect of authentic ecdysones on five insect species when fed in the diet, but several synthetic analogues were shown to disrupt larval growth and metamorphosis when ingested. According to Williams & Robbins (1968) ecdysones can penetrate the insect cuticle when topically applied in appropriate solvents so that uptake of phyto-ecdysones by this route in Nature is theoretically possible. These authors also cite evidence indicating that the uptake of large amounts of ecdysone can disrupt development often fatally in certain insect species. In this connection, it should be recalled that the paper factor must reach the tissues of pyrrhocorid bugs by absorption either from the gut or through the cuticle. Thus, although the evidence remains largely circumstantial, it seems not unlikely that phyto-ecdysones and plant-derived juvenile hormone analogues are, or were at some time in the past, utilized by plants as protective agents against attack by phytophagous insects.

CONCLUSIONS

From the foregoing discussion it is clear the pheromone-endocrine interactions play a primary role in the regulation of reproduction and the control of developmental events in a variety of insect species. Further research will doubtless reveal numerous additional examples of such interactions. It is reasonably clear from the examples discussed in what types of situations such interactions are likely to occur. In all cases it is obvious that a great deal more research, particularly of a biochemical and neurophysiological nature, is needed before a full understanding of these interactions can be achieved.

However, even in the light of our current limited understanding it is becoming increasingly evident that these pheromone–endocrine interactions have potentially important practical applications in insect control. Williams (1967) has discussed some of these possibilities with particular reference to plant products which mimic insect developmental hormones. As indicated above, such materials do show a certain degree of species-specificity, an important feature for the ideal insect control measure, and can fatally disrupt developmental patterns if administered in a suitable fashion at an appropriate stage in the life-cycle. An all-out search for such materials in the plant kingdom might reveal many compounds useful in insect control. Synthetic analogues of insect hormones also have potential uses in insect control. A case in point has been described by Masner, Slama & Landa (1968). A synthetic analogue of juvenile hormone, a di-hydrochloride of methyl farnesoate (see Fig. 4), causes permanent sterility in females of *Pyrrhocoris apterus* if 1 μg is applied to the body surface. Most remarkable is the fact that males treated with 100–1000 μg of this compound transfer enough of it to the females with which they mate that all are permanently sterilized. The possibilities of such compounds for insect control are immediately obvious. The sex pheromones of insects as well as such primer pheromones as the maturation pheromone of locusts, and the inhibitory pheromones involved in the caste determination of termites could also be potentially useful in insect control programmes once their chemistry and biology are more fully understood.

Clearly further research into pheromone–endocrine interactions in insects will yield results of both theoretical and practical value. Not only will our understanding of insect reproduction and development be enhanced but also selectively effective insect control measures operating with a minimum of ecological disruption will in all likelihood be developed.

REFERENCES

BARBIER, M. & LEDERER, E. (1960). Structure chimique de la substance royale de la reine d'Abeille (*Apis mellifera* L.). *C. r. hebd. Séanc. Acad. Sci., Paris* **250**, 4467–9.

BARTH, R. H. JR (1961). Hormonal control of sex attractant production in the Cuban cockroach. *Science, N.Y.* **133**, 1598–9.

BARTH, R. H. JR (1962). The endocrine control of mating behavior in the cockroach, *Byrsotria fumigata* (Guérin). *Gen. comp. Endocrinol.* **2**, 53–69.

BARTH, R. H. JR (1964). The mating behavior of *Byrsotria fumigata* (Guérin) (*Blattidae, Blaberinae*). *Behaviour* **23**, 1–30.

BARTH, R. H. JR (1965). Insect mating behavior: endocrine control of a chemical communication system. *Science, N.Y.,* **149**, 882–3.

BARTH, R. H. JR (1968). The comparative physiology of reproductive processes in cockroaches. I. Mating behaviour and its endocrine control. *Adv. Reprod. Physiol.* **3**, 167–207.

BELL, W. J. (1969). Dual role of juvenile hormone in the control of yolk formation in *Periplaneta americana. J. Insect Physiol.* **15**, 1279–90.

BIER, K. (1954). Über den Einfluss der Königin auf die Arbeiterinnen-Fertilität im Ameisenstaat. *Insectes Sociaux* **1**, 7–19.

BIER, K. (1956). Arbeiterinnfertilität und Aufzucht von Geschlechstieren als Regulationsleistung des Ameisenstaates. *Insectes Sociaux* 3, 177–84.

BIER, K. (1958). Die Regulation der Sexualität in den Insektenstaaten. *Ergebn. Biol.* 20, 97–126.

BOWERS, W. S., FALES, H. M., THOMPSON, M. J. & UEBEL, E. C. (1966). Juvenile hormone: identification of an active compound from balsam fir. *Science, N.Y.* 154, 1020.

BRIAN, M. V. (1953). Oviposition by workers of the ant *Myrmica*. *Physiol. Comp. Oecol.* 3, 25–36.

BRIAN, M. V. (1963). Studies of caste differentiation in *Myrmica rubra* L. 6. Factors influencing the course of female development in the early third instar. *Insectes Sociaux* 10, 91–102.

BRIAN, M. V. & HIBBLE, J. (1963). 9-oxodec-trans-2-enoic acid and *Myrmica* queen extracts tested for influence on brood in *Myrmica*. *J. Insect Physiol.* 9, 25–34.

BRONSON, F. H. (1968). Pheromonal influences on mammalian reproduction. In *Reproduction and Sexual Behavior*, pp. 341–61 (ed. M. Diamond). Bloomington: Indiana University Press.

BRUCE, H. M. (1959). An exteroceptive block to pregnancy in the mouse. *Nature, Lond.* 184, 105.

BUTLER, C. (1954). The method and importance of the recognition by a colony of honeybees (*A. mellifera*) of the presence of its queen. *Trans. R. ent. Soc. Lond.* 105, 11–29.

BUTLER, C. G. (1956). Some further observations on the nature of queen substance and its role in the organization of a honey-bee *Apis mellifera* community. *Proc. R. ent. Soc. Lond.* A 31, 12–16.

BUTLER, C. G. (1960). The significance of queen substance in swarming and supercedure in honey-bee (*Apis mellifera* L.) colonies. *Proc. R. ent. Soc. Lond.* A 35, 129–32.

BUTLER, C. G. (1961). The scent of queen honeybees (*A. mellifera* L.) that causes partial inhibition of queen rearing. *J. Insect Physiol.* 7, 258–64.

BUTLER, C. G. (1964). Pheromones in sexual processes in insects. *Symp. R. ent. Soc. Lond.* 2, 66–77.

BUTLER, C. G. (1965). Die Wirkung von Königinnen-Extrakten verschiedener sozialer Insekten auf die Aufzucht von Königinnen und die Entwicklung der Ovarien von Arbeiterinnen der Honigbiene. *Z. Bienenforsch.* 8, 143–7.

BUTLER, C. G. (1967a). Insect pheromones. *Biol. Rev.* 42, 42–87.

BUTLER, C. G. (1967b). A sex attractant acting as an aphrodisiac in the honeybee (*Apis mellifera* L.). *Proc. R. ent. Soc. Lond.* A 42, 71–6.

BUTLER, C. G., CALAM, D. H. & CALLOW, R. K. (1967). Attraction of *Apis mellifera* drones by the odour of the queens of two other species of honeybees. *Nature, Lond.* 213, 423–4.

BUTLER, C. G., CALLOW, R. K. & JOHNSTON, N. G. (1961). The isolation and synthesis of queen substance, 9-oxodec-trans-2-enoic acid, a honeybee pheromone. *Proc. Roy. Soc. Lond.* B 155, 417–32.

BUTLER, C. G. & FAIREY, E. M. (1963). The role of the queen in preventing oogenesis in worker honeybees (*A. mellifera* L.). *J. Apicult. Res.* 2, 14–18.

BUTLER, C. G. & PATON, P. N. (1962). Inhibition of queen rearing by queen honeybees (*Apis mellifera* L.) of different ages. *Proc. R. ent. Soc. Lond.* A 37, 114–16.

BUTLER, C. G. & SIMPSON, J. (1958). The source of the queen substance of the honey-bee (*Apis mellifera* L.). *Proc. R. ent. Soc. Lond.* A 33, 120–2.

CARLISLE, D. B., ELLIS, P. E. & BETTS, E. (1965). The influence of aromatic shrubs on sexual maturation in the desert locust, *Schistocerca gregaria*. *J. Insect Physiol.* 11, 1541–58.

CARLISLE, D. B. & ELLIS, P. E. (1968). Bracken and locust ecdysones: their effects on molting in the desert locust. *Science, N.Y.* **159**, 1472–4.

CHAUVIN, R. (1958). L'Action de groupement sur la croissance des grillons (*Gryllus domesticus*). *J. Insect Physiol.* **2**, 235–48.

CLEVELAND, L. R., BURKE, A. W. & KARLSON, P. (1960). Ecdysone-induced modifications in the sexual cycles of the protozoa of *Cryptocercus*. *J. Protozool.* **7**, 229–39.

ELLIS, P. E., CARLISLE, D. B. & OSBORNE, D. J. (1965). Desert locusts: sexual maturation delayed by feeding on senescent vegetation. *Science, N.Y.* **149**, 546–7.

EMMERICH, H. & BARTH, R. H. JR (1968). Effect of farnesyl methyl ether on reproductive physiology in the cockroach *Byrsotria fumigata* (Guérin). *Z. Naturf.* **23** B, 1019–20.

ENGELMANN, F. (1960). Hormonal control of mating behavior in an insect. *Experientia* **16**, 69–70.

ENGELMANN, F. (1965). The mode of regulation of the corpus allatum in adult insects. *Archs Anat. microsc. Morph. exp.* **54**, 387–404.

ENGELMANN, F. (1968). Endocrine control of reproduction in insects. *Ann. Rev. Ent.* **13**, 1–26.

ENGELMANN, F. (1969). Female specific protein: biosynthesis controlled by corpus allatum in *Leucophaea maderae*. *Science, N.Y.* **165**, 407–9.

ENGELMANN, F. & BARTH, R. H. JR (1968). Endocrine control of female receptivity in *Leucophaea maderae* (Blattaria). *Ann. Ent. Soc. Am.* **61**, 503–5.

GARY, N. E. (1962). Chemical mating attractants in the queen honey bee. *Science, N.Y.* **136**, 773–4.

GLEASON, K. K. & REYNIERSE, J. H. (1969). The behavioral significance of pheromones in vertebrates. *Psychol. Bull.* **71**, 58–73.

GREGG, R. E. (1942). The origin of castes in ants with special reference to *Pheidole morrisi* Forel. *Ecology* **23**, 295–308.

HALDANE, J. B. S. (1955). Animal communication and the origin of human language. *Sci. Progr.* **43**, 385–401.

HOFFMEISTER, H. (1966). Ecdysterone, ein neues Häutungshormon der Insekten *Angew. Chem.* **78**, 269.

ISHII, S. & KUWAHARA, Y. (1967). An aggregation pheromone of the German cockroach, *Blattella germanica* L. (Orthoptera: Blattellidae). I. Site of the pheromone production. *Appl. Ent. Zool.* **2**, 203–17.

ISHII, S. & KUWAHARA, Y. (1968). Aggregation of German cockroach (*Blattella germanica*) nymphs. *Experientia* **24**, 88–9.

JACOBSON, M. (1965). *Insect Sex Attractants.* New York: Wiley Interscience, 154 pp.

JOHNSTON, N. C., LAW, J. H. & WEAVER, N. (1965). Metabolism of 9-ketodec-2-enoic acid by worker honeybees (*Apis mellifera* L.). *Biochemistry* **4**, 1615–21.

KARLSON, P. & BUTENANDT, A. (1959). Pheromones (ectohormones) in insects. *Ann. Rev. Ent.* **4**, 39–58.

KARLSON, P., HOFFMEISTER, H., HUMMEL, H., HOCKS, P. & SPITELLER, G. (1965). Zur Chemie des Ecdysons. VI. Reaktionen des Ecdysonmoleküls. *Chem. Ber.* **98**, 2394–402.

LEA, A. D. & THOMSEN, E. (1969). Size independent secretion by the corpus allatum of *Calliphora erythrocephala*. *J. Insect Physiol.* **15**, 477–82.

VAN DER LEE, S. & BOOT, L. M. (1955). Spontaneous pseudopregnancy in mice. *Acta physiol. pharmac. néerl.* **4**, 442–3.

VAN DER LEE, S. & BOOT, L. M. (1956). Spontaneous pseudopregnancy in mice. II. *Acta physiol. pharmac. néerl.* **5**, 213–15.

LIGHT, S. F. (1943). The determination of the castes of social insects. *Q. Rev. Biol.* **18**, 46–63.

LIGHT, S. F. (1944). Experimental studies on ectohormonal control of the development of supplementary reproductives in the termite genus *Zootermopsis*. *Univ. Calif. Publ. Zool.* **43**, 413–54.

LOHER, W. (1960). The chemical acceleration of the maturation process and its hormonal control in the male of the desert locust. *Proc. Roy. Soc. Lond.* B **153**, 380–97.

LÜSCHER, M. (1960). Hormonal control of caste differentiation in termites. *Ann. N.Y. Acad. Sci.* **89**, 549–63.

LÜSCHER, M. (1961). Social control of polymorphism in termites. *Symp. R. ent. Soc. Lond.* **1**, 57–67.

LÜSCHER, M. (1962). Hormonal regulation of development in termites. *Sym. gen. Biol. Italica* **10**, 1–11.

LÜSCHER, M. (1963). Functions of the corpora allata in the development of termites. *Proc. XVIth Int. Cong. Zool.* **4**, 244–50.

LÜSCHER, M. (1964a). Die Elimination überzähliger Ersatzgeschlechtstiere bei der Termite *Kalotermes flavicollis* (Fabr.). *Rev. Suisse Zool.* **71**, 626–32.

LÜSCHER, M. (1964b). Die spezifische Wirkung männlicher und weiblicher Ersatzgeschlechtstiere auf die Entstehung von Ersatzgeschlechtstieren bei der Termite, *Kalotermes flavicollis* (Fabr.). *Insectes Sociaux* **11**, 79–90.

LÜSCHER, M. & SPRINGHETTI, A. (1960). Untersuchungen über die Bedeutung der Corpora Allata für die Differenzierung der Kasten bei der Termite, *Kalotermes flavicollis* F. *J. Insect Physiol.* **5**, 190–212.

LÜSCHER, M. & WALKER, I. (1963). Zur Frage der Wirkungsweise der Königinnenpheromone bei der Honigbiene. *Rev. Suisse Zool.* **70**, 304–11.

MASNER, P., SLAMA, K. & LANDA, V. (1968). Sexually spread insect sterility induced by the analogue of juvenile hormone. *Nature, Lond.* **219**, 395–6.

McFARLANE, J. E. (1966). Studies on group effects in crickets. I. Effect of methyl linolenate, methyl linoleate, and vitamin E. *J. Insect. Physiol.* **12**, 179–88.

McFARLANE, J. E. (1968). Fatty acids, methyl esters, and insect growth. *Comp. Biochem. Physiol.* **24**, 377–84.

MICHENER, C. D. (1961). Social polymorphism in Hymenoptera. *Symp. R. ent. Soc. Lond.* **1**, 43–56.

MILBURN, N., WEIANT, E. A & ROEDER, K. D. (1960). The release of efferent nerve activity in the roach *Periplaneta americana* by extracts of the corpus cardiacum. *Biol. Bull. mar. biol. Lab., Woods Hole* **118**, 111–19.

NAKANISHI, M., KOREEDA, S., SASAKI, M., CHANG, L. & HSU, H. Y. (1966). Insect hormones, the structure of ponasterone A, an insect moulting hormone from the leaves of *Podocarpus nakii* Hay. *Chem. Commun.* 915–17.

NORRIS, M. J. (1954). Sexual maturation in the desert locust (*Schistocerca gregaria* Forskal) with special reference to the effects of grouping. *Anti-Locust Bull.* no. 18, pp. 1–44.

NORRIS, M. J. (1964a). Accelerating and inhibiting effects of crowding on sexual maturation in two species of locusts. *Nature, Lond.* **203**, 784–5.

NORRIS, M. J. (1964b). Environmental control of sexual maturation in insects. *Symp. R. ent. Soc. Lond.* **2**, 56–65.

NORRIS, M. J. & PENER, M. P. (1965). An inhibitory effect of allatectomized males and females on the sexual maturation of young male adults of *Schistocerca gregaria* (Forsk.) (Orthoptera: Acrididae). *Nature, Lond.* **208**, 1122.

ÖZBAS, S. & HODGSON, E. S. (1958). Action of insect neurosecretion upon central nervous system in vitro and upon behavior. *Proc. natn. Acad. Sci. U.S.A.* **44**, 825–30.

REGNIER, F. E. & LAW, J. H. (1968). Insect pheromones. *J. Lipid Res.* **9**, 541–51.

RIDDIFORD, L. M. (1967). Trans-2-hexenal: mating stimulant for polyphemus moths. *Science, N.Y.* **158**, 139–41.

RIDDIFORD, L. M. & WILLIAMS, C. M. (1967a). Volatile principle from oak leaves: role in the sex life of the polyphemus moth. *Science, N.Y.* **155**, 589–90.

RIDDIFORD, L. M. & WILLIAMS, C. M. (1967b). Chemical signaling between polyphemus moths and between moths and host plant. *Science, N.Y.* **156**, 541.

RÖLLER, H., DAHM, K. H., SWEELY, C. C. & TROST, B. M. (1967). The structure of the juvenile hormone. *Angew. Chem.* (Eng. Ed.) **6**, 179–80.

RÖLLER, H., PIEPHO, H. & HOLZ, I. (1963). Zum Problem der Hormonabhängigkeit des Paarungsverhaltens bei Insekten, Untersuchingen an *Galleria mellonella. J. Insect Physiol.* **9**, 187–94.

ROMANUK, M., SLAMA, K., & SORM, F. (1967). Constitution of a compound with a pronounced juvenile hormone activity. *Proc. natn. Acad. Sci. U.S.A.* **57**, 349–52.

ROTH, L. M. & BARTH, R. H. JR (1964). The control of sexual receptivity in female cockroaches. *J. Insect Physiol.* **10**, 965–75.

ROTH, L. M. & STAHL, W. H. (1956). Tergal and cercal secretion of *Blatta orientalis* L. *Science, N.Y.* **123**, 798–9.

ROTHSCHILD, M. & FORD, B. (1964a). Breeding of the rabbit flea (*Spilopsyllus cuniculi* Dale) controlled by the reproductive hormones of the host. *Nature, Lond.* **201**, 103–4.

ROTHSCHILD, M. & FORD, B. (1964b). Maturation and egg-laying of the rabbit flea (*Spilopsyllus cuniculi* Dale) induced by the external application of hydrocortisone. *Nature, Lond.* **203**, 210–11.

ROTHSCHILD, M. & FORD, B. (1966). Hormones on the vertebrate host controlling ovarian regression and copulation of the rabbit flea. *Nature, Lond.* **211**, 261–6.

SCHILDKNECHT, H., SIEWERDT, R. & MASCHWITZ, U. (1966). A vertebrate hormone as defensive substance of the water beetle (*Dytiscus marginalis*). *Angew. Chem.* (Eng. Ed.) **5**, 421–2.

SCHNEIDER, D. (1966). Chemical sense communication in insects. *Symp. Soc. exp. Biol.* **20**, 273–97.

SCHNEIRLA, T. C. & ROSENBLATT, J. S. (1961). Behavioral organization and genesis of the social bond in insects and mammals. *Am. J. Orthopsychiat.* **31**, 223–53.

SCHMIALEK, P. (1961). Die Identifizierung zweier in Tenebriokot und in Hefe vorkommender Substanzen mit Juvenilhormonwirkung. *Z. Naturf.* **16** b, 461–4.

SLAMA, K. & WILLIAMS, C. M. (1965). Juvenile hormone activity for the bug, *Pyrrhocoris apterus. Proc. natn. Acad. Sci. U.S.A.* **54**, 411–14.

SLAMA, K. & WILLIAMS, C. M. (1966). The juvenile hormone. V. The sensitivity of the bug *Pyrrhocoris apterus* to a hormonally active factor in American paper-pulp. *Biol. Bull. mar. biol. Lab., Woods Hole* **130**, 235–46.

STAAL, G. B. (1967). Plants as a source of insect hormones. *Proc. K. ned. Akad. Wet.* **70**, 409–18.

TAKEMOTO, T., HIKINO, Y., NOMOTO, K. & HIKINO, H. (1967). Structure of cyasterone, a novel C_{29} insect-molting substance from *Cyathula capitata. Tetrahedron Lett.* **33**, 3191–4.

TAKEMOTO, T., OGAWA, S. & NISHIMOTO, N. (1967). Isolation of the molting hormones of insects from *Achyranthis radix. Yakugaku Zasshi (J. Pharm. Soc. Japan)* **87**, 325–7.

WEAVER, N. (1966). Physiology of caste determination. *Ann. Rev. Ent.* **11**, 79–102.

WHITTEN, W. K. (1957). Effect of exteroceptive factors on the oestrous cycle of mice. *Nature, Lond.* **180**, 1436.

WHITTEN, W. K. (1966). Pheromones and mammalian reproduction. *Adv. Reproductive Physiol.* **1**, 155–77.

WILLIAMS, C. M. (1967). Third-generation pesticides. *Scient. Am.* **217**, 13–17.

WILLIAMS, C. M., & ROBBINS, W. E. (1968). Conference on insect-plant interactions. *Bioscience* **18**, 791–2, 797–9.

WILLIS, E. R., RISER, G. R. & ROTH, L. M. (1958). Observations on reproduction and development in cockroaches. *Ann. Ent. Soc. Am.* **51**, 53–69.

WILLIS, J. H. & BRUNET, P. C. J. (1966). The hormonal control of colleterial gland secretion. *J. exp. Biol.* **44**, 363–78.

WILSON, E. O. (1963). The social biology of ants. *Ann. Rev. Ent.* **8**, 345–68.

WILSON, E. O. (1965). Chemical communication in the social insects. *Science, N.Y.* **149**, 1064–71.

WILSON, E. O. (1968). Chemical systems. In *Animal Communication*, pp. 75–102 (ed. T. A. Sebeok). Bloomington: Indiana University Press.

WILSON, E. O. & BOSSERT, W. H. (1963). Chemical communication among animals. *Recent Progr. Horm. Res.* **19**, 673–716.

DISCUSSION

EVANS: I would point out that there is an analagous mechanism to insect 'calling' behaviour in the marking behaviour of terrestrial mammals. These behaviours transfer pheromones either directly—through self-anointing (with skin-gland secretions, urine, saliva)—or indirectly via object-marking. Steroids can thus influence not only the synthetic, but also the distributional processes which regulate scent-signal transmission, as in the micturition patterns of the domestic dog.

DE WILDE: I would like to comment on the mode of action of queen-substance on worker bees. As you probably know, Mrs Verheyer-Voogd of Utrecht has done extensive work on the suppressing activity of queen-substance on worker ovaries. She has not been able to confirm the action of queen-substance by injection, but has found that this pheromone is only active in a 'pattern' situation, where it is present on small objects like pieces of wood, lumps of wax, or dead worker bees. The action of this pheromone seems to require a releasing situation.

As to the function of insect hormones in plants, Dr Staal and others in our institute have never been able to find any effect of ecdysterone in plants fed upon by lepidopterous larvae. In fact, it appeared that the intestine was rather impermeable to this hormone, and this was true also for insects not normally feeding on plants containing hormones, such as *Taxus*-feeders.

BARTH: Yes, I agree. Carlisle and his associates also have found that ecdysones are not taken up by the gut when you feed locusts on ecdysone-containing materials (see page 397 above).

BERN: Don't you think that there is very real danger in the notion of insect control based upon the use of hormone-like products? Claims of specificity are all very well until one finds out that the 'specific' hormone product has wiped out several species of insects that one did not want to destroy. It would seem to me that a lot more hope could be placed in the idea of insect control by means of pheromonal factors, for which there is much greater evidence of real specificity of biological effectiveness. I really am concerned about the potential dangers of hormonal DDTs.

BARTH: I quite agree. I think that without a doubt pheromones really offer much safer possibilities for insect control at the present time. We still must do a great deal of testing of all of these substances to be sure that they really are species-specific before I would be in favour of widespread use of them in this way.

BERN: We are still concerned with the whole environment. While I agree that it would be very reassuring to know that a hormone-mimetic substance is not lethal to us, what goes in also comes out. Would you like to say something about the possible fate of these compounds on the environment, after they leave the human (or other animal) body. I think it would be hard to predict that the effects will be limited to those species which happen to be annoying your grain.

HIGHNAM: I agree entirely with Professor Bern in this: a very great deal of work will have to be done. The specific example quoted by Dr Barth of DMF and its transference from male to female, making the female sterile for life, is really only an extension of the screw-worm type of control which has been very successfully applied in the United States, but which for economic and other reasons can't be used with insect pests in which the female mates more than once. A type of chemical control with juvenile hormone analogues has the specificity of the sterile male technique as used in the screw worm, but with the added advantage that it can be applied to insects in which the female will mate more than once during a fairly long reproductive life.

DE WILDE: I am assured that our colleagues involved in the elaboration of pheromones for insect control realize the danger of a non-selective hormonal insecticide, it is reassuring that at present non-persistent substances are already known with a high selective action in Diptera, Hemiptera and Coleoptera.

SESSION V
NEURO-ENDOCRINE MEDIATION

FEEDBACK MECHANISMS AND THE CONTROL OF THE SECRETION OF THE HYPOTHALAMIC RELEASING FACTORS

By MARCELLA MOTTA, F. PIVA, L. TIMA, MARIAROSA ZANISI and L. MARTINI

INTRODUCTION

It is now generally accepted that the central nervous system (CNS) plays an essential role in the regulation of many endocrine functions and that the hypothalamus is the crucial element of such regulation (Mangili, Motta & Martini, 1966; Martini, Fraschini & Motta, 1968; Mess & Martini, 1968; Motta, Fraschini & Martini, 1969). The key position of the hypothalamus is due to its strategic location in close contact with the anterior-pituitary gland, and to the fact that this region of the brain contains particular neurones which synthesize the chemical messengers necessary for stimulating or for inhibiting the anterior pituitary (the so-called hypothalamic releasing and inhibitory factors), as well as feedback receptors whose activity is modified by hormonal influences (Mangili et al. 1966; Martini et al. 1968; Mess & Martini, 1968; Motta et al. 1969). The hypothalamus then contains, simultaneously, the last element of an *afferent pathway* (feedback receptors) and the first element of an *efferent pathway* (the cells synthesizing the releasing and the inhibitory factors).

Two hypotheses can be put forward with regard to the simultaneous presence of *sensitive elements* and of an *executive component* in the hypothalamus: (*a*) the structures which synthesize the hypophysiotropic messengers are also able to receive and to evaluate feedback signals directly; or (*b*) the receptor function is the responsibility of completely different structures; if this is the case, the information collected by sensory elements must be transmitted to the executive area in order to enable this to adjust its function accordingly. The data obtained in the authors' laboratory which will be reviewed in this paper, support the possibility that, within the hypothalamus, separate structures might be responsible for receiving feedback signals and for transforming them into executive messages.

AFFERENT COMPONENT

The study of the localization of the elements sensitive to signals originated either in the peripheral target glands ('long' feedback mechanisms) (Mangili *et al.* 1966) or in the anterior pituitary ('short' feedback mechanisms) (Motta *et al.* 1969) has been performed using a stereotaxic approach, i.e. by placing microquantities of different active compounds directly in the brain or in the anterior pituitary.

Adrenocortical hormones

The localization of the feedback receptors sensitive to adrenocortical steroids has been investigated in normal male rats by implanting two potent ACTH-blocking steroids, cortisol and dexamethasone (Δ_1-9-α-fluoro-16-α-methylcortisol) in the basal region of the hypothalamus; animals were killed 1 week post-operatively and their plasma and adrenal corticosterone levels were evaluated. Dexamethasone was very effective in reducing adrenal weight as well as plasma and adrenal corticosterone levels, when placed in the median eminence (ME). Cortisol, when placed in the ME, was also inhibitory, but not to as great an extent as dexamethasone; ME implants of cortisol were actually able to reduce plasma and adrenal corticosterone levels, but were ineffective on adrenal weight (Corbin, Mangili, Motta & Martini, 1965). Both steroids were ineffective when implanted directly in the anterior pituitary or in the cerebral cortex (Table 1). These studies, which have been repeatedly confirmed in several laboratories (see Mess & Martini, 1968, for references) provide a good demonstration that the receptors sensitive to adrenal hormones are mainly located in the ME region of the hypothalamus.

Table 1. *Effect of implants of dexamethasone and cortisol in the median eminence (ME), the anterior pituitary (Pit) and the cerebral cortex (CC) of male rats*

Values are means ± S.E. Number of rats in parentheses.

Groups	Pituitary weight (mg)	Adrenal weight (mg)	Corticosterone μg/100 ml plasma	Corticosterone μg/100 mg adrenal
Controls (20)	8·25 ± 0·19	42·1 ± 3·9	18·1 ± 1·75	3·46 ± 0·32
ME-sham (10)	8·20 ± 0·37	39·1 ± 2·0	17·0 ± 1·54	3·72 ± 0·39
ME-cortisol (6)	6·90 ± 0·48	35·2 ± 4·2	9·4 ± 1·60*	1·98 ± 0·19*
ME-dexamethasone (17)	6·87 ± 0·24	27·4 ± 1·6*	6·4 ± 1·38*	1·05 ± 0·15*
Pit-dexamethasone (14)	7·50 ± 0·24	37·0 ± 2·8	17·1 ± 2·20	3·69 ± 0·42
CC-dexamethasone (13)	8·30 ± 0·45	37·5 ± 5·4	16·0 ± 2·37	3·59 ± 0·38

* $P \leqslant$ 0·01 *v.* ME-sham.

Sex hormones

The localization of the feedback elements which participate in the control of the secretion of gonadotropic hormones has also been investigated using a similar technique. For these experiments microquantities of oestrogen were placed either in the hypothalamus or directly in the anterior pituitary gland. Two different types of experiments were performed: one in adult and one in prepubertal animals.

For the first study, male rats were preferred to females; this choice was made, since it is still necessary to rely on indirect criteria in order to have some information on the amounts and on the type of gonadotropins present in the general circulation. Modifications of testicular weight represent a reliable index of plasma levels of follicle stimulating hormone (FSH), while changes of the weights of the prostates and of the seminal vesicles are strictly related to the amounts of circulating luteinizing hormone (LH); these parameters, although not ideal, are certainly more specific and sensitive than those provided by the study of the histological picture of the ovary (Martini *et al.* 1968).

It has been observed that the implantation of oestradiol in the ME of normal adult male rats causes, in 5 days time, a significant decrease in the weights of the testes, of the prostates, and of the seminal vesicles (Table 2).

Table 2. *Effect of implants of oestradiol in the median eminence (ME) of male rats (negative feedback)*

Values are means ± S.E. Number of rats in parentheses.

Groups	Body wt (g)	Pituitary wt (mg)	Adrenal wt (mg)
ME-sham (60)	231 ± 4·1	11·00 ± 0·39	42·80 ± 1·7
ME-oestradiol (60)	223 ± 2·6	13·07 ± 0·45*	45·20 ± 1·4

	Testes wt (g)	Prostates wt (mg)	Seminal vesicles wt (mg)
ME-sham (60)	2·71 ± 0·53	182·9 ± 7·8	288·8 ± 16·0
ME-oestradiol (60)	2·53 ± 0·43*	129·1 ± 7·5**	151·8 ± 9·2**

* $P \leqslant 0.01$.   ** $P \leqslant 0.001$.

These data suggest that, in adult animals, the release of FSH and of LH is reduced following placement of oestrogen in the ME. The evaluation of pituitary stores of the two hormones in animals having oestrogen implanted in this region of the brain has provided a more direct proof of this fact and has shown in addition that synthesis of both gonadotropins is also decreased; ME implants of oestradiol have actually been found to induce a significant reduction of the pituitary content of FSH and LH (Table 3) (Martini *et al.* 1968).

Table 3. *Effect of implants of oestradiol in the median eminence (ME) on pituitary FSH and LH levels of male rats (negative feedback)*

Number of rats in parentheses

Groups	Pituitary FSH (μg/mg†)	Limits, 95 %	Pituitary LH (μg/mg‡)	Limits, 95 %
ME-sham (20)	44·83	38·7–54·1	1·50	1·20–2·30
	54·31	47·2–62·1	1·37	1·00–2·40
ME-oestradiol (20)	26·73	19·1–32·3	0·92	0·71–1·10
	33·33	27·2–39·5	0·33	0·15–0·54

† Microgramme equivalents of NIH-FSH-S-3 ovine per milligramme wet weight of pituitary tissue.
‡ Microgramme equivalents of NIH-LH-S-11 ovine per milligramme wet weight of pituitary tissue.

Table 4. *Effect of implants of oestradiol in the median eminence (ME), the anterior pituitary (Pit) and the cerebral cortex (CC) of immature female rats (positive feedback)*

Values are means ± s.e. Number of rats in parentheses.

Groups	Body wt (g)	Pituitary wt (mg)	Ovaries wt (mg)	Uterus wt (mg)	Vaginal opening (days)
Controls (10)	133 ± 2·5	6·4 ± 0·53	49·5 ± 4·2	187·8 ± 27·7	36·3 ± 0·90
ME-sham (10)	110 ± 4·1	5·1 ± 0·28	47·3 ± 3·9	158·2 ± 10·1	36·0 ± 0·88
ME-oestradiol (10)	118 ± 5·9	10·9 ± 0·80**	44·3 ± 3·7	203·0 ± 6·5*	29·0 ± 0·62**
Pit-oestradiol (10)	123 ± 5·2	9·4 ± 0·82**	48·2 ± 4·0	154·4 ± 11·4	36·4 ± 0·82
CC-oestradiol (10)	117 ± 3·2	8·0 ± 0·45**	46·0 ± 3·7	168·6 ± 7·0	35·0 ± 0·85

* $P \leqslant$ 0·01 *v.* ME-sham.   ** $P \leqslant$ 0·001 *v.* ME-sham.

The experiments in perpubertal animals were performed following the work of Ramirez & Sawyer (1965) who reported that the systemic administration of a series of small doses of oestrogens to prepubertal female rats advances puberty and increases the secretion of LH and of the luteinizing hormone releasing factor (LH-RF); these data suggested that, under certain circumstances, oestrogen may facilitate LH release and that the hypothalamus may be a critical site for this effect. The data recorded when small amounts of oestradiol were placed in the ME of 26 days old prepubertal female rats are as follows: puberty was considerably advanced, as shown by the very precocious opening of the vagina (29th day); the difference from the controls (vagina opening around the 36th day) was highly significant (Table 4) (Motta, Fraschini, Giuliani & Martini, 1968). In addition, all animals bearing ME implants of oestrogen had ovulated at the time of autopsy (performed at 39 days of age) as shown by the presence of young corpora lutea; none of the control animals had ovulated by that time. Both phenomena (advanced puberty and ovulation) are apparently related to the enhanced release of LH induced by the presence of oestrogen in the ME; this is suggested by the decline of pituitary stores of LH (Table 5) and by the increase of plasma LH titres in the implanted animals (Table 6). The data here reported confirm then that oestrogen administered to immature animals may exert a 'positive' feedback effect on LH release, and indicate in addition that they do so through an effect on receptor elements located in the ME (Martini *et al.* 1968; Motta *et al.* 1968).

Table 5. *Effect of implants of oestradiol in the median eminence (ME), the anterior pituitary (Pit) and the cerebral cortex (CC) on pituitary LH levels of immature female rats (positive feedback)*

Values are means ± s.e. Number of rats in parentheses.

Groups	Pituitary LH	
	μg/mg*	μg/Pit
Controls (10)	1·24 ± 0·04	7·68
ME-sham (10)	1·54 ± 0·05	7·85
ME-oestradiol (10)	0·28 ± 0·01†	3·05
Pit-oestradiol (10)	1·42 ± 0·04	13·35
CC-oestradiol (10)	1·36 ± 0·07	10·88

* Microgramme equivalents of NIH-LH-B-1 bovine per milligram wet weight of pituitary tissue. † $P \leqslant 0.001$ v. controls.

Table 6. *Effect of implants of oestradiol in the median eminence (ME), the anterior pituitary (Pit) and the cerebral cortex (CC) on plasma LH levels of immature female rats (positive feedback)*

Values are means ± s.e. Number of donor rats in parentheses.

Groups	Ovarian ascorbic acid depletion (%)	Plasma LH
ME-sham (8)	1·3 ± 0·7	Absent
ME-oestradiol (8)	21·4 ± 1·3	Present
Pit-oestradiol (8)	4·4 ± 0·8	Absent
CC-oestradiol (8)	2·3 ± 1·2	Absent

The two groups of experiments summarized in the preceding paragraphs agree in indicating that the ME contains receptors sensitive to oestrogen; however, they differ considerably, since they show (*a*) that ME implants of oestrogen inhibit both synthesis and release of LH in adult animals; and (*b*) that similar implants stimulate LH release (with no concomitant activation of synthesis) in prepubertal ones. Several tentative suggestions have been put forward to explain this discrepancy (Motta *et al.* 1968; Martini *et al.* 1968; Martini, Fraschini & Motta, 1969).

Adrenocorticotrophic hormone

It has been recently discovered that, in addition to the 'long' feedback systems described in the preceding sections of this paper, feedback mechanisms in which the inhibitory signals are provided by pituitary hormones ('short' mechanisms) may also participate in the control of anterior pituitary functions (Motta *et al.* 1969). Consequently, it was deemed of interest to study whether the receptors for the 'short' effect of hormones are located in the same areas which contain the receptors for the 'long' feedback effect. Motta, Mangili & Martini (1965) have made stereotaxic implantations of solid ACTH (crude material and synthetic β-1-24-peptide) into several areas of the brain and into the pituitary of normal male rats. Only ACTH implants performed in the ME region modified plasma corticosterone levels; these were significantly reduced 5 days following ACTH implantation; there were no variations in adrenal weights (Table 7). ME implants of other pituitary hormones were completely ineffective in modifying the activity of the pituitary–adrenal axis. It was argued from these experiments that the ME might also contain receptors sensitive to changing levels of ACTH. This postulation has recently been supported by behavioural (De Wied, 1969) as well as by electrophysiological (Beyer & Sawyer, 1969) data (see Motta *et al.* 1969, for references).

Table 7. *Effect of implants of ACTH in the median eminence (ME),*
the anterior pituitary (Pit) and the cerebral cortex (CC) of male rats

Values are means ± S.E. Number of rats in parentheses.

Groups	Adrenals weight (mg)	Plasma corticosterone (μg/100 ml)
Controls (12)	43·1 ± 4·2	19·1 ± 1·25
ME-sham (18)	41·2 ± 2·2	18·4 ± 1·60
ME-ACTH (USP Reference Standard) (24)	44·0 ± 2·8	9·1 ± 0·78*
ME-ACTH (synthetic) (8)	43·1 ± 2·2	10·7 ± 0·84*
Pit-ACTH (USP Reference Standard) (12)	43·7 ± 2·8	18·1 ± 2·02
CC-ACTH (USP Reference Standard) (12)	43·4 ± 1·6	19·4 ± 2·00
ME-LH (NIH-B-1 bovine) (8)	42·2 ± 2·7	19·2 ± 1·04

* $P \leqslant$ 0·001 *v.* ME-sham

Gonadotrophic hormones

That the ME might also be a receptor area sensitive to gonadotrophins was suggested initially by the data of Dávid *et al.* (1966); they have found that the implantation of pure LH into the ME of normal or castrated rats results, in 5 days, in a significant decrease of both pituitary and plasma LH concentrations (Tables 8, 9). Implants of LH performed in the pituitary or in other regions of the brain were completely ineffective on both parameters. These data, as well as similar results described by Corbin (1966) and by Corbin & Cohen (1966), suggest that the ME contains receptors which, when loaded with LH, bring about an inhibition of both synthesis and release of LH. The specificity of these ME receptors is shown by the negative results obtained when FSH instead of LH was placed in this region (Tables 8, 9).

Table 8. *Effect of implants of LH in the median eminence (ME), the anterior pituitary (Pit) and the cerebral cortex (CC) on pituitary LH levels of castrated female rats*

Values are means ± s.e. Number of rats in parentheses.

Groups	Pituitary weight (mg)	Pituitary LH (μg/mg†)	Limits, 95 %
ME-sham (15)	10·60 ± 0·62	3·82	2·4–5·6
		4·68	3·1–7·2
ME-LH (14)	10·11 ± 0·45	1·94	1·2–2·6
		1·57	1·1–2·4
Pit-LH (8)	9·81 ± 0·74	4·11	2·6–6·3
CC-LH (8)	11·28 ± 0·60	3·06	1·4–4·1
		3·44	2·3–5·4
ME-FSH (6)	8·83 ± 0·37	4·36	2·8–7·2

† Microgramme equivalents of NIH-LH-B-1 bovine per milligramme wet weight of pituitary tissue.

Table 9. *Effect of implants of LH in the median eminence (ME), and the cerebral cortex (CC) on plasma LH levels of castrated female rats*

Values are means ± s.e. Number of donor rats in parentheses.

Groups	Ovarian ascorbic acid depletion (%)	Plasma LH
ME-sham (6)	20·3 ± 1·7	Present
ME-LH (6)	1·5 ± 0·4	Absent
CC-LH (6)	22·3 ± 2·8	Present
ME-FSH (6)	18·4 ± 1·9	Present

Recently data have appeared which indicate that the ME may act as a sensory element also for FSH (Corbin & Story, 1967) and Prolactin (Clemens & Meites, 1968); implants of these two hormones in the basal hypothalamus have been shown to result in a conspicuous reduction of their own secretion under different experimental conditions (see Motta et al. 1969, for references and details).

EFFERENT COMPONENT

It has already been mentioned that the hypothalamus controls the pituitary gland through the release of specific releasing and inhibitory factors (RFs and IFs) into the blood flow of the pituitary portal vessels. The chemical and purification work on these hypothalamic principles has been performed mainly using ME tissue as the starting material (McCann & Dhariwal, 1966). However, it occurred to us that the presence of these neurohumoral agents in high concentrations in the ME might not be a sufficient proof of their synthesis taking place in this region. The three groups of experiments here to be described were planned in order to provide some direct evidence in favour of the postulation that the synthesis of the RF's might take place in hypothalamic centres located far from the ME; according to this hypothesis the ME would be only the storage site for the different hypothalamic hypophysiotrophic mediators.

Effect of hypothalamic lesions on ME stores of releasing factors

The first technique to be used was that of placing separate electrolytic lesions in each of the hypothalamic areas which had been previously reported to play some role in the control of anterior-pituitary functions, and to study whether these lesions might modify the concentration of each RF at the ME level. It was expected that an RF would disappear from the ME (or that its concentration would be reduced) if a lesion were placed exactly in the area which synthesizes it. The following RF's were considered: follicle stimulating hormone-releasing factor (FSH-RF), luteinizing hormone-releasing factor (LH-RF), thyroid stimulating hormone-releasing factor (TSH-RF) and corticotropin releasing factor (CRF). Four independent areas of the brain were lesioned; they will be referred to as: (*a*) paraventricular area, (*b*) suprachiasmatic area, (*c*) arcuate-ventromedial area, and (*d*) ME area. Five days following placement of the lesions the animals were killed and their ME were collected in order to evaluate their content in FSH-RF, LH-RF, TSH-RF and CRF. For these assays the procedures described by Fraschini, Motta & Martini (1966) were used. Using this approach it has been possible to localize, within the hypothalamus of the rat, a circumscribed region in which FSH-RF is synthesized; it has actually been shown that lesions in or around the paraventricular nuclei are the only ones which reduce the concentrations of this RF in the ME (Fig. 1). The content of LH-RF in the ME is reduced when lesions are placed in the suprachiasmatic area; also lesions placed in the arcuate-ventromedial nuclei bring about some reduction of ME concentrations of LH-RF (Fig. 1) (Mess, Fraschini,

Motta & Martini, 1967; Martini *et al.* 1968). It is apparent from these data that separate hypothalamic structures are involved in the synthesis of the two RF's controlling gonadotrophin secretion: one single region (the paraventricular area) synthesizes FSH-RF; two different zones, one in the anterior hypothalamus (suprachiasmatic area) and one located more caudally (arcuate-ventromedial region) apparently secrete LH-RF. Suprachiasmatic, paraventricular and arcuate-ventromedial lesions (as well as

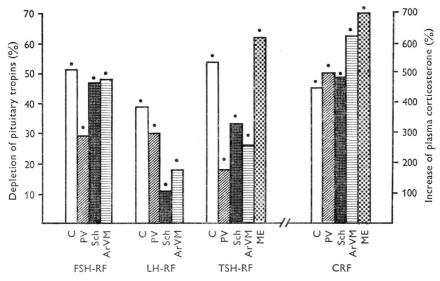

Fig. 1. Effect of lesions localized in the paraventricular (PV), suprachiasmatic (Sch), arcuate–ventromedial (ArVM) and median eminence (ME) areas on FSH-RF, LH-RF, TSH-RF and CRF activity of the median eminence of male rats. Columns represent either the depletion of pituitary trophins (left) or the increase of plasma corticosterone (right) induced by the intracarotid injection of hypothalamic extracts prepared from non-lesioned controls (C) or from animals with the different hypothalamic lesions.

lesions localized between these nuclei) have all reduced TSH-RF in the ME; apparently paraventricular lesions were more effective than the others (Fig. 1). Although other explanations might be put forward, these results may be interpreted as indicating that TSH-RF is synthesized over a large area located in the anterior hypothalamus; apparently each portion of the hypothalamus manufactures TSH-RF at a low rate; so that when one region is destroyed, the remaining ones are not able to compensate for the amount of tissue which is not functioning (Mess *et al.* 1967; Martini *et al.* 1968). None of the lesions proved successful in reducing CRF concentrations in the ME; the concentrations of this RF were actually unmodified following lesions located in the suprachiasmatic and paraventricular regions; a small increase in CRF content was found after lesioning the arcuate-ventromedial area; an even larger increase followed partial destruction of the ME (Fig. 1). These data might indicate either that CRF is synthesized in a zone which has not been lesioned so far, or that the structures which manufacture this mediator have a large distribution and can very rapidly compensate for the lesioned areas; this would explain why circumscribed lesions, like the ones which were used in these experiments, have not reduced its storage; the increase in CRF concentration observed following placement of lesions in the arcuate-ventromedial area or in the ME itself is probably due to the damage of the pituitary portal vessels, which prevents its delivery to the pituitary (Mess *et al.* 1967; Martini *et al.* 1968).

Effect of 'hypothalamic deafferentation' on ME stores of releasing factors

The second approach devised to study whether the RF's might be synthesized outside the ME was that of evaluating the concentration of FSH-RF in the *hypothalamic islands* of animals submitted to a complete

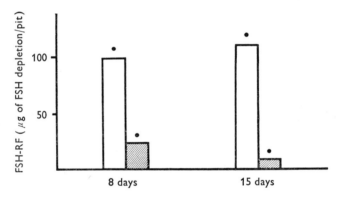

Fig. 2. Effect of *hypothalamic deafferentation* on the FSH-RF content of the *hypothalamic island* of adult normal male rats (8 and 15 days after the operation). Columns represent the depletion of pituitary FSH induced in normal male rats by the intracarotid injection of hypothalamic extracts prepared from controls (□) or from *deafferented* animals (▨), (Method of Dávid, Fraschini & Martini, 1965).

hypothalamic deafferentation (Halász, 1969). This operation permits a total separation of the paraventricular region from the rest of the hypothalamus. Consequently, if the hypothesis put forward by Mess *et al.* (1967) and by Martini *et al.* (1968) were correct, one would expect the complete disappearance of FSH-RF from the *island* a few days after the operation. The experiments were performed in adult male rats; FSH-RF content in the *deafferented island* was measured 8 and 15 days following the operation, using the technique described by Dávid, Fraschini & Martini (1965). The results summarized in Fig. 2 shows that FSH-RF stores are significantly reduced in the ME region 8 days after a complete *hypothalamic deafferentation*; FSH-RF disappears completely from the *isolated hypothalamus* 15 days after the operation (Tima, Motta & Martini, 1969). These data apparently confirm that FSH-RF is synthesized in the paraventricular region and provide additional support for the hypothesis that the RF's are synthesized outside the ME.

Effect of hypothalamic implants of inhibitors of protein synthesis on ME stores of releasing factors

In the third group of experiments, the hypothesis that FSH-RF might be synthesized in the paraventricular nuclei was tested by implanting cycloheximide (Actidione), an inhibitor of protein synthesis, into the paraventricular region of adult castrated male rats and by evaluating the effects

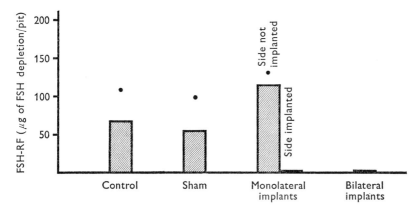

Fig. 3. Effect of implants of cycloheximide (Actidione) in the paraventricular region on the FSH-RF content of the median eminence of adult castrated male rats. See text for more details. Columns represent the depletion of pituitary FSH induced in normal male rats by the intracarotid injection of hypothalamic extracts prepared from controls or from implanted animals (Method of Dávid, Fraschini & Martini, 1965). Control: unimplanted animals. Sham: animals implanted with empty cannulae. Monolateral implants: animals implanted with cycloheximide in one paraventricular nucleus. Bilateral implants: animals implanted with cycloheximide in both paraventricular nuclei.

of such implants on FSH-RF stores in the ME. Cycloheximide was implanted either unilaterally or bilaterally; when unilateral implants were performed at time of autopsy ME tissue was collected in a way which permitted the separation of the half ME corresponding to the implanted side from the half corresponding to the non-implanted one; the two halves of the ME were then tested separately for their content in FSH-RF. The data shown in Fig. 3 indicate that, 5 days after unilateral implantation of the drug, FSH-RF disappears only from the ipsilateral half of the ME; complete disappearance of FSH-RF from the ME is induced by bilateral implants (Zanisi & Martini, 1969). It may be concluded that inhibition of protein synthesis in the cells of the paraventricular nuclei interfere with some biochemical process which is essential for the synthesis of FSH-RF in this region; these data, however, are not taken as a conclusive proof that FSH-RF itself is a protein or a polypeptide (McCann & Dhariwal, 1966). It is clear from the results that fibres originating in one paraventricular nucleus do not cross, and carry FSH-RF only to the ipsilateral half of the ME.

CONCLUSIONS

The data presented in the first part of this paper (see section 'Afferent component') indicate that, within the hypothalamus, the ME is probably the region which is particularly influenced by feedback messages, since it contains feedback receptors sensitive to the 'negative' effect of corticoids, of sex steroids, of ACTH and of LH as well as to the 'positive' effect of oestrogen. This list is certainly incomplete, since it is based only on the data obtained in the authors' laboratory. On the other hand, the results summarized in the second part of the paper (see section 'Efferent component') have validated the hypothesis that hypothalamic RFs are synthesized outside the ME. It has been demonstrated that lesions placed in hypothalamic structures situated far from the ME reduce the storage of several RFs in ME tissue; in addition, the disappearance of FSH-RF from the ME may be obtained either by separating the paraventricular region from the rest of the hypothalamus (*hypothalamic deafferentation*) or by implanting in this region an inhibitor of protein synthesis.

It is quite apparent from the data that the areas in which the RFs are synthesized do not overlap with those where the information from the periphery is received. Nervous interconnections between the *afferent component* (ME receptor area) and the *efferent executive component* (nuclei synthesizing the RFs) have to be postulated. Theoretically speaking, two types of circuits may be anticipated: (*a*) a direct one, from ME receptors (and possibly from other receptor areas located outside the hypothalamus) (Mangili *et al.* 1966; Mess & Martini, 1968) to the cell bodies where RFs are synthesized; (*b*) an indirect one, from the ME to the cells producing releasing factors via intermediate centres. These centres (located either in the hypothalamus or outside the hypothalamus) might be responsible for integrating the information received from feedback receptors with that provided by other brain structures (amygdala, hippocampus, olfactory bulbs, etc.), by the pineal gland, by exogenous influences (light, temperature, etc.), before translating it into a final message to be transmitted to the *executive elements* (Fig. 4). We believe that the elucidation of the pathways involved in these nervous circuits is of primary importance, and will be a fruitful field of research in the coming years.

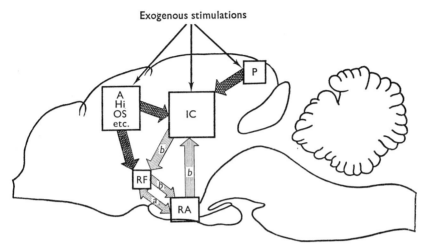

Fig. 4. Schematic representation of the pathways which might participate in transmitting the information received from feedback receptors to the sites where the releasing factors are synthesized. See text for more details on circuits (*a*) and (*b*). A, Amygdala; Hi, hippo-campus; OS, olfactory structures; P, pineal gland; RA, receptor area; RF, neurones synthesizing the releasing factors; IC, integrative centre.

The experimental work described in this paper has been supported by funds of the Department of Pharmacology, University of Milan, and by the following grants: AM-10119-01-03 and AM-11783/01-02 of the National Institutes of Health, Bethesda, Maryland, and 67–530 of the Ford Foundation, New York. Gifts of pituitary hormones were made by the National Institutes of Health, Bethesda, Maryland. All such support is gratefully acknowledged.

REFERENCES

BEYER, C. & SAWYER, C. H. (1969). Hypothalamic unit activity related to control of the pituitary gland. In *Frontiers in Neuroendocrinology*, pp. 255–87 (eds. W. F. Ganong and L. Martini). Oxford University Press.

CLEMENS, J. A. & MEITES, J. (1968). Inhibition by hypothalamic prolactin implants of prolactin secretion, mammary growth and luteal function. *Endocrinology* **82**, 878–81.

CORBIN, A. (1966). Pituitary and plasma LH of ovariectomized rats with median eminence implants of LH. *Endocrinology* **78**, 893–6.

CORBIN, A. & COHEN, A. I. (1966). Effect of median eminence implants of LH on pituitary LH of female rats. *Endocrinology* **78**, 41–6.

CORBIN, A., MANGILI, G., MOTTA, M. & MARTINI, L. (1965). Effect of hypo-thalamic and mesencephalic steroid implantations on ACTH feedback mecha-nisms. *Endocrinology* **76**, 811–18.

CORBIN, A. & STORY, J. C. (1967). 'Internal' feedback mechanism: response of pituitary FSH and of stalk-median eminence follicle stimulating hormone-releasing factor to median eminence implants of FSH. *Endocrinology* **80**, 1006–12.

DÁVID, M. A., FRASCHINI, F. & MARTINI, L. (1965). An *in vivo* method for evaluating the hypothalamic follicle stimulating hormone releasing factor. *Experientia* **21**, 483–4.

DÁVID, M. A., FRASCHINI, F. & MARTINI, L. (1966). Control of LH secretion: role of a 'short' feedback mechanism. *Endocrinology* **78**, 55–60.

DE WIED, D. (1969). Effects of peptide hormones on behaviour. In *Frontiers in Neuroendocrinology*, pp. 97–140 (eds. W. F. Ganong and L. Martini). Oxford University Press.

FRASCHINI, F., MOTTA, M. & MARTINI, L. (1966). Methods for the evaluation of hypothalamic hypophysiotropic principles. In *Methods in Drug Evaluation*, pp. 424–57 (eds. P. Mantegazza and F. Piccinini). Amsterdam: North-Holland Publishing Company.

HALÁSZ, B. (1969). The endocrine effects of isolation of the hypothalamus from the rest of the brain. In *Frontiers in Neuroendocrinology*, pp. 307–42. (eds. W. F. Ganong and L. Martini). Oxford University Press.

MANGILI, G., MOTTA, M. & MARTINI, L. (1966). Control of adrenocorticotropic hormone secretion. In *Neuroendocrinology*, vol. 1, pp. 297–370 (eds. L. Martini and W. F. Ganong). New York and London: Academic Press.

MARTINI, L., FRASCHINI, F. & MOTTA, M. (1968). Neural control of anterior pituitary functions. *Recent Progr. Hormone Res.* **24**, 439–96.

MARTINI, L., FRASCHINI, F. & MOTTA, M. (1969). Hypothalamic mechanisms and anterior pituitary functions. In *Advances in the Biosciences*, vol. 1, pp. 201–12 (ed. G. Raspé) Vieweg: Pergamon Press.

McCANN, S. M. & DHARIWAL, A. P. S. (1966). Hypothalamic releasing factors and the neurovascular link between the brain and the anterior pituitary. In *Neuroendocrinology*, vol. 1, pp. 261–96 (eds. L. Martini and W. F. Ganong). New York and London: Academic Press.

MESS, B., FRASCHINI, F., MOTTA, M. & MARTINI, L. (1967). The topography of the neurons synthesizing the hypothalamic releasing factors. In *Hormonal Steroids*, pp. 1004–13 (eds. L. Martini, F. Fraschini and M. Motta). Amster-dam: Excerpta Medica.

MESS, B. & MARTINI, L. (1968). The central nervous system and the secretion of anterior pituitary trophic hormones. In *Recent Advances in Endocrinology*, pp. 1–49 (ed. V. H. T. James). London: Churchill.

MOTTA, M., FRASCHINI, F., GIULIANI, G. & MARTINI, L. (1968). Hypothalamus, oestrogens and puberty. *Endocrinology* **83**, 1101–7.

MOTTA, M., FRASCHINI, F. & MARTINI, L. (1969). 'Short' feedback mechanisms in the control of anterior pituitary function. In *Frontiers in Neuroendocrinology*, pp. 211–53 (eds. W. F. Ganong and L. Martini). Oxford University Press.

MOTTA, M., MANGILI, G. & MARTINI, L. (1965). A 'short' feedback loop in the control of ACTH secretion. *Endocrinology* **77**, 392–5.

RAMIREZ, V. D. & SAWYER, C. H. (1965). Advancement of puberty in the female rat by estrogen. *Endocrinology* **76**, 1158–68.

TIMA, L., MOTTA, M. & MARTINI, L. (1969). Effect of 'hypothalamic deafferenta-tion' on hypothalamic follicle stimulating hormone releasing factor (FSH-RF) and on pituitary FSH. *Program 51st Meeting Endocrine Soc.*, p. 194.

ZANISI, M. & MARTINI, L. (1969). Effect of brain implants of cycloheximide on hypothalamic follicle stimulating hormone releasing factor (FSH-RF) and on pituitary FSH. *Program 51st Meeting Endocrine Soc.*, p. 202.

DISCUSSION

VAN OORDT: You told us about your experiments on the 'short circuit' feedback of pituitary hormones in which the pituitary hormones had been implanted into the median eminence. What is the physiological significance of the results of these experiments? Do the pituitary hormones also inhibit the output of releasers from the median eminence when they are administered via the circulatory system?

MARTINI: Thank you. I confined my presentation to data obtained by implanting hormones into the brain or into other areas like the pituitary, the cerebral cortex, etc. Another approach for studying both 'short' and 'long' feedback mechanisms, and for establishing whether they operate through the brain, is that of administering exogenous pituitary hormones and of measuring, a few days later, the modifications of the hypothalamic stores of the releasing factors induced by treatment. It is obvious that for assessing the effect of gonadotrophins on the gonadotrophin releasing factors, it is essential to operate in castrated animals; this in order to avoid a possible effect through the activation of the release of sex hormones. This type of experiment has provided the following results. Castration induces, after a week, a significant increase in the hypothalamic concentration of the follicle-stimulating hormone releasing factor (FSH-RF). If castrated animals are treated with exogenous FSH, the levels of FSH-RF in the hypothalamus are brought back to normal. Similar results have been obtained for other pituitary hormones. The data have been reviewed in the chapter by Motta *et al.* (1969). We believe that this approach provides a good validation of the results obtained by using the technique of implanting active materials into the brain.

CLEGG: You show a sex difference in the response of the median eminence to oestrogens, e.g. a 'negative' feedback in the male and a 'positive' one in the female. What happens in female rats androgenized shortly after birth?

MARTINI: I am sorry I did not make it clear that this is not a sexual difference. The 'negative' feedback effect of oestrogen may be observed also in female animals. The type of effect ('positive' or 'negative') depends probably more on the age of the animals than on their sex. There are preliminary data indicating that in female rats androgenized shortly after birth, oestrogen exerts a 'negative' effect. However, these animals appear less sensitive to oestrogen than normal ones (J. Szentágothai, B. Flerkó, B. Mess & B. Halász (1968), *Hypothalamic Control of the Anterior Pituitary* (Budapest: Akadémiai Kiado)).

FOLLET: Might I ask something about the integrative centre which you mentioned? Recently we have obtained evidence in the Japanese quail indicating that the gonadotrophin releasing factors are localized in the basal hypothalamus and median eminence, which suggests that they are synthesized within the region of the tuberal nucleus (probably corresponding with the arcuate nucleus of mammals.). However, lesions placed dorsal to this area abolish the release of pituitary gonadotrophins. This region might therefore be an example in lower vertebrates of such an integrative centre.

MARTINI: My last slide was only a diagram; the existence of 'integrative centres' is still hypothetical. However, we believe that this hypothesis may help to explain several data which are found in the literature. Those you quote may represent a valid example in favour of our hypothesis.

THE SIGNIFICANCE OF THE PINEAL

By W. B. QUAY

INTRODUCTION

The pineal complex of organs, derived from the posterior part of the diencephalon's roof, has had a complex structural evolution within vertebrates, including all major groups from cyclostomes to mammals. Structural and other characteristics of the evolving pineal complex suggest a gradual transformation from a primarily directly photoreceptive sensory system to an indirectly photoreceptive endocrine system with an increasing superimposed autonomic or sympathetic control. A recent resurgence in pineal research has provided us with a diversity of interesting and important results largely supporting this generalization. Our presentation here, however, is necessarily limited and selective. Recent reviews may be consulted for coverage of some areas and some points of view either not treated or only briefly mentioned here (Mess, 1968; Wurtman, Axelrod & Kelly, 1968; Quay, 1965 d, 1968 b, 1969 a, b).

The significance of the pineal as either a sensory or an endocrine organ might be expected to be revealed most simply by observing changes following its removal. Our emphasis here will be on such results of pineal removal, or pinealectomy, first in a brief survey of published comparative findings, and then in a presentation of new results from mammals. These last-mentioned results are of two kinds: (1) those showing pineal effects on brain chemistry, most likely involving homeostatic neuroglial mechanisms; and (2) those showing pineal effects on behavioural rhythmicity, with the definition of a probable pineal contribution in photoperiodic phase shifting. These two kinds of results are selected for presentation here for three reasons: (1) it is suggested that they are fundamental and specific effects; (2) it is believed that they may be directly related to each other, thus, effects on behavioural phase shifting are proposed as being likely to be dependent on related temporal events in brain chemistry; and (3) it is suggested that variable peripheral effects, such as those on the reproductive system, are indirect and secondary in relation to primary effects of the pineal on the nervous system.

The relations of pineal activity to environmental factors have been studied especially productively within the past two decades, with the most significant relationships established with environmental illumination and photoperiod in part and with stress-related or behavioural arousal conditions in part. Most of these investigations concern either effects near the time of puberty or those that might occur in adulthood. Less recognized are possibilities of environmental control of pineal development. Although continuous light has been shown to have such an effect, simpler and non-photic environmental factors can also affect pineal postnatal development in mammals, as will be demonstrated in a final experiment to be presented here.

COMPARATIVE SURVEY
Cyclostomes

Pioneering investigations by Young (1935) showed that in the ammocoete, or larva, of the lamprey (*Lampetra planeri*) the removal of the pineal complex caused interruption of the daily rhythm of the integumentary melanophores, which then remained in an expanded state under all levels of illumination. It was thought that the paling of an ammocoete when it passed from light to darkness resulted from an inhibition of pituitary melanophore hormone by nervous impulses set up by the change of illumination of the pineal complex. Prophetically for much of the subsequent and recent pineal research, it appeared that the active phase of the daily cycle occurred during the transition from light to darkness. Recently studies by Eddy & Strahan (1968) using two other genera (*Geotria* and *Mordacia*) have extended these results with emphasis on the likelihood of a pineal endocrine mediation of the photo-pigmentary response, such as by the secretion of melatonin. It was concluded that while the pineal complex has an endocrine as well as a photoreceptive role in the control of melanophores in *Geotria*, it is not involved in the pigmentary control in *Mordacia*. Pineal ultrastructure in *Lampetra planeri* is consistent with both photosensory and neurosynaptic capacities, with endocrine activity still remaining a possibility (Collin, 1968*b*; Collin & Meiniel, 1968).

Sharks

Electron-microscopic studies by Rüdeberg (1968, 1969) also support a photoreceptor capacity within the pineal of the dogfish, *Scyliorhinus canicula*. Unpublished electrophysiological results obtained by Hamasaki and Streck and cited by Rüdeberg (1969) showed the pineal in this species to be light responsive, with the spontaneous activity of the pineal nerve fibres being inhibited by light and with the duration of the inhibition proportional to light intensity. Effects of pineal extirpation have not been studied apparently in any of the cartilaginous fishes.

Bony fishes

Neurophysiological investigations have shown that the pineal of a number of different kinds of freshwater fishes is photoreceptive (de la Motte, 1963; Dodt, 1963; Morita, 1966 a). The fact that some anterior brain regions appear also to be photoreceptive in these forms makes difficult the interpretation of some of the most interesting experiments (Frisch, 1911 a, b; Scharrer, 1928; Breder & Rasquin, 1947). Effects of pineal removal likewise seem often to have been dependent on involvement of neighbouring brain regions (Pflugfelder, 1964, 1966, 1967). Nevertheless, available evidence suggests physiological roles for the pineal in at least some species in relation to photoperiodic phenomena on a daily basis in terms of melanophore responses and on a seasonal basis in terms of gonadotropic responses. Daily rhythms in teleost melanophore responses are species, area and compound specific. While the pineal compound melatonin is active and specific as a paling agent in some species, in some conditions it can act as a darkening agent (Reed, 1968; Reed, Finnin & Ruffin, 1969). Possible contributions of other amines, and species differences in response and innervation must be considered (Rasquin, 1958; Reed, 1968; Grove, 1969) as well as the possibility that melatonin may have possible sites of origin in addition to the pineal complex (Quay, 1965 b, 1968 b, c). In the light of recent experiments by Fenwick (1969) employing goldfish pinealectomized at different seasons, it is likely that the lack of effect in earlier studies was due to differences in timing of the experiments in relation to photoperiod or season (Schönherr, 1955; Pickford & Atz, 1957; Peter, 1968). Pinealectomy enhanced the gonadotropic response to light only during the seasonal period when such a response normally occurs (Fenwick, 1969).

Amphibians

Physiology of the pineal complex in amphibians has many similarities with that in bony fishes, particularly in relation to neurophysiological results (Dodt & Heerd, 1962; Dodt & Morita, 1967), ultrastructural implications (Eakin, 1961; Oksche & Vaupel-von Harnack, 1962; Eakin, Quay & Westfall, 1963; Kelly & Smith, 1964; Charlton, 1968; Ueck, 1968) and effects on melanophore activity (Bagnara, 1960, 1963, 1964). Again we have the problem of similar photosensory capacity in the forebrain as well, in this case demonstrated and mapped by electrophysiological recording (Dodt & Jacobson, 1963; Morita, 1965). Even with the latter possible limitation in pineal specificity as receptor site, recent implication of frog pineals in light-induced movement (Mrosovsky & Tress, 1966), metabolic regulation (Kasbohm, 1967) and inhibition of spontaneous electrical activity in the hypophyseal pars intermedia (Oshima & Gorbman, 1969) deserves more intensive study. An extra-optic photoreceptor, possibly in the pineal region, has been implicated in the mechanism for photic induction of phase shifts in daily locomotor rhythms in the salamander *Plethodon glutinosus* (Adler, 1969). Information concerning the possibility of a pineal effect on amphibian reproductive systems is scant and mostly negative. Thus while melatonin is reported to inhibit ovulation *in vitro* of mature *Rana pipiens* oocytes (O'Connor, 1969), it had no effect on the frog's spermatogenic response (Juszkiewicz & Rakalska, 1965), and pinealectomized larvae of *Alytes obstetricians* showed no significant modifications of the ovaries (Disclos, 1964).

Reptiles

The reptilian pineal complex, while including photoreceptoral capacities and afferent neuronal activities, especially in lacertilians, shows a marked increase in what appears to be endocrine secretory parenchyma. This has been abundantly documented by electron microscopy of different portions of the pineal complex in diverse genera (Eakin & Westfall, 1959; Steyn, 1959, 1960; Eakin, Quay & Westfall, 1961; Oksche & Kirschstein, 1966, 1968; Vivien, 1964; Lierse, 1965; Collin, 1967a, b; Lutz & Collin, 1967; Wartenberg & Baumgarten, 1968; Vivien & Roels, 1967; Petit, 1968, 1969; and others). Microelectrode records from the parietal (pineal) eye of two genera of lizards (*Anolis* and *Phrynosoma*) provide mutually comparable results with differences probably attributable to techniques of recording (Miller & Wolbarsht, 1962; Gruberg, Heath & Northcutt, 1968). One of the two electrically active units revealed in the parietal eye was spontaneously active in darkness but was inhibited by light. The physiological effects of covering or removing the parietal eye or the underlying pineal organ, or epiphysis, of lizards appear to be variable and difficult to define in quantitative terms. The effects as well as their inconstancy appear most likely to be due to species differences and critical factors of seasonal and photoperiod timing. In spite of these difficulties, promising results have been obtained relating to patterns and characteristics of daily locomotor activity and seasonal thyroid activity (Stebbins & Eakin, 1958; Glaser, 1958; Eakin, Stebbins & Wilhoft, 1959; Stebbins, 1960, 1963; Stebbins & Wilhoft, 1966).

Birds

Although attempts to detect avian pineal electrical responses to illumination have been unsuccessful (Morita, 1966b; Ralph & Dawson, 1968), other kinds of evidence suggest the probability of some photosensory capacity in some species. Thus pineal ganglion cells and rudimentary photoreceptoral cells have been demonstrated microscopically in various species (Quay & Renzoni, 1963; Oksche & Vaupel-von Harnack, 1965, 1966; Collin, 1966 a, b, 1968 a; Bischoff, 1967; Fujie, 1968; Quay, Renzoni & Eakin, 1968). Extra-retinal light perception by the sparrow (*Passer domesticus*), manifested either in reproductive effects or in entrainment of daily activity rhythm, has been studied recently by Menaker and co-workers (Menaker, 1968 a; Gaston & Menaker, 1968; Menaker & Keats, 1968). The most recent report from this group (Menaker, 1968 b) indicates that some region of the brain and not the pineal is involved in extraretinal reception of photic cues in this species. Extirpation of the pineal has continued to be investigated, usually in domestic fowl (*Gallus*) (Stalsberg, 1965; Zadura, Roszkowski & Cakala, 1969) or Japanese quail (*Coturnix coturnix japonica*) (Arrington, 1967; Homma, McFarland & Wilson, 1967; Renzoni, 1967; Sayler & Wolfson, 1967, 1968), and usually with negative results. However, some groups have reported results which might be taken to indicate 'progonadotropic' effects of the pineal in subadults of both species (Shellabarger, 1952; Sayler & Wolfson, 1967, 1968; Zadura *et al.* 1969). This stands in contrast with the evidence as usually reviewed from mammals and supporting an antigonadotropic role for the pineal.

Mammals

Diverse aspects of mammalian pineal physiology and problems in this field have been reviewed elsewhere recently (Wurtman, Axelrod & Kelly, 1968; Quay, 1969 a, b). To be recognized first of all is the early, frequent and continued proposal of a fundamentally antigonadotropic effect of the mammalian pineal (Kitay & Altschule, 1954; Thiebolt & Le Bars, 1955; Wurtman *et al.* 1968). Thus pinealectomy is expected to lead to increased gonadal size and activity and administration of pineal substance is expected to cause reduced gonadal size and activity. The strongest case for such a pineal action in any mammalian species has been provided by Reiter and co-workers studying the pineal-dependency of darkness-induced gonadal atrophy in the golden hamster (*Cricetus auratus*) (Reiter, Hoffman & Hester, 1966; Reiter, 1968 a, b, c). Confirmation (Quay, 1969 b) and appropriate quantitative analyses support what is now a much more convincing picture than had been available previously from laboratory rats. Nevertheless, in the ferret, Herbert (1969) has shown a quite different relationship between pineal and reproductive activity. Pinealectomy in this species did not affect the duration of the annual oestrous period, but it greatly increased the interval between onsets of successive postoperative annual oestrous periods. Accommodation of such differences in pineal actions seems possible if the mediation of pineal effects is through the central nervous system and if the potentially critical nature of pineal physiological timing in relation to daily photoperiod is appreciated.

PINEALECTOMY AND BRAIN CHEMISTRY

At an international conference on the epiphysis cerebri in July 1963 I presented results of studies showing that pinealectomized rats on a sodium-deficient diet have a lower cerebral potassium concentration than sham-operated or unoperated controls, while the sodium and potassium concentrations of their other examined organs are not significantly different (Quay, 1965a). An effect equivalent to pinealectomy was caused by continuous light, a treatment that in most respects is inhibitory as far as the rat pineal is concerned (Quay, 1962, 1963a). A review and discussion of these results has been provided recently (Quay, 1969b).

Another quantitative difference in the brain chemistry of pinealectomized rats has been suggested recently in collaborative studies with colleagues at Berkeley—Drs Edward L. Bennett and Mark R. Rosenzweig.

Methods

In this study sibling male rats of the S_1-strain were weaned, weighed, and operated upon when 22–29 days old. The three operation groups of pinealectomized, sham-operated and unoperated rats were maintained under identical conditions until autopsy 66–67 days later. All animals were born, raised and maintained in the same suite of windowless rooms with controlled temperature and daily photoperiod (lights on 4.00–18.00 Pacific Standard Time). Animals representing different groups were mixed in order throughout the experiment, and at autopsy, which occurred between 9.30 and 13.30 on three consecutive days. Following death by rapid decapitation brain regions were quickly dissected, weighed (Mettler balance) and frozen. Methods and 'landmarks' in the dissection have been described (Rosenzweig, Krech, Bennett & Diamond, 1962). Brain regions were analysed colorimetrically for activities of acetylcholinesterase (AChE) and cholinesterase (ChE) as described previously (Krech, Rosenzweig & Bennett, 1966; Rosenzweig, Krech, Bennett & Diamond, 1968). In the determination of AChE activity aliquots of brain region homogenates at pH 8 were incubated with acetylthiocholine and 5,5′-dithiobis-(2-nitrobenzoic acid). In the determination of ChE or 'non-specific' cholinesterase activity, butyrylthiocholine was substituted for acetylthiocholine and a specific inhibitor for AChE was added. Brain dissections, weighings and chemical analyses were done by personnel lacking knowledge of the specific treatments received by animals.

Results

Statistical analysis of the results showed a significant difference in enzyme activity only in the left occipital or 'visual' cortex (= 'V' area) (Fig. 1). The right occipital cortex, the site of the operation, was not analysed. Other cortical and brain divisions analysed bilaterally ('somesthetic cortex', 'remaining dorsal cortex', 'ventral cortex', cerebellum-pons-medulla,

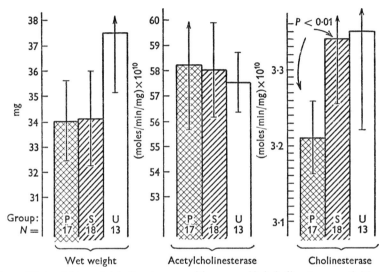

Fig. 1. Wet weight, acetylcholinesterase and 'non-specific' cholinesterase activities of the left occipital ('V' area) cerebral cortex of sibling male rats pinealectomized (P), sham-operated (S) or unoperated (U) when 22–29 days old and autopsied 66–67 days later. Tops of bars correspond to group means; vertical lines extend × 2 the standard error of the mean on each side of the means. N, number of individuals; P, probability from t based on pair-wise analysis of differences between siblings.

and remaining brain) showed no differences. Although at operation the meninges of the left 'V' area were not penetrated and although no visually detectable differences were observed postoperatively, the wet weight of the left 'V' area averaged about 3·5 mg less in both pinealectomized and sham-operated as compared with unoperated animals (Fig. 1). However, this difference could not be shown to have statistical significance. The tissue concentration of AChE was the same in all groups, but the concentration of ChE was significantly lower in the pinealectomized animals.

Discussion

It is not clear at this time whether the slight but significant difference in left occipital cortical ChE after pinealectomy is dependent on operative involvement of the contralateral cortex. Nevertheless, although equivalent operative haemorrhage and right occipital cortical trauma occurred in the sham-operated animals, their left cortical ChE activity was not distinguishable from that of their unoperated siblings. Thus the removal of the pineal appears to be a critical factor in the lowered ChE concentration. These results in conjunction with those presented earlier on brain electrolyte homeostasis (Quay, 1965a, 1969b) suggest that it is within the glial, and probably astrocytic, compartment, rather than the neuronal one, that a central homeostatic defect exists in the pinealectomized animal.

PINEALECTOMY AND BEHAVIOURAL RHYTHMS

Direct photoreceptoral capacity is not found in any mammalian pineal, at least according to currently available evidence. Nevertheless, the distal part of the pineal shows cytological responses to photoperiod (Quay, 1956) and daily biochemical rhythms of great amplitude and responsive to shifts in timing of environmental illumination within a 24 h period (Quay, 1963b, 1964a,b, 1966, 1967). The suggestion (Quay, 1963b) that the mediation of these photic pineal responses is by way of the lateral eyes, brain and superior cervical sympathetic system has been confirmed and extended in its essential relationships (Wurtman, Axelrod & Kelly, 1968). The thought that the pineal may therefore serve either as a transducer or as a central element in a 'biological clock' has remained, nevertheless, without any directly supporting evidence. Search for such direct evidence of pineal participation in physiologic and behavioural rhythms would seem to be most efficiently directed to study of 24 h or circadian rhythms in pinealectomized animals. In comparison with seasonal and reproductive rhythms, circadian rhythms are of more general occurrence, are less variable according to species and may be closer to central control factors in physiological rhythms of longer period length.

Our study of the circadian rhythm in spontaneous running activity of laboratory rats in relation to environmental photoperiod has failed to demonstrate any difference in the pinealectomized animal until early this year (1969). The free-running circadian rhythm in the male rat in both standard daily photocycle and continuous light is similar in normal and pinealectomized animals (Quay, 1965c). In the daily number and distribution of active 10 min periods the pinealectomized rat appears still to follow 'Aschoff's rule'. As a naturally nocturnal or dark-active species it shows a longer circadian period in constant light than in constant dark and a reduction in period length with decreasing intensity of illumination (Quay, 1968a). However, recent experimental analysis of photoperiod-induced phase shifting reveals a difference in the pinealectomized animal.

Methods

Seventeen female rats of the S_1-strain and from three litters were studied in activity wheels for almost 1 year, starting when they were 5 months old. Different aspects of the cycles in running activity were studied. As demonstrated earlier with adult male rats in the same room (Quay, 1968 *a*), at least the circadian rhythms of running activity of the females went out of phase with each other both in constant environmental conditions and also after they were blinded during the terminal 2 months of the study. Therefore, individuation of cycle length occurred even though the animals were in the same room, in adjacent cages and under identical environmental conditions. No other animals were in the room. After about 2 months in the activity cages matched siblings in adjacent cages were pinealectomized, sham-operated or unoperated according to procedures described previously (Quay, 1965*a*). There were no mortalities during the immediate post-operative period. Approximately 3 months post-operatively the circadian phase-shifting of the animals was tested by reversals of the daily photoperiod. Three 25 W red ceiling lamps were on continuously. The daily photoperiod, 12 h in length, was provided by three 100 W tungsten-filament white ceiling bulbs. Rotations of activity wheels were individually monitored by microswitches connected to separate channels of an Esterline–Angus event recorder. Conditions generally were the same as described previously (Quay, 1968 *a*).

Results

When the mean times of starting and ending the primary daily active period are plotted for the four most active pinealectomized animals and their four most active sham-operated siblings, an interesting and consistent difference appears (Figs. 2–5). Through four reversals of daily photoperiod, sham-operated animals lag the pinealectomized ones in phase shifting as far as the daily start of activity is concerned. The two groups are undistinguished in the shifting of the time of their daily termination of major activity. The precocity of the phase shifting of the starting times of the pinealectomized animals averages about 5 h after the first reversal (Fig. 2) and is about 1 h by the fourth. Computational analyses of these results will be presented elsewhere. The plots presented at this time, however, suggest that the essential difference between the pinealectomized and sham-operated animal during phase shifting may not be in the rate of the shift, but in the time lapse between photoperiod change and the change in the timing of the start of the daily active period. Results of the four consecutive phase shifts (Figs. 2–5) are informative not only in

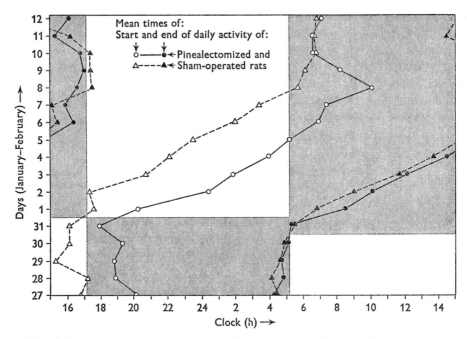

Fig. 2. First test of photoperiod-induced circadian phase-shifting in sibling pinealecto-mized and sham-operated female rats (four animals per group) running simultaneously in the same room. Advance in shift of mean daily starting time is seen in pinealectomized as compared with sham-operated animals. Red lights on continuously, white lights on only during clear (non-shaded) area of chart.

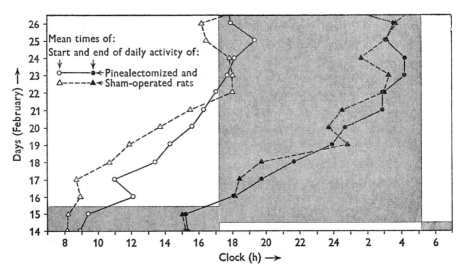

Fig. 3. Second test of photoperiod-induced circadian phase-shifting in sibling pinealecto-mized and sham-operated female rats (four animals per group); conditions as portrayed in Fig. 2 except that photoperiod phasing is returned to standard time relationship.

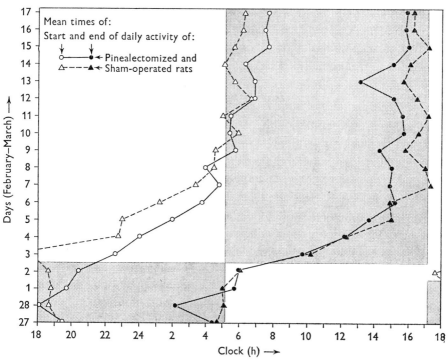

Fig. 4 Third test of photoperiod-induced circadian phase-shifting in sibling pinealectomized and sham-operated female rats (four animals per group); conditions as portrayed in
Figs. 2 and 3 except that 24 h of white light occurs at the time of photoperiod shift.

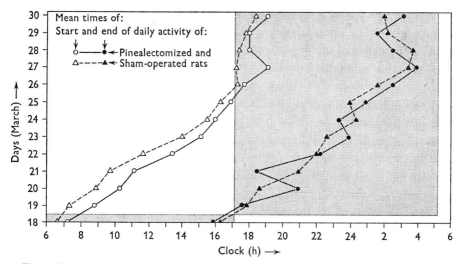

Fig. 5. Fourth test of photoperiod-induced circadian phase-shifting in sibling pinealectomized and sham-operated female rats (four animals per group); conditions as portrayed in
Fig. 4 except that photoperiod phasing is returned to standard time relationship.

regard to the reproducibility of the difference shown between the groups, but also in regard to (1) reversals through the two subdivisions of the day, (2) effects of dark and light transition periods, and (3) possible progressive attenuation of the difference in later phase reversals.

Discussion

These results suggest that the pinealectomized rat may respond more quickly in its start of daily activity in relation to modification of environmental photoperiod. It is tempting to extrapolate from this finding to speculate on how this might affect seasonal advance or reproductive activity in species that are reproductively stimulated by increase or advance of photoperiod, and how this might affect conversely the reproductive period in species that are stimulated by reduction or delay of photoperiod. Extensive additional experimentation and quantitative analyses are required before such thoughts can be offered with much hope of describing actual physiological relationships.

ENVIRONMENTAL CONTROL OF PINEAL DEVELOPMENT

Studies during the last two decades have succeeded especially in showing that the adult mammalian pineal is modified by various environmental factors, particularly those having to do with light and the sympathetic nervous system. It is important to realize also the probability of the critical nature of the early postnatal environment for pineal development and probable activity during later life. The profound effect of continuous light on the pineal organ of young rats was first shown by Fiske and co-workers (Fiske, Bryant & Putnam, 1960; Fiske, Pound & Putnam, 1962) and has been confirmed by others. In recent studies in collaboration with colleagues (Drs E. L. Bennett, M. R. Rosenzweig & D. Krech) in Berkeley, it has been shown that timing of weaning and early conditions of caging can have marked effects also. Isolation of young rats leads to a pineal hypertrophy in comparison to what is seen in siblings in various cage situations in which four or more animals are together (Quay, Bennett, Rosenzweig & Krech, 1969). In contrast with effects on other endocrines and brain, the isolation hypertrophy of the pineal is dependent on an earlier age of weaning and isolation (Figs. 6–8).

As described earlier, cerebral cortical changes are associated with having several animals per cage and having a higher level of environmental complexity. This occurs whether weaning and experimental conditions start at 20–21 or 27–28 days of age (Fig. 6). Similarly, with the pituitary, adrenal and thyroid, hypertrophy in isolated and unhandled animals occurs after either starting date of treatment (Figs. 7, 8). The restricted early period of mammalian pineal growth responses to environmental factors is likely to be due to the rapid postnatal attenuation of its mitotic activity (Quay & Levine, 1957). However, a restricted daily period of mitoses can be detected in adult rat pineals as well (Quay & Renzoni, 1966). Aside from this consideration of the pineal's difference in growth potential, recent studies of environmental effects on the pineal indicate (1) its dissociation from cerebro-cortical changes resulting from the stimulation of complex environments, and (2) a far from consistent or clear relationship with growth changes in other endocrines.

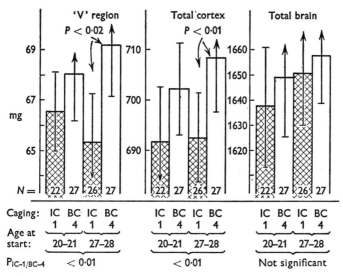

Fig. 6. Effect of isolation and lack of handling (IC-1) as compared with grouping four animals per cage and handling (BC-4) on post-weaning weight increases of brain regions in sibling male rats autopsied at 89–91 days of age. It is seen that the cerebro-cortical response is equivalent in animals weaned and started at 27–28 days of age as compared with siblings weaned and started at 20–21 days of age. Tops of bars correspond to group means; vertical lines extend × 2 the standard error of the mean on each side of the means N, Number of individuals; P, probability from t based on pair-wise analysis of differences between siblings.

CONCLUSIONS

1. The pineal complex of organs is active in vertebrates from cyclostomes to mammals.

2. Its major structural and physiological attributes are those of a photoreceptor in lower groups and progressively those of an endocrine organ in higher groups.

3. Throughout the vertebrate series pineal peripheral physiologic effects,

be they in chromatophore, reproductive or other systems, are often variable according to age, species, photoperiod, season and other environmental conditions, but are usually related to environmental illumination.

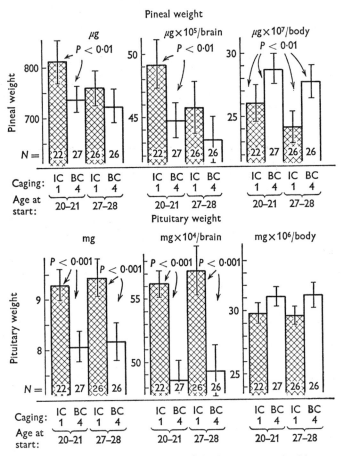

Fig. 7. Effect of isolation and lack of handling (IC-1) as compared with grouping four animals per cage and handling (BC-4) on pineal and pituitary weights in sibling male rats autopsied at 89–91 days of age. In contrast with pituitary weight, pineal weight is increased only when animals are weaned and isolated at the earlier age, 20–21 days as compared with 27–28 days. N, number of individuals; P, probability from t based on pair-wise analysis of differences between siblings.

4. Explanation of the varied effects of the pineal on peripheral systems is most readily provided by evidence relating the pineal's primary action to homeostatic mechanisms within the central nervous system.

5. Brain chemistry and homeostatic limitations in pinealectomized rats suggest that the astrocytic neuroglial system may be the primary central recipient of pineal regulation.

6. Endogenous physiological rhythms and species-specific responses to environmental photoperiods are likely to stem largely from the central nervous system, with the pineal's peripherally detectable contribution, at least usually, being secondary and intermittent in higher vertebrates.

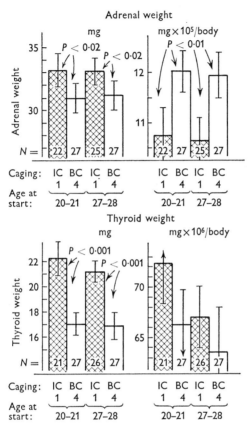

Fig. 8. Effect of isolation and lack of handling (IC-1) as compared with grouping four animals per cage and handling (BC-4) on adrenal and thyroid weights in sibling male rats autopsied at 89–91 days of age. Hypertrophy of these endocrines in the isolated animals of the two groups follows more closely the pattern of the pituitary than the pineal (Fig. 7) but differs proportionately when related to body weight. N, Number of individuals; P, probability from t based on pair-wise analysis of differences between siblings.

7. Differences in the timing of photoperiod-induced circadian phase-shifting in pinealectomized rats provides the first evidence of a general mechanism by which both daily and seasonal, species-specific, photoperiod effects may be modified according to different patterns by removal of the pineal.

8. Pineal postnatal development in mammals can be influenced by light

and other environmental factors, and the occurrence and timing of such pineal growth responses are differentiated from those of other studied endocrines and brain regions.

I am grateful to Emily Reid, Helen Sherry, Susan Evans, Marie Hebert, Hiromi Morimoto, Victor Gin, Peter Witte, Joe Wong and William Young, in addition to others mentioned earlier, for office, laboratory and animal-room aid. Some phases of these studies were supported in part by research grants (GM-05219, NB-06296) from the National Institutes of Health, U.S. Public Health Service.

REFERENCES

ADLER, K. (1969). Extraoptic phase shifting of circadian locomotor rhythm in salamanders. *Science, N.Y.* **164**, 1290–2.

ARRINGTON, L. C. (1967). Studies of the pineal function in Japanese quail (*Coturnix coturnix japonica.*) *Diss. Abstr.* **27**, 3253 B–3254 B.

BAGNARA, J. T. (1960). Pineal regulation of the body lightening reaction in amphibian larvae. *Science, N.Y.* **132**, 1481–3.

BAGNARA, J. T. (1963). The pineal and the body lightening reaction of larval amphibians. *Gen comp. Endocr.* **3**, 86–100.

BAGNARA, J. T. (1964). Independent actions of pineal and hypophysis in the regulation of chromatophores of anuran larvae. *Gen. comp. Endocr.* **4**, 299–303.

BISCHOFF, M. B. (1967). Ultrastructural evidence for secretory and photoreceptor functions in the avian pineal organ. *J. Cell Biol.* **35**, 13 A.

BREDER, C. M., JR & RASQUIN, P. (1947). Comparative studies in the light sensitivity of blind characins from a series of Mexican caves. *Bull. Am. Mus. nat. Hist.* **89**, 325–51.

CHARLTON, H. M. (1968). The pineal gland of *Xenopus laevis*, Daudin: a histological, histochemical, and electron microscopic study. *Gen. comp. Endocr.* **11**, 465–80.

COLLIN, J.-P. (1966a). Etude préliminaire des photorécepteurs rudimentaires de l'epiphyse de *Pica pica* L. pendant la vie embryonnaire et post-embryonnaire. *C. r. hebd. Séanc. Acad. Sci., Paris* D, **263**, 660–3.

COLLIN, J.-P. (1966b). Sur l'évolution des photorécepteurs rudimentaires épiphysaires chez la Pie (*Pica pica* L.). *C. r. Séanc. Soc. Biol.* **160**, 1876–80.

COLLIN, J.-P. (1967a). Structure, nature sécretoire, degenerescence partielle des photorecepteurs rudimentaires epiphysaires chez *Lacerta viridis* (Laurenti). *C. r. hebd. Séanc. Acad. Sci., Paris* D, **264**, 647–50.

COLLIN, J.-P. (1967b). Recherches préliminaires sur les propriétes histochimiques de l'epiphyse de quelques Lacertiliens. *C. r. hebd. Séanc. Acad. Sci., Paris* D, **265**, 1827–30.

COLLIN, J.-P. (1968a). Rubans circonscrits par les vésicules dans les photorécepteurs rudimentaires épiphysaires de l'Oiseau *Vanellus vanellus* (L) et nouvelles considérations phylogénétiques relatives aux pinéalocytes des Mammifères. *C. r. hebd. Séanc. Acad. Sci., Paris* D, **267**, 758–60.

COLLIN, J.-P. (1968b). L'épithélium sensoriel de l'organe pinéal de la larve âgée et de l'adulte de *Lampetra planeri*. *C. r. hebd. Séanc. Acad. Sci., Paris* D, **267**, 1768–71.

COLLIN, J.-P. & MEINIEL, A. (1968). Contribution à la connaissance des structures synaptiques du type ruban dans l'organe pinéal des vertébrés. Etude particulière en microscopic électronique des connexions de l'innervation efférente chez l'ammocète de lamproie de Planer. *Archs Anat. microsc. Morph. exp.* **57**, 275–96.

DISCLOS, P. (1964). Epiphysectomie chez le têtard d'Alytes. *C. r. hebd. Séanc. Acad. Sci. Paris* **258**, 3101–3.

DODT, E. (1963). Photosensitivity of the pineal organ in the teleost, *Salmos irideus* (Gibbons). *Experientia* **19**, 642–3.

DODT, E. & HEERD, E. (1962). Mode of action of pineal nerve fibers in frogs. *J. Neurophysiol.* **25**, 405–29.

DODT, E. & JACOBSON, M. (1963). Photosensitivity of a localized region of the frog diencephalon. *J. Neurophysiol.* **26**, 752–8.

DODT, E. & MORITA, Y. (1967). Conduction of nerve impulses within the pineal system of frog. *Pflügers Arch. ges. Physiol.* **293**, 184–92.

EAKIN, R. M. (1961). Photoreceptors in the amphibian frontal organ. *Proc. natn. Acad. Sci. U.S.A.* **47**, 1084–8.

EAKIN, R. M., QUAY, W. B. & WESTFALL, J. (1961). Cytochemical and cytological studies of the parietal eye of the lizard, *Sceloporus occidentalis. Z. Zellforsch. mikrosk. Anat.* **53**, 449–70.

EAKIN, R. M., QUAY, W. B. & WESTFALL, J. (1963). Cytological and cytochemical studies on the frontal and pineal organs of the treefrog, *Hyla regilla. Z. Zellforsch. mikrosk. Anat.* **59**, 663–83.

EAKIN, R. M., STEBBINS, R. C. & WILHOFT, D. C. (1959). Effects of parietalectomy and sustained temperature on thyroid of lizard, *Sceloporus occidentalis. Proc. Soc. exp. Biol. Med.* **101**, 162–4.

EAKIN, R. M. & WESTFALL, J. (1959). Fine structure of the retina in the reptilian third eye. *J. biophys. biochem. Cytol.* **6**, 133–4.

EDDY, J. M. P. & STRAHAN, R. (1968). The role of the pineal complex in the pigmentary effector system of the lampreys, *Mordacia mordax* (Richardson) and *Goetria australis* (Gray). *Gen. comp. Endocr.* **11**, 528–34.

FENWICK, J. C. (1969). The functions of the fish pineal organ. Ph.D. thesis, University of British Columbia,

FISKE, V. M., BRYANT, G. K. & PUTNAM, J. (1960). Effect of light on the weight of the pineal in the rat. *Endocrinology* **66**, 489–91.

FISKE, V. M., POUND, J. & PUTNAM, J. (1962). Effect of light on the weight of the pineal organ in hypophysectomized, gonadectomized, adrenalectomized or thiouracil-fed rats. *Endocrinology* **71**, 130–3.

FRISCH, K. VON (1911*a*). Beiträge zur Physiologie der Pigmentzellen in der Fischhaut. *Arch. ges. Physiol.* **138**, 319–87.

FRISCH, K. VON (1911*b*). Über das Parietalorgan der Fische als funktionierendes Organ. *Sber. Ges. Morph. Physiol. Münch.* **29**, 16–18.

FUJIE, E. (1968). Ultrastructure of the pineal body of the domestic chicken, with special reference to the changes induced by altered photoperiods. *Archvm histol. jap.* **29**, 271–303.

GASTON, S. & MENAKER, M. (1968). Pineal function: the biological clock in the sparrow. *Science, N.Y.* **160**, 1125–7.

GLASER, R. (1958). Increase in locomotor activity following shielding of the parietal eye in night lizards. *Science, N.Y.* **128**, 1577–8.

GROVE, D. J. (1969). The effects of adrenergic drugs on melanophores of the minnow, *Phoxinus phoxinus* (L.). *Comp. biochem. Physiol.* **28**, 37–54.

GRUBERG, E. R., HEATH, J. E. & NORTHCUTT, R. G. (1968). Photosensitivity of the parietal eye of the lizard, *Phrynosoma cornutum. Proc. Int. Union Physiol. Sci.* **7**, 171.

440 W. B. QUAY

HERBERT, J. (1969). The pineal gland and light-induced oestrus in ferrets. *J. Endocr.* **43**, 625–36.

HOMMA, K., McFARLAND, L. Z. & WILSON, W. O. (1967). Response of the reproductive organs of the Japanese quail to pinealectomy and melatonin injections. *Poultry Sci.* **46**, 314–19.

JUSZKIEWICZ, T. & RAKALSKA, Z. (1965). Lack of the effect of melatonin on the frog spermatogenic reaction. *J. pharm. Pharmac.* **17**, 189–90.

KASBOHM, P. (1967). Der Einfluss des Lichtes auf die Temperaturadaptation bei *Rana temporaria*. *Helgoländer wiss. Meeresunters.* **16**, 157–78.

KELLY, D. E. & SMITH, S. W. (1964). Fine structure of the pineal organs of the adult frog, *Rana pipiens*. *J. Cell Biol.* **22**, 653–74.

KITAY, J. I. & ALTSCHULE, M. D. (1954). *The Pineal Gland.* Harvard University Press.

KRECH, D., ROSENZWEIG, M. R. & BENNETT, E. L. (1966). Environmental impoverishment, social isolation and changes in brain chemistry and anatomy. *Physiol. Behav.* **1**, 99–104.

LIERSE, W. (1965). Elektronenmikroskopische Untersuchungen zur cytologie und angiologie des epiphysensteils von *Anolis carolinensis*. *Z. Zellforsch. mikrosk. Anat.* **65**, 397–408.

LUTZ, H. & COLLIN, J.-P. (1967). Sur la regression des cellules photo-receptrices épiphysaires chez la tortue terrestre: *Testudo hermanni* (Gmelin) et la phylogénie des photorécepteurs épiphysaires chez les vertébrés. *Bull. Soc. Zool. France* **92**, 797–801.

MENAKER, M. (1968 a). Extraretinal light perception in the sparrow. I. Entrainment of the biological clock. *Proc. natn. Acad. Sci. U.S.A.* **59**, 414–21.

MENAKER, M. (1968 b). Light perception by extra-retinal receptors in the brain of the sparrow. *Proc. Ann. Mtg Am. Psychol. Assoc.* no. 76, p. 299.

MENAKER, M. & KEATTS, H. (1968). Extraretinal light perception in the sparrow. II. Photoperiodic stimulation of testis growth. *Proc. natn. Acad. Sci. U.S.A.* **60**, 146–51.

MESS, B. (1968). Endocrine and neurochemical aspects of pineal function. *Int. Rev. Neurobiol.* **11**, 171–98.

MILLER, W. H. & WOLBARSHT, M. L. (1962). Neural activity in the parietal eye of a lizard. *Science, N.Y.* **135**, 316.

MORITA, Y. (1965). Extra- und intracelluläre Abteilungen einzelner Elemente des lichtempfindlichen Zwischenhirns anurer Amphibien. *Pflügers Arch. ges. Physiol.* **286**, 97–108.

MORITA, Y. (1966 a). Entladungsmuster pinealer Neurone der Regenbogenforelle (*Salmo irideus*) bei Belichtung des Zwischenhirns. *Pflügers Arch. ges. Physiol.* **289**, 155–67.

MORITA, Y. (1966 b). Absence of electrical activity of the pigeon's pineal organ in response to light. *Experientia* **22**, 402.

DE LA MOTTE, I. (1963). Untersuchungen zur vergleichenden Physiologie der Lichtempfindlichkeit geblendeter Fische. *Naturwissenschaften* **50**, 363.

MROSOVSKY, N. & TRESS, K. H. (1966). Plasticity of reactions to light in frogs and a possible role for the pineal eye. *Nature, Lond.* **210**, 1174–5.

O'CONNOR, J. M. (1969). Effect of melatonin on *in vitro* ovulation of frog oocytes. *Am. Zool.* **9**, 577.

OKSCHE, A. & KIRSCHSTEIN, H. (1966). Zur Frage der Sinneszellen im Pinealorgan der Reptilien. *Naturwissenschaften* **53**, 46.

OKSCHE, A. & KIRSCHSTEIN, H. (1968). Unterschiedlicher elektronenmikroskopischer Feinbau der Sinneszellen im Parietalauge und Pinealorgan (Epiphysis cerebri) der Lacertilia. Ein Beitrag zum Epiphysenproblem. *Z. Zellforsch. mikrosk. Anat.* **87**, 159–92.

OKSCHE, A. & VAUPEL-VON HARNACK, M. (1962). Elektronenmikroskopische Untersuchungen am Stirnorgan (Frontalorgan, Epiphysen endblase) von *Rana temporaria* und *Rana esculenta*. *Naturwissenschaften* **49**, 429–30.

OKSCHE, A. & VAUPEL-VON HARNACK, M. (1965). Über rudimentäre Sinneszellstrukturen im Pinealorgan des Hühnchens. *Naturwissenschaften* **52**, 662.

OKSCHE, A. & VAUPEL-VON HARNACK, M. (1966). Elektronenmikroskopische Untersuchungen zur Frage der Sinneszellen im Pinealorgan der Vögel. *Z. Zellforsch. mikrosk. Anat.* **69**, 41–60.

OSHIMA, K. & GORBMAN, A. (1969). Pars intermedia: unitary electrical activity regulated by light. *Science, N.Y.* **163**, 195–7.

PETER, R. E. (1968). Failure to detect an effect of pinealectomy in goldfish. *Gen. comp. Endocr.* **10**, 443–9.

PETIT, A. (1968). Ultrastructure de la rétine de l'œil pariétal d'un Lacertilien, *Anguis fragilis*. *Z. Zellforsch. mikrosk. Anat.* **92**, 70–93.

PETIT, A. (1969). Ultrastructure, innervation et fonction de l'épiphyse de l'orvet (*Anguis fragilis* L.). *Z. Zellforsch. mikrosk. Anat.* **96**, 437–65.

PFLUGFELDER, O. (1964). Wirkungen lokaler hirnläsionen auf hypophyse und thyreoidea von *Carassius gibelio auratus*. *Wilhelm. Roux Arch. Entw. Mech. Org.* **155**, 535–48.

PFLUGFELDER, O. (1966). Heterotopien vom Schilddrüsengewebe nach Störung des Hormonhaushalts. *Endokrinologie* **49**, 87–98.

PFLUGFELDER, O. (1967). Weitere Untersuchungen über Kypho-lordose und Scoliose nach Zerstörung der Epiphysenregion bei Fischen und Haushühnern. *Wilhelm. Roux Arch. Entw. Mech. Org.* **158**, 170–87.

PICKFORD, G. E. & ATZ, J. W. (1957). *The Physiology of the Pituitary Gland of Fishes*. *N.Y. Zool. Soc.*

QUAY, W. B. (1956). Volumetric and cytologic variation in the pineal body of *Peromyscus leucopus* (Rodentia) with respect to sex, captivity and day-length. *J. Morph.* **98**, 471–95.

QUAY, W. B. (1962). Metabolic and cytologic evidence of pineal inhibition by continuous light. *Am. Zool.* **2**, 550.

QUAY, W. B. (1963a). Cytologic and metabolic parameters of pineal inhibition by continuous light in the rat (*Rattus norvegicus*). *Z. Zellforsch. mikrosk. Anat.* **60**, 479–90.

QUAY, W. B. (1963b). Circadian rhythm in rat pineal serotonin and its modifications by estrous cycle and photoperiod. *Gen. comp. Endocr.* **3**, 473–9.

QUAY, W. B. (1964a). Circadian and estrous rhythms in pineal melatonin and 5-hydroxy indole-3-acetic acid. *Proc. Soc. exp. Biol. Med.* **115**, 710–13.

QUAY, W. B. (1964b). Circadian and estrous rhythms in pineal and brain serotonin. *Prog. Brain Res.* **8**, 61–3.

QUAY, W. B. (1965a). Experimental evidence for pineal participation in homeostasis of brain composition. *Prog. Brain Res.* **10**, 646–53.

QUAY, W. B. (1965b). Retinal and pineal hydroxyindole-O-methyl transferase activity in vertebrates. *Life Sci.* **4**, 983–91.

QUAY, W. B. (1965c). Photic relations and experimental dissociation of circadian rhythms in pineal composition and running activity in rats. *Photochem. Photobiol.* **4**, 425–32.

QUAY, W. B. (1965d). Indole derivatives of pineal and related neural and retinal tissues. *Pharmac. Rev.* **17**, 321–45.

QUAY, W. B. (1966). 24-Hour rhythms in pineal 5-hydroxytryptamine and hydroxyindole-O-methyl transferase activity in the macaque. *Proc. Soc. exp. Biol. Med.* **121**, 946–8.

QUAY, W. B. (1967). The significance of darkness and monoamine oxidase in the nocturnal changes in 5-hydroxytryptamine and hydroxyindole-O-methyl-transferase activity of the macaque's epiphysis cerebri. *Brain Res.* **3**, 277–86.

QUAY, W. B. (1968a). Individuation and lack of pineal effect in the rat's circadian locomotor rhythm. *Physiol. Behav.* **3**, 109–18.

QUAY, W. B. (1968b). Comparative physiology of serotonin and melatonin. *Adv. Pharmac.* **6**, A, 283–97.

QUAY, W. B. (1968c). Melatonin as a retinal neurohumor mediating light-induced cone contraction in the larval amphibian. *Proc. Int. Union Physiol. Sci.* **7**, 357.

QUAY, W. B. (1969a). The role of the pineal gland in environmental adaptation. In *Physiology and Pathology of Adaptation Mechanisms: Neural–Neuroendocrine–Humoral*, pp. 508–50 (ed. E. Bajusz). Oxford: Pergamon Press.

QUAY, W. B. (1969b). Evidence for a pineal contribution in the regulation of vertebrate reproductive systems. *Gen. comp. Endocr.* (Supp.) **2**, 101–10.

QUAY, W. B., BENNETT, E. L., ROSENZWEIG, M. R. & KRECH, D. (1969). Effects of isolation and environmental complexity on brain and pineal organ. *Physiol. Behav.* **4**, 489–94.

QUAY, W. B. & LEVINE, B. E. (1957). Pineal growth and mitotic activity in the rat and the effects of colchicine and sex hormones. *Anat. Rec.* **129**, 65–78.

QUAY, W. B. & RENZONI, A. (1963). Comparative and experimental studies of pineal structure and cytology in passeriform birds. *Riv. Biol.* **56**, 363–407.

QUAY, W. B. & RENZONI, A. (1966). Twenty-four hour rhythms in pineal mitotic activity and nuclear and nucleolar dimensions. *Growth* **30**, 315–24.

QUAY, W. B., RENZONI, A. & EAKIN, R. M. (1968). Pineal ultrastructure in *Melopsittacus undulatus* with particular regard to cell types and functions. *Riv. Biol.* **61**, 371–93.

RALPH, C. L. & DAWSON, D. C. (1968). Failure of the pineal body of two species of birds (*Coturnix coturnix japonica* and *Passer domesticus*) to show electrical responses to illumination. *Experientia* **24**, 147–8.

RASQUIN, P. (1958). Studies in the control of pigment cells and light reactions in recent teleost fishes. 1: Morphology of the pineal region. 2: Reactions of the pigmentary system to hormonal stimulation. *Bull. Am. Mus. nat. Hist.* **115**, 1–68.

REED, B. L. (1968). The control of circadian pigment changes in the pencil fish: a proposed role for melatonin. *Life Sci.* **7**, 961–73.

REED, B. L., FINNIN, B. C. & RUFFIN, N. E. (1969). The effects of melatonin and epinephrine on the melanophores of freshwater teleosts. *Life Sci.* **8**, 113–20.

REITER, R. J. (1968a). Morphological studies on the reproductive organs of blinded male hamsters and the effects of pinealectomy or superior cervical ganglionectomy. *Anat. Rec.* **160**, 13–24.

REITER, R. J. (1968b). Changes in the reproductive organs of cold-exposed and light-deprived female hamsters (*Mesocricetus auratus*). *J. Reprod. Fert.* **16**, 217–22.

REITER, R. J. (1968c). The pineal gland and gonadal development in male rats and hamsters. *Fert. Ster.* **19**, 1009–17.

REITER, R. J., HOFFMAN, R. A. & HESTER, R. J. (1966). The role of the pineal gland and of environmental lighting in the regulation of the endocrine and reproductive systems of rodents. *Edgewood Arsenal Tech. Report* no. 4032. Edgewood Arsenal, Maryland.

RENZONI, A. (1967). La fisiologia dell' epifisi negli uccelli. 1. Pinealectomia in *Coturnix coturnix japonica*. *Boll. Soc. ital. Biol. sper.* **43**, 585–8.

ROSENZWEIG, M. R., KRECH, D., BENNETT, E. L. & DIAMOND, M. C. (1962). Effects of environmental complexity and training on brain chemistry and anatomy. *J. comp. physiol. Psychol.* **55**, 429–37.

ROSENZWEIG, M. R., KRECH, D., BENNETT, E. L. & DIAMOND, M. C. (1968). Modifying brain chemistry and anatomy by enrichment or impoverishment of experience. In *Early Experience and Behaviour*, pp. 258–98 (eds. G. Newton and S. Levine). Springfield, Ill.: C. C. Thomas.

RÜDEBERG, C. (1968). Receptor cells in the pineal organ of the dogfish *Scyliorhinus canicula* Linné. *Z. Zellforsch. mikrosk. Anat.* **85**, 521–6.

RÜDEBERG, C. (1969). Light and electron microscopic studies on the pineal organ of the dogfish, *Scyliorhinus canicula* L. *Z. Zellforsch. mikrosk. Anat.* **96**, 548–81.

SAYLER, A. & WOLFSON, A. (1967). Avian pineal gland: progonadotropic response in the Japanese quail. *Science, N.Y.* **158**, 1478–9.

SAYLER, A. & WOLFSON, A. (1968). Influence of the pineal gland on gonadal maturation in the Japanese quail. *Endocrinology* **83**, 1237–46.

SCHARRER, E. (1928). Untersuchungen über das Zwischenhirn der Fische. I. Die Lichtempfindlichkeit blinder Elritzen. *Z. vergl. Physiol.* **7**, 1–38.

SCHÖNHERR, J. (1955). Über die Abhängigkeit der Instinkthandlungen vom Vorderhirn und Zwischenhirn (Epiphyse) bei *Gasterosteus aculeatus* L. *Zool. Jb.* Abt. 3, **65**, 357–86.

SHELLABARGER, C. J. (1952). Pinealectomy vs. pineal injection in the young cockerel *Endocrinology* **51**, 152–4.

STALSBERG, H. (1965). Effects of extirpation of the epiphysis cerebri in 6-day chick embryos. *Acta endocr., Copnh.* (Suppl.) **97**, 1–119.

STEBBINS, R. C. (1960). Effects of pinealectomy in the western fence lizard *Sceloporus occidentalis*. *Copeia* no. 4, pp. 276–83.

STEBBINS, R. C. (1963). Activity changes in the striped plateau lizard with evidence on influence of the parietal eye. *Copeia* no. 4, pp. 681–91.

STEBBINS, R. C. & EAKIN, R. M. (1958). The role of the 'third eye' in reptilian behavior. *Am. Mus. Novit.* no. 1870, pp. 1–40.

STEBBINS, R. C. & WILHOFT, D. C. (1966). Influence of the parietal eye on activity in lizards. In *The Galapagos; Proceedings of the Symposia of the Galapagos International Scientific Project*, pp. 258–68. (ed. R. I. Bowman). University of California Press.

STEYN, W. (1959). Ultrastructure of pineal eye sensory cells. *Nature, Lond.* **183**, 764–5.

STEYN, W. (1960). Electron microscopic observations on the epiphysial sensory cells in lizards and the pineal sensory cell problem. *Z. Zellforsch. mikrosk. Anat.* **51**, 735–47.

THIEBOLT, L. & LE BARS, H. (1955). *La glande pinéale ou épiphyse anatomie–histologie–physiologie–clinique*. Paris: Librairie Maloine S.A.

UECK, M. (1968). Ultrastruktur des pinealen Sinnesapparates bei einigen Pipidae und Discoglossidae. *Z. Zellforsch. mikrosk. Anat.* **92**, 452–76.

VIVIEN, J. H. (1964). Ultrastructure des constituants de l'épiphyse de *Tropidonotus natrix* L. *C. r. hebd. Séanc. Acad. Sci., Paris* **258**, 3370–2.

VIVIEN, J. H. & ROELS, B. (1967). Ultrastructure de l'épiphyse des Cheloniens. Présence d'un 'paraboloide' et de structures de type photorécepteur dans l'épithélium sécrétoire de *Pseudomys scripta elegans*. *C. r. hebd. Séanc. Acad. Sci., Paris* **264**, 1743–6.

WARTENBERG, H. & BAUMGARTEN, H. G. (1968). Elekronenmikroskopische Untersuchungen zur Frage der photosensorischen und sekretorischen Funktion des Pinealorgans von *Lacerta viridis* und *L. muralis*. *Z. Anat. Entw. Gesch.* **127**, 99–120.

WURTMAN, R. J., AXELROD, J. & KELLY, D. E. (1968). *The Pineal*. New York and London: Academic Press.

YOUNG, J. Z. (1935). The photoreceptors of lampreys. II. The functions of the pineal complex. *J. exp. Biol.* **12**, 254–70.

ZADURA, J., ROSZKOWSKI, J. & CAKALA, S. (1969). The influence of pinealectomy on the weight and histologic changes in the testes and adrenals of roosters. *Acta physiol. polon.* **20**, 117–24.

DISCUSSION

DODD: Dr Quay, in the early part of your paper you mentioned the work of Eddy and Strahan on the Australian genera of cyclostomes *Mordacia* and *Geotria*. Miss Eddy is now working with us in Bangor and she has three recent results that I think are very interesting, and perhaps important. The first is that if you take out the pineal in the ammocoete larva in the last year of its larval life it fails to metamorphose. Secondly, if you take out the pineal of *Lampetra fluviatius* at the start of its sexual maturation the males at any rate are ripe 6 weeks earlier than the controls. The third is that using *Xenopus* larvae a week after hatching, she can make extracts of both the ammocoete and the adult and show that they contain a melanophore contracting substance, possibly melatonin, but of course, we don't know for certain. These are, of course, results in the oldest group of vertebrates. It looks as though there might have been a pineal–pituitary axis perhaps even in them.

QUAY: I am very interested to hear these results and I hope she continues her interesting work.

VAN OORDT: Dr Quay, you have given us a wealth of information upon the significance of the pineal. You have told us about the effects of melatonin in non-mammalian vertebrates. To be certain that these effects of melatonin are concerned with the function of the pineal, one has to ascertain that the substance is exclusively formed within this organ. In this respect, it seems of interest to mention the recent work of van der Veerdonk in Utrecht. He succeeded in finding such an exclusive production of melatonin in the pineal of several non-mammalian vertebrates, i.e. the frog and the chicken. The methods he developed to accomplish this involved extraction of melatonin itself and the testing of its effect on the melanophores of *Xenopus* larvae.

QUAY: I am glad to hear of the recent work from Utrecht on this. There is perhaps a matter of minor disagreement. I myself feel as we go progressively lower in the vertebrates we find at least the melatonin-forming enzyme more broadly distributed in tissues. Actually in amphibians one finds more of the enzyme activity for making melatonin, acetylserotonin methyl transferase in the retina than in the pineal itself, which is rather a modest organ in the amphibian. In trout and other fishes the enzyme is even more widely distributed. This does not necessarily mean that the retina is an endocrine organ, secreting melatonin. At least we should keep in mind that the localization of the capacity to form melatonin is not necessarily restricted as we go down to lower vertebrates. I also refrain from saying much about melatonin because I am not really sure myself that melatonin is a hormone in a great many species. We must remember that the so-called biogenic amines, adrenaline and noradrenaline, serotonin and melatonin, are broad in distribution and vary markedly as to what they might do at the local tissue level.

In an evolutionary sense it appears that these things are more often local tissue humours and transmitters than hormones in the usual sense. It is true that the pineal contains proteins that are pineal specific and there are molecules of this nature which I think might be more likely contenders for the role as pineal hormones. There are some groups of workers in Europe, for example, and other places who feel that it is not melatonin or such small amines that are pineal hormones affecting eventually the reproductive organs, but it is indeed the pineal protein fractions which contain some thing or some things that are better contenders for roles as pineal hormones. However, I think that the most basic question is what happens when you remove the pineal. Hence my talking about this, and excluding some of the very interesting work that is being done on melatonin and on some of its effects, and on the effects of its congeners—such as the work by Wurtman, Martini, van der Veerdonk and others.

HERBERT: I was interested to see that rats which had been pinealecto-mized could alter their running rhythms in response to a change in environmental illumination. This surprised me because I thought perhaps this would not happen. But did you look at any other parameters—for example, the time of ovulation or the time of onset of oestrous behaviour?

QUAY: Over the years I have looked at many things and tried to reproduce many different experiments. I find that in rats reproductive effects of pinealectomy are not readily reproduced. A lot depends probably on the photoperiod conditions, the dietary conditions and seasonal conditions which are rarely reported in published literature. A vast array of publications is available showing, or not showing, or contradicting, pineal effects on reproductive or other systems. It is very hard to evaluate much of this since we don't know about the ambient conditions of photoperiod, how the animals were raised, etc., and I can only say that much goes unsaid.

THE NEURO-ENDOCRINE CONTROL OF WATER METABOLISM

By H. HELLER

The central regulation of body water loss has been intensively investigated in mammals and man and there is also a respectable body of evidence for its neural control in non-mammalian vertebrates. However, although the adjustment of water intake appears to be as accurately regulated as water excretion, much less attention has been paid to the regulatory mechanisms concerned with water intake. This certainly applies to mammals, and the evidence for 'lower' vertebrates is very scanty indeed.

I would like, therefore, to start this discussion with a short outline of recent inquiries into mechanisms concerned in the phenomena of drinking and thirst in mammals with the hope of providing a stimulus for the extension of such studies to non-mammalian species.

REGULATION OF WATER INTAKE

Central, blood-mediated stimuli

Many theories have been formulated to explain thirst and I can do no better than to reproduce a survey (Table 1) of them from the book by Wolf (1958). In this table no clear distinction is made between stimuli arising from water loss and the mechanisms integrating these stimuli. However, there is now good experimental evidence that certain regions of the central nervous system are particularly involved in the latter. Andersson (1953) and Andersson & McCann (1955) showed that the injection of 3–10 μl. of a 2–3% sodium chloride solution into the anterior hypothalamus of unanaesthetized goats produced drinking of 2–8 l. water beginning 30–60 s after the injection and lasting 2–5 m. In several experiments this effect could be repeated 3 or 4 times with a $\frac{1}{2}$ h interval. The most obvious effects were obtained in the region near the paraventricular nucleus. Injection of iso- or hypotonic saline had no effect (for experiments in dogs see Andersson & McCann, 1956).

Similar effects in goats were obtained by Andersson & McCann (1955) by electrical stimulation of a somewhat wider area of the same region. Drinking began 10–30 s after the onset of stimulation and ceased 2–3 s after discontinuing the current. These results, which have been confirmed and extended to other mammalian species (rat, Greer, 1955; dog, Andersson

[447]

& McCann, 1956) suggest the existence of a 'thirst centre' in the anterior hypothalamus responsive to osmotic stimuli. It would be logical to seek the osmoreceptors in the area responsive to the injection of hypertonic saline, since the wider area activated electrically may include fibres leading to the 'osmoreceptive centre'.

Table 1. *Theories of thirst: classification*

A. Instinctive desire (Magendie)

B. Anhydremia
    1. Acting on mouth and throat (Beaumont, Ludwig, Colin, Luciani)
    2. Acting on brain (Dumas, Müller)
    3. Acting generally (Schiff)

C. Peripheral (local) origin
    1. 'Dry mouth', etc. (Hippocrates, Tancredi, Haller, Deneufbourg, Luciani, Valenti, Cannon, Gregersen, Holmes)
    2. Oesophageal contraction (Müller)
    3. Heart and lungs (Galen)
    4. Viscera (Aristotle)

D. General origin
    1. General sensation
        (*a*) Depletion of water (Schiff)
        (*b*) Cellular dehydration (E. Bernard, Kerpel-Fronius, Gilman, Dill)
        (*c*) Diffuse (Wettendorff, Luciani, Carlson)
    2. General need of water (Deneufbourg, C. Bernard, Longet, Meigs)

E. Elevated osmotic pressure of blood (Mayer, Leschke, Nonnenbruch)

F. Osmometric
    1. Cellular dehydration (Wettendorff, Gilman, Dill)
    2. Central osmoreceptor (Wolf, Andersson)

G. Central
    1. Blood concentration (Dumas, Mayer, Leschke, Müller)
    2. Osmoreceptor (Wolf, Andersson)
    3. Association or conditioned reflex (Schiff, Wettendorff, Kourilsky, Wolf)
    4. Thirst hormones (Linazasoro, Adolph, Gilbert)
    5. Miscellaneous (Nothnagel, Voit, Bellows, Oehme, Kourilsky)

H. Multiple factor (Bellows, Adolph, Stellar)

From A. V. Wolf, 1958.

However, the osmolarity of the blood is clearly not the only type of thirst stimulus operating: hypovolaemia, i.e. decrease of the circulating plasmo-space in which neither the osmotic pressure of the body fluids nor the volume of the cellular space has been changed, seems to be of similar importance to cellular shrinkage (see e.g. Fitzsimons, 1961; Stricker & Wolf, 1967). This stimulus has long been thought to act directly on volume receptors in the hypothalamus but quite recent work by Fitzsimons (1969) shows that an indirect effect may also be involved. He found that rats in normal fluid balance started to drink water $\frac{1}{2}$–1 h after complete ligation of the inferior vena cava either above or below the renal veins and continued drinking for an hour or two. Constriction of the aorta above the renal arteries or constriction of both renal arteries also caused drinking, oliguria and the development of a positive fluid balance. Caval ligation was relatively ineffective as a stimulus to drinking after bilateral nephrectomy but was effective in rats made anuric by ligation of the ureters. Constriction of the aorta below the renal arteries or after nephrectomy was ineffective as a stimulus to drinking.

These results pointed to a humoral mediator arising in the kidney (see also Gutman, Benzakein & Chaimovitz, 1967) and this assumption was supported by the fact that simple saline extracts of renal cortex—but not of renal medulla—had a dipsogenic effect. The renal dipsogen, if not identical with renin, resembles it so closely that it proved impossible to separate the two factors during a fractionation procedure used to make renin. It could also be shown that renin itself stimulated drinking in normal rats as effectively as the intravenous injection of a hypertonic sodium chloride solution. A similar suggestion, namely that angiotensin acts as a stimulus for drinking, has been made by Fitzsimons & Simons (1969).

Fitzsimons points out that there are at least three ways in which the renin-angiotensin system could play a role in drinking: (a) activation of this system might increase the sensitivity of vascular stretch receptors in answer to a pre-existing hypovolaemia; (b) renin could act by causing increased capillary permeability and a fall in plasma volume. A vascular permeability factor has been found in the kidney (Asscher & Anson, 1963) which, according to Peart (1965), may be renin itself; (c) renin and/ or angiotensin may have a central action. The last possibility is supported by observations of Booth (1968) and Epstein, Fitzsimons & Simons (1968), who showed that small amounts of angiotensin (down to 5 ng = 1/250 the systemic dose needed to produce an effect on water uptake) placed in the medial preoptic and anterior hypothalamic regions caused rats to drink substantial amounts of water.

Mediation of the dipsogenic response through adrenocorticosteroids seems unlikely since in Fitzsimons' experiments adrenalectomy did not prevent drinking after caval ligation or injection of renal extracts. Moreover, aldosterone had no effect on water intake. However, against this view is a recent report by Gutman & Benzakein (1969) who studied the relation of the adrenals to the drinking response to hypovolaemia. They produced hypovolaemia by subcutaneous injection of 2 ml/100 g of polyethylene glycol (20% Carbowax, 20 M) in 0·9% sodium chloride solution, and found that the increased water uptake caused by the resulting iso-osmotic depletion of body fluid was significantly decreased after bilateral adrenalectomy.

According to Fitzsimons (1969) the mechanisms for the production of extracellular thirst may be as follows: a decrease in the extracellular fluid volume causes first a change in sensory information from stretch receptors in the low pressure side of the circulation; many of these receptors are in the atria and pulmonary vessels. This change in information, much of it carried in the vagi, directly activates the medial preoptic region which in turn stimulates the lateral hypothalamus and limbic structures to cause thirst and drinking. At the medial preoptic level this pathway seems to be distinct from the osmosensitive region, which appears to be more laterally situated in the hypothalamus (see also Andersson, Olsson & Warner, 1967). This pathway operates in the nephrectomized rat but when the kidneys are present the same afferent pathway leads also to a reflex release of renin (with the sympathetic nerves in the afferent arc, Vander & Luciano, 1967) which, after conversion to angiotensin, acts directly on the hypothalamus.

Involvement of central cholinergic and adrenergic mechanisms

In addition to these blood-mediated stimuli for drinking, i.e. changes in plasma osmotic pressure and activation of the stretch receptors, there is evidence (see e.g. Wolf, 1958, and Blass, 1968) that impulses for drinking can also be elicited from higher centres, including the cerebral cortex. It seems likely that—by analogy with the activation of supraoptic nuclear neurones subserving vasopressin release—cholinergic and adrenergic pathways may play a role in this mechanism. The existence of cholinergic circuits involved in the thirst drive has been convincingly demonstrated (Grossman, 1960; Fisher & Coury, 1962; Stein, 1963), whereas adrenergic pathways have been implicated primarily in the inhibition of drinking (Grossman, 1964a; Miller, 1965).

However, it has now been shown by Lehr and his co-workers (Lehr, Mallow & Krukowski, 1967) in the rat that the subcutaneous injection of the β-adrenergic compound isoprenaline invariably induced copious drinking which was not accompanied by an increase in urine volume. Water-loading before the injection did not abolish drinking and uncovered marked simultaneous antidiuresis. In contrast, rats injected with the α-adrenergic compound metaraminol, had increased urine flow 'in the absence of conspicuous attempts to recoup the water loss by drinking'. Pretreatment with the β-adrenergic blocker propranolol inhibited the isoprenaline-induced drinking and antidiuresis whereas α-adrenergic blockage with tolazoline caused isoprenaline-like effects, presumably due to β-adrenergic and/or cholinergic preponderance. Note also that in a study published before that of Lehr et al. (1967) Italian workers (Zamboni & Siro-Brigiani, 1966) had obtained very similar results, also in rats. They found that dopamine, noradrenaline and adrenaline inhibited thirst and increased diuresis while small doses of isoprenaline had a marked dipsogenic action accompanied by intense antidiuresis. Table 2 summarizes these results.

Table 2. *Effects of sympathomimetic compounds and their antagonists on spontaneous water intake (thirst) in the rat*

Name of compound	Type of action	Effect on water intake
Metaraminol	α-adrenergic	No effect
Isoprenaline	β-adrenergic	Copious drinking
Adrenaline	Both α and β	Inhibition of drinking
Tolazoline	α-adrenergic blocker	Increased water intake
Propranolol	β-adrenergic blocker	Inhibition of isoprenaline-induced drinking

Based on results of Lehr *et al.* (1967) and Zamboni & Siro-Brigiani (1966).

Reflex activation or inhibition of the hypothalamic 'thirst centre' by the contrasting haemodynamic effects of α- and β-adrenergic amines cannot be ruled out but—for example—the fact that tolazoline in large doses induced drinking without significant changes in blood pressure would not fit this interpretation. The evidence for the possible existence of β-adrenergic circuits involved in the thirst drive is much more appealing considering that (*a*) isoprenaline has been shown to stimulate drinking in rats made completely adipsic and unresponsive to the most intense osmotic stimuli by symmetrical lesions in the lateral hypothalamus, (*b*) Grossman (1964*b*) showed that the application of the α-adrenergic blocker dibenzyline to the central amygdala of the rat caused a pronounced increase in drinking and (*c*) as already mentioned, the significance of cholinergic mechanisms in the central nervous system in thirst has been well demonstrated.

Observations (*b*) and (*c*) suggest that we may be dealing with an interesting parallel to central mechanisms which regulate water loss (see below) since these findings indicate that adrenergic and cholinergic neurones from other parts of the brain act on the hypothalamic centre which also 'senses' osmotic and blood-borne chemical dipsogenic stimuli, possibly in analogy to the cholinergic innervation of the supraoptic nuclear neurones concerned with vasopressin release. In other words, one might postulate the type of cholinergic–adrenergic balanced control of fluid intake so familiar in the regulatory action of the autonomous nervous system on many peripheral organs. Fig. 1 attempts to incorporate this hypothesis into a general scheme of thirst-producing stimuli.

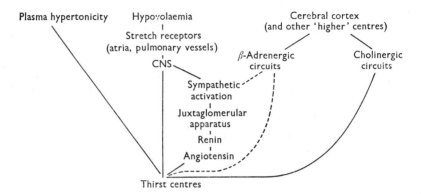

Fig. 1. Tentative scheme of thirst-producing stimuli (partly after Fitzsimons, 1970).

REGULATION OF WATER LOSS
Osmoreceptors

Like thirst, inhibition of urine flow by stimulation of vasopressin release can be elicited by electrical or osmotic activation of discrete hypothalamic sites. It is possible, therefore, that the same osmoreceptors subserve stimuli for increases in water intake and decreases in water loss, and this has in fact been postulated by Wolf in 1950. This concept was apparently supported by results of Andersson & McCann (1955), who found that both drinking and inhibition of water diuresis were evoked when the rostral portion of the 'drinking centre' was stimulated. However, against this identity is the observation of the same workers that stimulation of more caudal sites elicited drinking (though they may have only been stimulating afferent fibres converging on to the receptor area proper, see also Andersson et al. 1967). Moreover, in the same experiments stimulation within or adjacent to the paraventricular nucleus produced antidiuresis without drinking.

Hypovolaemia

Another observation suggesting that the control of water intake and water loss is functionally—though not necessarily anatomically—linked is the fact that hypovolaemia leads not only to increased drinking but also to increased renal water conservation. Ever since Ginsburg & Heller (1953) demonstrated in rats that the antidiuretic potency of blood rose progressively with the volume of blood removed, it has been accepted that iso-osmotic reduction of the blood volume by haemorrhage causes ADH (antidiuretic hormone) release (see also Ginsburg & Brown, 1956; de Wied, 1960; Weinstein, Berne & Sachs, 1960; Baratz & Ingraham, 1960 and Jahn, Stephen & Stahl, 1960). It has also been shown that more subtle procedures which alter blood volume (particularly in the thoracic circulation) or lead to the stimulation of atrial stretch receptors (for references see Ginsburg, 1968), may, dependent on the nature of the change, lead to increase or inhibition of antidiuretic hormone secretion. Receptors in the carotid sinuses seem also to be involved in the regulation of vasopressin secretion (see Perlmutt, 1963, and Share & Levy, 1966).

Since both hypovolaemic and osmotic stimuli are involved in the regulation of renal water loss by activating the hypothalamo–neurohypophysial system it might be asked whether, analogous to its role in the thirst mechanism, the renin-angiotensin system is also concerned in the control of water loss. The effect of systemically administered angiotensin on normal animals seems to vary from species to species. Intravenous infusion in man produces antidiuresis but in the rabbit, rat and dog, diuresis is the usual effect (for references see Peart, 1965). It has also been shown that both in man and in the cat (Barer, 1963) angiotensin causes a reduction in renal blood flow which suggests that even if antidiuretic, angiotensin acts by its effects on the renal vasculature and not by release of vasopressin. However, it would seem that the possibility of a release of vasopressin by a central (hypothalamic?) action of angiotensin on the neurohypophysis has, so far, not been investigated. That angiotensin may have *some* central effects is suggested by the reports of Bickerton & Buckley (1961) and Laverty (1963) that this peptide when injected into an internal carotid artery in doses which were ineffective intravenously, causes circulatory changes.

Cholinergic mechanisms

Continuing the comparison between dipsogenic and antidiuretic stimuli: what is the evidence that cholinergic and adrenergic neurones in the central nervous system are involved in controlling the release of the antidiuretic hormone? Participation of cholinergic neurones has, by now, been unequivocally demonstrated. Pickford (1947) has shown that acetylcholine injected into the caudal cells of a supraoptic nucleus produces marked antidiuresis. Even more convincingly, similar injections of anticholinesterases (physostigmine, di-isopropyl phosphorofluoridate) had the same effect (Pickford, 1947; Duke, Pickford & Watt, 1950), suggesting that endogenous acetylcholine is concerned. Moreover, true acetylcholinesterase has been found both in the region of the supraoptic and paraventricular nuclei (Abrahams, Koelle & Smart, 1957; Pepler & Pearse, 1957; Holmes, 1961; Lederis & Livingston, 1969) and in the neural lobe (Koelle & Geesey, 1961; Parmar, Sutter & Nickerson, 1961; Lederis & Livingston, 1969) of several mammalian species. Choline acetylase (choline acetyl transferase) was also found in both the hypothalamus and the neural lobe (Feldberg & Vogt, 1948; Fonnum, 1966; Lederis & Livingston, 1969).

The histochemical demonstration of the presence of cholinesterases in the neurohypophysis together with the occurrence of small subcellular organelles, resembling the 'synaptic vesicles' in peripheral autonomic nerves, led Koelle (1961) to propose the theory that the small neurohypophysial vesicles contain acetylcholine and are concerned in the intra-axonal release of the posterior pituitary hormones from their storage granules. This view of Koelle was supported by De Robertis and his colleagues (Gerschenfeld, Tramezzani & De Robertis, 1960; De Robertis, 1962). However, apart from the well-supported hypothesis of Douglas and his associates (for references see Ginsburg, 1968) that the arrival of the electric stimulus leads to depolarization and shifts of potassium which promote the entry of calcium into the neurosecretory terminals and causes hormone release without the mediation of a humoral transmitter, it would seem (Holmes, 1961) that in the neural lobe a positive histochemical reaction for cholinesterases can only be demonstrated around the blood vessels. Moreover, Lederis & Livingston (1968) found that, on density-gradient centrifugation, acetylcholine in rabbit neural lobe homogenates was differently distributed from vasopressin and oxytocin. In addition, experiments by three groups of workers (Douglas & Poisner, 1964; Dicker, 1966; Daniel & Lederis, 1967) who attempted to release hormones from isolated posterior pituitary lobes after the addition of acetylcholine were uniformly unsuccessful. Considering the distribution and quantities of acetylcholine and its co-ordinated enzymes in the neural lobe and the fact that the administration of acetylcholine has been shown to cause vaso-dilatation in the neurohypophysis (Konstantinova, 1967), it seems likely that the role of acetylcholine in the release of the neurohypophysial hor-mone is restricted to its release from cholinergic neurones in synapse with the neurosecretory fibres and that its presence in the neurohypophysis may indicate merely its participation in vascular events in this tissue.

Adrenergic mechanisms

At the present state of our knowledge it is difficult to obtain a clear picture of the involvement of adrenergic mechanisms in the regulation of water loss by the body for two main reasons. (1) It is frequently difficult to dis-tinguish between the central and peripheral effects of the sympathomime-tic agents which have been used experimentally. The permeability of the blood barrier to these substances has also to be considered in this context. (2) Such central effects which have been suggested to occur may or may not have been mediated by the hypothalamo–neurohypophysial system.

However, that adrenergic synapses with the neurosecretory fibres are concerned is very likely since it would seem the paraventricular and especially the supraoptic nuclei are richly innervated by aminergic neurones (Carlsson, Falk & Hillarp, 1962) and high concentrations of noradrenaline have been found in the hypothalamus (Vogt, 1954).

Table 3. *Effects of sympathomimetic compounds and their antagonists on urine flow and vasopressin (ADH) release in the rat*

Name of compound	Type of action	Effect on urine flow	Effect on ADH release
Metaraminol ⎱ Noradrenaline ⎰	α-adrenergic	Diuresis	No release
Isoprenaline	β-adrenergic	Anti-diuresis	?
Adrenaline	Both α and β	Unpredictable	Inhibits release
Phenoxybenzamine ⎱ Dihydro-ergotamine ⎰	α-blockers	No effect	Block release
Propranolol	β-blocker	No effect	?

Based mainly on results of Lehr *et al.* (1967) and Dyball (1968).

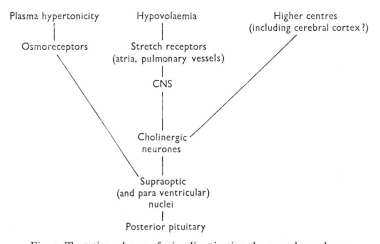

Fig. 2. Tentative scheme of stimuli activating the neurohypophyses.

Table 3 attempts to summarize some of the results achieved and, while not very informative, suggests at least that the confusing effects of adrenaline previously reported (for references see Ginsburg, 1968) can probably be ascribed to the affinity of this amine to both α and β-adrenergic receptors. It suggests also that, as in the case of dipsogenic stimuli, a role of β-adrenergic circuits in the regulation of water output may have to be envisaged.

Fig. 2 attempts to draw a scheme of stimuli which cause the secretion

of ADH and hence lead to water conservation. It is almost certainly incomplete and neglects other mechanisms (e.g. renal sodium retention by
aldosterone release) which counteract extracellular dehydration.

PHYLOGENETIC CONSIDERATIONS

There is a considerable variety of ways by which the various classes of
vertebrates replenish their extracellular fluid phase. Table 4 reviews the
phyletic distribution of sites of water intake. *Cyclostomes* in freshwater
take up large quantities of water osmotically through the gills (Bentley &
Follett, 1963); their skin seems to have a low permeability to water
(Bentley 1962*a*). When in the sea the pattern of osmoregulation of lampreys is much the same as that of marine teleosts (Morris, 1960); that is to

Table 4. *Phylogenetic distribution of sites of water uptake*

	Gills	Oral cavity	Skin	Urinary bladder	Gut,* cloaca (stomach?)
Cyclostomes					
Fresh water	+ +	+	(+)	?	−
Marine	−	−	(+)	?	+ +
Elasmobranchs	+ +	+	(+)	?	−
Teleosts					
Fresh water	+ +	+	(+)	?	−
Marine	−	−	−	?	+ +
Amphibians	−	?	+ +	+ +	−
Reptiles	−	−	−	(+)	+ +
Birds	−	−	−	−	+ +(−)
Mammals	−	−	−	−	+ +(−)

+ +, Main site; +, subsidiary site; (+), possibly subsidiary site; −, no absorption
(−), absent in some species.
* After ingestion by mouth.

say they drink sea water. By virtue of their peculiar ability to retain urea for
osmotic purposes, the blood of *elasmobranchs* is always hypertonic to their
environment and hence water flows in osmotically through the gills and,
presumably, also the mucous membranes of the mouth (see Krogh,
1939), though at a much higher rate in freshwater than in marine elasmobranchs. Urea retention for osmotic purposes appears to be a very ancient
mechanism since Pickford & Blake Grant (1967) found a high urea concentration in the blood of the Coelacanth. Having also determined the
main serum electrolytes these authors came to the conclusion that the
blood of *Latimeria* was probably slightly hypotonic to equatorial (Indian
Ocean) water. If so, they rightly point out that the fish must drink.
Smith (1941) has shown that sharks and rays do not drink habitually.
Again according to Smith (1930), *freshwater teleosts* do not normally
drink (some of these fishes seem to drink occasionally, see Allee & Frank,

1948, and Potts & Evans, 1967) and, since water diffuses only slowly through the skin (Krogh, 1937), the main water intake must be through the gills and the oral mucous membranes. In contrast, *marine teleosts* lose water through the gills and have been shown (Smith, 1930) to drink large quantities of water. *Amphibians*, as exemplified by observations on the frog, do not drink when immersed in water or dilute electrolyte solutions. Their main site of water intake is the skin (Townson, 1799). The extra-cellular fluid volume is also replenished by absorption of water stored in the urinary bladder (Townson, 1799; Darwin, 1839; Steen, 1930). The water metabolism of *reptiles* is a neglected subject. However, it is known (Buddenbrock, 1956) that lizards, tortoises, alligators and snakes drink. From blood and urine analyses Buddenbrock (1956) suspected that marine turtles ingest sea water, but this does not seem to have been demonstrated. There is some evidence for water absorption from the reptilian cloaca (for reference see Chew, 1961), and water transport from the mucosal to the serosal side has been demonstrated by Bentley (1962b) in the isolated tortoise bladder. Like their reptilian ancestors, most *birds* drink freely. Evidence for absorption of significant amounts of water from the cloaca is equivocal. There is some indication that marine birds drink sea water (Cade & Bartholomew, 1959; Schmidt-Nielsen, Jørgensen & Osaki, 1957), which they would be well adapted to deal with since excess salt can be excreted by their salt glands. The drinking of sea water by marine *mammals* is comprehensively discussed by Wolf (1958) in his monograph on *Thirst*. Many species of desert mammals never drink, others, like the camel, are adapted to go without water for long periods of time (for references see Schmidt-Nielsen & Schmidt-Nielsen, 1952).

It is thus evident (see Table 4) that many types of lower vertebrates drink. But at which point in evolution does the need for water ingestion become a conscious sensation so that one is entitled to speak of 'thirst'? At present there seems to be no answer to this question. The problem is obviously open to experimentation and it may be permissible to stress the need for such work.

While the involvement of centres 'higher' than those of the hypothalamus in the control of water loss in lower vertebrates has still to be clarified, there is much suggestive evidence that—at least in all tetrapods—the hypothalamo-neurohypophysial system is concerned in the regulation of water metabolism. It is now known that all vertebrate classes (including the bony and the cartilaginous fishes, see Sawyer, 1968; Heller & Pickering, 1969; Sawyer, Freer & Tseng, 1967; Pickering & Heller, 1970), with the exception of mammals, elaborate arginine vasotocin (AVT), a hormone which when injected into a tetrapod does usually cause antidiuresis and/or prevents water loss by an effect on extrarenal mechanisms. The minimum doses in which vasotocin acts on the kidneys of such animals as the domestic fowl, the water snake *Natrix siphedon* and the toad *Bufo marinus* is of the order of a few nanogrammes ($= 1 \times 10^{-9}$ g) of the pure peptide (for references see Heller, 1969). Since this is only a small fraction of the amount of hormone stored in the neurohypophysis of these species, there is the possibility that these doses are physiological.

There is additional, though rather incomplete evidence indicating that the hypothalamo-neurohypophysial system is involved in the control of the water metabolism of the vertebrate classes mentioned. For instance, neurohypophysectomy (Shirley & Nalbandov, 1956) or lesions of the supraoptic region (Ralph, 1960) have been shown to produce a diabetes insipidus-like condition in birds and the hormone content of the neurohypophysis has been found to be decreased in dehydrated frogs and toads (Levinsky & Sawyer, 1953; Tramezzani & Uranga, 1954; Jørgensen, Wingstrand & Rosenkilde, 1956). However, this type of evidence is missing for reptiles, and the effects of neurohypophysectomy do not appear to have been studied in this vertebrate class. Several studies on the effects of the removal of the pituitary in anuran amphibians are available but the results are contradictory (for references and discussion see Morel & Jard, 1968). Little is known about endogenous hormone levels in non-mammalian tetrapods though a good beginning has recently been made by Bentley (1969), who studied the effect of dehydration of *Bufo marinus* and *Rana catesbiana* on the water transfer (toad bladder) activity of the blood plasma.

The evidence for a physiological role of the neurohypophysis in the regulation of the water metabolism of fish is even more fragmentary. Elasmobranchs have hardly been investigated. All that seems to be known is that the neurohypophysis of elasmobranchs produces similar hormones to that of other vertebrates (see e.g. Heller & Pickering, 1970; Pickering & Heller, 1970), and that elasmobranch neurohypophysial extracts injected into the shark *Ginglymostoma cirrhatum* did not produce water retention (Heller & Bentley, 1965).

There is some information about actinopterygians and dipnoans as several workers have been able to show that neurohypophysial hormones given in small doses may influence the renal function of such fishes (goldfish, Maetz, Bourguet, Lahlou & Houdry, 1964; eel, Chester-Jones, Chan, Rankin & Poniah, 1969; African lungfish, Sawyer, 1967). The renal effects were uniformly an increase in glomerular filtration and an increase in urine flow. If these effects are 'physiological' they would suggest that in bony fishes in fresh water neurohypophysial hormones may promote water loss (and would thus fulfil a useful osmoregulatory function by preventing the fish from being swamped by water) rather than water conservation as in terrestrial vertebrates.

Finally, may I stress that I am well aware that, divorcing the multiple interactions between water and salt metabolism, as I have done in this survey, has introduced a somewhat artificial division which, however, was inevitable for reasons of time and space. Fortunately, many aspects of salt metabolism have been discussed on the first day of this symposium.

REFERENCES

ABRAHAMS, V. C., KOELLE, G. B. & SMART, P. (1957). Histochemical demonstrations of cholinesterases in the hypothalamus of the dog. *J. Physiol. Lond.* **139**, 137–44.

ALLEE, W. C. & FRANK, P. (1948). Ingestion of colloidal material and water by goldfish. *Physiol. Zool.* **21**, 381–90.

ANDERSSON, B. (1953). The effect of injections of hypertonic NaCl solutions into different parts of the hypothalamus of goats. *Acta physiol. scand.* **28**, 188–201.

ANDERSSON, B. & McCANN, S. M. (1955). A further study of polydipsia evoked by hypothalamic stimulation in the goat. *Acta physiol. scand.* **33**, 333–46.

ANDERSSON, B. & McCANN, S. M. (1956). The effect of hypothalamic lesions on the water intake of the dog. *Acta physiol. scand.* **35**, 312–20.

ANDERSSON, B., OLSSON, K. & WARNER, R. G. (1967). Dissimilarities between the central control of thirst and the release of antidiuretic hormone (ADH). *Acta physiol. scand.* **71**, 57–64.

ASSCHER, A. W. & ANSON, S. G. (1963). A vascular permeability factor of renal origin. *Nature, Lond.* **198**, 1097–9.

BARATZ, R. A. & INGRAHAM, R. C. (1960). Renal hemodynamics and antidiuretic hormone release associated with volume regulation. *Am. J. Physiol.* **198**, 565–70.

BARER, G. R. (1963). The action of vasopressin, a vasopressin analogue (PLV$_2$), oxytocin, angiotensin, bradykinin and theophylline ethylene diamine on renal blood flow in the anaesthetized cat. *J. Physiol. Lond.* **169**, 62–72.

BENTLEY, P. J. (1962a). Permeability of the skin of the cyclostome *Lampetra fluviatilis* to water and electrolytes. *Comp. Biochem. Physiol.* **6**, 95–7.

BENTLEY, P. J. (1962b). Studies on the permeability of the large intestine and urinary bladder of the tortoise (*Testudo graeca*) with special reference to the effect of neurohypophysial and adrenocortical hormones. *Gen. comp. Endocr.* **2**, 323–8.

BENTLEY, P. J. (1969). Neurohypophysial function in amphibia: hormone activity in the plasma. *J. Endocr.* **43**, 359–69.

BENTLEY, P. J. & FOLLETT, B. K. (1963). Kidney function in a primitive vertebrate, the cyclostome *Lampetra fluviatilis*. *J. Physiol., Lond.* **169**, 902–18.

BICKERTON, R. K. & BUCKLEY, J. P. (1961). Evidence for a central mechanism in angiotensin induced hypertension. *Proc. Soc. exp. Biol. Med.* **106**, 834–6.

BLASS, E. M. (1968). Separation of cellular from extracellular controls of drinking in rats by frontal brain damage. *Science, N.Y.* **162**, 1501–3.

BOOTH, D. A. (1968). Mechanism of action of norepinephrine in eliciting an eating response on injection into the rat hypothalamus. *J. Pharmac. exp. Ther.* **160**, 336–48.

BUDDENBROCK, W. VON (1956). *Vergleichende Physiologie*, vol. 3, pp. 536–7. Basel & Stuttgart: Birkhäuser.

CADE, G. J. & BARTHOLOMEW, G. A. (1959). Use of sea water by Savannah sparrows. *Physiol. Zool.* **32**, 320–8.

CARLSSON A., FALK, B. & HILLARP, N. A. (1962). Cellular localization of brain monoamine. *Acta physiol. scand.* (Suppl.) **196**, 1–28.

CHESTER-JONES, I., CHAN, D. K. O., RANKIN, J. C. & PONIAH, S. (1969). Renal function in the European eel (*Anguilla anguilla*, L.): effects of the corpuscles of *Stannius*, caudal neurosecretory system, neurohypophysial peptides and vasoactive substances. *J. Endocr.* **43**, 21–31.

CHEW, R. M. (1961). Water metabolism of desert-inhabiting vertebrates. *Biol. Rev.* **36**, 1–31.

DANIEL, A. R. & LEDERIS, K. (1967). Release of neurohypophysial hormones *in vitro*. *J. Physiol., Lond.* **190**, 171–87.

DARWIN, C. (1839). *Journal of Researches into the Natural History and Geology of the Countries visited during the voyage of H.M.S. Beagle around the World.* London: Colburn.

DE ROBERTIS, E. (1962). Ultrastructure and function in some neurosecretory systems. In *Neurosecretion*, pp. 3–20 (eds. H. Heller and R. B. Clark). New York and London: Academic Press.

DE WIED, D. (1960). A simple automatic and sensitive method for the assay of antidiuretic hormone with notes on the antidiuretic potency of plasma under different experimental conditions. *Acta physiol. pharmac. néerl.* **9**, 69–81.

DICKER, S. E. (1966). Release of vasopressin and oxytocin from isolated pituitary glands of adult and newborn rats. *J. Physiol., Lond.* **185**, 429–44.

DOUGLAS, W. W. & POISNER, A. M. (1964). Stimulus-secretion coupling in a neurosecretory organ and the role of calcium in the release of vasopressin from the neurohypophysis. *J. Physiol., Lond.* **172**, 1–18.

DUKE, H. N., PICKFORD, M. & WATT, J. A. (1950). The immediate and delayed effects of diisopropyl fluorophosphate injected into the supraoptic nuclei of dogs. *J. Physiol., Lond.* **111**, 81–8.

DYBALL, R. E. J. (1968). The effects of drugs on the release of vasopressin. *Br. J. Pharmac. Chemother.* **33**, 329–41.

EPSTEIN, A. N., FITZSIMONS, J. T. & SIMONS, B. J. (1968). Drinking caused by the intracranial injection of angiotensin into the rat. *J. Physiol., Lond.* **200**, 98–100P.

FELDBERG, W. & VOGT, M. (1948). Acetylcholine synthesis in different regions of the central nervous system. *J. Physiol., Lond.* **107**, 372–81.

FISHER, A. & COURY, J. M. (1962). Cholinergic tracing of a neural circuit underlying thirst drive. *Science, N.Y.* **138**, 691–3.

FITZSIMONS, J. T. (1961). Drinking by rats depleted of body fluid without increase in osmotic pressure. *J. Physiol., Lond.* **159**, 297–309.

FITZSIMONS, J. T. (1969). The role of a renal thirst factor in drinking induced by extracellular stimuli. *J. Physiol., Lond.* **201**, 349–68.

FITZSIMONS, J. T. (1970). The renin–angiotensin system in the control of drinking. In *Integration of Endocrine and Non-endocrine Mechanisms in the Hypothalamus.* Proc. of Int. Conf. at Stresa, Italy, 20–28 May, 1969. New York and London: Academic Press. (In the Press).

FITZSIMONS, J. T. & SIMONS, B. J. (1969). The effect on drinking in the rat of intravenous infusion of angiotensin, given alone or in combination with other stimuli of thirst. *J. Physiol., Lond.* **203**, 45–57.

FONNUM, F. (1966). A radiochemical method for the estimation of choline acetyltransferase. *Biochem. J.* **100**, 479–84.

GERSCHENFELD, H. M., TRAMEZZANI, J. H. & DE ROBERTIS, E. (1960). Ultrastructure and function in neurohypophysis of the toad. *Endocrinology* **66**, 741–62.

GINSBURG, M. (1968). Production, release, transportation and elimination of the neurohypophysial hormones. *Handbook of Experimental Pharmacology* vol. 23, pp. 286–371.

GINSBURG, M. & BROWN, L. M. (1956). Effect of anaesthetics and haemorrhage on the release of neurohypophysial antidiuretic hormone. *Br. J. Pharmac.* **14**, 327–33.

GINSBURG, M. & HELLER, H. (1953). Antidiuretic activity in blood obtained from various parts of the cardiovascular system. *J. Endocr.* **9**, 274–82.

GREER, M. A. (1955). Suggestive evidence of a primary 'drinking center' in hypothalamus of the rat. *Proc. Soc. exp. Biol. Med.* **77**, 59–62.

GROSSMAN, S. P. (1960). Eating and drinking elicited by direct adrenergic and cholinergic stimulation of the hypothalamus. *Science, N.Y.* **132**, 301–2.

GROSSMAN, S. P. (1964a). Behavioural effects of direct chemical stimulation on central nervous system structures. *Int. J. Neuropharmac.* **3**, 45–58.

GROSSMAN, S. P. (1964b). Some neurochemical aspects of central regulations of thirst. In *Thirst, First Int. Symp. on Thirst in the Regulation of Body Water*, pp. 487–514 (ed. M. J. Wayner). Oxford: Pergamon Press.

GUTMAN, Y. & BENZEKEIN, F. (1969). Relation of kidneys and adrenal glands to hypovolemic thirst. *Israel J. med. Sci.* **5**, 411–13.

GUTMAN, Y., BENZAKEIN, F. & CHAIMOVITZ, M. (1967). Kidney factor(s) affecting water consumption in the rat. *Israel J. med. Sci.* **3**, 910–11.

HELLER, H. (1969). Class and species-specific actions of neurohypophysial hormones. *C.R.N.S. Symposium*, no. 177, Paris, 1968, pp. 35–43.

HELLER, H. & BENTLEY, P. J. (1965). Phylogenetic distribution of the effects of neurohypophysial hormone on water and sodium metabolism. *Gen. comp. Endocr.* **5**, 96–108.

HELLER, H. & PICKERING, B. T. (1970). The distribution of vertebrate neurohypophysial hormones and its relation to possible pathways to their evolution. In *The Neurohypophysis*. International Encyclopedia of Pharmacology and Therapeutics, Section 41, pp. 59–79 (eds. H. Heller and B. T. Pickering). Oxford: Pergamon Press. (In the Press).

HOLMES, R. L. (1961). Phosphatase and cholinesterase in the hypothalamo–hypophysial system in the monkey. *J. Endocr.* **23**, 63–7.

JAHN, H., STEPHAN, F. & STAHL, J. (1960). Diuretic activity in blood and urine after bleeding in the dog. *Archs Sci. physiol.* **14**, 421–33.

JØRGENSEN, C. B., WINGSTRAND, K. G. & ROSENKILDE, P. (1956). Neurohypophysis and water metabolism in the toad *Bufo bufo* (L.). *Endocrinology* **59**, 601–10.

KOELLE, G. B. (1961). A proposed dual neurohumoral role of acetylcholine: its function at pre- and post-synaptic sites. *Nature, Lond.* **190**, 208–24.

KOELLE, G. B. & GEESEY, C. N. (1961). Localisation of cholinesterase in the neurohypophysis and its functional implications. *Proc. Soc. exp. Biol. Med.* **106**, 625–8.

KONSTANTINOVA, M. (1967). The effect of adrenaline and acetylcholine on the hypothalamic–neurohypophysial neurosecretion in the rat. *Z. Zellforsch. mikrosk. Anat.* **83**, 549–67.

KROGH, A. (1937). Osmotic regulation in freshwater fishes by active absorption of chloride ions. *Z. vergl. Physiol.* **24**, 656–66.

KROGH, A. (1939). *Osmotic Regulation in Aquatic Animals*. Cambridge University Press.

LAVERTY, R. (1963) A nervous-mediated action of angiotensin in anaesthetized rats. *J. Pharm. Pharmac.* **15**, 63–8.

LEDERIS, K. & LIVINGSTON, A. (1968). Subcellular localization of acetylcholine in the posterior lobe of the rabbit. *J. Physiol. Lond.* **196**, 34–6P.

462 H. HELLER

LEDERIS, K. & LIVINGSTON, A. (1969). Acetylcholine and related enzymes in the neural lobe and anterior hypothalamus of the rabbit. *J. Physiol., Lond.* **201**, 695–709.

LEHR, D., MALLOW, J. & KRUKOWSKI, M. (1967). Copious drinking and simultaneous inhibition of urine flow elicited by beta-adrenergic stimulation and contrary effect of alpha-adrenergic stimulation. *J. Pharmacol. exp. Ther.* **158**, 150–63.

LEVINSKY, N. G. & SAWYER, W. H. (1953). Significance of the neurohypophysis in regulation of fluid balance in the frog. *Proc. Soc. exp. Biol. Med.* **82**, 272–4.

MAETZ, J., BOURGUET, J., LAHLOU, B. & HOUDRY, J. (1964). Peptides neurohypophysaires et osmorégulation chez *Carassius auratus. Gen. comp. Endocr.* **4**, 508–22.

MILLER, N. (1965). Chemical coding of behaviour in brain. *Science, N.Y.* **148**, 328–38.

MOREL, F. & JARD, S. (1968). Actions and functions of the neurohypophysial hormones and related peptides in lower vertebrates. In *Handbook of Experimental Pharmacology.* vol. 23, pp. 655–716. Berlin: Springer.

MORRIS, R. (1960). General problems of osmoregulation with special reference to cyclostomes. *Symp. zool. Soc. Lond.* **1**, 1–16.

PARMAR, S. S., SUTTER, M. C. & NICKERSON, M. (1961). Localization and characterization of cholinesterase in subcellular fractions of rat brain and beef pituitary. *Can. J. Biochem. Physiol.* **39**, 1335–45.

PEART, W. S. (1965). The renin–angiotensin system. *Pharmac. Rev.* **17**, 143–82.

PEPLER, W. J. & PEARSE, A. G. E. (1957). The histochemistry of the esterase of rat brains with special reference to those of the hypothalamic nuclei. *J. Neurochem.* **1**, 193–202.

PERLMUTT, J. H. (1963). Reflex antidiuresis after occlusion of common carotid arteries in hydrated dogs. *Am. J. Physiol.* **204**, 197–201.

PICKERING, B. T. & HELLER, H. (1970). Oxytocin as a neurohypophysial hormone in the holocephalian elasmobranch fish *Hydrolagus colliei. J. Endocr.* (In the Press.)

PICKFORD, M. (1947). The action of acetylcholine in the supraoptic nucleus of the chloralosed dog. *J. Physiol., Lond.* **106**, 264–70.

PICKFORD, G. E. & BLAKE GRANT, F. (1967). Serum osmolality in the coelocanth, *Latimeria cholumnae*: Urea retention and ion regulation. *Science, N.Y.* **155**, 568–70.

POTTS, W. T. W. & EVANS, D. H. (1967). Sodium and chloride balance in the killifish *Fundulus heteroclitus. Biol. Bull. mar. Biol. Lab., Woods Hole* **133**, 411–25.

RALPH, C. L. (1960). Polydipsia in the hen following lesions in the supraoptic hypothalamus. *Am. J. Physiol.* **198**, 528–30.

SAWYER, W. H. (1967). Evolution of antidiuretic hormones and their functions. *Am. J. Med.* **42**, 678–86.

SAWYER, W. H. (1968). Phylogenetic aspects of the neurohypophysial hormones. *Handbook of Experimental Pharmacology*, vol. 23, pp. 717–47.

SAWYER, W. H., FREER, R. J. & TSENG, T. C. (1967). Characterisation of a principle resembling oxytocin in the pituitary of the holocephalian ratfish (*Hydrolagus colliei*) by partition chromatography on Sephadex columns. *Gen. comp. Endocr.* **9**, 31–7.

SCHMIDT-NIELSEN, K., JØRGENSEN, C. B. & OSAKI, H. (1957). Secretion of hypertonic solution in marine birds. *Fedn Proc. Fedn. Am. Socs exp. Biol.* **16**, 113–4.

SCHMIDT-NIELSEN, K. & SCHMIDT-NIELSEN, B. (1952). Water metabolism of desert mammals. *Physiol. Rev.* **32**, 135–66.

SHARE, L. & LEVY, M. N. (1966). Carotid sinus pulse pressure as a determinant of plasma antidiuretic hormone activity. *Am. J. Physiol.* **211**, 721–4.

SHIRLEY, H. V. & NALBANDOV, A. V. (1956). Effects of neurohypophysectomy in domestic chickens. *Endocrinology* **58**, 477–83.

SMITH, H. W. (1930). The absorption and excretion of water and salts by marine teleosts. *Am. J. Physiol.* **93**, 480–505.

SMITH, H. W. (1931). The absorption and excretion of water and salts by the elasmobranch fishes. II. Marine elasmobranchs. *Am. J. Physiol.* **98**, 296–310.

STEEN, W. B. (1930). On the permeability of the frog's bladder to water. *Anat. Rec.* **43**, 218–20.

STEIN, L. (1963). Anticholinergic drugs and the central control of the thirst drive. *Science, N.Y.* **139**, 46–7.

STRICKER, E. M. & WOLF, G. (1967). The effect of hypovolemia on drinking in rats with lateral hypothalamic damage. *Proc. Soc. exp. Biol. Med.* **124**, 816–20.

TOWNSON, R. (1799). *Tracts and Observations in Natural History and Physiology.* London: White.

TRAMEZZANI, J. H. & URANGA, J. V. (1954). Variaciones de la substancia Gomori positiva y activitad antidiuretica en la neurohypofisis de sapos hidratos y deshidratos. *Rev. Soc. argent. Biol.* **30**, 148–51.

VANDER, A. J. & LUCIANO, J. R. (1967). Neural and humoral control of renin release in salt depletion. *Circulation Res.* Suppl. to vols. 20 and 21, 69–75.

VOGT, M. (1954). The concentration of sympathin in different parts of the central nervous system under normal conditions and after the administration of drugs. *J. Physiol., Lond.* **123**, 451–81.

WEINSTEIN, H., BERNE, R. M. & SACHS, H. (1960). Vasopressin in blood. Effect of hemorrhage. *Endocrinology* **66**, 712–18.

WOLF, A. V. (1950). *The Urinary Function of the Kidney.* New York: Grune and Stretton.

WOLF, A. V. (1958). *Thirst.* Springfield, Ill.: Charles C. Thomas.

ZAMBONI, P. & SIRO-BRIGIANI, G. (1966). Effeto delle catecolamine sulla sete e la diuresi nel ratto. *Boll. Soc. ital. Biol. sperim.* **42**, 1657–9.

DISCUSSION

MADERSON: What is the evidence for your contention that the skin is a major site of water uptake in amphibians?

HELLER: Amphibians do not drink when kept in fresh water, but their body weight increases steadily when the cloaca is tied. Moreover, water influx can also be demonstrated in isolated pieces of skin. In anurans this process is enhanced, both *in vivo* and *in vitro*, by the neurohypophysial hormones, the so-called hydrosmotic effect.

MADERSON: Yes, but are you classifying this as a potential mechanism, or are you suggesting that this really does happen in nature on a regular basis?

HELLER: Indeed, I am. As long ago as 1799 it was observed by an English physician, R. Townson (see my paper), that the body weight of frogs or toads increases if they are placed on a piece of moist blotting paper. More recently, the influx of water into the skin of toads has been accurately measured by Ussing and his co-workers (see e.g. V. Koefoed-Johnsen & H. H. Ussing (1953), *Acta physiol. scand.* **28**, 60).

ACTIVE SUBSTANCES IN THE
CAUDAL NEUROSECRETORY SYSTEM OF
BONY FISHES

By K. LEDERIS

The search for active principles in the urophysis is of relatively recent origin although the neurosecretory nature of the large neurones, the 'Dahlgren' cells, in the caudal spinal cord of bony and cartilaginous fishes has been known for many years (Dahlgren, 1914; Speidel, 1919; 1922). In a study of a system, the physiological function of which is not clearly understood, approaches on different parameters may be attempted, e.g. 'functional' morphology, which may be aimed at detecting physiological events from morphologically recognizable changes. Alternatively, experimental procedures aimed at changing the 'milieu interior or exterior' may be instigated with the aim of detecting altered morphology or function of the system under investigation.

Thus differences were noted in nuclear morphology of the caudal neurosecretory cells of *Albula vulpes* taken from oceanic ponds and those caught in the open sea (Fridberg, Bern & Nishioka, 1966). Similarly, a more intense staining reaction was noted by Sano (1958) in this system in freshwater fish as compared with that of marine fish. Higher content of 'cholinergic substance' was found in the urophysis of freshwater than in sea fish (Kobayashi, Uemura, Oota & Ishii, 1963; Uemura, Kobayashi & Ishii, 1963). Electrophysiological investigations on single-unit responses have indicated that injections of distilled water in *Paralichthys dentatus* increased the rate of firing by the caudal neurosecretory cells (Bennett & Fox, 1962). Yagi & Bern (1963) showed that the frequency of spontaneous discharges from the caudal neurosecretory neurones could be changed drastically by altering the 'milieu exterior' (fresh to sea water). Pronounced morphological changes (light and electron microscopical) were reported to occur in the urophysis of *Tilapia mossambica* after electrical stimulation *in vivo* and *in vitro* (Fridberg *et al.* 1966).

Attempts during the last decade to find active substances which may be produced by the caudal secretory neurones and stored in (and released from) the urophysis have resulted in a considerable number of reports on a variety of biologically measurable activities. Table 1 shows a summary of the activities observed by various groups of investigators.

Perusal of the descriptions of experimental procedures employed, and of

Table 1. *Summary of effects of urophysial preparations*

Effect	Test preparation	Suggested identity of active substance	Authors
Balance, Mobility	Goldfish	?	Enami (1959)
Locomotor (tailfin)	Teleosts (various)	?	Honma (1966), Honma & Tamura (1967)
Pressor (blood pressure) kidney effects	Eel	Peptides, peptide-protein complex	Bern et al. (1967)
Depressor (blood pressure)	Rat	?	Kobayashi et al. (1968)
'Cholinergic' substance	Bivalve heart	ACh ?	Kobayashi et al. (1963)
Stimulation of Na$^+$ fluxes (gill), Na$^+$ retention (kidney)	Goldfish	?	Maetz et al. (1964)
ACTH release	Various Indian fishes	Zinc (Contamination with AVT)	Roy (1962)
Antidiuretic	Rat		Sawyer & Bern (1963)
Stimulation of uterus	Rat	AVT, Isotocin	
Stimulation of oviduct	Hen		Sterba et al. (1965)
Water retention	Frog		
Water retention	Frog, toad	?	Bern et al. (1967), Nishioka (1968)
'Hydrosmotic'	Toad bladder	?	Lacanilao (1969)
Smooth muscle stimulating:			
Bladder	Trout		
Intestine	Trout, mudsucker	'Low molecular' (mol.wt < 1,000) substance(s)	Lederis (1969a, c)
Ovary-oviduct	Guppy		

the findings obtained in these investigations, exposes a number of problems and apparent controversies, the solutions of which would require 'a common denominator' to be used on more than one parameter:

(1) The availability of an 'assay' procedure which would provide the means for the estimation of one particular biological activity. Such a procedure should be technically simple, rapid and precise.

(2) A urophysial standard preparation which would offer the means for quantitative comparison and evaluation of the known activities and others yet to be described.

(3) Biological, pharmacological and chemical identity of the active principle or principles with regard to all, or at least some, of the known activities.

(4) In view of the difficulties encountered so far in identifying the physiological role of the caudal neurosecretory system and of its products, elucidation of the above problems (e.g. knowledge of the biological and chemical identity of the neurosecretory products) could lead to a definition of the physiological importance of this system in bony and elasmobranch fishes.

The experiments to be described here may provide the basic requirements for the solution of some of the problems raised above.

ASSAY OF A SMOOTH MUSCLE-CONTRACTING ACTIVITY ON THE ISOLATED URINARY BLADDER OF THE TROUT (*SALMO GAIRDNERI*)

This assay which was evolved and described recently (Lederis, 1969 a, b; Geschwind, Lederis, Bern & Nishioka, 1968) is technically simple, requiring only a suitable strain-gauge pressure transducer–amplifier–chart recorder system. The bladder is attached to the transducer by a suitable connecting piece, the internal space (transducer–connector–bladder) is filled with dilute (1:5) Ringer's solution (at a hydrostatic pressure of 15–25 cm), the bladder is immersed in a 10 ml organ bath containing trout Ringer's (Nandi & Bern, 1965) to which urophysial extracts and other test substances are added. The responses (rhythmic contractions of the bladder) are recorded on paper and expressed as number of contractions/minute. Fig. 1 shows the dose–response relationships of aqueous and acid extracts of urophyses (Fig. 1 a) and also a comparison of dose–response slopes between urophysial extracts, acetylcholine (ACh) and 5-hydroxytryptamine (5-HT, Fig. 1 b).

Pharmacological comparison of the effects on the trout bladder, trout or mudsucker (*Gillichthys mirabilis*) intestine, rat uterus, rat blood pressure (Lederis 1969 a, c) of urophysial extracts and a number of biologically

30-2

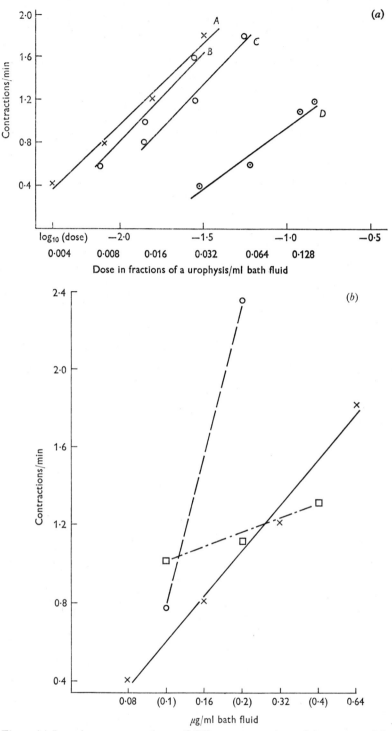

Fig. 1. (a) Log. dose–response slopes of different preparations of the trout caudal neuro-secretory system: (A) acetic acid extracts of trout urophysis; (B) whole homogenate (in distilled water) of trout urophysis; (C) extract in Ringer's solution; (D) whole homogenate (in Ringer's) of caudal spinal cord (immediately rostral to urophysial area). Responses expressed as contractions/min over a 5 min period after additions of geometrically increas-ing doses of extracts (expressed as fractions of a urophysis/ml of bath fluid). (From Lederis, 1969b). (b) Comparison of bladder responses to Urophysial Laboratory Standard (UHS), acetylcholine chloride (ACh) and 5-hydroxytryptamine creatinine sulphate (5-HT) (mean responses from three experiments). ×——×, UHS; ○ – – ○, 5-HT; □——□, ACh. (From Lederis, 1969c.)

active substances which might occur in the urophysis indicated that the bladder-contracting activity in urophysial extracts could not be due to any of the substances tested (Table 2). These tests and the comparison between the relative activity of urophysial extracts and the bladder-contracting activity of, for example, ACh and 5-HT indicated the presence in the urophysis of an extremely potent smooth-muscle-stimulating principle. To produce comparable responses on the trout bladder, approximately equal amounts (wt/wt) of either ACh-chloride or 5-HT-creatinine sulphate and of urophysial acetone-dried tissue were needed. If the concentration of the active principles in the urophysis in relation to dry tissue weight is assumed to be comparable, for example, to the active peptide to tissue weight in the neurohypophysis (in the latter the active peptides may amount to 1–2 % of dry neural lobe tissue weight), then the bladder-contracting principle of the urophysis may be calculated to be 50–100 times more potent on the trout bladder than ACh-chloride or 5-HT (wt/wt).

LABORATORY STANDARD AND REFERENCE PREPARATION

(a) Urophysial Laboratory Standard preparation (UHS)

When the trout-bladder preparation was found to satisfy the basic requirements for a useful and usable bioassay procedure, and before the investigation of the biological, pharmacological and chemical properties of the active principle(s), it was thought desirable to introduce a urophysial laboratory standard preparation. Availability of such a standard would provide means for a quantitative evaluation and comparison of all experimental data, and it would also ensure reproducibility of findings. Urophyses of *Gillichthys mirabilis* were readily available in the laboratory, and the trout-bladder contracting activity was found not to vary greatly when the assay data from a large series of experiments, with 2–20 *Gillichthys* urophyses in each, were compared. The *Gillichthys* urophysis was adopted as a basis for a laboratory standard which was prepared by drying batches of urophyses (20–300 organs/batch) in acetone, extracting the acetone-dried powder in 0·25 % acetic acid and heating the extracts for 3 min in a boiling water bath. Each batch of a Urophysial Laboratory Standard (UHS) was assayed against the previous one, the activity in the first UHS being taken as 100 %. Table 3 shows the comparison between six batches of UHS prepared and used over a period of about 12 months. Analysis of variance showed that the activity between batches, and the mean weight of acetone-dried urophyses in each batch, did not vary significantly. Moreover, since multiple assays (2 × 2 restricted block design) were done to 'standardize'

Table 2. *Summary of effects of urophysial laboratory standard preparation (UHS) and of some other compounds on the trout bladder, trout or Gillichthys intestine, rat uterus and rat blood pressure; and some inactivation tests on the UHS (from Lederis, 1969c)*

++, Highly active; +, active; (+), active at high doses; ↑/↓, increase/decrease in muscle tone; (↑), increase in muscle tone at high doses (inconsistent); N.T. = not tested.

| | Test preparation or procedure | | | | | Inactivation tests | | |
| Compound | Trout bladder | Trout–*Gillichthys* intestine | | Rat uterus | Rat blood pressure | Heating at pH 10 | Inactivation with thioglycollate | Treatment with trypsin (1) and chymotrypsin (2) |
		Isotonic	Isometric					
UHS	++(↑)	.	(+)↑	.	.	Decreased	.	Abolished by (1) and (2) in 5 expts; decreased in 2 expts
Acetylcholine	+(↑)	++(↑)	++(↑)	+	N.T.	Abolished	N.T.	N.T.
5-Hydroxytryptamine	+	++	++	+	N.T.	N.T.	N.T.	N.T.
Histamine	.	+	N.T.	(+)→	N.T.	N.T.	N.T.	N.T.
Adrenaline	.	↓	→	(+)→	N.T.	N.T.	N.T.	N.T.
Noradrenalin	Transient inhibition	N.T.	N.T.	N.T.	N.T.	N.T.	N.T.	N.T.
Arginine vasotocin	.	N.T.	N.T.	++	++	N.T.	N.T.	N.T.
Isotocin (ichthyotocin)	.	N.T.	N.T.	++	N.T.	N.T.	N.T.	N.T.
Arginine vasopressin	.	N.T.	N.T.	(+)→	++	N.T.	Abolished	N.T.
Oxytocin	.	N.T.	N.T.	++	N.T.	N.T.	N.T.	N.T.

each new batch against the previous one, an evaluation of the precision of the assays was possible. The assay precision was found to be satisfactory ($\lambda = 0.06 - 0.1$).

Table 3. *Summary of mean weights of urophyses and of bladder-contracting activity in different batches of Urophysial Laboratory Standard Preparation (batches UHS_2–UHS_7)*

(From Lederis, 1969 b)

Batch no.	No. of urophyses/ batch	Mean dry weight/UH (ng)	Activity (% of UHS_2)	Index of precision (λ)
UHS_2	20	22.0	100	
				0.067
UHS_3	100	21.1	76 (57.2–94.4)*	
				0.066
UHS_4	200	23.2	117 (87.6–146.0)	
				0.057
UHS_5	220	19.3	123 (101.6–148.7)	
				0.053
UHS_6	300	21.8	93 (75.4–111.0)	
				0.095
UHS_7	249	22.3	106 (76.2–151.6)	
		21.5 ± 1.2†	104.7 (85.7 ± 144.1)	

* 95 % fiducial limits. † Weighted mean ± S.E.M.

(b) *Urophysial reference preparation ('urotensin')*

The uncertainty prevailing at present about the identity and significance of the various activities found in urophysial homogenates, breis and extracts in various laboratories is mainly due to the lack of means of evaluation of each of the activities claimed against a common denominator. To overcome this and to make future studies more meaningful, it was recently proposed (Bern & Lederis, 1969; Lederis & Bern, 1969) that the UHS described here (and in Lederis, 1969b) be adopted as Urophysial Reference Preparation under the general name of urotensin. It was suggested that this preparation (or a local laboratory standard related to and standardized against urotensin) be employed in further studies of the active principle(s) of the caudal neurosecretory system. For convenience of use, a unit of activity was defined as 'the trout bladder-contracting-(or equivalent) activity present in an extract of 1 mg of acetone dried urophysis powder of *Gillichthys mirabilis* homogenized in 0.25% acetic acid and heated for 3 min in a boiling water bath'.

Effects of urotensin on other tissues

The use of doses up to 20 times higher than those necessary to induce contractions in the bladder failed to affect either the blood pressure of the rat, or the isolated rat uterus; at the same time stimulation (not dose-related) of the isolated intestine of the trout or *Gillichthys* was observed (Lederis, 1969 *b*). The rat blood pressure and the rat uterus experiments were done in an attempt to test once more the likelihood of the occurrence in the urophysis of substances similar to, or identical with, the neurohypophysial hormones. Sawyer & Bern (1963) came to the conclusion that substances with the pharmacological characteristics of the neurohypophysial hormone peptides did not occur in the teleost urophysis other than as trace-contaminants. However, Sterba, Luppa & Schuhmacher (1965) reported the presence in the carp urophysis of substances compatible, on the basis of their chromatographic and pharmacological properties, with arginine vasotocin and isotocin. The present findings, which were based on dosages (in the rat experiments) of UHS 10–20 times higher than the average bladder-contracting doses, do not support the conclusions reached by Sterba *et al.* (1965). The use of considerably higher dosages by these workers (up to 15,000 times the bladder-contracting doses) in their assays may explain the differences in findings.

Effects on blood pressure

The interesting, but opposite, effects of urophysial extracts on the blood pressure of the eel (Bern, Nishioka, Chester-Jones, Chan, Rankin & Ponniah, 1967) and of the rat (Kobayashi, Matsui, Hirano, Iwata & Ishii, 1968) were studied in relation to the trout-bladder contracting activity of urotensin. It could be shown that small doses of urotensin raised the eel blood pressure by 10–15 mm Hg (ventral aorta) and 5–10 mm Hg respectively (dorsal aorta) (Fig. 2). Considerably higher doses were needed to produce a pronounced and prolonged fall of the blood pressure of the anaesthetized rat (Fig. 3*a*). The above observations thus confirmed the findings of Bern *et al.* (1967) and those of Kobayashi *et al.* (1968). Treatment of the extracts (UHS and carp urophyses) with trypsin resulted in observations of two kinds, each leading to different conclusions about the number of vasoactive substances present in urophysial extracts.

In one series of experiments (3 expts) treatment of carp urophysial extracts with trypsin resulted in the abolition of the pronounced and prolonged fall in blood pressure in the rat, at the same time demonstrating more clearly another hypotensive effect in this species: a sudden fall in blood pressure (15–30 mm Hg) followed by a quick recovery (Fig. 3*b*).

The latter effect on rat blood pressure appeared to be pharmacologically similar to that produced by ACh but it was not abolished by atropine. The trout-bladder-contracting activity (in the same trypsin-treated material) was abolished, but the hypertensive effect in the eel persisted. In a second series of experiments done according to Hummel (1959) the eel blood-pressure-raising activity was also destroyed. In the same series of experi-

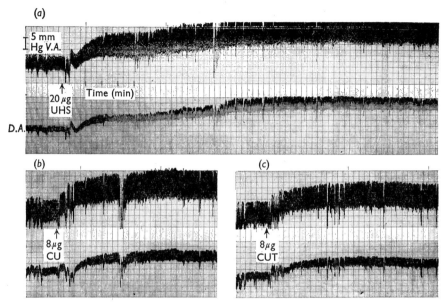

Fig. 2. Effects of urophysial extracts on the blood pressure of the free swimming eel. *V.A.*, Ventral aorta; *D.A.*, dorsal aorta. (*a*) At arrow 20 µg urophysial laboratory standard extract (UHS) injected i.v. Note a decrease in stroke volume immediately after injection (gill-vasoconstriction), followed by a slow increase in stroke volume, in both aortae, simultaneously with rising blood pressure. The duration of response is dose-dependent; may last 20–90 min (in this instance blood pressure returned to normal in 55 min). (*b*) Response to carp urophysis (CU) extract (= 8 µg UHS). (*c*) Response to trypsin-treated (see text, this page) carp urophysis (CUT) extract (original activity = 8 µg UHS).

ments, further samples of UHS and of carp urophyses were incubated also with chymotrypsin, the results were the same as those after trypsin treatment.

Subject to experiments designed to investigate further the effects of trypsin and of other enzymes on crude urophysial extracts and on separated active principles, the existence of up to three vasoactive substances in the urophysis can be postulated: (1) the rat hypotensive substance, which is not trypsin- or atropine-sensitive, (2) the rat-hypotensive, eel-hypertensive, trout bladder-contracting activity, which is destroyed by trypsin, (3) the eel-hypertensive effect, which remains active after (2) has been destroyed by trypsin. It cannot be excluded that the eel-hypertensive effect, remaining after incubation with trypsin in the earlier series of experiments, may be due to incomplete destruction of only one principle (which is rat-hypotensive and eel-hypertensive). As can be seen from

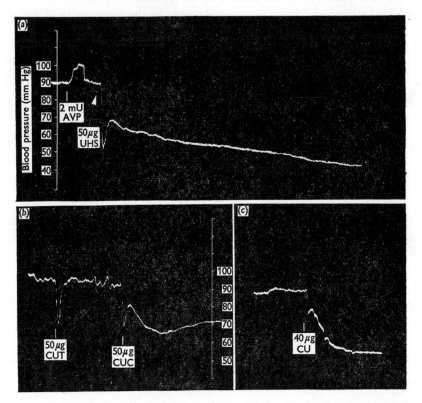

Fig. 3. Effects of urophysial extracts on rat blood pressure. (*a*) 2 mU arginine vasopressin (AVP). At arrow 50 μg UHS. Note a biphasic fall in blood pressure; the second fall slow and long lasting (210 min); maximum net fall > 40 mm Hg. (*b*) 50 μg CUT (trypsin-treated carp urophysis extract, initial activity = 50 μg UHS). Note only a sudden fall in blood pressure with immediate return to normal. 50 μg CUC (carp urophysis control, incubated in buffer without the enzyme, initial activity = 50 μg UHS). (*c*) 40 μg CU (carp urophysis extract = 40 μg UHS).

Figs. 2 and 3, higher doses are needed to produce an effect on rat blood pressure than those which are necessary to raise the blood pressure of the eel. The destruction by trypsin of 75% of the initial vasoactive principle, which would abolish the rat-hypotensive effect, could still leave sufficient activity in the extract to raise the blood pressure of the eel.

Effects on the smooth muscle of the ovary of Lebistes reticulatus

The ovary (with or without the attached oviduct) of this live-bearing species was found to be extremely sensitive to urotensin when used immediately after delivery of the young (Fig. 4*a*). Pronounced contrac-

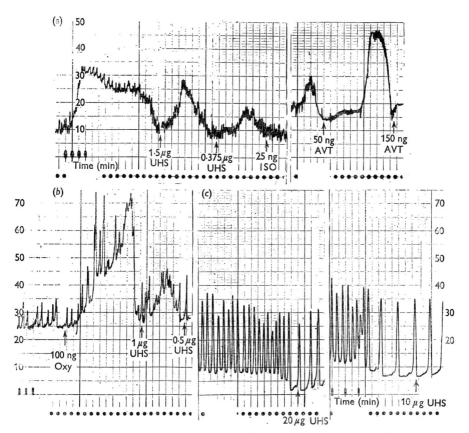

Fig. 4. Responses of the isolated ovary of *Lebistes reticulatus* to urophysial laboratory standard (UHS) and to neurohypophysial hormones (AVT = arginine vasotocin; ISO = isotocin; Oxy = oxytocin). Please read from right to left (*a*) 'Post-partum' fish. Note well-defined contractions. (*b*) Fish 'at term' but also some ova. Frequent rhythmic contractions superimposed on more prolonged responses; no response to 100 ng oxytocin. (*c*) 'Immature' fish (ovary filled with small ova). Only rhythmic rapid contractions with an indication of dose-dependence (shorter latent period after 20 µg UHS than that after 10 µg UHS).

tions were elicited by 0·25–1 µg UHS. As compared to the responses of the *Lebistes* ovarian muscle to neurohypophysial peptides (25 ng of AVT, Oxy, etc., or more), the 'urotensin' contractions appeared to be of a greater magnitude with a slow onset and longer duration. When the

Lebistes ovary was used from fishes in earlier reproductive stages (Fig.
4*b*, *c*) rhythmic and frequent contractions were elicited by urotensin; on
some occasions these were superimposed on more powerful contractions
of longer duration. These preliminary observations do not permit a
speculation on the possibility of an involvement of the urophysis in
reproduction in fishes.

Chemical nature of urotensin

In addition to the pharmacological tests and the attempts at inactivation
shown earlier (Table 2), experiments were initiated with the ultimate aim
of establishing the chemical nature and the molecular structure of urotensins (Geschwind *et al.* 1968).

The findings after chromatography on Sephadex G-25 are in agreement
with those of Bern *et al.* (1967) and Chan (1969); that is, the separation
of the initial activity into one or two 'low molecular' and one 'high
molecular' peaks was indicated. Crude extracts (UHS), and the active
'low molecular' peaks (trout bladder-contracting), were subjected to ultrafiltration through Amicon membranes (Geschwind *et al.* 1968; Geschwind
et al., unpublished observations). The results indicated that, in addition
to a low level of activity retained on the 'high molecular' filter (about
mol.wt 50,000), the bulk of activity was associated with substances which

Table 4. *Ultrafiltration of urophysial extracts through Amicon membranes*

	Activity in fractions (μg UHS)	
Filters	Whole urophyses	G25 IV and V
XM-50 (> 50,000)	4·5	5·9
UM-10 (> 10,000)	.	.
UM-2 (> 1,000)	9·6	16·3
Not retained (< 1,000)	16·1	5·0

were not retained, or only partially retained, on membranes of a molecular
weight cut-off of about mol.wt 1,000 (Table 4). In one experiment, in
which the 'low molecular' fraction obtained from chromatography on
Sephadex G-25 had been applied to the ultrafiltration membranes, a small
proportion of bladder-contracting activity was 'retained' at the 'high-
molecular' cut-off (Table 4*b*). The significance of this observation cannot
be determined at this stage. The possibility cannot be excluded that this
observation was due to an experimental artifact; otherwise a complex
formation, after a previous fractionation, would have to be postulated.

Subcellular localization of urotensin in the urophysis

Distribution of the urotensin activity (trout bladder-contracting) was estimated in cell fractions obtained by differential centrifugation of homogenates of urophyses of the trout or the *Gillichthys* according to a modification of the procedure of Lederis & Heller (1960). Of the total activity in the homogenate, 40–90% could be sedimented depending on the degree of homogenization and on the density of the suspension medium employed (hand- or motorized homogenization; 0·25–0·44 M sucrose; Bern, Nishioka & Lederis, unpublished observations). Table 5 shows the distribution of activity in fractions in a series of experiments in which urophyses had been homogenized in 0·44 M sucrose using a Waring blender (2000 rev/min for 10 s). Bern *et al.* (1967) had found that, after

Table 5. *Differential centrifugation of trout urophysis homogenates in* 0·44 M *sucrose* (*means* ± S.E. *from* 8 *experiments*)

	Activity in fractions	
	Distribution (%)	Range
Fraction 1. 8,000 *g* for 10 min	18·9 ± 2·7	5·6 − 29·7
Fraction 2. 40,000 *g* for 1 h	43·4 ± 4·3	29·7 − 63·3
Fraction 3. Supernatant	38·0 ± 4·0	20·9 − 52·5

similar centrifugation procedures had been employed, all the eel hypertensive activity remained in the final supernatant (i.e. was not sedimentable). It remains to be seen whether the difference in the subcellular distribution of the urotensin activity in these two series of experiments was due to variation in experimental procedures in the two laboratories or whether the eel hypertensive principle(s) are not associated with subcellular particles. It is of considerable importance to establish conclusively which activities are, and which are not, associated with ('stored in') subcellular particles. So far the active principles (hormones) in all endocrine glands have been found to be lodged in subcellular storage organelles. If the active principles in the urophysis are the products of secretory activity by the caudal neurosecretory neurones, then they are likely to be lodged in subcellular particles which are produced in the cell bodies, transported to, and stored in, the neurohaemal organ—the urophysis. The possibility arises, therefore, that any biologically active principles which cannot be sedimented from homogenates of the urophysis under the usual conditions of differential centrifugation, occur in the cell sap and may be produced by processes other than the accepted neurosecretory mechanisms.

The experimental data on the chemical nature of urotensins available to date, after the preliminary attempts at purification of the active principles of the urophysis, do not warrant any definite conclusions on the identity of the active substances. Nor can it be decided with certainty whether two, three, or more active principles occur in the urophysis. It seems likely that the bladder-contracting activity, and one or more of the vasoactive principles, may be peptidic in nature and of a relatively low molecular weight (mol.wt 1,000 or less). Furthermore the molecules may either contain no structural components with SH-groups, or, if such groups are present, they are of no importance for biological activity (activity not destroyed by treatment with sodium thioglycollate).

REFERENCES

BENNETT, M. V. L. & FOX, S. (1962). Electrophysiology of caudal neurosecretory cells in the skate and fluke. *Gen comp. Endocr.* **2**, 77–96.

BERN, H. A. & LEDERIS, K. (1969). A reference preparation for the study of active substances in the caudal neurosecretory system of teleosts. *J. Endocr.* **45**, xi–xii.

BERN, H. A., NISHIOKA, R. S. and CHESTER JONES, I., CHAN, D. K. O., RANKIN, J. C. & PONNIAH, S. (1967). The urophysis of teleost fish. *J. Endocr.* **37**, xl–xli.

CHAN, D. K. O. (1969). Pressor substances from the caudal neurosecretory system of teleost and elasmobranch fish. *Gen comp. Endocr.* (In the Press).

DAHLGREN, U. (1914). On the electric motor nerve-center in skates (Rajidae). *Science, N.Y.* **40**, 862.

ENAMI, M. (1959). The morphology and functional significance of the caudal neurosecretory system of fishes. In *Comparative Endocrinology*, pp. 697–724 (ed. A. Gorbman). New York: John Wiley.

FRIDBERG, G., BERN, H. A. & NISHIOKA, R. S. (1966). The caudal neurosecretory system of the isospondylous teleost, *Albula vulpes*, from different habitats. *Gen. comp. Endocr.* **6**, 195–212.

FRIDBERG, G., IWASAKI, S., YAGI, K., BERN, H. A., WILSON, D. M. & NISHIOKA, R. S. (1966). Relation of impulse conduction to electrically induced release of neurosecretory material from the urophysis of the teleost fish. *Tilapia mossambica. J. exp. Zool.* **161**, 137–50.

GESCHWIND, I. I., LEDERIS, K., BERN, H. A. & NISHIOKA, R. S. (1968). Purification of bladder-contracting principle from the urophysis of the teleost, *Gillichthys mirabilis. Am. Zool.* **8**, 758.

HONMA, Y. (1966). Some evolutionary aspects of neural control of internal secretion in the ichthyoform animals. *Rep. Jap. Soc. Syst. Zool.* **2**, 6–11.

HONMA, Y. & TAMURA, E. (1967). Studies on the Japanese chars of the genus *Salvelinus*. IV. The caudal neurosecretory system of the Nikko-iwana, *Salvelinus leucomaenis pluvius* (Higendorf). *Gen. comp. Endocr.* **9**, 1–9.

HUMMEL, B. C. W. (1959). A modified spectrophotometric determination of chymotrypsin, trypsin and thrombin. *Can. J. Biochem. Physiol.* **37**, 1393–5.

KOBAYASHI, H., MATSUI, T., HIRANO, T., IWATA, T. & ISHII, S. (1968). Vasodepressor substance in the fish urophysis. *Annot. Zool. Jap.* **41**, 154–8.

KOBAYASHI, H., UEMURA, H., OOTA, Y. & ISHII, S. (1963). Cholinergic substance in the caudal neurosecretory storage organ of fish. *Science, N.Y.* **141**, 714–16.

LACANILAO, F. (1969). Teleostean urophysis: stimulation of water movement across the bladder of the toad *Bufo marinus. Science, N.Y.* **163**, 1326–7.

LEDERIS, K. (1969*a*). Teleostean urophysis: stimulation of rhythmic contractions of the bladder of the trout (*Salmo gairdneri*). *Science, N.Y.* **163**, 1327–8.

LEDERIS, K. (1969*b*). Teleost urophysis: I. Bioassay of an active principle on the isolated urinary bladder of the rainbow trout (*Salmo gairdneri*). *Gen. comp. Endocr.* (In the Press.)

LEDERIS, K. (1969*c*). Teleost urophysis: II. Biological characterization of the bladder-contracting activity. *Gen. comp. Endocr.* (In the Press.)

LEDERIS, K. & BERN, H. A. (1969). Teleost urophysis: A reference preparation ('urotensin') and its biological activity spectrum. *Gen. comp. Endocr.* (In the Press.)

LEDERIS, K. & HELLER, H. (1960). Intracellular storage of vasopressin and oxytocin. *Proc. Ist Int. Congr. Endocr.*, Copenhagen, pp. 115–16.

MAETZ, J., BOURGUET, J. & LAHLOU, B. (1964). Urophyse et osmorégulation chez *Carassius auratus. Gen comp. Endocr.* **4**, 401–14.

NANDI, J. & BERN, H. A. (1965). Chromatography of corticosteroids from teleost fishes. *Gen. comp. Endocr.* **5**, 1–15.

NISHIOKA, R. S. (1968). In G. Fridberg and H. A. Bern: The urophysis and the caudal neurosecretory system of fishes. *Biol. Rev.* **43**, 175–99.

ROY, B. B. (1962). Action of zinc on the hypothalamo–pituitary–adrenal axis of the fish, guinea-pig and dog and on different experimental preparations. *Proc. natn. Acad. Sci. India* B, **28**, 13–28.

SANO, Y. (1958). Weitere Untersuchungen über den Feinbau der Neurophysis spinalis caudalis. *Z. Zellforsch. mikrosk. Anat.* **48**, 236–60.

SAWYER, W. H. & BERN, H. A. (1963). Examination of the caudal neurosecretory system of *Tilapia mossambica* (Teleostei cichlidae) for the presence of neuro-hypophyseal-like activity. *Am. Zool.* **3**, 555–6.

SPEIDEL, C. C. (1919). Gland-cells of internal secretion in the spinal cord of the skates. *Papers Dep. mar. Biol. Carnegie Instn Wash.* **13**, 1–31.

SPEIDEL, C. C. (1922). Further comparative studies in other fishes of cells that are homologous to the large irregular glandular cells in the spinal cord of the skates. *J. comp. Neurol.* **34**, 303–17.

STERBA, G., LUPPA, H. & SCHUHMACHER, U. (1965). Untersuchungen zur Funktion des kaudalen neurosekretorischen Systems beim Karpfen. *Endokrinologie* **48**, 25–39.

UEMURA, H., KOBAYASHI, H. & ISHII, S. (1963). Cholinergic substance in the neurosecretory storage-release organs. *Zool. Mag.* **72**, 204–12.

YAGI, K. & BERN, H. A. (1963). Electrophysiologic indications of the osmoregulatory role of the teleost urophysis. *Science, N.Y.* **142**, 491–3.

DISCUSSION

VAN OORDT: Dr Lederis, among the fascinating results of your work I was particularly struck by the effects you obtained with urophysial extracts upon the ovary of the guppy. You observed a contraction of the ovary when the extract was administered after parturition.

Would you be willing to speculate on the possible physiological significance of this finding? In the guppy the ovary shows a clear gestational cycle of approximately 4 weeks and thus parturition is a cyclical phenomenon. Cyclical contraction of the ovary could be due to either a cyclical activity of the urophysis or to a cyclical change in sensitivity of the ovarian muscle cells to the urophysial hormone. In Utrecht, Lambert (thesis, University of Utrecht) has shown that during this cycle ovarian steroid production is mainly restricted to the period of some days before to some days after parturition. It is not impossible that the presence of these steroids might render the ovary sensitive to urotensin and that in the absence of a strong steroid production in the ovary during gestation the organ is less sensitive to urotensin and will thus not show contraction during that period.

LEDERIS: I'm glad you use the word 'speculate'. I could not do more than just speculate about the urophysis in reproduction at this stage. We would like to believe that the urophysis may have a physiological role in reproduction in fishes, but we have not as yet got any firm evidence to support it. In addition to the experiments I have described here, we have some preliminary findings on the reproductively migrating Pacific salmon, that is, salmon before they move into brackish water from the Pacific; the fish in the brackish water, and thirdly salmon at the spawning grounds. We found that the urotensin activity (urotensin is the name Dr Bern and I suggested for the proposed reference preparation) was considerably higher in the ocean fish than in the delta fish. We are not yet certain about the urotensin activity in the spawning fish. It would appear that the activity present in the urophysis of the fish that are about to spawn is lower than that in the fish that are still assembling outside the Golden Gate. Whether these changes are related to reproductive stages or whether they are simply due to changes in the osmotic environment we cannot say at this stage. Equally, I cannot offer an explanation about the effects, or the likely importance of the effects, of urophysial extracts on the *Lebistes* ovary at different stages of reproduction. I would like to make it clear that we have just started scraping the surface of the whole problem of the biological actions or of the possible physiological role of the urophysis. We hope to be able to answer questions such as yours in the not too distant future.

VAN OORDT: Thank you for this clarification. However, may I suggest that others who may follow you in studying this most interesting subject should follow your example in clearly stating the physiological situation of the target organ, for the urotensin preparation they are using?

LEDERIS: Yes, this is the main reason, or at least one of the several reasons, why we think that the adoption of a common reference preparation (a common laboratory standard) would serve a very important role in standardizing further studies on the urophysis.

CHAN, D. K. O.: It is always a surprise that when two groups of people working independently on the same problem tackle it using almost identical methods—I refer to your chemical studies on the urophysial principle. Recently in Hong Kong I had the opportunity to repeat some of our early investigations done in Sheffield on the chemical nature of the urophysial substances. I was able to confirm that there are indeed three biologically active fractions separable by gel filtration through Sephadex G-25 and G-10. One fraction was excluded from the columns together with large-molecular-weight proteins. Two other fractions were retained by both gel columns (Sephadex G-10 has an exclusion molecular weight of 700). The smaller fractions could also pass through Amicon membranes by ultrafiltration showing that their molecular weights are smaller than 1,000. Digestion by proteolytic enzymes produced some interesting results (Table 1). Trypsin, chymotrypsin and papain, which attack peptide, amide or ester bonds of amino acids, destroyed the pressor activity of the urophysial substances. However, pepsin, which breaks only peptide bonds, especially those of aromatic amino acids, and leucine aminopeptidase which attack the peptide bond of the amino acid carrying a free α-amino group at the N-terminal, had no effect. Granted that the molecular weight estimations by molecular sieving through Sephadex gels be accurate, our results seem to indicate that we might be dealing with a small amide or ester rather than a polypeptide. Do you have any data to give us further insight into the probable molecular structure of the urophysial active principles?

Table 1

Enzyme	Ability to destroy urophysial pressor activity	Bonds attacked
Trypsin	Yes	Peptide, amide or ester of arginine/lysine
Chymotrypsin	Yes	Peptide, amide or ester of aromatic and other amino acids
Papain	Yes	Peptide, amide or ester
Pepsin	No	Peptide bonds (only) of aromatic and other amino acids
Leucine aminopeptidase	No	Peptide bond of amino acid at N-terminal and contains free α-amino group.

LEDERIS: Thank you very much for your very interesting slide Dr Chan. I would prefer not to comment any further on the chemical nature of the active substances or on our thoughts on it, because I feel we have not reached the stage where more speculation could be fruitful. I would only like to add one gentle word of caution on your interpretation of the molecular weights, namely, that I think those who are familiar with gel filtration in the Sephadex systems will agree that molecular size is not necessarily given by the rate at which a given substance is eluted from the column. One has to consider a number of physico-chemical properties of the substances in question before one can speculate on molecular weights on the basis of chromotography in G-25. While I have no evidence against your speculation, I would prefer to remark that this should not be used as more than a speculation, because gel filtration may be affected by many factors, such as electrical, aromatic, etc., properties of the molecules in question. I would be inclined to exercise caution at this stage on the molecular size and chemical identity of the urophysial principles.

MARTINI: Has thioglycollate inactivation been attempted on your substance?

LEDERIS: Yes, Professor Martini, we attempted that at a time when we were searching for a neurohypophysial peptide-like activity in urophysial extracts. From incubation experiments with trypsin, chymotrypsin and sodium thioglycollate we came to the conclusion that (a) we might be dealing with a peptide, (b) that if it was a peptide, it is a peptide not like a neurophysial peptide, at least it appears not to have sulphydryl groups in the molecule which are of importance for biological activity. Thats about all we can say.

BERN: I think one important point has to be brought out again and this is the problem of multiple factors in a broader sense than Dr Chan and Dr Lederis have been discussing. We still have a few additional effects to cope with. One of them is the very important effect on sodium uptake described by Dr Maetz and his colleagues a few years ago, although he may wish to say something about this himself. I know something of his thinking and I imagine that maybe one can account for this gill effect on the basis of haemodynamic changes. Maybe this is not necessarily going to prove to be a different factor, but the hydrosmotic influence that Lacanilao has described remains outside the pale as far as the discussion up to now is concerned. The fractions which have been isolated, the ones to which Dr Lederis referred, don't have hydrosmotic activity. At the moment— the moment being about 3 weeks ago—Lacanilao was prepared to say that there was nothing that he had done at the present juncture which would distinguish his hydrosmotic principle from arginine vasotocin. What arginine vasotocin is doing down in the urophysis none of us at the moment has any idea, except some extremely wild ones. Let me add one more thing. Just before I left Berkeley, Professor Fleming told me that he and Professor Potts had been injecting some of our materials into *Fundulus kansae* and were getting major changes in ion fluxes, which they were then preparing to examine again using our several fractions. I would emphasize that whereas we may be arriving at some common feelings about blood pressure and bladder contracting, and indeed, ovary/oviduct-contracting activities, there is still the issue of possible effects on water and ion balance, which may be due to a different principle.

LEDERIS: May I say a few words in reply to Dr Bern. I simply did not mention the effects on ionic and water fluxes because they are not pertinent to the work on which I reported today. Because of limitations on the time available, I neither mentioned the work of Dr Maetz and his colleagues nor did I mention the observations made by Drs Chan and Chester Jones and their co-workers on ionic fluxes.

MAETZ: I would simply like to add that the salt transport activity of the urophysis in the goldfish is lost after treatment with acetic acid and heat. This may mean that this activity is independent of haemodynamic activity on the gill. I wonder if the preparation of Professor Geschwind is based only on the acetic extract. If it is, this may mean that important activity is lost.

LEDERIS: The batch which I mentioned to you earlier on was in acetic acid, but we have used fresh urophyses in some experiments done together with Dr Geschwind. This is an important point; in fact, this subject has cropped up from time to time in discussions with Dr Bern. The possibility remains that the sodium flux-influencing factor may be the same as the Lacanilao hydrosmotic factor, which is also, as far as I recollect, not acid-stable. The earlier findings of Lacanilao indicated that the activity was not acid-stable, but in his more recent experiments Lacanilao has found, I think, that it is acid-stable. Dr Maetz's findings indicated that treatment of extracts with weak acids destroyed the sodium flux-affecting activity.

CHAN, D. K. O.: One of the main concerns for those working on the urophysis is whether we are dealing with one or more classes of compounds. My own work with 12 species of teleost fishes and one species of elasmobranch fish showed that the caudal neurosecretory system always contains substances that are vasopressor in the eel. Does anyone have any data to show whether elasmobranch caudal spinal cord also contains vasotocin?

LEDERIS: I would like to quote the report of Sterba and his colleagues (Sterba *et al.* 1965, see my paper), who claim to have found arginine vasotocin-like and isotocin-like peptides, plus neurophysin-like substances in the carp urophysis. The dosages used by Sterba and co-workers are 10,000–30,000 times higher than the ones we use on the bladder, or, if you like, at least 1,000–3,000 times higher than the dosages you use on the eel. The answer to the question on the presence of neurohypophysial-like peptides might therefore be self-explanatory.

CHAN, D. K. O.: Our recent work in fact shows that the threshold dose for the pressor activity in the Japanese eel, *Anguilla japonica*, is in the order of 0·1 μg/kg of the partially purified compound (gel-filtration through Sephadex G-25, lyophilized and chromatographed through Sephadex G-10).

One other question I wish to raise is this. When we talk about stability in acid, what do we actually mean? Do we mean stability in acetic acid (pH 2) or acid hydrolysis? Our material is definitely stable in an acid pH at room temperature.

LEDERIS: When I referred to acid stability I was only referring to 0·25 % acetic acid but Dr Kobayashi used 1 N hydrochloric acid, and when I first read his paper I had the thought that acid hydrolysis could not be excluded in this case. At present I am inclined to think that this is not so, because the weak acid extracts and also watery extracts that we have used on the rat blood pressure produced similar effects to those described by Kobayashi *et al.* (1968) (see my paper).

CHAN, D. K. O.: In our hands, extraction of urophyses with acetic acid, 0·1 N-HCl, distilled water or saline, all gave identical results.

LEDERIS: Our observations agree with that in principle.

WEIR: Is it possible to ablate the urophysis, and if so, what effect does it have on the fish?

LEDERIS: Yes. However, Dr Bern is in a much better position to comment on that because he has done such experiments.

BERN: It is a very easy thing to ablate initially but it is not easy to determine the usefulness of this procedure because of the tendency, at least in some species of teleosts, for it to regenerate almost immediately. I'm just not sure what urophysectomy means, physiologically speaking. However, Ireland (*J. Endocr.* **43**, 133) has recently reported effects on survival after ablation of the urophysis. My former colleague Dr Fridberg tells me that the regeneration phenomenon, which he described in such detail, after complete removal of the system from *Tilapia*, does not apply to the fish he uses in Sweden, namely *Leuciscus rutilus*. So I do not know how generalized the phenomenon of neogenesis may be when the whole system is removed. In summary, then, the urophysis is an easy thing to take out, but removal does not necessarily allow one to conclude that one has thereby dispensed with urophysial factors.

SESSION VI
TEMPORAL CHANGES IN ENDOCRINE SECRETIONS

HORMONES AND INSECT DIAPAUSE

By J. de WILDE

INTRODUCTION

As insects depend on plants and animals for their subsistence, geographical and temporal coincidence of their active stages with their host plant, host animal or prey is a primary condition for each species. No less important is their adaptation to the latitudinal climatic zonation within their areas of dispersal, as well as to regional climatic differences within the zones.

The main problem with which the insect is faced is how to cover the risks presented by periodic cold and arid conditions and by absence of food. As these conditions mostly follow a seasonal pattern, the problem is often one of synchronization of the active stage with the favourable season. Next to this, there is the problem of surviving the immediately adverse conditions presented by winter or dry, hot periods; a problem of resistance and protection.

Hibernation and aestivation in insects often takes place in a state of diapause, a state of arrest in reproduction and morphogenesis, during which the resistance to adverse climatic conditions is increased by physiological and sometimes morphological features fitting together into an 'adaptation syndrome' varying from species to species.

The arrest can be complete or partial, and may even be limited to individual organ systems. Ovarian diapause in some mosquitos is known as 'gonotrophic dissociation'. Intraspecific adaptation to climatic zones is generally not effected by variation of cold resistance but by variation of responses to environmental tokens. This has recently been borne out by extensive experiments in the USSR (Danilevskii & Kuznetsova, 1968). With five species of Lepidoptera, transfer experiments were made from mediterranean to boreal climates. Geographic races from the Black Sea coast and from the Leningrad region differed very little in lethal minimum temperature in the diapausing stage, but largely in photoperiodic response (Fig. 2).

Seasonal tokens (signals) seldom consist of the adverse factors (cold, drought) themselves, but are mostly stimuli correlated with seasonal time preceding the adverse conditions: photoperiod, temperature, plant factors. They do not act directly on the effector system, but rather have a semantic effect: they are 'translated' by the organism to induce the required adaptation syndrome.

Most remarkably these stimuli often by-pass the sense organs and have a direct effect on the central nervous system (Williams, 1946; Lees, 1964; Williams & Adkisson, 1964). This 'extrasensory information' is apparently of basic importance to survival.

Synchronization partly implies time measurement and partly adjustment to critical environmental factors: availability of adequate food and appropriate temperatures. Time measurement generally takes place on account of the daily photoperiod synchronizing the insect's activity with the astronomic season. Corrections are provided by temperature and food. High temperatures annihilate the effect of short-day and thus tend to prolong the period of growth and reproductive activity. The insect's activity thus coincides with the climatic season.

Ageing of the host plant often provides a third type of token stimulus which may abruptly bring growth and reproduction to a standstill. The termination of the insect's seasonal activity is thus synchronized with the end of the vegetative season of the host plant (Müller, 1962; de Wilde & Ferket, 1967).

While photoperiod is the major ecological factor inducing diapause, break of diapause is generally brought about by a more or less prolonged stay at relatively low temperatures. However, more and more cases are known where photoperiod is involved in the break of diapause. In such cases, photoperiods preventing the onset of diapause may also bring about its termination.

It may seem strange that such extremely different stimuli as photoperiod (photodynamic reaction) temperature (thermodynamic effect) and chemical plant substances may induce one and the same syndrome. This can only be understood if we assume with Ernst Scharrer (1966) that the syndrome is governed by a centre which integrates the environmental impact and leads it into a common path. It will be shown that this centre is located in the brain, and that the common path is provided by the cerebral neurosecretory cells and tracts (Fig. 1).

Aestivation diapause, hibernation diapause, quiescence, wintersleep and their subdivisions have been subjected to detailed nomenclature and elaborate definitions, mainly by ecologists (Way, 1962; Müller, 1965; Ushatinskaya, 1966).

To the physiologists, guided by the thoerem of Claude Bernard ('On ne definit pas les phénomènes naturelles, on ne definit que les produits de l'esprit humaine'), this tendency to undue definition has little appeal. The analysis and description of the syndromes observed in different types of dormancy, their endocrine basis and the mechanism of their induction by environmental tokens have long been a fascinating field of study.

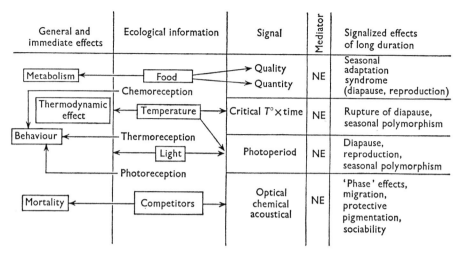

Fig. 1. Scheme of the dual effect of ecological factors. Signal or 'token' information is channeled through the neuroendocrine system (NE) to provide long-term effects of an adaptive nature, e.g. diapause.

EXTRINSIC CONTROL OF DIAPAUSE

Diapause occurs in eggs, larvae, pupae and adults of a great many insect species. Different, but even closely related species may have a different diapause stage, but in each species it usually occurs in one characteristic stage or larval instar.

In some insect species diapause occurs during each generation. In these 'univoltine' species diapause has long been considered as being obligatory. But the list of such cases is continuously being reduced, as highly specific conditions are often required to prevent diapause, and it has taken precise experimentation to detect the critical factors. Sometimes only a very narrow band of photoperiods will allow the active condition to occur. In cases of facultative diapause, which form the great majority, diapause is known to be induced by environmental conditions.

Such insects may have several generations per year, and are indicated as bi-, tri- or multi-voltine. Aphids may pass through seven or eight generations before entering diapause in the winter egg stage.

Environmental information is received during a sensitive stage, which may be more or less remote from the responsive stage. Adult diapause in the Colorado beetle is mainly governed by photoperiods applied to the adult itself, but egg diapause in the commercial silkworm depends on photoperiods received during the egg stage in the previous generation. This delayed response implies the storage of environmental information, a process which at present has barely been touched by physiologists.

Extrinsic control of diapause is, in a sense, extrinsic control of growth, differentiation, reproduction. The role of the endocrine system in this control can only be understood on the basis of a thorough knowledge of the endocrine control of these functions in general. And, as this knowledge is still very incomplete, the study of diapause has often been at the base of new discoveries in insect endocrinology and, for that matter, in insect physiology in general.

DIAPAUSE TOKENS

The specific environmental conditions inducing diapause have the character of 'token' stimuli: chemical or physical energies of a specific nature or pattern 'translated' by the insect into highly specific physiological responses, and in this respect comparable with 'releasing stimuli' or 'releasing situations' known in ethology (Tinbergen, 1951).

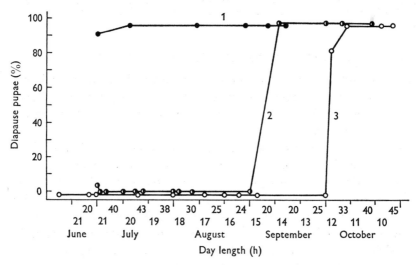

Fig. 2. Photoperiodic response curves of three geographic races of the Noctuid *Acronycta rumicis*, and lethal minimum temperatures of the hibernating (pupal) stage. Note that, while cold resistance does not differ appreciably, photoperiodic response shows adaptive differences between the races (after Danilevskii & Kuznetsova, 1968). 1, Leningrad population ($-22 \cdot 0$ °C); 2, Bielgorod populaton ($-22 \cdot 6$ °C); 3, Suchumi population ($-23 \cdot 6$ °C).

As diapause is often a seasonal phenomenon, photoperiod is perhaps the most basic factor, as it varies with unfailing precision during the seasons and is thus a very accurate expression of seasonal time.

Photoperiodic response follows a response curve characterized by a steep change at an intermediate value named critical photoperiod (Fig. 2). The position of this critical photoperiod is often a racial character of the insect, and is then related to the geographical latitude at which the specific geographical race occurs. It may further be modified by temperature. As follows from the curves given in Fig. 2, in some insects diapause is induced by shorter photoperiods than those allowing for growth and reproduction. These 'long-day insects' are the prevailing types. In the inverse

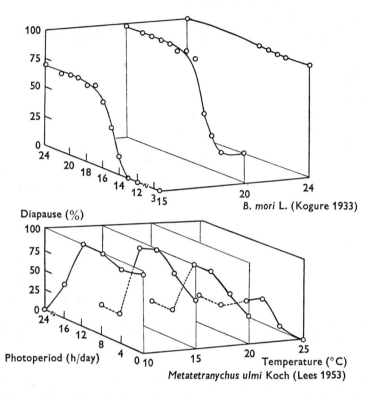

Fig. 3. The combined effect of photoperiod and temperature in a short-day insect (*Bombyx mori*), and in a long-day species, the fruit-tree red spider, *Metatetranychus ulmi*. Note that high temperatures abolish the effect of short-day and tend to shift the critical photoperiod towards lower values.

case, the term 'short-day insect' is applied. There are also cases where all photoperiods, with the exception of a very narrow band, induce diapause. In this case, an insect may be univoltine and still not have obligatory diapause.

Temperature may modify photoperiodic response in such a way that high temperatures tend to annihilate the effect of short day, and may shift the position of the critical photoperiod of long-day insects towards lower values (Fig. 3).

Temperature may also be a diapause-inducing factor by itself, as is the case in the Khapra beetle, *Trogoderma granarium* (Burges, 1959), where even a slight increase in temperature of 5 °C may induce the beetles to leave their food and enter diapause.

The quantitity and condition of the food is a third category of factors providing tokens in diapause induction. In the fruit-tree red spider, *Panonychus ulmi*, severe leaf damage (bronzing) induces the formation of winter females (Kuenen, 1946; Blair & Groves, 1952). In the Colorado beetle adult, feeding for less than 10 h per day results in diapause (de

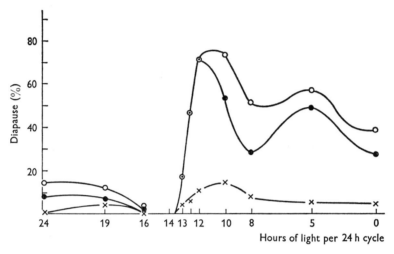

Fig. 4. Effect of food quality on photoperiodic response in *Pectinophora gossypiella* (after Adkisson, 1961). ○——○, Cottonseed meal (5 % fat); ●——●, cottonseed meal (1 % fat); ×——×, wheat germ (0·3 % fat).

Wilde, Duintjer & Mook, 1959) even under long photoperiods. This condition is also induced when the beetle is fed with physiologically aged potato leaves (de Wilde & Ferket, 1967). In *Aleurochiton complanatus*, a long-day insect, diapause is entered when the host plant is ageing, even under photoperiods above 16 h (Müller, 1962).

In the red bollworm, *Pectinophora gossypiella*, diapause only occurs if the fat content of the diet reaches a value of ± 5%, a situation prevailing in cotton seed in an initial phase of maturation (Adkisson, 1961) (Fig. 4).

Break of diapause is generally effected by a more or less prolonged stay at relatively low temperatures (5–10 °C), a treatment comparable with 'jarovization' of plants. On the contrary, the Colorado beetle needs a temperature of 30 °C for diapause termination, while low temperatures delay this process (de Wilde, 1957). There are also cases where, after low-temperature treatment, long photoperiods are needed to obtain activation, as is the case in pupal diapause of *Antheraea pernyi* (Williams & Adkisson, 1964). And, finally, there is a growing list of insects in which diapause is broken only by day length. Examples are the Lepidoptera, *Dendrolimus pini* and *Euproctis chrysorrhoea* (Geyspitz, 1949, 1953), *Ostrinia nubilalis* (McLeod & Beck, 1963) and *Pectinophora gossypiella* (Adkisson, 1964). In these cases photoperiods which, when applied permanently, would prevent diapause promote its termination.

Other signals from the environment can also be utilized, correlated with food condition or other essential environmental requisites. The hypopus of the mite *Histiosoma* (Perron, 1954) is activated by the smell of yeast. The dormant larva of the Tachinid parasite *Eucarcelia rutilla* is activated by the moulting hormone of its host (Schoonhoven, 1962).

THE MECHANISM OF PHOTOPERIODIC INDUCTION OF DIAPAUSE

In an earlier paper (de Wilde, 1962) a distinction has been made between photoperiodic induction (a reversible, partly photodynamic process) and photoperiodic determination (the induced state of the over-all controlling centres of growth and reproduction). Although in an increasing number of cases diapause appears to be under constant photoperiodic control, the distinction still seems to be useful in those cases where delayed photo-periodic responses occur, and the sensitive stage is remote from the res-ponsive stage.

It is a striking feature that the ocelli or compound eyes have not been found to be the site of photoperiodic reception (Tanaka, 1950; de Wilde *et al.* 1959; Lees, 1964). Though there are indications that in some insect species the midbrain is the region where this reception takes place (Lees, 1964; Williams, 1969), it is not widely believed that the medial proto-cerebral neurosecretory cells are directly involved in this reception, for, in the case of delayed responses, these cells are supposed to undergo one or more cycles of activity before the quiescent state is entered.

As reported by Lees (1968), in the photoperiod-sensitive aphid *Megoura vitiae*, the medial neurosecretory cells (MNSC) do not reveal clear differ-ences in secretory activity after long and short day, as judged by histological and ultrastructural criteria.

In the case of *Megoura*, photoperiodic response is not an arrest but a modification of morphogenesis. It is therefore not to be expected that the MNSC will be inactivated or activated by antagonistic photoperiods, but rather that a differential change in the secretion of some NS components will take place. In this respect, aphids, with only one pair of MNSC, are not favourable material, as one cell is supposed to regulate a series of functions.

In the Colorado beetle, investigated by Schooneveld (in de Wilde, 1969a), the situation is different. Here the alternative responses are diapause and reproduction, accompanied by remarkable changes in the histological picture of the neurosecretory (NS) A-cells, but not in some of the other five types of NS cells distinguished by this author. Under short-day (SD) treatment, inducing diapause, the changes occur very gradually in the course of about 2 weeks, and become most expressive when diapause has started.

To obtain diapause in *Leptinotarsa*, a number of 8–10 SD cycles is required. In the Lepidopteron *Diataraxia oleracea*, between the last larval moult and the fifth day of the last larval instar, two or three short day cycles suffice to induce pupal diapause.

Neither the nature nor the location of the photosensitive pigment transferring light energy to the neurones in photoperiodic reception have been elucidated. In the European corn borer, *Ostrinia nubilalis*, where diapause occurs in the last larval instar, according to Beck (1964) two centres are involved in the photoperiodic termination of diapause: one of them located in the brain, the other in the posterior intestine. The intestinal light-detectors are identified as epithelial ridges projecting from the inner line of the ileum. These cells would also form an endocrine centre producing a hormone 'proctodone', the function of which would be to activate the cerebral neurosecretory cells after these have been made receptive by long photoperiods. No confirmation of this situation in any other insect has as yet been obtained.

The mechanism of time measurement is a most controversial aspect of photoperiodic induction. The crucial problem is how insects distinguish between photoperiods differing by no more than 15–30 min.

It has been known for many years that neither the total amount of light nor the absolute length of the photophase nor the ratio of the durations of photophase and skotophase are specifically determining photoperiodic response. As neither the ultrastructure of the NS nor their histochemical properties give any clue as to their cyclic reaction to light and darkness, indirect methods have been used, deriving conclusions from the effects of independent variation of photo- and skotophase and of total cycle length.

In this manner, attempts have been made to prove or disprove the following two hypotheses: (a) photoperiodic induction involves a two-phase timing process, measuring the length of photophase and skotophase with reference to an 'innate' response to a critical photoperiod. The results of Lees (1953) with *Panonychus ulmi*, of Dickson (1949) with *Grapholitha molesta*, and of de Wilde (1965) with *Leptinotarsa* have been interpreted in this manner. The fact that the dark period is especially temperature-sensitive has been considered as an argument in favour of this idea: (*b*) photoperiodic induction is related to an endogenous circadian rhythm

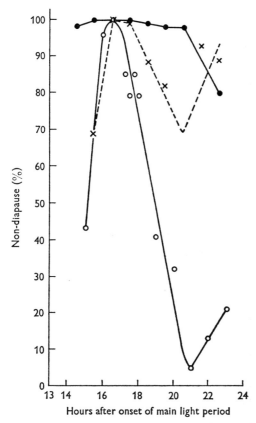

Fig. 5. Effect of light interruptions during the skotophase in *Adoxophyes reticulana*. ———, Basic photoperiod 13 h; – – –, basic photoperiod 14 h. Note that a 1 h light break already saturates the long-day induction system. Normal response to short-day is diapause. Light interruptions: ●———●, 1 h; ○———○, 2 min; ×– – –×, 2 min. After Ankersmit, 1968.

which is present in all cellular functions of an organism, and which determines the receptivity of the central nervous system to the 'on' or 'off' switch of light. The fact that in some cases a flash of light, applied at a critical moment in the skotophase, may change the effect of SD into one of long-day (LD), and may, for example, prevent the induction of diapause, renders some probability to this supposition. It is not possible in the frame of this discussion to sum up all the evidence which has been considered in favour of either of the two hypotheses. If the method of 'light breaks' provides a crucial test in deciding between one or the other hypothesis, it would seem that there is place for each of the two mechanisms, and that some insects rely on the circadian apparatus and others on the 'two-phase timer'. This may be illustrated with two examples.

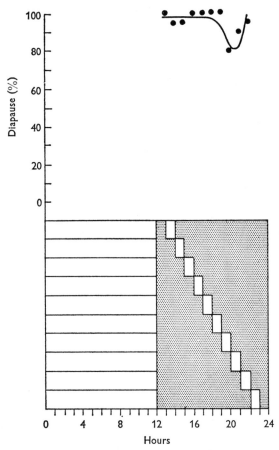

Fig. 6. Absence of effect of 1 h light interruptions during the skotophase in *Leptinotarsa decemlineata*. After de Wilde, 1965.

In *Adoxophyes reticulana* the fruit-tree leafroller, diapause occurs in the last larval instar and is induced by photoperiods below 16 h. The effect of short photoperiods of very different duration is annihilated by a light break of 2 min and even by a photoflash of 10 msec applied precisely 16 h after the beginning of the photophase (Ankersmit, 1968) (Fig. 5). This seems difficult to explain by a two-phase timing mechanism measuring the length of phases.

At the other hand, in *Leptinotarsa decemlineata* light breaks applied during a long skotophase are only effective if their duration, added to the preceding or following photophase, would provide a LD, and if they are applied in immediate connection with the photophase (de Wilde, 1965; Hodek & de Wilde, unpublished) (Fig. 6). It would seem highly unlikely that the basic mechanism of photoperiodic time measurement would be a different one in different species. But we have to realize that an 'internal clock' is required in both cases to account for the existence of a critical photoperiod.

On the other hand, even in two related insect species, photoperiodic response and the incidence of diapause can be so entirely different that there is every reason to assume that diapause has been evolved independently many times in the evolutionary history of insects. The same could be true for photoperiodic response. In this respect it should be considered that the general sensitivity to light is an elementary property of all neurones which has lost its functional importance in metazoan evolution in almost all cases excepting photoperiodic response. And, as the nervous system displays chemical and electrical activity, photoperiodic induction may predominantly involve neurohumours in some insect species and nervous impulses in others.

THE DIAPAUSE SYNDROME

It goes without saying that the diapause syndrome differs greatly between embryonic, larval, pupal and adult diapause. But apart from these differences between developmental stages, the diapause syndrome varies in intensity and extent even between related species in the same stage. This is often due to preparatory 'prediapause' phenomena sometimes even performed in a former stage of development. Winter eggs from *Panonychus ulmi* differ morphologically from summer eggs (Beament, 1951). In the codling moth larval diapause is preceded by the spinning of a 'hibernaculum' cocoon, which is repaired when damaged during winter. And in the Colorado beetle a dramatic change in behaviour takes place before entering diapause, the beetle responding to granular substrates with characteristic 'burrowing' movements. This only stops at a depth of ± 25 cm in the soil, where diapause is entered. These examples suffice to demonstrate that the diapause syndrome is not simply an arrest of growth, development or reproduction.

Although diapause is in many respects an 'all-or-none' phenomenon, its intensity may differ in geographical races of an insect species (Danilevskii, 1965) and different photoperiods may induce diapause of different intensity in one and the same geographic race (Masaki, 1965).

Egg diapause may occur in different stages of embryonic development. In *Bombyx mori* it occurs at the 'germ band' stage, in *Lymantria dispar* at the end of embryonic development just prior to hatching.

Larval diapause may occur in any instar (*Dendrolimus pini*, Geyspitz, 1958), but this is usually the first (*Euproctis phaeorrhoea, Choristoneura fumiferana*) or the last (*Enarmonia pomonella, Adoxophyes reticulana*).

There may be some increase in volume, but never an ecdysis, with the exception of the Khapra beetle larva, where an occasional moult may occur leading to a larva decreasing in size (Burges, 1961). Inanition is generally prevailing during larval diapause, and though motility is arrested, at favourable temperatures stimulation may result in some locomotion. It is therefore not a matter of surprise that spontaneous electrical activity, levels of cholinesterase and cholinacetylase activity, and a cholinergic substance were demonstrated throughout diapause in the larval brain of *Ostrinia nubilalis* (Mansingh & Smallman, 1967).

Pupal diapause takes place before the pharate adult is being developed. Motility is retained in the abdominal segments, indicating the intact state of the intersegmental abdominal musculature.

Spontaneous electrical activity and cholinesterase and choline acetylase activity have been demonstrated in the brain of diapausing pupae of *Hyalophora cecropia, Antheraea polyphemus* and *A. pernyi* (Schoonhoven, 1963; Shappirio & Harvey, 1965; Mansingh & Smallman, 1967).

Adult diapause is characterized by inversion of the ovaria in female and the accessory glands in male insects. Where the ovaries had been active before diapause, oosorption is observed. In *Leptinotarsa* it has furthermore been shown that the flight muscles are atrophied. Motility has come to a complete standstill, but when transferred to favourable temperature and light conditions locomotion may occur within minutes. Needless to say, electrical and neurohumoral activity are retained in the brain (Schoonhoven, 1963).

The metabolic component of the diapause syndrome has attracted the interest of workers from early times. Low water and high fat content were considered to be essential features connected with cold resistance and the need for energy reserves. Both points of view have recently been criticized. It is now known that cold resistance is due to the changed composition of haemolymph and tissue fluids, especially the presence of rather high concentrations of glycerol (Wyatt & Meyer, 1959), and that the fat body contains, next to lipids, considerable amounts of protein reserves.

The occurrence of glycerol in the body fluids was first detected in the diapausing pupa of *Hyalophora cecropia*. It was shown to be due to special regulatory mechanisms (Harvey & Haskell, 1966) and not to largely anaerobic metabolism, as was first concluded from the reduced rate of oxygen consumption and the insensitivity to some metabolic inhibitors.

In the egg of the grasshopper *Melanoplus differentialis*, diapause is characterized by an extremely low rate of oxygen consumption. After break of diapause, respiratory intensity increases more than tenfold. During diapause the eggs are remarkably insensitive to cyanide and carbon monoxide, indicating a very reduced significance of the cytochrome system. Sensitivity increases steadily as cytochrome oxidase is synthesized during postdiapause (Bodine, 1934; Bodine & Boell, 1934; Boell & Bodine, 1938).

In the diapausing pupa of *Hyalophora cecropia* metabolism is also of the carbon monoxide and cyanide-insensitive type. Spectroscopical and enzymological studies showed that cytochrome oxidase was present in low levels, but that cytochrome *b* and *c* could not be detected (Schneidermann & Williams, 1954).

It was originally concluded that the 'metabolism of morphogenesis' was specifically coupled with the cytochrome system, and that a distinction could be made between this metabolism and the 'metabolism of maintenance'. Resumption of growth in postdiapause would depend on the resynthesis of cytochromes (Schneiderman & Williams, 1953). Subsequent studies have, however, revealed that there is a co-ordinated decrease of metabolic activity in diapausing *H. cecropia* pupae, and that the pathways do not differ qualitatively from those in active tissues. The actual rate of respiration is not limited by the levels of respiratory enzymes but by the low energy demand as a consequence of the low rate of protein synthesis (Harvey, 1962; Mitchell, 1962). This again is related to a much reduced rate of RNA synthesis (Oberländer & Schneiderman, 1966).

The standstill of morphogenesis characteristic of pupal diapause was thereby brought back to the level of transcription in protein biosynthesis. Recent experiments (Bowers & Williams, 1964; Krishnakumaran & Schneiderman, 1964) suggest that the replication of DNA is the *primum movens* after break of pupal diapause.

The considerable reduction of respiratory intensity observed in adult insects during diapause has been related to the state of involution of some specific organ systems: the ovaries in the case of the heteropterous bug *Pyrrhocoris*, the flight muscles in *Leptinotarsa* (Slama, 1964; Stegwee, 1964). In the latter case it has even been shown that during the onset of diapause, the activities of mitochondrial and extramitochondrial enzymes reflect the rate of degeneration of separate elements of the flight muscle syncytium; the sarcosomes degenerating earlier than the muscle fibrils (de Kort, 1969). In the haemolymph of diapausing beetles, three specific SD proteins are observed which are absent in active, LD treated beetles. These proteins contribute to a high total protein concentration in the haemolymph during diapause—three times as high as observed in active long-day beetles (de Loof, 1969). It follows that in the adult *Leptinotarsa* also, protein biosynthesis is at the base of the diapause syndrome.

ENDOCRINE CONTROL OF DIAPAUSE

Hierarchical aspects of endocrine control

It is generally agreed that in insects the brain is the seat of neuroendocrine integration, and that the cerebral neurosecretory cells form an endocrine centre superordinated above other endocrine centres directly controlling growth and reproduction (Scharrer & Scharrer, 1963).

The cerebral neurosecretory cells are generally divided into a medial and a lateral group, the tracts of which leave the brain by the medial and lateral cardiac nerves.

As first proposed by Johansson (1958), the MNSC may be divided into several distinct categories on the basis of their staining characteristics when treated with Gomori chrome-haemotoxylin and with paraldehyde-fuchsin. In a detailed histological study with the bug *Oncopeltus fasciatus*, Johansson distinguished between A, B, C and D cells. The A cells appeared to be related to the stimulation of corpus allatum activity.

Subsequent authors have followed this distinction, though not always using the same criteria. Recently, the alcian blue–yellow technique and ultrastructural measurements have provided useful criteria in following the transport of NS granules from different cell types along NS fibres. Although histological distinction points to the existence of several NS hormones, these are generally indicated by the unified term 'brain hormone'.

The corpora cardiaca (CC) are considered to be partly 'neurohaemal organs' consisting of the endings of NS fibres transmitting the brain hormone to the haemolymph. Partly they have their own secretory and NS cells. Partly also, reflectory and NS nerve fibres pass through the CC to the corpora allata (CA). It is thought that the brain hormone extruded from the CC activates the prothoracic glands via the haemolymph, perhaps in a specific way (Williams, 1952), perhaps by a general stimulating effect on protein synthesis (Thomsen, 1952). It is also generally accepted that the brain hormone stimulates the CA. Most authors relate this function to the NS A-cells of the midbrain, but in the locust *Schistocerca paranensis*, according to Strong (1965), it would reside in the lateral group of cerebral NS cells. Inhibitory effects on CA activity would be provided by innervation from the brain. The chemical nature of the brain hormone is as yet uncertain (Williams, 1967).

The prothoracic glands are believed to secrete the steroid hormone ecdysone, the CA produce the terpenoid juvenile hormone (JH). Both hormones have been chemically defined.

It is generally believed that an ecdysial cycle is started in the brain, and that it involves the periodic release of ecdysone and, in the case of a pre-adult moult, of JH. Each wave of secretion ultimately results in a moult.

In some instances the JH has been shown to have a stimulating effect on the prothoracic glands (Williams, 1959; Gilbert & Schneiderman, 1959; Röller & Dahm, 1968). Also, it has been claimed that the CA exert a stimulating effect on the cerebral NSC (Highnam, 1962; Lea & Thomsen 1962, 1969).

Oogenesis would require the joint release of brain hormone and JH (Thomsen, 1952, with *Calliphora*; Highnam, Lusis & Hill, 1963, with *Schistocerca*), but other authors (e.g. Karlinsky, 1967, with *Pieris*) have obtained oogenesis merely on account of CA activity. It is believed that a reproductive cycle results from the cyclic release of the above hormone(s) (Strangways-Dixon, 1962).

Endocrine activity in relation to diapause

Egg diapause

In *Bombyx mori* egg diapause is induced by a maternal 'diapause hormone' released by the sub-oesophageal ganglion (SOG). Partly it inhibits embryogenesis by direct penetration into the ova, partly it changes the composition of yolk and the pigmentation of the egg by an effect on the follicle (Hasegawa, 1963; Hasegawa & Yamashita, 1965). Under LD conditions inducive to diapause, the brain promotes the secretion of the diapause hormone. Already during the last larval instars two caudal neurosecretory cells of the SOG are activated, which apparently secrete this hormone (Fukuda & Takeuchi, 1967). Under diapause-preventing conditions an inhibitory effect is exerted by the brain on the SOG, and the diapause factor is not released.

Larval diapause

The interesting experiments of Fukaya & Mitsuhashi (1961) indicate that during larval diapause in the rice-stem borer, *Chilo suppressalis*, the prothoracic glands are inactive, but that the corpora allata secrete throughout the diapause period (Fig. 7). The result is a high JH titre. Implantation of active prothoracic glands or brains results in break of diapause and pupation, but the pupa shows many metathetelic characters. The authors conclude that 'the hormone released from the corpora allata keeps the brain or prothoracic glands inactive' (Fukaya & Mitsuhashi, 1961). No experiments were made with extra-implantation of active CA to induce diapause. The authors refer to a discussion by Bodenstein (1953) assuming that a delay of the time of moulting observed after implanting extra CA must indicate a depression of prothoracic gland activity. It would seem that this observation could also be explained by assuming that high JH titres diminish the response of the tissues to ecdysone.

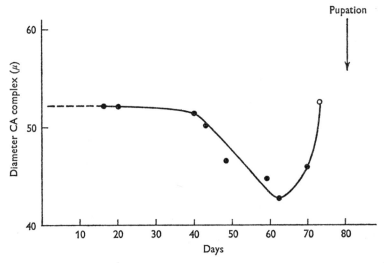

Fig. 7. Size of corpora allata (CA) during larval diapause in *Chilo suppressalis*. Note that the glands are involuted when diapause is broken, but increase in size shortly before pupation (after Fukaya & Mitsuhashi, 1961).

The prothoracotropic effect of JH observed in Lepidoptera seems to render an inhibitory effect in *Chilo* on the level of the prothoracic glands rather improbable. This is also borne out by the fact that implantation of active prothoracic glands resulted in break of diapause. In view of the fact that during diapause the neurosecretory B-cells (*sensu* Kobayashi, 1957) are crammed with fuchsinophile material, it would seem probable that these cells are related to the activation of the prothoracic glands. This would, however, leave the activity of the CA during larval diapause in *Chilo*, also observed in *Plodia interpunctella* (Waku, 1960) and *Ostrinia nubilalis* (Mitsuhashi & Fukaya, 1960), unexplained.

In this respect it should be remembered that recent studies by Riddiford (1969) have revealed that the JH and analogues may influence the programming of development. Also a special role of the CA in fat body metabolism may not be excluded. Other studies on larval diapause in different insect orders have mainly pointed to a temporary failure of brain hormone release (Sellier, 1949, in *Gryllus*, Church; 1955, in *Cephus cinctus*; Rahm, 1952, in *Sialis lutaria*).

Pupal diapause

The experiments of Williams and his school have greatly contributed to our understanding not only of the endocrine situation during pupal diapause and its break, but of the sequence of endocrine events during insect growth in general.

'The dormant condition is, in fact, a syndrome of endocrine deficiency due to a failure of the brain to secrete a hormone prerequisite for the initiation of adult development' (Williams & Adkisson, 1964). The experiments upon which these results were based were performed with the pupae of several Saturniid species, in which diapause is induced during the larval stage (Mansingh & Smallman, 1967). The evidence was based on the effect of chilling on break of diapause. Under the influence of low temperature the diapausing brain would become a 'competent' brain which, at a temperature of 25 °C, becomes an 'active' brain and secretes a hormone-activating the prothoracic glands. The microsurgical experiments by which this course of events was proven at the same time provided a base of our understanding for the regulation of insect growth in general.

As stated before (de Wilde, 1969*b*), these experiments only indicate that the inactive state of the brain *maintains* the state of diapause. It has not been shown that decerebration, NSC cautery or extirpation of the prothoracic glands induces a complete diapause syndrome. In fact, the prothoracotropic effect of ecdysone renders it improbable that the prothoracic glands, once activated, are brought to silence by inactivity of the brain. Some inhibitory effect of an unknown nature exerted on the prothoracic glands must accompany the induction of diapause. But as yet our knowledge of the inhibition of prothoracic gland activity in general is insufficient to provide any clue.

Highnam (1958) has found that in *Mimas tiliae* the CA are secretory during diapause but become inactive as the brain has gained 'competence' by chilling (Fig. 8). Removal of the CA, however, did not result in break of diapause. Also, by implantation of a chilled brain, diapause was broken despite the CA activity. In this respect, the situation in pupal diapause of *Mimas* resembles the state of affairs prevailing during larval diapause in the rice-stem borer.

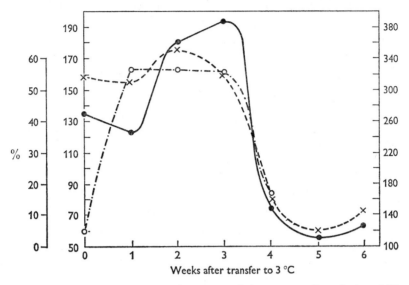

Fig. 8. Glandular activities in the brain and the corpora allata during chilling in the diapausing pupa of *Mimas tiliae;* showing that important changes take place not only in the cerebral neurosecretory cells (NSC) but also in the postcerebral glands. —·—, Percentage of post-secretory NSC; ———, total volume (CC) allata ($\times 10^{-3}$ μ^3); – – –, cell volumes (CC) allata ($\times 10^{-3}$ μ^3). After Highnam (1958).

Adult diapause

Though endocrinological experiments on adult diapause are few, the results point to a central role of the CA. In *Macrodytes* (Joly, 1945), *Pyrrhocoris* (Slama, 1964) and *Leptinotarsa* (de Wilde *et al.* 1959; de Wilde & de Boer, 1961) many characteristic elements of the diapause syndrome are produced by allatectomy. However, in *Pyrrhocoris* the rate of oxygen consumption is only depressed to the diapause level if the corpora cardiaca are also removed. Behavioural and reproductive elements of the diapause syndrome are also produced by removal of the MNSC of the brain (Geldiay, 1967; de Wilde & de Boer, 1969).

In *Leptinotarsa* the complete syndrome of diapause, including the change of behaviour, the degeneration of the flight muscles, the standstill of oogenesis and the much reduced rate of oxygen consumption are produced by allatectomy alone (de Wilde & Stegwee, 1958; de Wilde & de Boer, 1961; Stegwee, Kimmel, de Boer & Henstra, 1963). Subsequent implantation of one or two pairs of active CA completely restore the active situation.

Histological evidence by Geldiay (1967), Siew (1965) and Schooneveld (unpublished) points to an inability of the neurosecretory A cells to release their hormone during adult diapause, and this may well be the primary endocrine failure. Break of diapause involves the return of a high titre of JH activity accompanied by regeneration of the flight muscles. Reproductive activity is only resumed after feeding has again taken place.

It has proved possible to measure the JH level in the haemolymph of *Leptinotarsa* (Fig. 9). After a brief increase following adult emergence, the titre falls sharply in SD treated beetles and remains at levels below the accuracy of measurement during diapause (de Wilde *et al.* 1968).

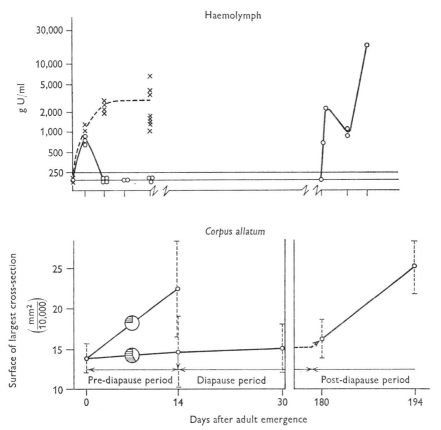

Fig. 9. Juvenile hormone (JH) titre in the haemolymph, and dimensions of the corpora allata as a function of photoperiodic treatment in *Leptinotarsa decemlineata*. Note that the JH titre during diapause is below the error of determination. In the upper graph: × – – – ×, long-day treatment; ○———○, short-day treatment. After de Wilde *et al.* (1968).

Extrinsic control of endocrines in relation to diapause

Photoperiod

The relation between photoperiodic treatment and the induction of egg diapause in *Bombyx mori* is as yet obscure. As diapause may already be induced by treating the embryonic stage of the previous generation, it is clear that photoperiod interferes with a programming event ultimately leading to endocrine consequences. Neither do we know how photoperiod induces larval diapause in *Chilo suppressalis*, a LD insect in which the endocrine situation was described in the foregoing chapter.

In the case of pupal diapause in *Antheraea pernyi*, which is broken after chilling by photoperiods above 16 h, the analysis has been carried a good deal further. We now know that the photoperiodic receptor is located in

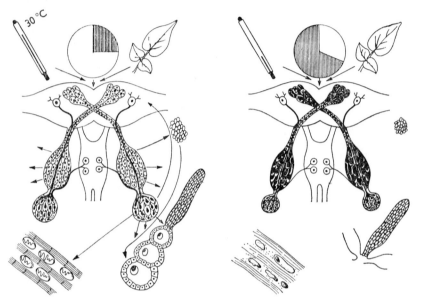

Fig. 10. Elements of the diapause syndrome (right) and of the reproductive state (left) induced by photoperiod, temperature, and food condition in the Colorado beetle, *Leptinotarsa decemlineata*.

the brain, probably in the side lobes. When a chilled brain is excised and transplanted to the tip of the abdomen, and is illuminated through a plastic window, break of diapause ensues. Normally, illumination reaches the brain through a translucent region of the anterior cuticle. In *Antheraea* the photoperiodic response curve of diapause break is the inverse of that of diapause induction, the critical photoperiod being the same (Williams & Adkisson, 1964).

It is now known that *A. pernyi*, *Hyalophora cecropia* and *A. polyphemus* all have a diapause governed by photoperiod applied during the larval stage (Mansingh & Smallman, 1967). It would be very interesting indeed to know how the endocrine events in the pupa are programmed by this photoperiodic induction process.

In the adult *Leptinotarsa decemlineata,* photoperiods below 15 h induce diapause. In an initial stage of diapause reversion to long photoperiods may reactivate the beetles (Hodek & de Wilde, unpublished), and successive inactivation and activation of CA activity occurs by successively applying short and long photoperiods.

There are some indications pointing to the NS A-cells of the brain as the primary site of endocrine photoperiodic response (de Wilde & de Boer, 1969, Schooneveld, unpublished). SD treatment results in a specific pattern of haemolymph proteins. LD treatment abolishes the synthesis of the specific SD proteins and instead leads to the induction of selective vitellogenic proteins in female beetles. The abolition of SD protein synthesis by LD is apparently due to the increase in JH titre, as the same effect is produced under SD conditions by application of pure JH (de Loof, 1969).

Temperature

Low-temperature treatment conducive to break of diapause may produce its effect at different rates. In *Hyalophora cecropia* chilling apparently affects primarily the brain. Brains obtained from chilled pupae cannot immediately activate brainless pupae, but only do so after being 'activated' by a stay at 25 °C. In *Mimas tiliae* competence is gained by a process during which the neurosecretory cells elaborate intercellular material which is passed on to the CC and apparently released during chilling. Thus, in *Mimas,* both 'competence' and 'activity' of the brain are attained during chilling.

Food

In *Leptinotarsa,* feeding with physiologically aged potato leaves results in arrest of oogenesis, and, ultimately, in diapause. This is not due to lack of materials necessary for oogenesis. After inhibition of oogenesis, implantation of one or two pairs of active CA results in a significant increase in egg production. Apparently, food condition provides a 'diapause token' leading to inhibition of CA activity (de Wilde & Ferket, 1967).

THE CONTRIBUTION OF DIAPAUSE RESEARCH TO GENERAL THEORY

Apart from its contribution to insect ecology, diapause research has had important implications for the general theory of endocrine control. First of all, our knowledge of the hierarchy of endocrine functions determining growth has been greatly advanced by the work of C. M. Williams and his co-workers with *Hyalophora cecropia*, and this has also stimulated much research on the mode of action of insect hormones. Furthermore, diapause research has made it abundantly clear that our knowledge of the regulation of endocrine activity and of the control of growth and metamorphosis is still very incomplete. The much simplified concept of two successive endocrine activities stimulated by a wave of NS activity has been changed into one which includes programmed cell growth, interactions between endocrines and even autostimulation of endocrine glands.

And, above all, the role of a hormone as a 'message' in the *milieu inter-ieur* has been shown to depend on the species not unlike the effect of environmental messages with relation to diapause.

Low-temperature treatment is conducive to break of diapause in *H. cecropia*, and retards the process in *Leptinotarsa*. LD's may induce reproduction in some species, and diapause in others.

High titres of JH are prevailing during larval diapause in *Chilo* but would break diapause in *Leptinotarsa*. Similarly, high titres of JH promote flight-muscle synthesis in *Leptinotarsa* but induce flight-muscle degeneration in *Acheta* (Chudakova, 1969) and in some Scolytidae (Chapman, 1956; Ticheler, 1961).

If anything, diapause research has underlined that the response to hormones is a function of the genetic constitution of the insect in a detailed sense.

REFERENCES

ADKISSON, P. A. (1961). Effect of larval diet on the seasonal occurrence of diapause in the Pink Bollworm. *J. econ. Ent.* **54**, 1107–12.

ADKISSON, P. A. (1964). Action of the photoperiod in controlling insect diapause. *Am. Nat.* **98**, 337–74.

ANKERSMIT, G. W. (1968). The photoperiod as a control agent against *Adoxophyes reticulana* (Lepidoptera, Tortricidae). *Ent. exp. appl.* **11**, 231–40.

BEAMENT, J. W. L. (1951). The structure and formation of the egg of the fruit tree red spider mite, *Metatetranychus ulmi* Koch. *Ann. appl. Biol.* **38**, 1–24.

BECK, S. D. (1964). Time-measurement in insect photoperiodism. *Am. Nat.* **98**, 329–45.

BLAIR, C. A. & GROVES, J. R. (1952). Biology of the fruit tree red spider mite *Metatetranychus ulmi* (Koch) in S.E. England. *J. hort. Sci.* **27**, 14–43.

BODENSTEIN, D. (1953). Studies on the humoral mechanisms in growth and meta-morphosis of the cockroach, *Periplaneta americana*. III. Humoral effects on metabolism. *Biol. Bull. mar. biol. Lab., Woods Hole* **124**, 105–15.

BODINE, J. H. (1934). The effect of cyanide on the oxygen consumption of normal and blocked embryonic cells. *J. cell. comp. Physiol.* **4**, 397–404.

BODINE, J. H. & BOELL, E. J. (1934). Respiratory mechanisms of normally develop-ing and blocked embryonic cells. *J. cell. comp. Physiol.* **5**, 97–113.

BOELL, E. J. & BODINE, J. H. (1938). Effect of nitrophenols on the respiratory metabolism of Orthopteron embryos. *Proc. Iowa Acad. Sci. Des Moines* **44**, 207.

BOWERS, B. & WILLIAMS, C. M. (1964). Physiology of insect diapause. XIII. DNA synthesis during the metamorphosis of the Cecropia silkworm. *Biol. Bull. mar. biol. Lab., Woods Hole* **126**, 205–19.

BURGES, H. D. (1959). Studies on the Dermestid beetle, *Trogoderma granarium* Everts. II. The occurrence of diapause larvae of a constant temperature and their behaviour. *Bull. ent. Res.* **50**, 407–22.

BURGES, H. D. (1961). The effect of temperature, humidity and quantity of food on the development and diapause of *Trogoderma parabile* Beal. *Bull. ent. Res.* **51**, 685–96.

CHAPMAN, J. A. (1956). Flight-muscle changes during adult life in a scolytid beetle. *Nature, Lond.* **177**, 1183.

CHUDAKOVA, I. (1969). Endocrine regulation of degeneration of cricket (*Acheta domestica*) deafferented flight muscles. *Abstracts short communications of Vth Conference of European Comparative Endocrinologists*, Utrecht, 1969.

CHURCH, N. S. (1955). Hormones and the termination and reinduction of diapause in *Cephus cinctus* Nort. *Can. J. Zool.* **33**, 339–69.

DANILEVSKII, A. S. (1965). *Photoperiodism and Seasonal Development of Insects.* London: Oliver & Boyd.

DANILEVSKII, A. S. & KUZNETSOVA, J. A. (1968). The intra-specific adaptations of insects to the climatic zonation. In *Photoperiodic Adaptation*, pp. 5–51. Comm. Leningrad University. (Russian.)

DICKSON, R. C. (1949). Factors governing the induction of diapause in the Oriental fruit moth. *Ann. ent. Soc. Am.* **42**, 511–32.

FUKUDA, S. & TAKEUCHI, S. (1967). Diapause factor producing cells in the sub-oesophageal ganglion of the silkworm, *Bombyx mori* L. *Proc. Japan. Acad.* **43**, 51–6.

FUKAYA, M. & MITSUHASHI, J. (1961). Larval diapause in the rice stem borer with special reference to its hormonal mechanism. *Bull. natn. Inst. agric. Sci.* C **13**, 1–32.

GELDIAY, S. (1967). Hormonal control of adult reproductive diapause in the Egyptian grasshopper, *Anacridium aegyptium* L. *J. Endocr.* **37**, 63–71.

GEYSPITZ, K. F. (1949). Light as a factor regulating the cycle of development of the pine *Lasiocampid dendrolimus pini* L. *Dokl. Akad. Nauk. SSSR* **68**, 781–4. (Russian.)

GEYSPITZ, K. F. (1953). The reaction of univoltine Lepidoptera to day-length. *Entomol. Obozr.* **33**, 17–31. (Russian.)

GEYSPITZ, K. F. (1958). The adaptational significance of the photoperiodic reaction and its role in the ecology of the pine moth (*Dendrolimus pini* L.). *Sci. Mem. of LSU*, **240**, 21–33. (Russian.)

GILBERT, L. I. & SCHNEIDERMANN, H. A. (1959). Prothoracic gland stimulation by juvenile hormone extracts of insects. *Nature, Lond.* **184**, 171–3.

HARVEY, W. R. (1962). Metabolic aspects of insect diapause. *Ann. Rev. Ent.* **7**, 57–80.

HARVEY, W. R. & HASKELL, J. A. (1966). Metabolic control mechanisms in insects. *Adv. Insect Physiol.* **3**, 133–205.

HASEGAWA, K. (1963). Studies on the mode of action of the diapause hormone in the silk worm, *Bombyx mori* L. 1. The action of diapause hormone injected into pupae of different ages. *J. exp. Biol.* **40**, 517–29.

HASEGAWA, K. & YAMASHITA, O. (1965). Studies on the mode of action of the diapause hormone in the silk worm, *Bombyx mori* L. VI. The target organ of the diapause hormone. *J. exp. Biol.* **43**, 271–7.

HIGHNAM, K. C. (1958). Activity of the corpora allata during pupal diapause in *Mimas tiliae* (Lep.) *Q. J. microsc. Sci.* **99**, 171–80.

HIGHNAM, K. C. (1962). Neurosecretory control of ovarian development in *Schistocerca gregaria*. *Proc. IIIrd Int. Symp. Neurosecr. Mem. Soc. Endocr.* no. 12, pp. 379–90.

HIGHNAM, K. C., LUSIS, O. & HILL, L. (1963). Factors affecting oöcyte resorption in the desert locust *Schistocerca gregaria* (Forskål). *J. Insect Physiol.* **9**, 827–37.

JOHANSSON, A. S. (1958). Hormonal regulation of reproduction in the milkweed bug, *Oncopeltus fasciatus* (Dallas). *Nature, Lond.* **181**, 198–9.

JOLY, P. (1945). La fonction ovarienne et son central humoral chez les Dystiscides. *Archs Zool. exp. gen.* **84**, 49–164.

KARLINSKY, A. (1967). Physiologie des insectes. Influence des corpora allata sur le fonctionnement ovarien en milieu mâle de *Pieris brassicae* L. (Lepidoptera). *C. r. hebd. Séanc. Acad. Sci., Paris* D, **265**, 2040–2.

KOBAYASHI, M. (1957). Studies on the neurosecretion in the silk worm, *Bombyx mori* L. *Bull. seric. Exp. Stn Japan* **15**, 181–273.

DE KORT, C. A. D. (1969). Hormones and the structural and biochemical properties of the flight muscles in the Colorado beetle. *Meded. LandbHoogesch. Wageningen* **69** (2), 1–63.

KRISHNAKUMARAN, A. & SCHNEIDERMANN, H. A. (1964). Developmental capacities of the cells of an adult moth. *J. exp. Zool.* **157**, 293–306.

KUENEN, D. J. (1946). Het fruitspint en zijn bestrijding. *Meded. LandbVoorlDienst, Den Haag* no. 44.

LEA, A. O. & THOMSEN, E. E. (1962). Cycles in the synthetic activity of the medial neurosecretory cells of *Calliphora erythrocephala* and their regulation. *Proc. IIIrd Int. Symp. Neurosecr. Mem. Soc. Endocr.* **12**, 345–7.

LEA, A. O. & THOMSEN, E. E. (1969). Size independent secretion by the corpus allatum of *Calliphora erythrocephala*. *J. Insect Physiol.* **15**, 477–82.

LEES, A. D. (1953). The significance of the light and dark phases in the photoperiodic control of diapause in *Metatetranychus ulmi* Koch. *Ann. appl. Biol.* **40**, 487–97.

LEES, A. D. (1964). The location of the photoperiodic receptors in the aphid *Megura viciae* Buckton. *J. exp. Biol.* **41**, 119–33.

LEES, A. D. (1968). Photoperiodism in insects. In *Photophysiology*, pp. 47–137. New York and London: Academic Press.

DE LOOF, A. (1969). Causale mechanismen bij de vitellogenese van de Coloradokever (*Leptinotarsa decemlineata* Say.). Thesis, University of Ghent.

MANSINGH, A. & SMALLMAN, B. N. (1967). The cholinergic system in insect diapause. *J. Insect Physiol.* **13**, 447–67.

MANSINGH, A. & SMALLMAN, B. N. (1967). Neurophysiological events during larval diapause and metamorphosis of the European corn borer, *Ostrinia nubilalis* Hübner. *J. Insect Physiol.* **13**, 861–8.

MANSINGH, A. & SMALLMAN, B. N. (1967). Effect of photoperiod on the incidence and physiology of diapause in two saturniids. *J. Insect Physiol.* **13**, 1147–62.

MASAKI, S. (1965). Geographic variation in the intrinsic incubation period: a physiological dine in the Emma-field cricket. *Bull. Fac. Agirc. Hirosaki Univ.* **11**, 59–90.

McLEOD, D. G. R. & BECK, S. D. (1963). Photoperiodic termination of diapause in an insect. *Biol. Bull. mar. biol. Lab., Woods Hole* **124**, 84–96.

MITCHELL, P. (1962). Metabolism, transport and morphogenesis; which drives which? *J. gen. Microbiol.* **29**, 25–37.

MITSUHASHI, J. & FUKAYA, M. (1960). The hormonal control of larval diapause in the rice stem borer *Chilo suppressalis*. III. Histological studies on the neurosecretory cells of the corpora allata during diapause and postdiapause. *Jap. J. appl. Entom. Zool.* **4**, 127–34.

MÜLLER, H. J. (1962). Über die Induktion der Diapause von der Ausbildung der Saisonformen bei *Aleurochiton complanatus* (Baerensprung) (Homoptera, Aleurodidae). *Z. Morph. Ökol. Tiere* **51**, 575–610.

MÜLLER, H. J. (1965). Probleme der Insektendiapause. *Verh. dt. zool. Ges.*, 7–13 *Jena* 192–222.

OBERLANDER, H. & SCHNEIDERMANN, H. A. (1966). Juvenile hormone and RNA synthesis in pupal tissues of saturnid moths. *J. Insect Physiol.* **12**, 37–41.

PERRON, R. (1954). Untersuchungen über Bau, Entwicklung und Physiologie der Milbe, *Histiosoma laboratorium* Hughes. *Acta. zool. Stockh.* **35**, 71–112.

RAHM, U. H. (1952). Die innersekretorische Steuerung der post-embryonalen Entwicklung von *Sialis lutaria* L. *Revue suisse Zool.* **59**, 173–237.

RÖLLER, H. & DAHM, K. H. (1968). The chemistry and biology of juvenile hormone. *Recent Prog. Horm. Res.* **24**, 651–80.

RIDDIFORD, L. M. (1969). The action of the juvenile hormone on silk worm embryos. *Proc. Int. Symp. Insect Endocr., Brno* 1966. (In the Press.)

SCHARRER, E. (1966). Principles of neuroendocrine integration. In *Endocrines and the Central Nervous System*, pp. 1–35. Baltimore: Williams and Wilkins.

SCHARRER, E. & SCHARRER, B. (1963). *Neuroendocrinology*. Columbia University Press.

SCHNEIDERMANN, H. A. & WILLIAMS, C. M. (1953). The physiology of insect diapause. VII. The respiratory metabolism of the Cecropia silkworm during diapause and development. *Biol. Bull. mar. biol. Lab., Woods Hole* **105**, 320–34.

SCHNEIDERMANN, H. A. & WILLIAMS, C. M. (1954). The physiology of insect diapause. VIII. Qualitative changes in the metabolism of the Cecropia silkworm during diapause and development. *Biol. Bull. mar. biol. Lab., Woods Hole* **106**, 210–29.

SCHOONHOVEN, L. M. (1962). Diapause and the physiology of host–parasite synchronization in *Bupalus piniarius* L. and *Eucarcelia rutilla* Vell. *Archs néerl. Zool.* **15**, 111–74.

SCHOONHOVEN, L. M. (1963). Spontaneous electrical activity in the brain of diapausing insects. *Science, N.Y.* **141**, 173–4.

SELLIER, R. (1949). Diapause larvaire et macroptérisme chez *Gryllus compestris* (Orth.). *C. r. hebd. Séanc. Acad. Sci., Paris* **228**, 2055–6.

SHAPPIRIO, D. G. & HARVEY, W. R. (1965). The injury metabolism of the Cecropia silkworm. II. Injury induced alterations in oxidative enzyme systems and respiratory metabolism of the pupal wing epidermis. *J. Insect Physiol.* **11**, 305–27.

SIEW, Y. C. (1965). The endocrine control of adult reproductive diapause in the Chrysomelid beetle *Galeruca tanaceti* (L.). II. *J. Insect Physiol.* **11**, 463–79.

SLAMA, K. (1964). Hormonal control of haemolymph protein concentration in the adult of *Pyrrhocoris apterus* L. *J. Insect Physiol.* **10**, 773–82.

STEGWEE, D. (1964). Respiratory chain metabolism in the Colorado beetle. II. Respiration and oxidative phosphorylation in 'sarcosomes' from diapausing beetles. *J. Insect Physiol.* **10**, 97–102.

STEGWEE, D., KIMMEL, E. C., DE BOER, J. A. & HENSTRA, S. (1963). Hormonal control of reversible degeneration of flight muscle in the Colorado beetle, *Leptinotarsa decemlineata* Say. *J. Cell Biol.* **19**, 519–27.

STRANGWAYS-DIXON, J. (1962). The relationship between nutrition, hormones and reproduction in the blowfly *Calliphora erythrocephala* (Meig.). III. The corpus allatum in relation to nutrition, the ovaries, innervation and corpora cardiaca. *J. exp. Biol.* **39**, 293–306.

STRONG, L. (1965). The relationships between the brain, corpora allata and oocyte growth in the Central American locust, *Schistocerca* sp. II. The innervation of the corpora allata, the lateral neurosecretory complex and oöcyte growth. *J. Insect Physiol.* **11**, 271–80.

TANAKA, S. (1950). Studies on hibernation with special reference to photoperiodicity and breeding of the Chinese tussar silkworm. III. *J. seric. Sci. Japan*, **19**, 580–90.

THOMSEN, E. E. (1952). Functional significance of the neurosecretory brain cells and the corpus cardiacum in the female blow-fly, *Calliphora erithrocephala* Meig. *J. exp. Biol.* **29**, 137–72.

TICHELER, J. (1961). Etude analytique de l'épidémiologie du Scolyte des graines de café, *Stephanoderes hampei* Ferr., en Côte de l'Ivoire. *Meded. Landbouwgeschool. Wageningen*, **61** (1), 1–49.

TINBERGEN, N. (1951). *The Study of Instinct*. Oxford: Clarendon Press.

USHATINSKAYA, R. (1966). Diversity of forms of physiological rest in the Colorado potato beetle as one of the causes of extension of its range. In *Ecology and Physiology of Diapause in the Colorado Potato Beetle* (*Leptinotarsa decemlineata* Say.). *Dokl. Akad. Nauk. SSSR*, pp. 5–22. (Russian.)

WAKU, Y. (1960). Studies on the hibernation and diapause in insects. IV. Histological observations of the endocrine organs in the diapause and non-diapause larvae of the Indian mealmoth, *Plodia interpunctella* Hübner. *Sci. Rep. Tôhoku Univ.*, 4th ser. (Biol.), **26**, 327–40

WAY, M. J. (1962). Definition of diapause. *Ann. appl. Biol.* **50**, 595–6.

DE WILDE, J. (1957). Breeding the Colorado beetle under controlled conditions. *Z. Pflkrankh.* **64**, 589–93.

DE WILDE, J. (1962). Photoperiodism in insects and mites. *Ann. Rev. Ent.* **7**, 1–26.

DE WILDE, J. (1965). Photoperiodic control of endocrines in insects. *Archs Anat. microsc. Morph. exp.* **54**, 547–64.

DE WILDE, J. (1969a). Diapause and seasonal synchronization in the adult Colorado beetle (*Leptinotarsa decemlineata* Say.). In *Dormancy and Survival. Symp. Soc. exp. Biol.* no. XXIII, pp. 263–84.

DE WILDE, J. (1969b). Hormones and diapause. Proc. Third *Int. Congr. Endocr.* Mexico, pp. 356–65.

DE WILDE, J. & DE BOER, J. A. (1961). Physiology of diapause in the adult Colorado beetle. II. Diapause as a case of pseudo-allatectomy. *J. Insect Physiol.* **6**, 152–61.

DE WILDE, J. & DE BOER, J. A. (1969). Humoral and nervous pathways in photoperiodic induction of diapause in *Leptinotarsa decemlineata* Say. *J. Insect Physiol.* **15**, 661–75.

DE WILDE, J., DUINTJER, C. S., MOOK, L. (1959). Physiology of diapause in the adult Colorado beetle (*Leptinotarsa decemlineata* Say.). I. The photoperiod as a controlling factor. *J. Insect Physiol.* **3**, 75–85.

DE WILDE, J. & FERKET, P. (1967). The hostplant as a source of seasonal information. *Meded. Rijksfac. Landbwet. Gent.* **32**, 387–92.

DE WILDE, J., STAAL, G. B., DE KORT, C. A. D., DE LOOF, A. & BAARD, G. (1968). Juvenile hormone titer in the haemolymph as a function of photoperiodic treatment in the adult Colorado beetle (*Leptinotarsa decemlineata* Say.). *Proc. K. ned. Akad. Wet. C* **71**, 321–6.

DE WILDE, J. & STEGWEE, D. (1958). Two major effects of the corpus allatum in the adult Colorado beetle (*Leptinotarsa decemlineata* Say.). *Archs néerl. Zool.* (Suppl.) **13**, 277–89.

WILLIAMS, C. M. (1946). Physiology of insect diapause: the role of the brain on the production and termination of pupal dormancy in the giant silkworm, *Platysamia cecropia. Biol. Bull. mar. biol. Lab., Woods Hole*, **90**, 234–43.

WILLIAMS, C. M. (1952). Physiology of insect diapause. IV. The brain and pro-thoracic glands as an endocrine system in the cecropia silkworm. *Biol. Bull. mar. biol. Lab.*, *Woods Hole* **103**, 120–38.

WILLIAMS, C. M. (1959). The juvenile hormone. I. Endocrine activity of the corpora allata of the adult Cecropia silkworm. *Bull. Biol. mar. biol. Lab.*, *Woods Hole* **116**, 323–38.

WILLIAMS, C. M. (1967). The present status of the brain hormone. In *Insects and Physiology* pp. 133–43. London: Oliver and Boyd.

WILLIAMS, C. M. (1969). Photoperiodism and endocrine aspects of insect diapause. In *Dormancy and Survival. Symp. Soc. exp. Biol.* no. XXIII, pp. 285–300.

WILLIAMS, C. M. & ADKISSON, P. L. (1964). Physiology of diapause. XIV. An endocrine mechanism for the photoperiodic control of pupal diapause in the oak silkworm, *Anthereya pernyi. Biol. Bull. mar. biol. Lab.*, *Woods Hole* **127**, 511–25.

WYATT, H. G. R. & MEYER, W. L. (1959). The chemistry of insect haemolymph. III. Glycerol. *J. gen. Physiol.* **42**, 1005–11.

DISCUSSION

QUAY: I wish to enquire about the possibility of the involvement of 'biogenic amines' in the mediation of environmental effects on diapause and related internal events. Recalling especially the work several years ago of Shaaya, Karlson, Sekeris and Marmaras, on developmental changes in *Calliphora* of DOPA decarboxylase, 5-hydroxytryptophan decarboxylase and the catecholamines and indoleamines produced, it appeared that peaks and falls in these were correlated with other developmental events. What is our understanding at the present time of the participation of catecholamines and indoleamines in the onset of diapause, or related events?

DE WILDE: There is practically no information at all, I am sorry to say. Any study of diapause should begin with the analysis of the over-all induction process. All the work on *Cecropia*, for instance, pertains to break of diapause and thus has to do with diapause development rather than diapause induction. Experiments where people have in fact analysed the induction of diapause are extremely few. The transfer of environmental information at the biochemical level is still *terra incognita*.

HIGHAM: It is a well-known fact, I think, that in insects the sensitive stage for the photoperiodic stimulus may occur some considerable time before the actual induction of diapause. In the case of the eggs of the silk moth a generation intervenes between the sensitive stage and the resultant effect on the endocrine system to induce diapause. Does Professor de Wilde have any idea what happens to the stimulus after being received at one stage, that the endocrine system can react normally for several stages and then will react in a different way to induce diapause? In other words, what does happen to the stimulus between the sensitive stage and the actual diapause induction.

DE WILDE: What you are asking is: how can the environmental information be stored for such a long time without interfering with the activities of the endocrine system, which occur in between? I think this is the strongest argument against the idea that the neuroendocrine system is directly affected by photoperiodic stimuli. It remains a riddle how this environmental information is stored. The only thing we may perhaps assume is that it will not be directly within the neurosecretory cells. We know that photoperiodic induction takes place in the mid-brain, and, according to Williams, not in the neuropile but in the perikarya. There must occur a 'memory' in the midbrain, the nature of which is as yet unknown.

INTERSEXUALITY IN FISHES

By RUDOLF REINBOTH

INTRODUCTION

The term 'intersex' is defined in many different ways. According to Bacci (1965), who adopts the classical definition of Goldschmidt (1915), an intersex is an individual of unisexual species whose reproductive organs and/or secondary sex characters are partly of one sex and partly of the other, although it does not show genetically different parts. In Beatty's opinion the 'narrowest definition' of the term requires 'that an intersex must produce both male and female gametes' and 'the definition of intersex as equivalent to a true hermaphrodite, i.e. an animal that possesses simultaneously both male and female gonadal tissue' is regarded by him as being of wider meaning (Beatty, 1964). Atz (1964), however, claims that intersexuality means the presence of both male and female characteristics in a single individual and that hermaphrodites are intersexes, but some intersexes are not hermaphrodites. Even from these few examples it is apparent that ambiguity surrounds the term. In this paper, in order to emphasize the occurrence of simultaneous hermaphroditism and consecutive appearance of both sex characteristics as being a normal phenomenon in certain species, it is proposed to revive the term 'ambosexuality' (amphisexuality), a term that has been used occasionally by a few investigators (e.g. Coe, 1938; Gallien, 1965). In an ambosexual animal both male and female characteristics are associated normally in a single individual—either simultaneously or in a temporal succession.

Table 1 presents a short summary of our present knowledge on ambosexual teleosts. Its comprehensiveness is debatable for several reasons—limited number of specimens on which conclusions are based, considerable differences in the thoroughness of investigations—but, nevertheless, it gives the picture that normal intersexuality (ambosexuality) seems to be rather widespread among teleosts. It is probable also that the numbers will increase in the future, in particular as far as wrasses (Labridae), parrot fishes (Scaridae), porgies (Sparidae) and the family group of sea basses (Serranidae) and their relatives are concerned.

This paper does not consider the many cases of abnormal intersexuality (accidental hermaphroditism) which have been reported in a large number of species and readers are referred to the comprehensive review by Atz (1964) on this topic.

Table 1. *Normally intersexual teleosts*

Systematic group*	Type of intersexuality No. of species			Habitat		References†
	⚥	♂→♀	♀→♂	Marine	Fresh water	
Clupeiformes‡						
Gonostomatidae	.	1	.	+	.	11
Scopeliformes						
Sudidae	8	.	.	+	.	22, 23
Evermannellidae‡	1	.	.	+	.	23
Alepisauridae	2	.	.	+	.	8
Cypriniformes‡						
Cobitidae	.	1	.	.	+	18
Cyprinodontiformes						
Cyprinodontidae	1	.	.	.	+	9
Polynemiformes‡						
Polynemidae	.	2	.	+	.	10, 19
Synbranchiformes						
Synbranchidae	.	.	3	.	+	3, 14, 27
Perciformes						
Serranidae (including	12	.	.	+	.	5, 33, 37, 39
Grammistidae and	.	.	16	+	.	13, 21, 26, 34,
Grammidae)						39, 40
Sparidae	.	7	.	+	.	1, 6, 7, 12, 15, 16, 17, 25, 30, 33
	.	.	3	+	.	7, 33, 35
Emmelichthyidae§	.	.	3	+	.	20, 33, 43
Cepolidae	.	.	1	+	.	42
Labridae	.	.	12	+	.	24, 29, 31, 32, 33, 36, 37, 38, 41
Scaridae‡	.	.	3	+	.	4, 36, 37
Platycephalidae	.	3	.	+	.	2, 28

* The classification system of Berg (1958) has been adopted.

† (1) Aoyama (1955), (2) Aoyama *et al.* (1963) and personal communication by Aoyama, (3) Chan & Phillips (1967), (4) Choat (1966), (5) Clark (1959), (6) D'Ancona (1940/41), (7) D'Ancona (1949), (8) Gibbs (1960), (9) Harrington (1961, 1967), (10) Hida (1967), (11) Kawaguchi & Marumo (1967/68), (12) Kinoshita (1936), (13) Lavenda (1949), (14) Liem (1963, 1968), (15) Lissia-Frau (1966 a, 1968), (16) Lissia-Frau & Casu (1968 b), (17) Lissia-Frau & Pala (1968), (18) Lodi (1967 a, b), (19) Longhurst (1965), (20) Lozano Cabo (1953), (21) McErlean & Smith (1964), (22) Mead (1960), (23) Mead *et al.* (1964), (24) Okada (1964 a, b), (25) Okada (1965 a, b), (26) Okada (1965 c, d), (27) Okada (1966 a), (28) Okada (1966 b, 1968), (29) Oliver & Massuti (1952), (30) Pasquali (1940/1), (31) Quignard (1966), (32) Reinboth (1957, 1967 c, 1970), (33) Reinboth (1962 a), (34) Reinboth (1963 b, 1964, 1965, 1967 a), (35) Reinboth (1967 b), (36) Reinboth (1968), (37) Reinboth (unpublished), (38) Roede (1966), (39) Smith (1959), (40) Smith (1965), (41) Sordi (1962, 1964, 1967), (42) Vives *et al.* (1959), (43) Zei (1949).

‡ In these groups normally intersexual representatives were still unknown at the time of Atz's (1964) review.

§ According to Greenwood *et al.* (1966) the genus *Spicara* (= *Maena*) has to be included into this family.

METHODS OF DIAGNOSIS

Although a considerable literature on ambosexual fishes has accumulated in the past few years, most conclusions are based on rather indirect evidence. The vast majority of ambosexual teleosts are marine species that are difficult to breed in the laboratory and hence it has been difficult to obtain information both of an experimental kind and also in recognizing and describing a fish's particular intersexual pattern.

The presence of ovulated oocytes and masses of sperm in the gonads of a species can tell us that such animals are simultaneous hermaphrodites, but we may run into trouble if the fish we are looking at is not ready to spawn. Especially in the serranids, among which most of the simultaneous hermaphrodites do occur, the gonads of functional females of a proto-gynous species (e.g. *Sacura*) may have striking similarities with the ovo-testis of a true hermaphrodite, at least during certain periods of the repro-ductive cycle. The observation of spawning behaviour of a simultaneous hermaphrodite (e.g. *Serranellus subligarius* (Clark, 1959)) or successful artificial fertilization of eggs with sperm that is taken from the same individual (Reinboth, 1962a; Salekhova, 1963) may furnish additional support for the identification of simultaneous hermaphrodites. But again, these methods are restricted to a particular season of the year and they have not been employed more often than in the cases referred to.

For studies on sex-inversion* all workers rely on statistical data in which size and sex of fishes are correlated. All results that are obtained in this way require an extremely careful evaluation, for, as long as our knowledge on the biology of a species is insufficient, one can neither exclude the existence of sex-specific differences in growth rate nor the possibility of selective catching. It is obvious that such events would greatly influence any result.

Histological examination of the gonads should be an indispensable pre-requisite for diagnosis. For example, Bacci (1965) and his student Zunarelli Vandini (1965) drew incorrect conclusions on sex-inversion in wrasses because both authors constantly overlooked striking differences in the morphology of the testes—differences that can tell us whether a male functioned previously as a female or not. Finally, it is highly desirable to have gonads from different phases of the reproductive cycle. This is particularly true when we try to get an idea about transformation processes. Partial gonadectomy, followed by histological investigation of the sample and later of the gonad that remained in the operated specimen, may reveal gonadal changes and provide the evidence that sex-inversion had taken place. But this type of experimentation has been done so far only in *Coris julis* (Reinboth, 1962).

It is relatively easy to recognize sex-inversion when all individuals of a species are starting as males or females respectively and change to the

---

* As the term 'sex-reversal' may imply the meaning of reverting to a primary sex condition it is abandoned in favour of the more neutral term 'sex-inversion'.

opposite sex at a later stage. But the difficulties increase considerably when functional males and functional females are present at the same size (and age) and when sooner or later a certain percentage of them may still invert to the other sex.

TYPES OF INTERSEXUALITY

There are three main types of ambosexual organization among teleosts: simultaneous hermaphroditism, protandry and protogyny. In protandrous and protogynous species we encounter a considerable number of varieties in morphology of the gonads, in the proportion of inverting animals and in the ratio of inverted specimens which contribute to the total stock of one sex.

In the schemes A and B of Fig. 1 an attempt is made to summarize our present knowledge of the different patterns of gonad development that can be found in single individuals of ambosexual species. The scheme provides a brief and convenient way of describing all the subtypes of ambosexual organization manifested in different species by using the various letters as references. Analogous cases in protogynous and protandric species are characterized by corresponding letters of the Latin and Greek alphabet.

The other graphs of Fig. 1 show how a particular sex type can be illustrated in a brief way. (A similar presentation has been used previously by Lissia-Frau, 1968b). For instance, the sexual organization of the protogynous *Thalassoma bifasciatum* (no. V) could be described as (a, b, m) and that of the protandric *Pagellus acarne* (no. II) as (ϕ, γ, λ).

Simultaneous hermaphroditism

Since the review by Atz (1964) on the subject, only two papers by Harrington (1967, 1968) on the unique self-fertilizing *Rivulus marmoratus* have enlarged our knowledge of simultaneous hermaphroditism beyond morphological description. Apart from one exceptional primary male, all speciments that Harrington caught in the wild proved to be hermaphrodites. Under laboratory conditions, however, he obtained 97% hermaphrodites and 3% primary male gonochorists from a total of 384 eggs. Carefully planned experiments on the influence of a variety of environmental factors (temperature, light, salinity) revealed that exposure to low temperature alone may bring about an increase of the number of primary males (Harrington, 1967). In his most recent paper Harrington (1968) defined the 'thermolabile phenocritical period of sex determination' in terms of developmental stages. Methodological limitations, however, did not permit an exact estimation of the point within the critical stage where thermolability begins nor how brief a low-temperature pulse may produce an effect. The temperature threshold lies near 20 °C. Subthreshold temperatures yielded an average of 63·8 % primary males, whereas temperatures as low as 21·1 ± 0·2 °C during the phenocritical period produced 100 % hermaphrodites. Climatic conditions during the presumed spawning

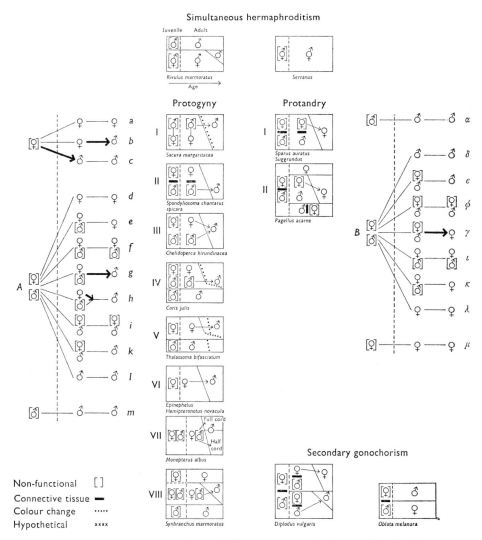

Fig. 1. Schematic representation of different patterns of ambosexuality. Further explanations in the text.

season of *Rivulus* in Florida do not favour the differentiation of primary male gonochorists. Unpublished observations on histological sections of gonads from serranids of the Gulf of California provided by Dr R. H. Rosenblatt of Scripps Institute: (*Serranus aequidens* Gilbert, 1 specimen;

Serranus fasciatus (Jenyns), 17 specimens; an undetermined species of the genus *Diplectrum* Holbrook, 12 specimens) indicate clearly that these species are simultaneous hermaphrodites (Plate 1 *a*). The topographical arrangement of ovarian and testicular tissue is only slightly different from the situation which is known from their close relatives in the genus *Serranus* (Reinboth, 1962 *a*).

Protandry

Protandric species have been reported from five families (cf. Table 1). Hida's recent paper on *Polydactylus sextarius* gives no histological data (Hida, 1967), whereas the publication on *Gonostoma gracile* by Kawaguchi & Marumo (1967/8) and the data on *Cobitis taenia* by Lodi (1967) suggest the absence of preformed ovarian tissue during the functional male phase. Lodi's micrographs demonstrate a complete mixture of both germinal elements during the period of sexual transformation in *Cobitis* and Kawaguchi & Marumo (1967/8) indicate that they found smaller oocytes in mature testes about to undergo sex-inversion. Further information on this problem, however, seems to be desirable. As far as *Cobitis taenia* is concerned, the question remains to be answered whether protandric sex-inversion does normally occur in this freshwater inhabiting species or whether Lodi's findings are valid only for the one population which he investigated (Lodi, 1967 *a, b*).

Protandric porgies (Sparidae) and flatheads (Platycephalidae) have the peculiarity of possessing ovarian tissue from the very beginning of the functional male phase. But the oocytes in the ovarian lamellae do not pass a certain growth stage (early perinucleolus stage), the size of which corresponds to about the diameter of oocytes in functional ovaries between two successive spawning periods. Sooner or later the testicular portion becomes more and more reduced whilst the development of the ovarian part advances until it reaches maturity. This basic process is apparently the same in all species. There are, however, some structural differences between flatheads and porgies. Since Aoyama, Kitajima & Mizue (1963) do not mention them in the English summary of their Japanese publication, additional comments on material of *Cociella crocodila* (Tilesius), *Suggrundus meerdervoorti* (Bleeker) and *Rogadius asper* (Cuv. et Val.), kindly provided by Dr Aoyama, are appropriate. As may be seen from Aoyama's and his fellow workers' paper (cf. Fig. 1) in the flatheads both germinal tissues are surrounded by the common gonadal wall, the testicular portion lying ventrally to the ovarian part. A similar dorso-ventral arrangement of both germinal parts exists in all ambosexual serranids, sparids and emmelichthyids but the inclusion of the heterologous tissues into a common gonadal cavity is restricted to sea basses and flatheads. In ambosexual gonads that had been identified as functional testes by Aoyama the ovarian portion may be much larger than the testicular part. The testis consists of extremely wide tubules the walls of which are formed by a very distinct germinal epithelium which contains the gonocytes (Plate 1*b*). As opposed to all other known ambosexual males (and simultaneous hermaphrodites), the vas deferens is limited to the ventral part of the gonad wall and no-where surrounds the gonadal lumen. According to the data by Aoyama *et al.* (1963) the juvenile gonad of *C. crocodila* is exclusively male and does not contain female gonocytes. If this is true, the flatheads would be unique among all other ambosexual fishes in which topographical separa-tion of the heterologous germinal tissues occurs. The observations by the Japanese workers would imply that ovarian tissue develops secondarily from testicular tissue, yet without becoming functionally active at that time, Okada (1966*b*) denies the existence of a testicular phase without ovarian elements for *Inegocia* (= *Suggrundus*) *meerdervoorti*. Instead of this he claims the occurrence of two different bisexual phases in this species and an ovarian phase between both of them. According to him the testis in the second bisexual phase is formed from a primordium different from that in the first bisexual phase. Nevertheless the evidence for such an assumption seems insufficient, especially since he indicates in a later paper (Okada, 1968) that only a part of the smaller specimens (45–50 mm) proved to be clearly bisexual (13 out of 81). The large majority of his specimens in this size-group were either purely male or female. All these problems should be followed up by investigators in the Pacific area.

Among the porgies we find the best-known examples of protandric sex-inversion (Pasquali, 1941; D'Ancona, 1941; 1950; Reinboth, 1962*a*, Lissia-Frau, 1966*a, b*; Lissia-Frau & Casu, 1968*b*; Lissia-Frau & Pala, 1968). The testis which disappears completely during sex-transformation does not participate in the structural formation of the functional ovary. In spite of this basic uniformity there seems to exist a considerable species-specific variety with regard to (i) the extent of the time-range during which sex-inversion takes place within the species' average life-span, (ii) the proportion of inverting animals, (iii) the quantitative ratio of heterologous germinal tissue in functional males (and females?) and (iv) the eventual existence of 'pure' males and 'pure' females besides ambosexual functional males that may, or may not, develop secondarily to functional females. But as long as we have no other methods available than those that have been mentioned above it is rather difficult to describe such differences in a quantitative way. Lissia-Frau (1966*a*), Lissia-Frau & Casu (1968*b*) and Lissia-Frau & Pala (1968) have tried to summarize the various conditions, mainly from D'Ancona's (1950), Reinboth's (1962*a*) and their own work.

Most of their recent arguments are based on the distinction between (i) exclusively male gonads, (ii) exclusively female gonads, (iii) gonads with a prevailing ovarian portion, (iv) gonads with a prevailing testicular portion and (v) ovotestes in which male and female parts are about the same size. But unless a regular sequence of transformation stages during a certain period of the reproductive cycle is traceable, such a classification can hardly contribute to answer the question whether sex-inversion occurs or not. The proportion of both germinal tissues does not by itself reveal whether an animal is a functional male or female or an inverting specimen. In *Lithognathus mormyrus*, which is considered protandric by Lissia-Frau & Casu (1968*b*), has been found several ambosexual gonads which must be considered as functional testes (onset of spermatogenesis, entirely quiescent ovarian part with smaller oocytes than females caught the same day) although the ovarian portion is larger than the testicular one (Plate 1*c*, *d*; Reinboth, unpublished). The Italian workers themselves mention in their paper (p. 8) that in 49 out of 65 specimens with a prevailing ovarian part the testicular crest is of considerable size and of normal histological appearance. Moreover, it should be emphasized that the presence of 'brown bodies' within the testis (which can be stained with the usual dyes for lipids and which remain unchanged during the passage through fat solvents for paraffin-embedding) cannot be taken as an indication of the beginning of testicular regeneration. They have been found frequently in fully active purely male gonads of many species (Reinboth, unpublished). The blurring of the outlines of the gonocytes, frequently mentioned by the Italian investigators, is also an unreliable criterion: rapid autolysis of gonadal tissue may occur when the material is coming from the market and has not been preserved immediately after the animal's death. Consequently one must consider the problem of protandry in *L. mormyrus* as still an open question (cf. Reinboth, 1962*a*).

Similar considerations apply to the paper by Lissia-Frau & Pala (1968) on the genus *Diplodus*. If, for example, in *D. sargus* the ratio 107 ♀♀:40 ♂♂ among the smaller specimens (14–23 cm) changes to 87 ♀♀:55 ♂♂ among the larger ones (23·1–40 cm) it seems difficult to argue for protandric sex-inversion. A similar situation applies to *D. annularis*.

The functional significance of the great individual variation in the ratios of female to male tissue remains unexplained. In species in which 'pure' males (males without ovarian tissue) exist it should be worth while looking for the ratio ambosexual:pure ♂♂ in different size classes as one of the parameters that could have some diagnostic value. But it should be taken into consideration that a decrease in the number of ambosexual males may come about either by γ or by ϵ (cf. Fig. 1*b*).

Two papers on *Mylio macrocephalus* by Okada (1965*a*, *b*) magnify the problems. Okada makes the following significant points. (i) *M. macrocephalus* is becoming sexually mature by the third year after hatching. (ii) More than 50% of 3-year-old specimens are functional males—(is there a significant shift in the ratio ♂♂:♀♀ corresponding to increasing size?). (iii) The earliest appearance of functional females is in the fourth year. (iv) Among the 3-year-old specimens distinct ovary-bearing females without any testicular tissue can be found. (v) Differentiated ovaries never contain testicular tissue. Interestingly the same species had been investigated previously as *Sparus longispinis* by Kinoshita (1936). He observed that by the third year the fishes were becoming gonochoristic by degeneration of the heterologous tissue. In spite of the small number of fishes available to Kinoshita (1936) and the lack of figures in Okada's (1965*a*, *b*) papers Okada's data suggest that *M. macrocephalus* might be a secondary gonochorist (ε, λ) (Fig. 1*b*). From the present information a definite statement on the sexual pattern of *M. macrocephalus* seems to be premature. The species should be reinvestigated as well as *Sparus aries* and *S. latus* examined by Kinoshita (1939).

Protogyny

In the animal kingdom protandric sex-inversion is much more widespread than a change from female to male (Hartmann, 1956). Among teleosts, however, protogyny is clearly preponderant. The process of gonad transformation may vary from species to species. In certain cases testis formation starts from preformed areas at the periphery of the gonad (e.g. *Spondyliosoma, Spicara, Coris julis* (Reinboth, 1962*a*), *Sacura* (Reinboth, 1963*b*)). In other fishes no trace of potential male tissue is identifiable until the ovary starts to degenerate and nests of gonia within the ovarian tissue enter the spermatogenetic cycle (e.g. some serranids (Smith, 1959, 1965), *Thalassoma bifasciatum* (Reinboth, unpublished)). There is no evidence for taxonomic correlations. But all protogynous species have in common a massive degeneration of the former ovarian tissue during gonad transformation. Atz (1964) has reviewed the literature up to 1963. Subsequently further papers on this subject have been published, on labrids by Quignard (1966), Reinboth (1967*c*), Roede (1966), Sordi (1964, 1967), Zunarelli Vandini (1965), on serranids by McErlean & Smith (1964), Okada (1965*c, d*), Reinboth (1964, 1965, 1967*a, b*), on synbranchids by Chan & Phillips (1967), Liem (1968), Okada (1966*a*). Choat (1966) and Reinboth (1968) have good evidence for the occurrence of protogyny in parrot fishes (scarids), which represent the most recently reported ambosexual family (cf. Rosenblatt & Hobson, 1969).

As in protandric teleosts the diagnosis of sex-inversion is easy when all representatives or the great majority of one sex belong to particular size groups (e.g. *Spondyliosoma, Spicara, Sacura*). The reverting animals have an intermediate size and can be recognized in the histological preparation by the striking breakdown of the ovarian part whilst the testicular portion starts developing. In some species with secondary sex characters the stage of sexual transformation is marked by an intermediate colour pattern (e.g. *Coris julis* (Reinboth, 1962*a*), *Sacura* (Reinboth, 1963*b*)).

Labrids

In wrasses the sexual organization may be particularly complicated because in several species (e.g. *Coris julis* (Reinboth, 1961, 1962*a*), *Thalassoma bifasciatum* (Reinboth, 1967), *T. lucasanum, Halichoeres poecilopterus, H. bivittatus, H. dispilus* (Reinboth, unpublished)) a relatively large number of males occur even among the smallest specimens. Fortunately, in these fishes the structure of the testis reflects clearly whether a male originated by sex-inversion (secondary male) or has been born as a male (primary male) (Reinboth, 1961, 1962*a*). In secondary males the testicular tissue surrounds the former ovarian cavity and is sometimes arranged in the same way as the ovarian lamellae (cf. Plate 2). Moreover, the structure of the vas deferens is entirely different in both male types. The vas deferens of a secondary male arises secondarily and surrounds the persisting ovarian lumen like a double-walled cylinder whereas in the primary male the vas deferens is a simple tube (Fig. 2). The morphologic difference between primary and secondary males has been corroborated by Harrington (1967, and personal communication) in his experiments on *Rivulus* and by Liem (1968) on several species of the synbranchids. Consequently I consider Zunarelli Vandini's (1965) opinion on the sexual pattern of *Coris julis* as untenable. She neglected the differences between primary and secondary testes and mixed both of them up in one group. Apart from body size and the morphologic characters of the secondary testes, the frequent presence of small oocytes or remnants of degenerated larger ones point to the origin of these male organs. Furthermore, in many hundreds of primary male wrasses of many species examined no trace of ovarian elements was seen (Reinboth, unpublished).

As far as can be seen from the present literature the protogynous wrasses should be subdivided into two main categories: (1) diandric species with primary and secondary males (e.g. *Coris julis, Thalassoma bifasciatum, Halichoeres bivittatus,* etc.); (2) monandric species in which all males are secondary ones (*Labrus turdus, L. merula* (Sordi, 1962), *L. bergylta* (Quignard, 1966), *Hemipteronotus novacula* (Oliver & Massuti, 1952; Sordi,

1967, Reinboth, unpublished)). The case of *L. bimaculatus* needs further investigation. According to Sordi (1964) and Quignard (1966) the occurrence of males is restricted to larger specimens (secondary males?), whereas Lönnberg & Gustafson (1936) found smaller ones too and provided a micrograph which seems to show a primary testis.

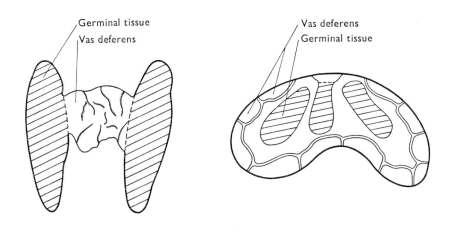

Primary testis   Secondary testis

Fig. 2. Schematic drawing of primary and secondary testes. Further explanations in the text.

The puzzling question whether all females of protogynous wrasses or only a proportion of them change sex has been often discussed but no satisfactory explanation has been found. Sordi (1962), for instance, states that in *Labrus merula* only about one half of the females become male whereas in *L. turdus* all of them undergo sex-inversion. Quignard (1966), however, considers Sordi's material as insufficient and concludes from his own observations that in *L. turdus*, too, all females finally become male. More data and comments on this problem will be published elsewhere (Reinboth, 1970).

Serranids

In protogynous serranids all males have a secondary testis. This does not exclude, however, the possibility that some specimens are already functional males when they become sexually mature for the first time. It should also be noted that the testes of certain secondary gonochorists have a secondary structure (Smith & Young, 1966). In some species a few females are as big as the largest males that have been caught (Smith, 1959), and it is unknown whether such females must be considered as 'pure' females or whether they are still able to undergo sex-inversion.

Synbranchids

In a recent paper Liem (1968) reported for this family that primary and secondary males may occur in the same species. According to Liem (1963) and also from the investigations on *Monopterus albus* by Chan & Phillips (1967) either all or at least the majority of female synbranchids become male at a later stage. Liem's most exciting result is the discovery of marked geographical differences in the sexuality of certain species (Liem, 1968). In *Synbranchus marmoratus* from Brazil he found a small percentage of primary males in the smallest individuals, whereas 'the secondary male is invariably represented by fish larger than 25 cm'. In a population of the same species on Pearl Island, just off the coast of Panama, primary males were absent. Unfortunately, he gives no data about the ratio of both male types and he never mentions primary males among the larger specimens. Liem's Plate 2, fig. *c* remains somewhat obscure. It would be interesting to know which functions he attributes to the lumina at the right and left of the cavity which he denotes as the sperm duct. Liem (1968) reinvestigated *Monopterus albus* and he concludes that the 'full cord males', described by Chan & Phillips are primary males but his conclusion finds no support from Chan & Phillips (1967). In their micrograph of the gonad of such a full cord male both gonadal cavities that are characteristic for the ovaries are clearly visible in the male organ.

Sparids

Our knowledge of protogynous sparids has advanced little since Atz's (1964) review except for some fragmentary evidence for protogynous sex-inversion in *Calamus pennatula* from the Western Atlantic (Reinboth, 1967b).

The only arguments in favour of protogynous sex-inversion in *Pagellus erythrinus* remain the differences of the numeric ratio ♀♀/♂♂ in smaller and larger size groups and the presence of male germ cells in the gonadal wall of females. Beyond this, *P. erythrinus* has nothing in common with other protogynous porgies (Reinboth, 1962a). The most peculiar feature is the presence of ovarian tissue in many testes, even at the height of the spawning season. Such a phenomenon is unknown in any other protogynous teleost and occurs in protandric species and secondary gonochorists only. In addition to this there is no evidence for ovarian degeneration while testicular tissue is developing and the rather large numbers of specimens covering all periods of the year makes it unlikely that such stages have been missed (Reinboth, 1962, unpublished). There are, however, some indications that a part of the ambosexual testes develops into 'pure' testes in the course of time. A random sampling revealed 15 ambosexual testes and 30 'pure' ones among males up to 230 mm in length, but only 7 ambosexual and 29 purely male gonads in fish of larger size. In terms of Fig. 1 a the sexual organization of *Pagellus erythrinus* could eventually be described as a combination of *e*, *g*, *k* and *l*. But whether *g* is realized seems still questionable. *P. erythrinus* is the most striking example demonstrating the difficulties in judging the sexual organization of an intersexual fish from examination of gonads only.

Origin of the heterologous germinal tissues

In the protandric porgies and flatheads the functional ovarian part derives from preformed non-functional ovarian tissue. Based on histological observations of juvenile undifferentiated gonads in sparids (Pasquali, 1941; D'Ancona, 1941, 1950; Lissia-Frau & Casu, 1968 a; Okada, 1965 a) it may be concluded that this tissue develops very early during gonad formation. In the flatheads the situation appears to be quite similar, although the problem of the alleged 'pure males' and the occurence of two bisexual phases (cf. p. 523) has still to be solved. In *Cobitis taenia* and *Gonostoma gracile* the origin of oocytes has not yet been established. But it seems that a particular ovogenous area does not exist.

Among a certain number of protogynous teleosts (e.g. *Sacura*, *Spondyliosoma*, *Spicara*) during the female stage non-functional testicular areas are present which commence growing when the degeneration of the ovary begins. At least in protogynous porgies the testicular part seems to arise early during gonad development. Although no protogynous sparid has been investigated from this standpoint, such an assumption appears to be reasonable because of the basic similarity of gonad structure in this family. From all other protogynous teleosts we have no data whatsoever on gonad development. A paper by Arru (1966) dealing with gonad formation in wrasses is based on very limited material. Nevertheless, it seems possible that where testicular tissue is present in well-defined areas then this region arises early in the gonad primordium. *Coris julis* is at present the only case among wrasses in which the origin of the secondary testis can be traced back to particular cell groups attached to or enclosed in the ovarian wall (Reinboth, 1962 a). Testis formation starts from these cells and in many intermediate gonads there is no evidence for the origin of male germ cells within the ovarian lamellae. This explains why the secondary testis of *Coris* never has a lamellar structure (Plate 2) in the arrangement of the male germinal tissue. In *Thalassoma bifasciatum* and in other wrasses too potential male cells could not be identified during the female stage (Reinboth, 1970).

According to the first paper by Liem (1963) on *Monopterus* the testicular tissue originates from clusters of cells placed at irregular distances along the connective tissue capsule of the gonad. Thus the situation seemed to be quite similar to that described earlier for *Coris* (Reinboth, 1962a). Chan & Phillips (1967), however, did not confirm Liem's observations. They claim the outer gonad wall to be sterile and found along the inner edges bordering the inner gonadal cavities in both gonadal lamellae gonocytes and cells similar to spermatogonia. In his most recent paper on synbranchids Liem (1968) refers to Chan & Phillips's (1967) observations and reports that he found a similar situation in *Synbranchus marmoratus*, passing over his earlier results and reporting Okada's (1966) findings about male gonocytes inside the fibrous connective tissue of the capsule; contrasting with the data of Chan & Phillips (1967). In the serranids (epinephelids) that have been studied by Smith (1959), specific male areas are lacking during the female stage. Sooner or later, however, small clusters of germ cells with spermatogenetic activity scattered within the ovarian tissue become discernible. The testis tissue is formed from such cell groups and hence the testis has ovary-like lamellae. The serranid fish *Chelidoperca hirundinacea* shows an intermediate pattern (Reinboth, 1967a).

It is an open question whether the resting deutogonia that are identified as potentially male by their specific location only, are qualitatively different from the germ cells lying within the ovarian lamellae. In other words, since oogonia and spermatogonia do not have differential cytological characters (Reinboth, 1967b) it is not possible to say whether male tissue originates from early determined male germ cells or from a stock of sexually bipotent gonocytes. The implications of this problem will be discussed later.

EXPERIMENTS ON INTERSEXUALITY IN FISHES

Experiments on ambosexual teleosts are extremely scanty. Apart from the few papers that have been listed by Atz (1964), only Harrington (1967, 1968), Okada (1964a, b) and Reinboth (1967b, c, 1970) have reported additional data.

Influence of environmental factors

Harrington's (1967, 1968) experiments on the effect of low temperature on the production of primary males in *Rivulus* are the only contribution to this topic. He emphasizes that low temperature must not be the only external factor which influences primary male induction; that is, the suppression of ovarian tissue in a normally hermaphroditic species. In order to explain his experimental results he suggests that certain embryonic processes might be uncoupled by differential effects of low temperature on two or more constituent rates of development. According to Harrington's unpublished observations environmental factors may also bring about a conversion of hermaphrodites to secondary males. Consequently *Rivulus* possesses in one and the same species features in common with several families of fishes containing hermaphroditic species (Harrington, personal communication). The effect of temperature stress during ontogenetic development on sexual differentiation is known from other animals, both from vertebrates (cf. Foote's (1964) review on amphibia) and invertebrates (experiments by Anderson & Horsfall (1963) and Brust & Horsfall (1965) on subarctic species of the genus *Aedes* in which exposure to larval stages to high temperatures suppressed the development of male tissues but activated female ones in heterozygous individuals). Recent experiments on *Betta splendens* revealed an extremely high variation of sex ratios under the influence of several extrinsic factors (Lucas, 1968) but the author has not been able to demonstrate a special significance of one particular variable. In the protogynous *Monopterus albus*, Liem (1963) has observed the appearance of males and a higher frequency of intersexes among starved individuals. At present the mechanisms of the effects of environmental factors on sex differentiation and development are entirely unknown. But apart from this it is doubtful whether environmental factors lend themselves to be generally taken as responsible for the induction of sex-inversion. Since in the ocean in which most of the intersexual teleosts are living the environmental conditions are rather stable, it is difficult to imagine how— for example, in a tropical reef—environmental factors could account for the change of sex in a few specimens only that share the same biotope with the large majority of members of the same species.

The influence of sex steroids on ambosexual teleosts

Okada (1964 *a, b*) reports on the work on a wrasse performed by Yamasu who injected female *Halichoeres poecilopterus* with testosterone propionate and male specimens with oestradiol benzoate. Unfortunately, the experimental conditions are so ill-defined that only general statements seem to be possible. Repeated injections of oestradiol (0·1 mg/injection) caused complete degeneration and final disappearance of male germ cells. Administration of testosterone to females induced degeneration of the oocytes and the onset of spermatogenesis. According to Okada (1964 *a*) the oogonia are stimulated to grow and produce male germ cells. Such an interpretation evidently contradicts our current knowledge. The secondary sex characters of the treated females did not change to the male pattern, whereas in a former experiment (Okada, 1962) on the same species methyltestosterone produced the male colour pattern without affecting the gonads. It is unknown whether the different hormones or other experimental conditions account for these contradictory results. Nevertheless, Okada's findings agree with the observations that administration of androgens causes precocious sex-inversion in protogynous species (Reinboth, 1962a, b, 1963a, 1967c, 1970).

In a short note on *Mylio macrocephalus* Okada (1965b) reports that subcutaneous implantation of a small amount of methyltestosterone elicits a marked stimulation of the testicular part of the gonads, seminiferous tubules being filled with spermatozoa within a few days. Okada used fish with presumptive ovaries in which the testicular part is greatly reduced in size but the author does not say in which way he determined the initial condition of the gonads of his experimental animals. In the absence of experimental details the evaluation of Okada's data is rather difficult and as pointed out above it appears to be doubtful whether *M. macrocephalus* may be considered as a protandric species. On the other hand it would be most interesting if androgens can bring about a masculinization of female gonads in a protandric fish. A previous experiment to elicit precocious ovarian development by injection of oestradiol into juvenile protandric *Sparus auratus* has not been successful (Reinboth, 1962a). Generally the effect of oestrogens on ambosexual teleosts seems to be much less specific than that of androgens (Reinboth, 1962a). In a few pilot experiments which included both sexes of *Diplodus annularis*, *Spicara maena*, *Thalassoma pavo* and *Coris julis* the injection of oestradiol considerably damaged both types of gonadal tissues.

Artificial sex-inversion in freshwater teleosts

Because of the difficulties in obtaining and handling marine teleosts for experimental investigations interest has been focussed on freshwater species that might be suitable for the study of sex-inversion. Yamamoto's (1953, 1955, 1958, 1959, 1961) pioneering work on artificial sex-inversion in *Oryzias* pointed the way to the endocrinological approach to the problem of sex-differentiation in teleosts. The cyprinodonts, however, represent a more primitive group, whereas most of the normally ambosexual species belong to the higher Perciform fishes. Among these the Cichlid fishes and the Belontiids (former Anabantids) proved to be good experimental animals.

Two groups of research workers (Noble & Kumpf, 1937; Kaiser &

Schmidt, 1951) reported that gonadectomy of female *Betta splendens* may lead to regeneration of testicular tissue. Becker (1969) succeeded in producing males with ripe sperm by incompletely gonadectomizing females (26 out of 87 specimens). The percentage of inverting fishes could be increased considerably by squashing the ovary between two slides and reimplanting the tissue-'brei' into the gonadal cavity (31 out of 48 specimens). Analogous experiments with *Macropodus opercularis* yielded similar results: out of 49 female specimens 27 individuals produced testicular tissue. But whereas in *Betta* mostly 'pure' males and a few hermaphrodites were obtained, in *Macropodus* all the males had more or less large amounts of ovarian tissue. Yet all these fishes acquired the male secondary sex characters within two months and some of them spawned successfully with females. The offspring of such couples were exclusively female. Consequently it must be assumed that in *Macropodus* the male is the heterogametic sex. The same surgical procedure applied to other species (*Tilapia mossambica*—Cichlidae; *Lepomis cyanellus*—Centrarchidae) did not cause a change of sex. Injection of testosterone into juvenile females of *Betta*, *Tilapia* and *Lepomis* had no masculinizing effects on the gonads. Becker found some indication that follicular cells give rise to the new germinal tissue. By the use of grafting experiments on *Betta splendens* Eckert (unpublished) succeeded in demonstrating that heterologous gonad regeneration proceeds from the former gonadal tissue even in a heterotopic position.

Experimental work with Cichlid fishes in our laboratory yielded some interesting and rather surprising results. The addition of either testosterone, methyltestosterone or oestradiol (mostly 0·5 mg/l.) to the water in which juvenile *Hemihaplochromis multicolor* were reared caused feminization of the gonads (up to 100%) although the androgens elicited male secondary sex characters (Müller, 1969). Hackmann (unpublished) has been able to delimit a sensible period of sex differentiation with a duration of 48 h at the most. As opposed to Clemens & Inslee (1968), androgens likewise had a feminizing effect on *Tilapia mossambica*. Administration of 11-ketotestosterone elicited male secondary sex characters in *Hemihaplochromis* but did not affect gonad differentiation. From a preliminary experimental series there is no evidence that the paradoxical effect might depend on hormone concentration but the method of steroid administration may have some directing influence under particular circumstances (Reinboth, 1969b). From such data it is difficult to agree with Yamamoto's (1962) hypothesis that sex steroids are acting as specific sex inductors (termones). Furthermore, in Cichlid fishes we have no supporting evidence for Ashby's (1965) suggestion that hormonal sex inversion might be the indirect result of intensification of a juvenile intersexual phase.

34-2

In the genera of *Pseudotropheus* and *Labeotropheus* from Lake Nyassa Peters (personal communication) has observed a frequent occurrence of oocytes in the midst of testicular tissue of functional males. On the other hand, some juvenile females seem to have spermatogenetic tissue. Simon (unpublished) has discovered a particular population of *Lepomis cyanellus* (Centrarchidae) in which more than 20% of the fish had gonads with both types of germinal tissue. The expectation of finding more freshwater teleosts with marked deviations from the usual gonochoristic pattern seems to be probable.

SOME COMMENTS ON THE GENETICAL APPROACH TO THE PROBLEM OF INTERSEXUALITY IN TELEOSTS

In the past the problem of sex differentiation in fishes has been looked upon mainly from a genetical point of view and the growing literature on normally intersexual species has enhanced the speculation in this approach. Bacci (1965) states that 'the study of a hereditary determination of any biological character requires in fact a precise and extended knowledge of its variability and any theory that is not based upon such knowledge is inadequate from a genetical point of view'. But he deviates from this principle when he is dealing with sex inversion in wrasses. The cyprinodonts are the only teleosts on which careful genetical studies have been carried out. But in spite of many ingenious experiments by different authors, sometimes on the same species, the interpretations are not always consistent (cf. reviews by Atz (1964), Bacci (1965), Gordon (1957), Hartmann (1956)).

Such a situation should discourage any premature and purely formal classification of ambosexual teleosts when they have never been propagated in the laboratory. These reservations apply to the work of Bacci (1965), Sordi (1962, 1964, 1967) and Zunarelli Vandini (1965). Harrington's (1967, 1968) meticulous work on *Rivulus* is the only case in which an ambosexual teleost has been subjected to genetical studies. Kallmann & Harrington (1964) have been able to show that *Rivulus* is homozygous and there seems to be good evidence that the experimental animals which they used were 'homogametic as well, both hermaphrodites and experimentally produced primary male gonochorists' (Harrington, 1967, p. 195). According to Harrington (1967) in *Rivulus* the hermaphroditic constitution is the epistatic one and the male constitution the hypostatic. Kallmann & Harrington (1964) consider certain strains of *Rivulus* as representing a 'clone'. According to Rieger, Michaelis & Green (1968) a 'clone' is 'a population of cells or organisms derived from a single cell or common ancestor by mitoses. The mode of reproduction giving rise to a clone is asexual'. Since *Rivulus* reproduces sexually the term 'clone' should be avoided in this case.

Okada (1964*b*) erroneously ascribed to Reinboth the opinion that the colour dimorphism of males of *Coris julis* would be genetically determined. Sordi's (1962) pretension of having demonstrated the existence of sexual digamety in *Labrus merula* can hardly be considered as valid because the methods which he used are inadequate for a statement of that kind. In his paper on synbranchids Liem (1968) reports that in wrasses 'the alternative of primary versus secondary male is decided by a genetic shift mechanism instead of by environmental conditions' whereas the original author (Harrington, 1967) discusses that topic merely as a hypothetical possibility. Liem (1968) is veiling our ignorance when he discounts Harrington's (1967) very cautious considerations on the eventual existence of a homogamety–heterogamety switch mechanism for the production of protogynous hermaphrodites versus primary males and then applies a part of them—word by word but without quotation—to the synbranchids.

Ohno (1967) is of the opinion that the state of sex differentiation in synchronous hermaphroditism 'reflects the total lack of chromosomal sex-determining factors. The protogynous and protoandrous types of functional hermaphroditism, on the other hand, may be regarded as most primitive because here sex-determination is the result of the aging process.' But this view is difficult to substantiate on our present knowledge.

Even when we are unable to ascribe the formation of ambosexual gonads and the regular succession of heterologous sex function to the existence of a chromosomal switch mechanism or to the actions of some kind of a polyfactorial gene system, it cannot be doubted that specific sexual constitutions are fixed in some way within the genome of the species. The combination of intrinsic and extrinsic factors that bring about the particular phenotypic expression of sex is a problem which extends, perhaps far, into adult life.

In ambosexual fishes, the current alternative of genotypic versus phenotypic sex determination should be dismissed, perhaps, as an over-simplification, since it frequently invites a one-sided look for sex-determining factors in the animal's environment. The search for intrinsic processes prior to certain developmental stages in the animal's gonad would automatically include both genetic and environmental parameters. The interplay of a well-established genetical constitution and somatic elements of the gonad in the development of sex is an outstanding but not the only example for the relativity of seemingly contrasting principles. The undoubted role of either hormones or inductor substances in the process of gonad formation should be a stimulus for the comparative endocrinologist to interest himself in the intersexuality problem of fishes.

CONCLUSION

Since Witschi's pioneering report on sex-differentiation in amphibia the discussion of the role of hormones and hypothetical inductor substances on the development of sex in vertebrates has continued (reviews by Burns (1961), Dodd (1960), Gallien (1965), Witschi (1957)). But it is frequently neglected that the distinction between cortex and medulla does not apply to the largest group of vertebrates, the teleosts. D'Ancona's (1950) efforts to adapt Witschi's concept to the particular situation in ambosexual sparids raises considerable problems (Reinboth, 1962a, 1967b). Lissia-Frau & Casu (1968a) emphasize that the process of gonadogenesis in sparids which leads to a spatial segregation of heterologous gonad tissues 'cannot be included among the factors responsible for the great variety of herm-aphrodite manifestations in these species'. In addition to this, our review has demonstrated that in a fairly large number of normally ambosexual teleosts the capacity of sex-inversion during adult life has no morphological counterpart in terms of a visible spatial zonation. Although the topo-graphical separation of male and female tissues is very distinct in certain ambosexual species its significance should not be overestimated with regard to such cases in which sex transformation is performed by the same tissues. Moreover, we are unable to describe specific differences between the gonia of both heterologous tissues (Reinboth, 1967b).

The most interesting results obtained by Charniaux-Cotton (1965) on sex development in crustaceans demonstrate that spatial segregation of germ cells plays a minor role in directing gametogenesis. Berreur-Bonnen-fant & Charniaux-Cotton (1965) have shown that even in species with topographical zonation of the germinal crest the titre of hormone pro-duced by the androgenic gland decides whether the gonia are entering spermatogenesis or oogenesis. This would confirm the generally accepted opinion of the sexual bipotentiality of gonia, even in adults, although it remains a puzzling fact that in many protandric and protogynous species the heterologous germinal tissues do not participate in the development of the second sex phase. Occasional exceptions have been reported by Lissia-Frau (1966b) and Lissia-Frau & Pala (1968).

The close functional relationship between gonads and pituitary, the conflicting and often fragmentary data on the effects of sex steroids on gametogenesis in a relatively small number of lower vertebrates (cf. reviews by Pickford & Atz (1957), Dodd (1960), Forbes (1961), Hoar (1965), Barr (1968), Lofts (1968)), make it hardly possible to use our information on crustaceans as a model for explaining certain mechanisms of intersexuality in fish. A comparison of observations like those by Lofts,

Pickford & Atz (1966) on the spermatokinetic effects of methyltestosterone in hypophysectomized *Fundulus heteroclitus* with experiments by Basu & Nandi (1965) on the suppression of spermatogenesis by administration of testosterone to hypophysectomized *Rana pipiens* should caution against undue generalizations about specific effects of specific steroids on oogenesis and spermatogenesis in lower vertebrates (cf. Ashby, 1965). The incompleteness of our present knowledge makes difficult the formulation of a hypothesis to serve as a guideline for subsequent elucidation of the complicated sex patterns in ambosexual teleosts.

The induction of complete artificial sex-inversion by hormone treatment in certain gonochoristic teleosts (Yamamoto, 1953–62; Yamamoto & Kajishima (1968); Dzwillo, 1962; Clemens & Inslee, 1968; Müller, 1969) has been obtained in undifferentiated animals only. It is questionable whether the mechanisms involved are equivalent to those which occur during spontaneous sex-inversion of ambosexual fish. It cannot be excluded that the morphogenetic effects of steroids during gonad differentiation are different from the processes that lead to a drastic change in the activities of functional sex organs.

We have ample information that among vertebrates androgens and oestrogens are produced simultaneously by the same individual. In spite of technical difficulties in determining the production rate of these substances there is general agreement that in males the androgens outweigh the oestrogens whilst in females the oestrogens preponderate. Normal sexual functions seem to depend on a delicate quantitative and qualitative balance of androgenic and oestrogenic hormones, and some cases of teratological intersexuality are explained as a disturbance of that equilibrium. Since ambosexuality must be considered as a normal phenomenon in many teleosts knowledge of the role of sex steroids is an intriguing problem. This is particularly true for simultaneous hermaphrodites. At present we have one single paper only which deals with this question. Lupo di Prisco & Chieffi (1965) have identified oestradiol, oestrone, oestriol, progesterone, testosterone and androstenedione in the gonadal extracts of *Serranus scriba*. But the investigation was certainly not exhaustive and should be repeated, taking into consideration intrinsic gonadal changes during the reproductive cycle. For a final evaluation of any results the additional investigation of related gonochorists seems indispensable. The same arguments apply to the first paper on *in vitro* steroid synthesis by the ovaries of a protogynous fish (Reinboth, Callard & Leathem, 1966). Recently Chan & Phillips (1969) investigated the biosynthesis of steroids in both functional sex stages of the protogynous *Monopterus albus*. Using labelled pregnenolone as a precursor they have been able to demonstrate that in both sexes androstenedione, testosterone, oestradiol-17β and oestrone are produced. But during sex-inversion the balance between androgens and oestrogens shifts towards a marked increase in androgen production and a decline of the amount of oestrogens produced. These results seem to support the former assumption that androgens might inhibit or even suppress ovarian development (Reinboth, 1962b). On the other hand it remains an open question whether the shift in the hormonal balance should be regarded as causative or whether it only reflects primary processes of a different kind. In transforming gonads of the protogynous *Coris julis* considerable amounts of cholesterol have been found (Reinboth, 1969a), but it is not yet known whether this has to be related to a particular steroidogenic activity.

In spite of the recent contributions just mentioned the *in vivo* processes which keep an animal hermaphroditic or which cause the succession of two entirely different sexual functions await clarification. Age-dependent differential responsiveness of the heterologous germinal tissues towards hormones (both steroids and gonadotrophins), mutual chemical (hormonal) inhibition of male and female gametogenesis, abnormal development of one gonadal element at least in its initial stages, selective chemical (hormonal) stimulation of either spermatogenesis or oogenesis—these could be some of the mechanisms that might be interrelated in specific combinations. It is up to endocrinologists to elucidate their eventual significance or to bring forward new ideas in order to explain a most puzzling phenomenon that resists, at present, satisfactory integration into the current concepts of vertebrate reproduction.

REFERENCES

ANDERSON, J. F. & HORSFALL, W. R. (1963). Thermal stress and anomalous development of mosquitoes (*Diptera: Culicidae*). I. Effect of constant temperature on dimorphism of adults of *Aedes stimulans*. *J. exp. Zool.* **154**, 67–107.

AOYAMA, T. (1955). On the hermaphroditism in the yellow sea bream, *Taius tumifrons*. *Jap. J. Ichthyol.* **4**, 119–29. (Japanese with English summary).

AOYAMA, T., KITAJIMA, T. & MIZUE, K. (1963). Study of the sex reversal of inegochi, *Cociella crocodila* (Tilesius). *Bull. Seikai Region. Fish. Res. Lab.* **29**, 13–33.

ARRU, A. (1966). Prime indagini sulla gonadogenesi nei Labridi. *Boll. Zool.* **33**, 327–33.

ASHBY, K. R. (1965). The effect of steroid hormones on the development of the reproductive system of *Salmo trutta* L. when administered at the commencement of spermatogenetic activity in the testes. *Riv. Biol.* **58**, 139–65.

ATZ, J. W. (1964). Intersexuality in fishes. In *Intersexuality in Vertebrates including Man*, pp. 145–232. (eds. C. N. Armstrong and A. J. Marshall). London and New York: Academic Press.

BACCI, G. (1965). *Sex Determination*. pp. 306. Oxford: Pergamon Press.

PLATE I

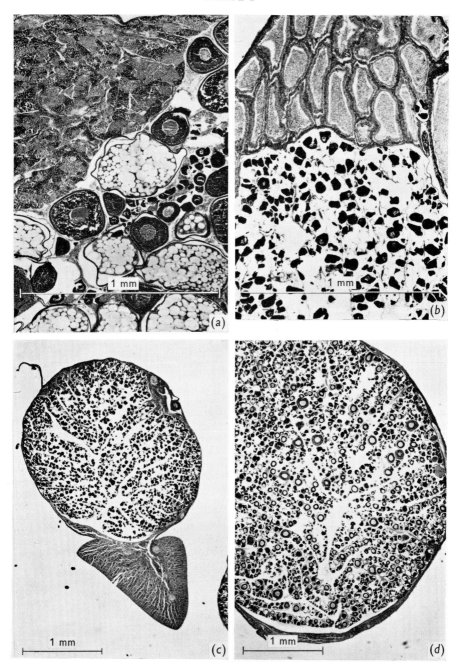

(*a*) Part of a cross-section through the gonad of the simultaneous hermaphrodite *Serranus fasciatus* (Jenyns); 85 mm, July.

(*b*) Part of a cross-section through the functional male gonad of the protandric platycephalid *Rogadius asper* (Cuv. et Val.). 173 mm, November.

Cross-sections through ambosexual gonads of *Lithognathus mormyrus* (L.). (*c*) Functional male with large non-functional ovarian portion. 222 mm, April; (*d*) Functional female. Testicular rudiment at the lower left of the gonad wall. 232 mm, April.

PLATE 2

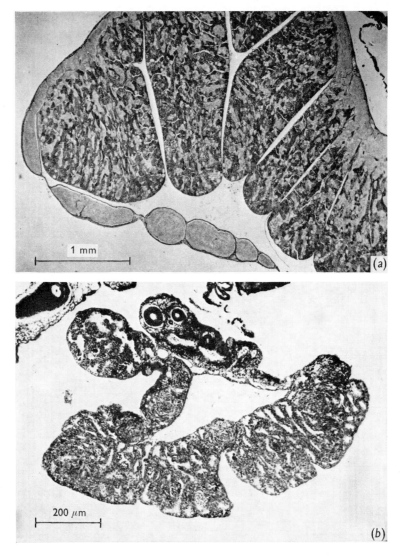

Cross-sections through secondary testes of labrids. (a) *Halichoeres poecilopterus* with testicular lamellae protruding into the former ovarian cavity: 196 mm, June. (b) *Coris julis* (L.) without lamellae: 142 mm, October.

BARR, W. A. (1968). Patterns of ovarian activity. In *Perspectives in Endocrinology*, pp. 164–238. (eds. E. J. W. Barrington and C. Barker Jørgensen). London: Academic Press.

BASU, S. L. & NANDI, J. (1965). Effects of testosterone and gonadotropins on spermatogenesis in *Rana pipiens* Schreber. *J. exp. Zool.* **159**, 93–111.

BEATTY, R. A. (1964). Chromosome deviations and sex in vertebrates. In *Intersexuality in Vertebrates Including Man*, pp. 17–144 (eds. C. N. Armstrong and A. J. Marshall). London: Academic Press.

BECKER, P. (1969). Experimentell induzierter Geschlechtswechsel bei Anabantiden. pp. 88, Dissertation, Mainz.

BERG, L. S. (1958). *System der rezenten und fossilen Fischartigen und Fische*, p. 310. Berlin: VEB Deutscher Verlag Wissenschaften.

BERREUR–BONNENFANT, J. & CHARNIAUX-COTTON, H. (1965). Hermaphrodisme protérandrique et fonctionnement de la zone germinative chez la crevette *Pandalus borealis* Kröyer. *Bull. Soc. zool. Fr.* **90**, 243–59.

BRUST, R. A. & HORSFALL, W. R. (1965). Thermal stress and anomalous development of mosquitoes (Diptera: Culicidae). IV. *Aedes communis. Can. J. Zool.* **43**, 11–53.

BURNS, R. K. (1961). Role of hormones in the differentiation of sex. In *Sex and Internal Secretions*, vol. 1, pp. 76–158 (ed. W. C. Young). Baltimore: Williams and Wilkins.

CHAN, S. T. H. & PHILLIPS, J. G. (1967). The structure of the gonad during natural sex reversal in *Monopterus albus* (Pisces: Teleostei). *J. Zool., Lond.* **151**, 129–41.

CHAN, S. T. H. & PHILLIPS, J. G. (1969). The biosynthesis of steroids by the gonads of the ricefield eel *Monopterus albus* at various phases during natural sex reversal. *Gen. comp. Endocr.* **12**, 619–36.

CHARNIAUX-COTTON, H. (1965). Hormonal control of sex differentiation in invertebrates. In *Organogenesis*, pp. 701–40 (eds. R. L. DeHaan and H. Ursprung). New York: Holt Rinehart and Winston.

CHOAT, H. (1966). Parrot fish. *Aust. Nat. Hist.* **15**, 265–8.

CLARK, E. (1959). Functional hermaphroditism and self-fertilization in a serranid fish. *Science, N.Y.* **129**, 215–16.

CLEMENS, H. P. & INSLEE, T. (1968). The production of unisexual broods by *Tilapia mossambica* sex-reversed with methyl testosterone. *Trans. Am. Fish. Soc.* **97**, 18–21.

COE, W. R. (1938). Influence of association on the sexual phases of gastropods having protandric consecutive sexuality. *Biol. Bull. mar. biol. Lab., Woods Hole* **75**, 274–85.

D'ANCONA, U. (1940/41). Ulteriori osservazioni e considerazioni sull' ermafroditismo ed il differenziamento sessuale dell' orata (*Sparus auratus* L.). *Pubbl. Staz. zool. Napol.* **18**, 313–36.

D'ANCONA, U. (1949). Il differenziamento della gonade e l'inversione sessuale degli Sparidi. *Arch. Oceanogr. Limnol.* **6**, 97–163.

D'ANCONA, U. (1950). Détermination et différenciation du sexe chez les poissons. *Archs Anat. microsc. Morph. exp.* **39**, 274–94.

DODD, J. M. (1960). Genetic and environmental aspects of sex determination in cold-blooded vertebrates. In *Sex Differentiation and Development*, pp. 17–44 (ed. C. R. Austin). *Mem. Soc. Endocr.* no. 7. Cambridge University Press.

DZWILLO, M. (1962). Über künstliche Erzeugung funktioneller Männchen weiblichen Genotyps bei *Lebistes reticulatus*. *Biol. Zbl.* **81**, 575–84.

FOOTE, C. L. (1964). Intersexuality in amphibians. In *Intersexuality in Vertebrates Including Man*, pp. 233–72 (eds. C. N. Armstrong and A. J. Marshall). New York and London: Academic Press.

FORBES, T. R. (1961). Endocrinology of reproduction in cold-blooded vertebrates. In *Sex and Internal Secretions*, vol. II, pp. 1035–87 (ed. W. C. Young). Baltimore: Williams and Wilkins.

GALLIEN, L. (1965). Genetical control of sexual differentiation in vertebrates. In *Organogenesis*, pp. 583–610 (eds. R. L. DeHaan and H. Ursprung). New York: Holt, Rinehart and Winston.

GIBBS, R. H. JR (1960). *Alepisaurus brevirostris*, a new species of lancet fish from the western North Atlantic. *Breviora* no. 123, 14 pp.

GOLDSCHMIDT, R. (1915). Vorläufige Mitteilung über weitere Versuche zur Vererbung und Bestimmung des Geschlechts. *Biol. Zbl.* **35**, 565–70.

GORDON, M. (1957). Physiological genetics of fishes. In *The Physiology of Fishes*, vol. 2, pp. 431–501. (ed. M. Brown). New York: Academic Press.

GREENWOOD, P. H., ROSEN, D. E., WEITZMAN, S. H. & MYERS, G. S. (1966). Phyletic studies of teleostean fishes, with a provisional classification of living forms. *Bull. Am. Mus. Nat. Hist.* **131**, 341–455.

HARRINGTON, R. W. JR (1961). Oviparous hermaphroditic fish with internal self-fertilization. *Science, N.Y.* **134**, 1749–50.

HARRINGTON, R. W. JR (1967). Environmentally controlled induction of primary male gonochorists from eggs of the self-fertilizing hermaphroditic fish, *Rivulus marmoratus* Poey. *Biol. Bull. mar. biol. Lab., Woods Hole* **132**, 174–99.

HARRINGTON, R. W. JR (1968). Delimitation of the thermolabile phenocritical period of sex determination and differentiation in the ontogeny of the normally hermaphroditic fish *Rivulus marmoratus* Poey. *Physiol. Zoöl.* **41**, 447–59.

HARTMANN, M. (1956). *Die Sexualität.* 463 pp. Stuttgart: Fischer.

HIDA, T. S. (1967). The distribution and biology of polynemids caught by bottom trawling in Indian seas by the R/V *Anton Bruun*, 1963. *J. Mar. biol. Ass. India*, **9**, 281–99.

HOAR, W. S. (1965). Comparative physiology: Hormones and reproduction in fishes. *A. Rev. Physiol.* **27**, 51–70.

KAISER, P. & SCHMIDT, E. (1951). Vollkommene Geschlechtsumwandlung beim weiblichen siamesischen Kampffisch *Betta splendens*. *Zool. Anz.* **146**, 66–73.

KALLMAN, K. D. & HARRINGTON, R. W. JR (1964). Evidence for the existence of homozygous clones in the self-fertilizing hermaphroditic teleost *Rivulus marmoratus* Poey. *Biol. Bull. mar. biol. Lab., Woods Hole* **126**, 101–14.

KAWAGUCHI, K. & MARUMO, R. (1967). Biology of *Gonostoma gracile* Günther (Gonostomatidae). I. Morphology, life history, and sex reversal. *Informat. Bull. on Planctonol. Japan*. Commem. no. Dr Y. Matsue (included in *Coll. Repr. Oceangr. Lab. met. Res. Inst. Univ. Mabashi* **6** (1968), 53–67).

KINOSHITA, J. (1936). On the conversion of sex in *Sparus longispinis* (Temminck & Schlegel) (Teleostei). *J. Sci. Horishima Univ.* (B-1), **4**, 69–80.

KINOSHITA, J. (1939). Studies on the sexuality of the genus *Sparus* (Teleostei). *J. Sci. Hiroshima Univ.* (B-1), **7**, 25–37.

LAVENDA, N. (1949). Sexual differences and normal protogynous hermaphroditism in the Atlantic sea bass, *Centropristes striatus*. *Copeia* **3**, 185–94.

LIEM, K. F. (1963). Sex reversal as a natural process in the synbranchiform fish *Monopterus albus*. *Copeia* 303–12.

LIEM, K. F. (1968). Geographical and taxonomic variation in the pattern of natural sex reversal in the teleost fish order Synbranchiformes. *J. Zool. Lond.* **156**, 225–38.

LISSIA-FRAU, A. M. (1966a). Ricerche sul differenziamento sessuale di *Boops salpa* (L.) (Teleostei, Sparidae). *Atti Accad. gioenia Sci. nat.* 6 ser., **18**, 165–74.

LISSIA-FRAU, A. M. (1966b). Sulla presenza di oviciti nell' area testicolare delle gonadi ermafrodite della Boga (*Boops boops* (L.)). *Boll. Zool.* **33**, 343–9.

LISSIA-FRAU, A. M. (1968). Le manifestazioni della sessualità negli Sparidi (Teleostei, Perciformes). *Studi Sassaresi* 2, 19 pp.

LISSIA-FRAU, A. M. & CASU, S. (1968a). Il processo gonadogenetico in alcune specie di Sparidi (Teleostei, Perciformes). *Studi Sassaresi* 1, 23 pp.

LISSIA-FRAU, A. M. & CASU, S. (1968b). Il differenziamento sessuale di Lithognathus mormyrus (L.) e di Oblada melanura (L.). *Studi Sassaresi* 2, 19 pp.

LISSIA-FRAU, A. M. & PALA, M. (1968). Ricerche sull'ermafroditismo nei Saraghi: *Diplodus sargus* (L.), *Diplodus vulgaris* (Geoffr.), *Diplodus annularis* (L.) e *Puntazzo puntazzo* Cetti. *Studi Sassaresi* 2, 20 pp.

LODI, E. (1967a). Sex reversal of *Cobitis taenia* L. (Osteichthyes, Fam. Cobitidae). *Experientia*, 23, 446–7.

LODI, E. (1967b). Inversione sessuale in *Cobitis taenia* L. (Cobitidae, Osteichthyes). *Arch. zool. ital.* 52, 129–35.

LOFTS, B. (1968). Patterns of testicular activity. In *Perspectives in Endocrinology*, pp. 239–304. (eds. E. J. W. Barrington and C. Barker Jørgensen). New York and London: Academic Press.

LOFTS, B., PICKFORD, G. E. & ATZ, J. W. (1966). Effects of methyl testosterone on the testes of a hypophysectomized cyprinodont fish, *Fundulus heteroclitus*. *Gen. comp. Endocr.* 6, 74–88.

LONGHURST, A. R. (1965). The biology of West African polynemid fishes. *J. Conseil* 30, 58–74.

LÖNNBERG, E. & GUSTAFSON, G. (1936). Contributions to the life history of the striped wrasse, *Labrus ossifagus* Lin. *Ark. Zool.* A, 29, no. 7, 16 pp.

LOZANO CABO, F. (1953). Monografia de los centracántidos mediterráneos con un estudio especial de la bimetría, biología y anatomía de *Spicara smaris* (L.). *Mem. R. Acad. Cienc. exact. fis. nat. Madr.* 17, 128 pp.

LUCAS, G. A. (1968). Factors affecting sex determination in *Betta splendens*. *Genetics*, 60, 199–200. (Abstr.)

LUPO DI PRISCO, C. & CHIEFFI, G. (1965). Identification of steroid hormones in the gonadal extract of the synchronous hermaphrodite teleost, *Serranus scriba*. *Gen. comp. Endocr.* 5, abstr. no. 71.

MCERLEAN, A. & SMITH, C. L. (1964). The age of sexual succession in the protogynous hermaphrodite *Mycteroperca microlepis*. *Trans. Am. Fish. Soc.* 93, 301–2.

MEAD, G. W. (1960). Hermaphroditism in archibenthic and pelagic fishes of the order Iniomi. *Deep Sea Res.* 10, 251–7.

MEAD, G. W., BERTELSEN, E. & COHEN, D. M. (1964). Reproduction among deep-sea fishes. *Deep Sea Res.* 11, 569–96.

MÜLLER, R. (1969). Die Einwirkung von Sexualhormonen auf die Geschlechtsdifferenzierung von *Hemihaplochromis multicolor* (Hilgendorf) (Cichlidae). *Zool. Jb. Physiol.* 74, 519–62.

NOBLE, G. K. & KUMPF, K. F. (1937). Sex reversal in the fighting fish, *Betta splendens*. *Anat Rec.* 70, (Suppl. 1), 97.

OHNO, S. (1967). *Sex Chromosomes and Sex-linked Genes*, p. 192. Heidelberg: Springer.

OKADA, Y. K. (1962). Sex reversal in the Japanese wrasse, *Halichoeres poecilopterus*. *Proc. Japan Acad.* 38, 508–13.

OKADA, Y. K. (1964a). A further note on sex reversal in the wrasse, *Halichoeres poecilopterus*. *Proc. Jap. Acad.* 40, 533–5.

OKADA, Y. K. (1964b). Effects of androgen and estrogen on sex reversal in the wrasse, *Halichoeres poecilopterus*. *Proc. Jap. Acad.* 40, 541–4.

OKADA, Y. K. (1965a). Bisexuality in sparid fishes. I. Origin of bisexual gonads in *Mylio macrocephalus*. *Proc. Jap. Acad.* 41, 294–9.

OKADA, Y. K. (1965 b). Bisexuality in sparid fishes. II. Sex segregation in *Mylio macrocephalus*. *Proc. Jap. Acad.* **41**, 300–4.

OKADA, Y. K. (1965 c). Sex reversal in the serranid fish, *Sacura margaritacea*. I. Sex characters and changes in gonads during reversal. *Proc. Jap. Acad.* **41**, 737–40.

OKADA, Y. K. (1965 d). Sex reversal in the serranid fish, *Sacura margaritacea*. II. Seasonal variations in gonads in relation to sex reversal. *Proc. Jap. Acad.* **41**, 741–5.

OKADA, Y. K. (1966 a). Observations on the sex reversal in the symbranchoid eel, *Fluta alba* (Zuiew). *Proc. Jap. Acad.* **42**, 491–6.

OKADA, Y. K. (1966 b). Sex reversal in *Ingeocia meerdervoorti* with special reference to repetition of hermaphroditic state. *Proc. Japan Acad.* **42**, 497–502.

OKADA, Y. K. (1968). A further note on the sex reversal in *Inegochia meerdervoorti*. *Proc. Japan Acad.* **44**, 374–8.

OLIVER, M. & MASSUTI, M. (1952). El raô, *Xyrichthys novacula* (fam. Labridae). Notas biologicas y biometricas. *Boll. Inst. esp. Oceanogr.* **48**, 1–14.

PASQUALI, A. (1940/1). Contributo allo studio dell'ermafroditismo e del differenziamento della gonade nell'orata (*Sparus auratus* L.). *Pubbl. Staz. zool. Napoli* **18**, 282–312.

PICKFORD, G. E. & ATZ, J. W. (1957). *The Physiology of the Pituitary Gland of Fishes*, p. 613, New York: *N. Y. Zool. Soc.*

QUIGNARD, J.-P. (1966). Recherches sur les Labridae (poissons téléostéens perciformes) des côtes européennes: systématique et biologie. *Naturalia monspel. sér. Zool.* **5**, p. 247.

REINBOTH, R. (1957). Sur la sexualité du Téléostéen *Coris julis* (L.). *C. r. hebd. Séanc. Acad. Sci., Paris* **245**, 1662–5.

REINBOTH, R. (1961). Natürliche Geschlechtsumwandlung bei adulten Teleosteern. *Zool. Anz.* (Suppl.) **24**, 259–62.

REINBOTH, R. (1962 a). Morphologische und funktionelle Zweigeschlechtlichkeit bei marinen Teleostiern (Serranidae, Sparidae, Centracanthidae, Labridae). *Zool. Jb.* (Physiol.), **69**, 405–80.

REINBOTH, R. (1962 b). The effects of testosterone on female *Coris julis* (L.), a wrasse with spontaneous sex-inversion. *Gen. comp. Endocr.* **2**, abstr. no. 39.

REINBOTH, R. (1963 a). Experimentell induzierter Geschlechtswechsel bei Fischen. *Verh. dt. zool. Ges. München*, pp. 67–73.

REINBOTH, R. (1963 b). Natürlicher Geschlechtswechsel bei *Sacura margartiacea* (Hilgendorf) (Serranidae). *Annotnes zool. jap.* **36**, 173–8.

REINBOTH, R. (1964). Inversion du Sexe chez *Anthias anthias* (L.) (Serranidae). *Vie Milieu* (Suppl.), **17**, 499–503.

REINBOTH, R. (1965). Sex reversal in the black sea bass *Centropristes striatus*. *Anat. Rec.* **151**, 403. (abstr.).

REINBOTH, R. (1967 a). Protogynie bei *Chelidoperca hirundinacea* (Cuv. et Val.) (Serranidae). Ein Diskussionsbeitrag zur Stammesgeschichte amphisexueller Fische. *Annotnes zool. jap.* **40**, 181–6.

REINBOTH, R. (1967 b). Zum Problem der amphisexuellen Fische. *Verh. dt. zool. Ges. München*, pp. 316–25.

REINBOTH, R. (1967 c). Biandric teleost species. *Gen. comp. Endocr.* **9**, abstr. no. 146.

REINBOTH, R. (1968). Protogynie bei Papageifischen (Scaridae). *Z. Naturf.* **23 b**, 852–5.

REINBOTH, R. (1969 a). Histochemische Beobachtungen zum Geschlechtswechsel bei *Coris julis* L. *Verh. dt. zool. Ges. Würzburg*, (in the Press).

REINBOTH, R. (1969 b). Varying effects of testosterone with different ways of hormone administration on gonad differentiation of a teleost fish. *Gen. comp. Endocr.* **13**, 527–8.

REINBOTH, R. (1970). Der Geschlechtswechsel bei Labriden unter besonderer Berücksichtigung von *Thalassoma bifasciatum. Z. Morph.* (in preparation).

REINBOTH, R., CALLARD, I. P. & LEATHEM, J. H. (1966). *In vitro* steroid synthesis by the ovaries of the teleost fish, *Centropristes striatus* (L.). *Gen comp. Endocr.* **7**, 326–8.

RIEGER, R., MICHAELIS, A. & GREEN, M. M. (1968). *A Glossary of Genetics*, 3rd ed. Berlin: Springer.

ROEDE, M. J. (1966). Notes on the labrid fish *Coris julis* (Linnaeus, 1758) with emphasis on dichromatism and sex. *Vie Milieu* A, **17**, 1317–33.

ROSENBLATT, R. H. & HOBSON, E. S. (1969). Parrotfishes (Scaridae) of the eastern Pacific, with a generic rearrangement of the Scarinae. *Copeia* 434–453.

SALEKHOVA, L. P. (1963). *Vop. Ikhtiol.* **3**, 275–87 (Russian).

SMITH, C. L. (1959). Hermaphroditism in some serranid fishes from Bermuda. *Pap. Michigan Acad. Sci.* **44**, 111–19.

SMITH, C. L. (1965). The patterns of sexuality and the classification of serranid fishes. *Am. Mus. Novit.* **2207**, 20 pp.

SMITH, C. L. & YOUNG, P. H. (1966). Gonad structure and the reproductive cycle of the kelp bass, *Paralabrax clathratus* (Girard), with comments on the relationships of the serranid genus *Paralabrax. Calif. Fish Game* **52**, 283–92.

SORDI, M. (1962). Ermafroditismo proteroginico in *Labrus turdus* L. e in *L. merula* L. *Monitore zool. ital.* **69**, 69–89.

SORDI, M. (1964). Ermafroditismo proteroginico in *Labrus bimaculatus* L. *Monitore zool. ital.* **72**, 21–30.

SORDI, M. (1967). Ermafroditismo proteroginico in *Xyrichthys novacula* (L.). *Archo zool. ital.* **52**, 305–8.

VIVES, F., SUAU, P. & PLANAS, A. (1959). Sobre la biología de la cinta. *Investigación pesq.* **14**, 3–23.

WITSCHI, E. (1914). Experimentelle Untersuchungen über die Entwicklungsgeschichte der Keimdrüsen von *Rana temporaria. Arch. mikrosk. Anat. EntwMech.* **85**, 9–113.

WITSCHI, E. (1957). The inductor theory of sex differentiation. *J. Fac. Sci. Hokkaido Univ.* ser. 6 (zool.), **13**, 428–39.

YAMAMOTO, T. (1953). Artificially induced sex-reversal in genotypic males of the medaka (*Oryzias latipes*). *J. exp. zool.* **123**, 571–94.

YAMAMOTO, T. (1955). Progeny of artificially induced sex-reversals of male genotype (*XY*) in the medaka (*Oryzias latipes*) with special reference to *YY* male. *Genetics* **40**, 406–19.

YAMAMOTO, T. (1958). Artificial induction of functional sex-reversal in genotypic females of the medaka (*Oryzias latipes*). *J. exp. zool.* **137**, 227–64.

YAMAMOTO, T. (1959). A further study on induction of functional sex reversal in genotypic males of the medaka (*Oryzias latipes*) and progenies of sex reversals. *Genetics* **44**, 739–57.

YAMAMOTO, T. (1961). Progenies of sex-reversal females mated with sex-reversal males in the medaka *Oryzias latipes. J. exp. zool.* **146**, 163–79.

YAMAMOTO, T. (1962). Hormonic factors affecting gonadal sex-differentiation in fish. *Gen. comp. Endocr.* suppl. **1**, 341–5.

YAMAMOTO, T. & KAJISHIMA, T. (1968). Sex hormone induction of sex reversal in the goldfish and evidence for male heterogamety. *J. exp. zool.* **168**, 215–22.

ZEI, M. (1949). Typical sex-reversal in teleosts. *Proc. zool. Soc. Lond.* **119**, 917–20.

ZUNARELLI VANDINI, R. (1965). Il problema dei maschi primari di *Coris julis. Monitore zool. ital.* **73**, 102–10.

DISCUSSION

HOAR: I might make a comment concerning differentiation of the testes of the guppy that may be of interest. S. Pandey, working in our laboratory, compared guppies which had been hypophysectomized as adults with individuals hypophysectomized as juveniles prior to differentiation of gonads. As expected, the gonads in adults showed regression of gametogenetic tissue to the juvenile condition, while gonads of hypophysectomized juveniles failed to differentiate. When the two groups were then treated with methyltestosterone the adults showed some differentiation of the spermatogenetic tissues while the juveniles, now grown to adult size, showed no testicular response. It seems that prior pituitary stimulation of the gonad may be essential to the later response which followed the application of exogenous androgens.

REINBOTH: The unpublished experiments by Hackmann on the delimitation of a critical period of sex determination in *Hemihaplochromis* are certainly another example which demonstrates the existence of qualitatively different physiological stages with regard to the effectiveness of exogenous steroids on gonadal tissue. But at present we have no idea whether the pituitary plays an essential role in this particular case.

CHAN, S. T. H: At present there exists two schools of thought concerning the evolution of hermaphroditism in fishes, namely the evolution of hermaphroditic form (e.g. secondary males) from gonochoristic females and the evolution of gonochorism from hermaphroditism. I don't know whether Dr Reinboth would have any comments on these.

REINBOTH: My answer should be, perhaps, 'no comment'. My own investigations on a great amount of material so far provides no evidence to support more strongly either one or the other idea. But I am well aware that more competent ichthyologists (e.g. C. L. Smith (1967), Contribution to a theory of hermaphroditism, *J. Theoret. Biol.* **17**, 76–90) favour the idea of a polyphyletic origin of hermaphroditism from the more usual vertebrate gonochorism.

CHAN, S. T. H.: With regard to the impact of environment on natural selection of sexuality, isolated habitat has been suggested to be a selective factor in the tooth carps (*Rivulus marmoratus*) for synchronous hermaphroditic forms and internal self-fertilization before the eggs were laid, provided significant survival value. However, many marine hermaphroditic fishes occur in some sort of open environment apparently without any isolation. I wonder whether you would have any suggestion on the type of selective pressure on these marine hermaphroditic fishes.

REINBOTH: At present I don't know of any particular environmental factor that we may hold responsible for favouring the development of simultaneous hermaphroditism. I would like to point out once again that the vast majority of ambosexual teleosts occur in the oceans, in which environmental conditions are usually more even and stable than in any other biotope. Mead *et al.* (1964) speculated on the occurrence of simultaneous hermaphrodites among deep-sea fishes as a means to meet the difficulties in finding a sex partner. On the other hand one should not forget that observations on the spawning behaviour of marine hermaphroditic teleosts do not suggest the occurrence of self-fertilization as a normal phenomenon.

HYDER: I wonder if Professor Reinboth has found that there is a critical dosage for the feminizing effect of methyltestosterone in his cichlid experiments?

REINBOTH: In Müller's (1969) experiments on *Hemihaplochromis* all doses higher than 0·25 mg testosterone/l. water had a feminizing effect, whereas a concentration of 0·05 mg hormone/l. did not affect the normal 1:1 sex ratio. For methyltestosterone the critical dosage has not yet been determined since 0·05 mg methyltestosterone/l. water also produced 100 % females.

TEMPORAL CHANGES IN THE PITUITARY-GONADAL AXIS

BY B. LOFTS, B. K. FOLLETT
AND R. K. MURTON

The title of this symposium is 'Hormones and the Environment' and most of the papers have demonstrated the manner whereby endocrine mechanisms enable an organism to survive in its environment. We shall in this contribution, in effect, reverse the symposium title because we visualize endogenous hormone mechanisms as adaptations, albeit often compromise adaptations, to environmental selection pressures. Our theme will be the hormonal mechanisms regulating reproduction. Evolution has taken place in an environment subject to regular and cyclic fluctuations both of short duration, such as the diurnal cycle of night and day, and of much longer periodicities, as exemplified by the changing seasons. It is not surprising therefore that the survival of a species has required it to adapt many of its physiological processes to these cyclic phenomena, and in particular, that reproduction be synchronized so as to ensure the maximal survival chances for both parents and offspring. Natural selection will strongly favour those individuals producing offspring at the most propitious season; and progeny reared at less favourable times will suffer a high and wasteful mortality. Such a differential survival rate will rapidly define the characteristic breeding season of any particular species, and in many of the higher vertebrates at least, has resulted in the establishment of annual breeding cycles.

The differential survival of both adults and young will favour genotypes with appropriate reproductive abilities and Baker (1938) has distinguished the selective factors involved as the *ultimate* factors ensuring that populations breed at the optimal season. In birds, but not necessarily in other taxa, the availability of suitable food supplies, including such influences as the distance from the food source, is the most important ultimate factor determining the number of young that can be reared to a reproductive age (Lack, 1950; Murton & Isaacson, 1962; Lack, 1967; Lofts & Murton, 1968). In order to anticipate the arrival of suitable seasons for breeding, animals have come to respond to a host of environmental signals which govern the behavioural and physiological processes leading to the physical maturation of the gonads. These factors Baker distinguishes as *proximate* factors, their ability to serve as effective timing mechanisms being no less subject to natural selection. Ultimate factors ensure that the gonads will recrudesce in response to a specific gonadotrophin secretion; that such secretion occurs when the animal experiences a particular photoperiodic stimulation is a consequence of proximate adaptations. It is likely that proximate adaptations are more amenable to selective change than the more deep-rooted physiological mechanisms underlying ultimate responses. For instance, there is little latitude for most species between conception and the date of hatch or birth, and in order to breed rapidly and to make use of highly seasonal foods in spring and early summer many temperate mammals, e.g. sheep and deer, pair and copulate in autumn so that the long gestation period does not impair the early production of young in the following spring. To achieve a suitable breeding season all that has been required is a responsiveness to declining autumn daylengths, a photoperiodic response unnecessary for those tropical ungulates with otherwise comparable gestation periods.

Many kinds of information could be used as proximate synchronizers, but species have tended to evolve responses to those environmental changes which constitute the most stable source of predictive information. At mid and high latitudes the annual variable which is uniquely consistent is the change in daylength and it is therefore not surprising that photoperiodic controlling mechanisms are found in many species of plants and animals (for reviews see Aschoff, 1955; Farner, 1961; Bünning, 1963; Beck, 1968). It should be emphasized, however, that while daylength changes may regulate breeding seasons in the temperate zones the same is not necessarily true for tropical and equatorial species, or for animals living in particularly arid areas of the world. Here other proximate factors such as seasonal rains often serve to time reproduction (Lofts & Murton, 1968; Marshall, 1970).

Among the vertebrates the most highly evolved photoperiodic mechanisms occur in birds and more is known about photoperiodism for these than for any other class (for reviews see Marshall, 1959; Farner, 1964; Wolfson, 1966; Farner & Follett, 1966; Lofts & Murton, 1968): this is true both from the viewpoint of time measurement and also regarding the means whereby photic information is translated into the complex endocrinological changes associated with the development of the reproductive organs. For this reason our discussion will largely be confined to avian data.

So far about 60 avian species from widely separate Orders have been proved experimentally to depend on photo-stimulation for gonadal development, and although other secondary factors, such as a sudden cold spell or an unseasonal period of warm sunny weather, may retard or accelerate the rate of gonadal recrudescence (Lofts & Murton, 1966), there is no experimental evidence that these factors by themselves can initiate testicular growth in the absence of stimulatory daylengths. There are a few temperate-zone birds which come into breeding condition regardless of daylength but this usually seems to be a secondary adaptation resulting from domestication as in the feral pigeon *Columba livia* (Lofts, Murton & Westwood, 1967*a*) and certain strains of the Japanese quail *Coturnix c. japonica* (Ablanalp, 1966), or due to an endogenous cycle of gonadal growth such as apparently exists in the Pekin duck *Anas platyrhynchos* (Benoit, Assenmacher & Brard, 1956*a*, *b*).

It is clear from the wide diversity of breeding patterns displayed by different species that both intra- and inter-specific differences in the photoperiodic response must exist (Lofts & Murton, 1968). For example, Dolnik (1963) has shown that for two populations of chaffinches (*Fringilla coelebs*), living respectively at 55° and 60 °N, there is a difference in their photosensitivity to the annual cycles of daylength experienced at these latitudes. Thus, the more southerly population responds to daylengths of about 10 h while the other requires 11 h of daylight to commence gonadal growth. Similar differences in photosensitivity have also been established for three closely related species of British pigeons (Lofts *et al.* 1967*a*). In this case the results of experimental photostimulation can be correlated directly with field observations on the time of testicular recrudescence. Spermatogenesis can be induced in the stock dove (*Columba oenas*) with artificial daylengths increasing from 9·1 to 10·8 h, the equivalent of natural daylengths at 52 °N in late January and mid-February, the time of year when these birds naturally show a spermatogenetic recrudescence in their normal environment. The same photoperiod, however, does not stimulate the testes of wood pigeons (*C. palumbus*), which require daylengths of from 11·0 to 12·5 h to initiate spermatogenesis. In the wild, stock doves begin gonadal recrudescence approximately 1 month before wood pigeons living in the same area (Lofts *et al.* 1966). It is to be expected that such quantitative and qualitative differences in the photoperiodic response mechanisms may be quite widespread, and reflect variations in the optimal breeding seasons for different species, as they do in pigeons. It is also possible that species might differ in the threshold of the response and yet breed at the same time. In this case one species may show a greater rate of gonadal growth for a given increment in daylength, so that it may start somewhat later in the year but take less time to reach maturity. An example of this may be the chaffinches of Dolnik (1963), the more northerly race showing the faster rate of recrudescence (see fig. 2 in Farner, 1964).

In the final analysis the annual variation in the size and functioning of the gonads is dependent upon changes in the output of the adenohypophysial gonadotrophins, which in turn are under the control of the hypo-thalamus and central nervous system. To a large extent the problem of avian photoperiodism resolves itself physiologically into a knowledge of how the bird translates information received from its environment into the secretion of such hormones. The questions which we shall attempt to answer in this discussion are therefore: first, how is the duration of the daily photoperiod measured and what constitutes a stimulatory daylength, and secondly, how is this reflected in the activity of the hypothalamic–pituitary and pituitary–gonadal axes?

THE PHOTOPERIODIC GONADAL RESPONSE

Under experimental conditions the gonads of males can normally be brought to a state of complete spermatogenesis by light stimulation. Farner & Wilson (1957) demonstrated that when male white-crowned sparrows (*Zonotrichia leucophrys gambelii*) were subjected to constant stimulatory daylengths, the growth of the testes up to a combined weight of about 250 mg (max. wt 500–600 mg) followed a log-linear function of the time of photo-stimulation. The rate of testicular growth during this period could thus be represented by the following formula:

$$k = (\log W_t - \log W_0)/t,$$

where W_0 is the resting testicular weight (in mg), W_t is the testicular weight after t days of photostimulation and k is the logarithmic growth-rate constant. Similar relationships have been found in the brambling, greenfinch and chaffinch (Dolnik, 1963), and in the Japanese quail (Follett & Farner, 1966a).

Experimentally the value of such log-linear growth curves derives from calculating k, which is a quantitative measure of the rate of gonadal growth, and, indirectly, of gonadotrophin secretion. Valid comparisons can then be drawn between the effectiveness of different photoperiodic treatments in stimulating the pituitary–gonadal axis. In the species examined to date k has been shown to depend on the length of the daily photoperiod, p, and at least over part of the range, k and p are directly proportional (Fig. 1). For most species there is a complete daily photoperiod below which no growth occurs and also one which induces the maximum rate of develop-ment. All these features of the response mechanism must be satisfied by any theory of time-measurement.

In *Coturnix* the rate of growth is particularly great (Fig. 2), the testes increasing in weight from 8 to 3,000 mg within 3 weeks of exposure to long (20L/4D) daylengths. The size increase results primarily from growth in both the length and diameter of the seminiferous tubules. These

changes are accompanied by a steady advance in the stages of spermatogenesis so that in *Coturnix* about one-half of all males are fully developed
when the combined testicular weight is 1,500–2,000 mg. Fig. 2 also stresses

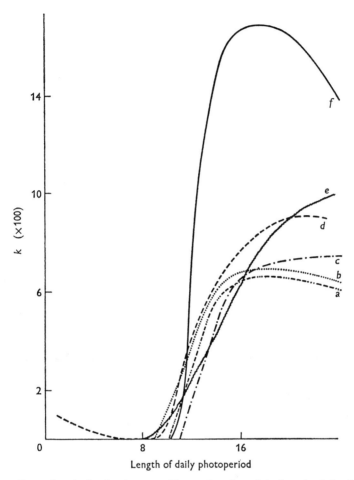

Fig. 1. Rate of testicular development (*k*) as a function of the length of the daily photoperiod in: (*a*) chaffinch; (*b*) greenfinch; (*c*) brambling (Dolnik, 1963); (*d*) house sparrow
(see Middleton, 1965); (*e*) white-crowned sparrow (Farner & Wilson, 1957); (*f*) Japanese
quail (Follett, unpublished). Redrawn from Farner (1964).

the importance of long daylengths, since males replaced on to short daily
photoperiods show a rapid regression soon after experiencing the reduction
in daylengths. Histologically, signs of degeneration appear within a few
days of changing the photoperiod. A complete explanation for the log-
linear growth phase is not yet forthcoming but it may well reflect a relatively constant daily output of gonadotrophins over this period.

Towards maturity the growth rate slows down perhaps as a result of a decrease in gonadotrophin secretion resulting from negative feedback caused by testicular androgens. Alternatively there may be a shift in the relative output of LH (luteinizing hormone) and FSH (follicle stimulating hormone). Adequate physiological evidence on these points is still lacking,

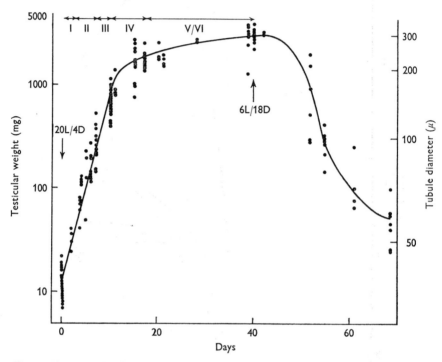

Fig. 2. Photoperiodically induced testicular growth in Japanese quail submitted to long (20 L/4 D) daily photoperiods for 40 days and then replaced on a short daylength of 6 L/18 D. All birds were 28 days of age at the start of the experiment and had been reared under short daylengths. Log$_{10}$ testicular weight is plotted for individual birds against time in days. The line to 1,000 mg was plotted by the method of least squares ($k = 0.176$), the rest of the curve being fitted by eye. The right-hand abscissa indicates the tubule diameter for testes of a given weight. The range of spermatogenetic stages (Bartholomew, 1949) is as follows: I, spermatogonia only; II, spermatogonia dividing, a few spermatocytes; III, many spermatocytes; IV, spermatocytes with spermatids; V, VI, spermatids and spermatozoa. Data from Follett & Sharp (unpublished).

but we do know that photo-stimulation results in a marked over-all increase in the level of pituitary gonadotrophins (Benoit, Assenmacher & Walter, 1950; Follett & Farner, 1966 b), which, at the cytological level, is best interpreted as an increase in FSH content and turnover, for it is the beta cells (presumptive FSH producing cells) which show the most marked changes during testicular growth (Tixier-Vidal, Follett & Farner, 1968), increasing in both size and number. The presumptive LH or gamma cells show evidence of increased activity throughout testicular growth but never become particularly abundant. The problem of steroid feedback and its possible physiological role remains unresolved. That such a mechanism exists is suggested by the fact that castration combined with photo-stimulation leads to a very great increase in the size and gonadotrophin content of the pituitary (e.g. Benoit *et al.* 1950; Tixier-Vidal *et al.* 1968), and the fact that implants of testosterone completely block photo-induced testicular growth and suppress pituitary gonadotrophins in *Coturnix* is also in agreement with this idea. Moreover, hypothalamic implants of testosterone in the duck cause testicular regression (Gogan, 1968). Castration has been shown to lead to an increased activity of hypothalamic acid phosphatases in *Emberiza rustica latifascia* (Uemura, 1964) and *Zonotrichia leucophrys gambelii* (Kobayashi & Farner, 1966). Finally, there is clear evidence that testicular growth is accompanied by rises in plasma testosterone (Jallageas & Attal, 1968; Boddy, unpublished). Thus, although conclusive evidence is lacking, these data taken together imply that negative feedback effects may be present in birds as in mammals. However, the situation is complicated by results such as those of Lofts (1962), which show that spermatogenetic activity can be both initiated and maintained in weaver finches (*Quelea quelea*) by androgen injections. The effect of testosterone treatment in this species (1 mg/day) depends on when the steroid is administered. Testicular collapse, as a result of the birds entering a refractory phase, can be prevented by exogenous androgen administration, and testosterone can induce growth in the regressed gonads of such birds. Nevertheless it had a negative feedback effect when given to males undergoing normal photostimulated testicular growth. Hamner (1968) has similarly been able to stimulate the quiescent testes of photorefractory house finches, *Carpodacus mexicanus* with exogenous androgen. Unlike weaver finches, the mature testes of androgen-injected (1 mg/day) feral pigeons (*C. livia*) show signs of regression after 3 weeks administration (Plate 1 *a*, *b*). At present it is difficult to know whether all these scattered observations result from pharmacological rather than physiological effects and more information is required on the levels of androgens in the blood, their half-lives and the rate of secretion from the testis.

The early stages of ovarian development show growth curves similar to those of the testis (white-crowned sparrow, Farner, Follett, King & Morton, 1966; *Coturnix*, Follett & Farner, 1966a). However, in most species (an exception is *Coturnix*) the ovary will not mature under experimental conditions, the block to growth occurring at the onset of vitellogenesis (Polikarpova, 1940; Phillips & van Tienhoven, 1960; Lehrman, 1961). Captivity itself is adequate to stop gonadal development, for white-crowned sparrows caught during their spring migration undergo little further ovarian growth if maintained in large outdoor aviaries. After a month in such conditions the mean ovarian weight only increases from 18·6 to 30 mg while the pituitary gonadotrophin content falls from 11·1 to 2·7 µg/gland (King, Follett, Farner & Morton, 1966). Males, on the other hand, reach sexual maturity during the same period, testicular weights increasing from 47·7 to 353 mg and pituitary gonadotrophins from 11·1 to 16·9 µg/gland. Phillips (1964), investigating a similar case of inhibition in captive mallards, found that growth took place when lesions were placed in the ventral medial archistriatum or the occipito-mesencephalic tract. He concluded that 'fear' was involved in the original blockade since 'escape' behaviour was also abolished in the lesioned animals. Whether sensory stimuli other than daylength are required in the wild female to induce the vitellogenic stages of ovarian development is not yet established although nest building and ovulation certainly seem to require other proximate factors (for review see Lehrman, 1961). The possible involvement of extra-hypothalamic structures in inhibiting gonadotrophin secretion is of particular interest because of the renewed interest in such mechanisms in mammals (e.g. Sawyer, 1967; Eleftheriou, Zolovick & Norman, 1967).

TIME-MEASUREMENT IN THE PHOTOPERIODIC GONADAL RESPONSE

During the past 50 years a great number of papers have described photoperiodic effects in a wide range of plants, insects and higher vertebrates but relatively few have been devoted to the specific problem of how daylength is measured. Nevertheless, a number of theories have been advanced of which two receive the most attention. The first, proposed by Bünning in 1936, suggested that endogenous daily rhythms were somehow involved in estimating the length of the photoperiod. The hypothesis envisaged a circadian rhythm of cellular function consisting of two half-cycles each of approximately 12 h duration, one of which was 'light requiring' (the photophil) while the other was 'dark-requiring' (the scotophil). Photoperiodic induction of a process requiring long days occurred only when the duration of the natural photoperiod extended into the scotophilic part of the cycle (Bünning, 1960). In its simplest form the model is essentially a coincidence device between light and some direct or indirect rhythm of 'photosensitivity', the phase of which is locked on to dawn. Recently, Pittendrigh (1966) and Pittendrigh & Minis (1964) have represented a more explicit version of this hypothesis stressing the dual action of the *whole* light cycle in acting both as the photoperiodic inducer and as the entraining agent for the circadian oscillation. The latter point is important because it means that the phase of the circadian rhythm is not fixed by the onset of the daily photoperiod (i.e. 'dawn') alone but is dependent on the over-all characteristics of the driving light cycle for determining its position in the 24 h period. Pittendrigh therefore proposes the more restrictive term 'photoperiodically inducible phase' to refer to that portion of the daily oscillation which, if illuminated, leads to photoperiodic induction. The second hypothesis of time-measurement envisages it to be based on an hourglass principle. This model, which has never been explicitly formulated, considers that a physiological process, in the present case gonadotrophin secretion, is dependent on the duration of the dark period.

A large body of experimental evidence now supports the Bünning hypothesis in both plants and animals (for reviews see Bünning, 1960; Bunsow, 1960; Hamner & Takimoto, 1964; Pittendrigh & Minis, 1964; Pittendrigh, 1966), but only recently has it been implicated in the avian photoperiodic response. In a series of elegant experiments W. H. Hamner (1963, 1964, 1966) showed a circadian rhythm to be involved in photo-induced testicular growth in the house finch (*Carpodacus mexicanus*). By exposing birds with undeveloped gonads to light cycles consisting of a 6 h light period followed by either 6, 18, 30, 42, 54 or 66 h of darkness, and maintaining them under these ahemeral cycles for about a month, he was able to demonstrate that testicular growth did not occur in birds exposed to 24 h cycles or multiples thereof (i.e. 6 L–18 D, 6 L–42 D, 6 L–72 D), but cycles of 12, 36 or 60 h induced rapid development. It is clear that these data cannot support any hypothesis which stresses a requirement for a given duration of light or darkness in order to trigger a response; they are best interpreted by assuming that a rhythm of sensitivity exists within the bird with a periodicity of about 24 h. It is not the quantity of light which is important but where it falls relative to the photo-inducible phase of the rhythm (Fig. 3).

Using schedules generally similar to those of Hamner, it is now apparent that circadian rhythms are involved in the photoperiodic testicular response of the white-crowned sparrow (Farner, 1965), the slate-coloured junco (*Junco hyemalis*) and the bobolink, *Dolichonyx oryzivorus* (Wolfson, 1966) as well as controlling testicular and ovarian growth in the Japanese quail (Follett & Sharp, 1969). Fig. 3 illustrates the results of exposing *Coturnix* males to ahemeral light-cycles. Gonadal maturation under cycles of 12, 36 and 60 h is significantly greater than under the other schedules.

Although such experiments illustrate the presence of a circadian rhythm in a response mechanism, they have a drawback in that they do not give much information on the shape and duration of the photo-inducible phase.

For this we depend upon another class of experiments, in which a skeleton schedule of two light periods per day is used to simulate a complete photoperiod. The design may be symmetrical such that the light pulses are of the same duration, but more usually it is asymmetrical, with the first being a short non-stimulatory photoperiod (e.g. 6 L–18 D) while the second, which is of much shorter duration, is used to interrupt the dark period. By exposing different experimental groups to an interruption of the dark period at different times of the night, the dark period is effectively 'scanned' by the

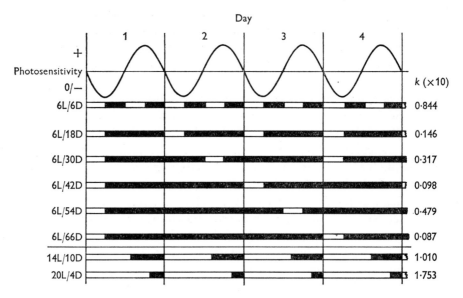

Fig. 3. Testicular responses of Japanese quail to ahemeral cycles. The theoretical circadian rhythm is shown as a sine curve. k refers to the testicular growth-rate constant. Values of k for normal (i.e. 24 h) stimulatory schedules are illustrated in the lower panel for comparison. Data from Follett & Sharp (1969).

light pulse, and an analysis of the gonads of birds kept under such conditions for several weeks will indicate the duration of the photo-inducible phase. This experimental approach has been used with a number of species including the house finch (Hamner, 1964), the white-crowned sparrow (Farner, 1965), the house sparrow *Passer domesticus* (Menaker, 1965; Menaker & Eskin, 1967; Murton, Lofts & Orr, 1970), and the greenfinch *Chloris chloris* (Murton, Bagshawe & Lofts, 1969) and generally the light pulse used has been of 1 or 2 h duration. Fig. 4 summarizes the photoperiodic sensitivity curves in different species. The results of an experiment carried out on *Coturnix* (Follett & Sharp, 1969) are shown in Fig. 5. In this investigation a much shorter 15 min light pulse was used as an interruption of the dark period so that birds were exposed to a total of only 6·25 h of light per day, an amount that would be non-stimulatory if given as a complete photoperiod (Fig. 1). This, however, was adequate to induce gonadal growth if the light pulse fell between 11·5 and 16 h after the onset of the

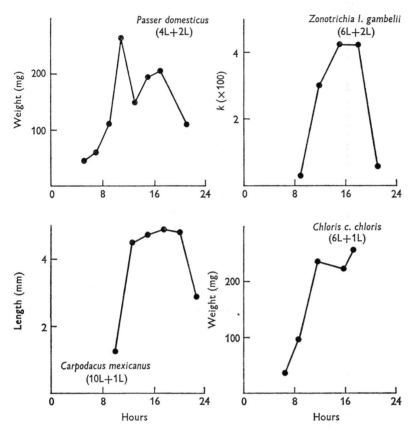

Fig. 4. Summary diagram to show photoperiodic testicular responses in night-interruption experiments. Note that the length of the primary and secondary light periods varies between experiments. The data are drawn from: *Passer domesticus* (Menaker, 1965), *Zonotrichia leucophrys gambelii* (Farner, 1965), *Carpodacus mexicanus* (Hamner, 1968), *Chloris c. chloris* (Murton & Lofts, unpublished).

primary photoperiod, whereas light pulses at other times failed to stimulate significant development. In this species therefore, by using a much shorter light pulse the photoinducible phase is much more precisely outlined and appears to last about 4 h. A similar curve is also produced by the females.

The nature of these curves explains many of the properties of the response mechanism under complete daily photoperiods. As daylengths increase, so the degree of coincidence between light and the photo-inducible

phase will increase, causing a greater secretion of gonadotrophins. In
Bünning's original model the rhythm was entrained to dawn and it would
necessarily follow that the light-sensitive phase must occur *early* in the

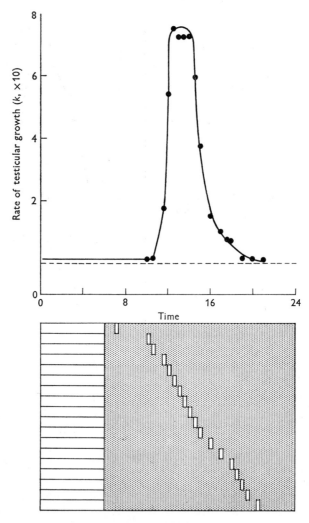

Fig. 5. The effect of asymmetric skeleton photoperiods on the rate (*k*) of testicular development in *Coturnix*. Each point represents the mean value of *k* for a schedule consisting
of 6 L/18 D together with a light pulse of 15 min given at the times indicated in the night
period. Data from Follett & Sharp (1969).

subjective night, as indicated in model A of Fig. 6. Contemporary thought,
however, stresses that phase relationships are determined by the characteristics of the whole light cycle and that a change in the length of the

daily photoperiod will cause a shift in the position of the photo-inducible phase. In this case a second model (Fig. 6B) can be visualized in which the photo-inducible phase might lie late in the subjective night, and the effect of a light pulse given at 14 h may be to cause a phase-delay of the

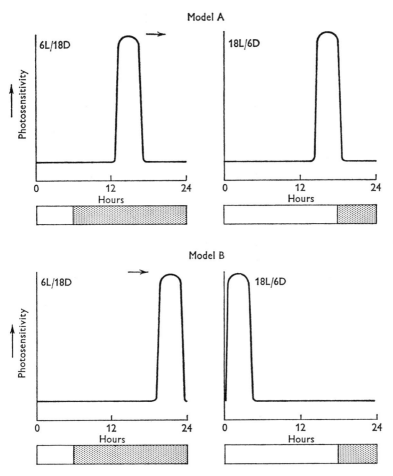

Fig. 6. Two possible models of how increasing daylengths could illuminate the photo-inducible phase and stimulate the release of gonadotrophins.

rhythm so that the photo-inducible phase now falls in the subsequent early subjective day and thus be illuminated by the 6 h light period. We have some evidence of the phase shift being induced by alteration of the primary photoperiod in male quail, and these results are summarized in Table 1. In this investigation groups of birds were exposed to daily photoperiods of 3, 6 or 9 h, followed by 15 min pulses in the night (Follett & Sharp, 1969).

As can be seen from a comparison of the rates of testicular growth at 12, 15 and 16 h, this produced a small but significant phase shift in the position of the photo-inducible phase, involving a delay as the primary photoperiod is increased. Nevertheless, photoperiodic experiments cannot by themselves distinguish between these two possibilities and also give no information as to the location of the photo-inducible phase under complete daily photoperiods.

Table 1. *Testicular growth rates (k) in 'night-interruption' schedules in* Coturnix *(Follett & Sharp, 1969)*

Time of light pulse[†]	Duration of 'initial' photoperiod (h)		
	3	6	9
10·5	0·0371 (0·0240–0·0502)	0	0
12	0·0808* (0·0697–0·0919)	0·0521* (0·0375–0·0667)	0·0249 (0·0092–0·0406)
13	0·0817 (0·0684–0·0950)	0·0718 (0·0595–0·0841)	—
14	0·0830 (0·0720–0·0940)	0·0726 (0·0576–0·0876)	0·0751 (0·0632–0·0870)
15	0·0209 (0·0109–0·0309)	0·0415* (0·0298–0·0532)	0·0802* (0·0658–0·0916)
16	0	0·0148* (0·003–0·0298)	0·0353 (0·0241–0·0465)

k Values are given with their 95 % Confidence Limits.
* $P < 0.05$ compared with relevent values at other photoperiod lengths.
† The time of the 15 min light pulse after the onset of the main photoperiod is given in hours.

THE RELEASE OF GONADOTROPHINS UNDER STIMULATORY PHOTOPERIODIC SCHEDULES

The ultimate response to any stimulatory light schedule is the secretion of gonadotrophins; perhaps this property could be used to assess the position and duration of the photo-inducible phase. In birds, as in mammals, the hypothalamus exerts absolute control over the gonadotrophic functions of the adenohypophysis so that hypophysectomy, transplantation of the pituitary to a distal site or sectioning of the portal vessels causes testicular growth to cease immediately (e.g. Assenmacher, 1958; Benoit & Assenmacher, 1959; Ma & Nalbandov, 1963; Baylé & Assenmacher, 1967). Electrolytic lesions in the basal hypothalamus which destroy the nucleus infundibularis (= *n. tuberis*) also block photo-induced testicular growth in a number of species (Wilson, 1967; Stetson 1969; Sharp & Follett, 1969), and cause testicular or ovarian regression in the domestic fowl (Graber, Frankel & Nalbandov, 1967; Lepkovsky & Yasuda, 1966; Kanematsu, Sonoda, Kii & Kato, 1966). In the male Pekin duck, however, gonadotrophic function is apparently controlled by anterior hypothalamic areas which may involve the 'classical' neurosecretory system (Benoit & Assenmacher, 1959; Gogan, Kordon & Benoit, 1963). There may also be other areas outside the basal hypothalamus which are involved in some photoperiodic species. Thus, Sharp & Follett (1969) have reported that lesions in the n. hypothalamicus posterior medialis, i.e. dorsal to the n. tuberis, equally block testicular growth in *Coturnix* even though the tuberohypophysial system is left intact.

The hypothalamic neurohormones or releasing factors which control gonadotrophin secretion in mammals are well established but comparable information in lower vertebrates has been lacking. Recently, however, the quail hypothalamus has been shown to contain an activity which stimulates gonadotrophin release *in vitro* from chicken or quail pituitaries (Follett, 1969), the rate of release depending on the logarithmic dose of hypothalamic extract. Jackson & Nalbandov (1969) have similarly found an LH-releasing factor in the chicken hypothalamus as judged by its ability to release LH *in vitro* from rat pituitaries. Such techniques provide us with a means of measuring the gonadotrophin releasing factors in the hypothalamus, and together with information on the gonadotrophin content of the donor's pituitary, can provide a system in which it might be possible to determine the time of hormone release during any photoperiodic cycle. The results of such an experiment carried out on male quail are summarized in Figs. 7 and 8. The basis of the gonadotrophin assay is ^{32}P-uptake into the testes of day-old chicks (Follett & Farner, 1966*b*). This method is not specific to LH or FSH and is therefore a 'total' gonadotrophin assay, although it probably has a strong bias towards avian FSH (Furr & Cunningham, 1970). In this experiment groups of 60 male quail from the same hatch were reared to maturity under a short (6L–18D) photoperiod. They were then exposed to long (20 L/4 D) daylengths to induce a high rate of growth (Fig. 1) and killed after 10 day's treatment when the testicular weight was about 1,000 mg. Generally, five to six batches of eight to ten birds were taken at approximately 4 h intervals through the 24 h cycle. Tissues from the birds were pooled and the pituitary gonadotrophin content (PG) measured directly. The gonadotrophin-releasing activity (GRA) was measured indirectly after an *in vitro* incubation of the hypothalamic extracts with cockerel pituitaries. Four such experiments have been completed with essentially similar results although there have been some minor differences in the over-all levels which makes pooling of data difficult. Fig. 7 shows the results from one such experiment while Fig. 8 contains the mean values from all four. It can be seen that up to 16 h after dawn the hypothalamic GRA remains relatively constant but the level falls significantly (average 32%) by 19·5 h. In all cases this was followed by a rapid rise in content so that by 23 h the level had been restored. The pattern of changes in PG content is generally similar except

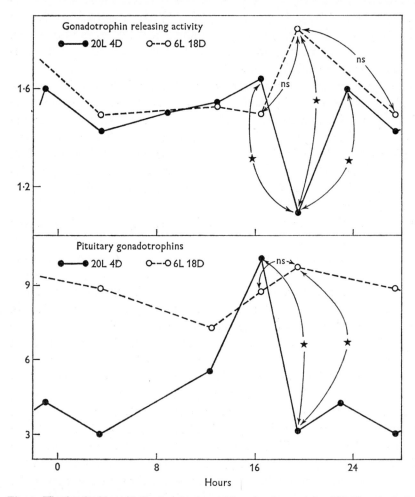

Fig. 7. The level of hypothalamic gonadotrophin releasing activity (GRA) and pituitary gonadotrophins (PG) in male *Coturnix* exposed to either long (20 L/4 D) daily photoperiods for 10 days or short-day lengths (6 L/18 D) following 10 days of photostimulation. The level is plotted against the time in hours at which a sample was taken. The values on the ordinates are expressed as follows: GRA, μg eqvts. NIH-LH-S 11 released/mg pituitary tissue/h; PG, μg eqvts. NIH-LH-S 11/gland. ns, Not significantly different; * $P < 0.05$. Data from Follett & Sharp (1969).

that the decrease from 16 to 19·5 h was even more marked (average 48%) and the level does not rise significantly by 23 h. Subsequently the content seemed to increase steadily and reached a peak 16 h after dawn. If the

PLATE I

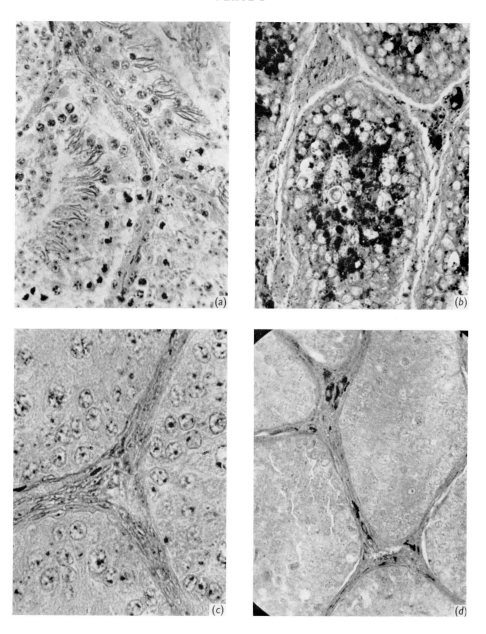

For explanation see p. 575

(facing p. 560)

PLATE 2

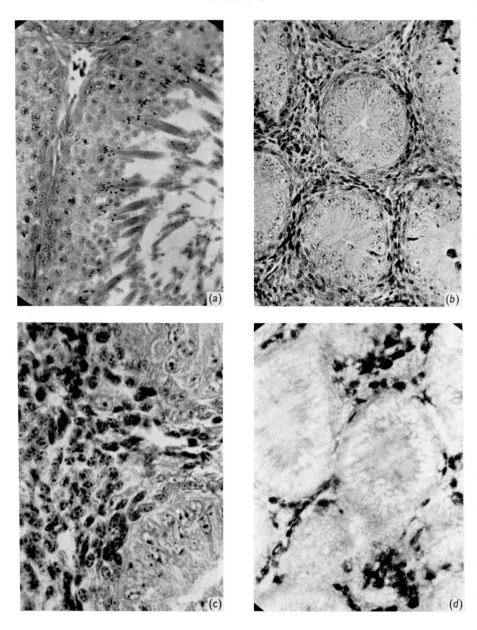

For explanation see p. 575

PLATE 3

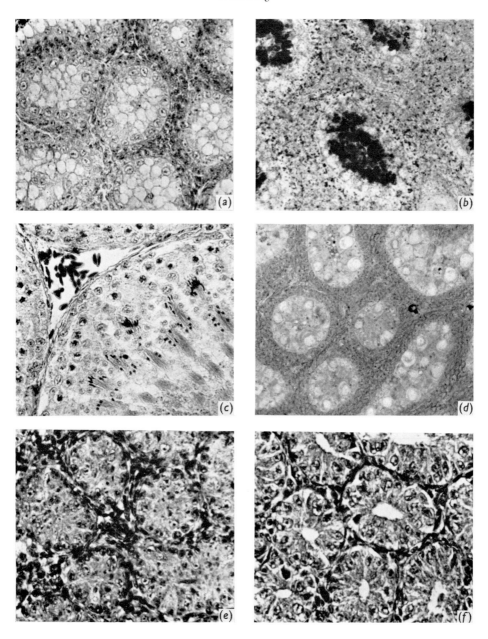

For explanation see p. 575

changes are due to hormone release they should be abolished when birds are moved from long stimulatory days to short daily photoperiods. In two experiments, therefore, 120 birds were placed on 20 L/4 D. One-half of the group was killed after 10 days and the remainder replaced on 6 L/18 D.

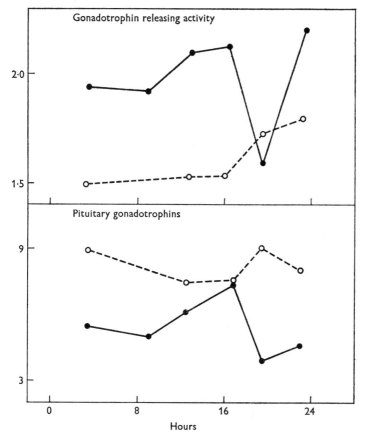

Fig. 8. The mean levels, from four experiments, of hypothalamic gonadotrophin releasing activity (GRA) and pituitary gonadotrophins (PG) in male quail exposed to either 10 days of long daily photoperiods or short-day lengths following 10 days photostimulation. For other details see Fig. 7 and text (Follett, unpublished).

These were killed either on the following day or 2 days later. The results are also plotted in Figs. 7 and 8. No changes in the levels of GRA or PG could be found through the cycle.

A parallel may be drawn between these results and those obtained for hypothalamic LH-RF in the rat (Ramirez & Sawyer, 1965; Chowers & McCann, 1965). In the latter, a sharp decrease in hypothalamic LH-RF content has been associated with a surge release of LH for ovulation. It seems possible, therefore, that the decrease in content shown in Fig. 7 might also reflect an increased hormone secretion under a stimulatory schedule. If so, the release of GRA in the quail photoperiodic response seems to be a discrete and relatively rapid event occurring over a 4 h period each day. This would agree with the duration of the photo-inducible phase as determined using skeleton photoperiods (Fig. 5). Furthermore, the pituitary gonadotrophins are apparently secreted immediately in response to this neurohormonal stimulation. However, the gonadotrophic level does not show a subsequent rise by 23 h as does the GRA, which suggests either continued secretion of gonadotrophins for a longer period than the hypothalamic neurohormones, or perhaps an inability to replete the gland rapidly.

The time of release under 20 L/4 D seems to lie in the subjective night rather than early in the day, which would support the first of the hypotheses outlined above (Fig. 6A) and suggests that in asymmetric skeleton photoperiods (Fig. 5) it was the 15 min light pulse which actually illuminated the photo-inducible phase. This conclusion makes the important assumption that secretion of the hypothalamic releasing factors occurs rapidly following coincidence of light and the photo-inducible phase. At present there is no method of testing this but intuitively one might not expect a long delay in a system operating through the central nervous system. The slight difference in the position of the rhythm determined by the two methods (Figs. 5, 8) may be a further example of phase delay caused by the long daily photoperiod (cf. Table 1).

So far we have only considered changes during the early phases of testicular growth in *Coturnix* when the predominant gonadotrophin being secreted is probably FSH. Moreover, the response mechanism in *Coturnix* appears to be relatively simple in that LH and FSH may well be secreted simultaneously at one circadian phase. Rather more complex systems may exist, however, in which the two hormones might be secreted at different times in the 24 h cycle. This suggestion, first proposed by Lofts & Murton (1968), would solve many problems relating to behavioural and ecological aspects of the normal reproductive cycles in different species and evidence for it has recently been produced by Murton *et al.* (1969, 70) in experiments on greenfinches (*Chloris c. chloris*). Males which had been caught in late December when their testes were regressed (Plate 1*c, d*), were divided into groups and placed on asymmetric light skeletons containing 6 h of light followed by an additional hour given in the dark period. The treatments are shown in Fig. 9. After 19 days all birds were killed (3–4 h after the onset of the primary 6 h photoperiod) and the testes analysed histologically. Pooled plasma samples were assayed for gonadotrophins using a radioimmunoassay developed for measuring human chorionic gonadotrophin (Wilde, Orr & Bagshawe, 1967). In this assay the dose response curves for HCG and human LH are parallel (Bagshawe, Wilde & Orr, 1966), while chicken gonadotrophins also gives a parallel response line (Bagshawe, Orr & Godden, 1968). It is probable therefore that the activity

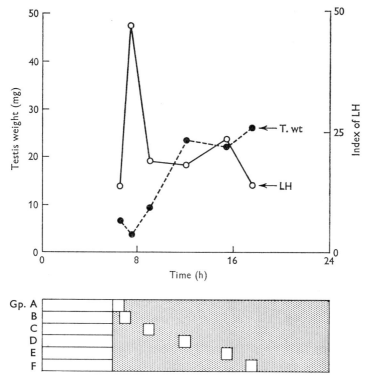

Fig. 9. The effect of asymmetric skeleton photoperiods on testicular development and plasma LH levels in *Chloris*. A high peak of LH occurs in the group exposed to a light pulse 7·5 h from 'dawn', but this is not accompanied by testicular development. From Murton, Bagshawe & Lofts (1969).

in greenfinch plasma is LH rather than FSH, although cross-reactions cannot be excluded. The method is extremely sensitive and a 5 % difference between samples may be considered statistically significant. The results are shown in Fig. 9, Plate 1 c, d and Plate 2 a, b).

As in previous experiments, when the size of the testes has been employed as an index of the effectiveness of a stimulatory schedule (Fig. 4), testicular growth was induced most strongly in group F, which received the light pulse from 17 to 18 h after dawn. This was confirmed histologically (Plate 2 a) for subjects in groups D, E and F all displayed gametogenetic activity with the production of spermatids. This indicates the secretion of FSH in these groups. More interesting, however, was group A, where little obvious testicular growth had occurred and the seminiferous tubules were completely inactive (Plate 2b) with undividing spermatogonia. On the other hand the interstitium of this group was highly stimulated, with numerous Leydig cells heavily charged with lipids (Plate 2c, d) and with swollen and prominent nuclei (Plate 2c). The LH titre in the plasma of this group was nearly three times as great as in the other groups (Fig. 9). The slight elevation of plasma activity in group E may reflect either a smaller second peak of LH release or a cross-reaction of greenfinch FSH in the assay. This demonstration that LH may be released independently from FSH is important in many respects, not the least of which is to emphasize that estimates of testicular weight are not necessarily a complete reflection of hormone secretion. Furthermore, they also indicate that when endogenous LH (and possibly androgen) titres are high this is not accompanied, in the greenfinch at least, by a concomitant development of spermatogenesis, whereas much of the evidence derived from the *exogenous* administration of androgens have led to the belief that these steroids have a stimulatory effect and may be involved in the seasonal recrudescence of gonadal activity. The results also imply that in the hypothalamus there may be separate centres controlling LH and FSH release, each perhaps receiving information from discrete rhythms of photoperiodic sensitivity. Anatomical evidence for such mechanisms in birds is virtually non-existent, as yet, but Stetson (personal communication) has found an area within the hypothalamus of *Coturnix* where lesions cause a cessation of FSH release (testicular collapse) and yet the interstitium remains active. Lesions in the more basal hypothalamus appear to block both FSH and LH release.

We have earlier laid emphasis on the differences in the photoperiodic response mechanisms which lead to the differences in the patterns of annual breeding cycles among avian species. In terms of the endogenous hormonal mechanisms, this phenomenon must obviously have its basis in differences of pattern in the release of the two gonadotrophic hormones regulated, in part at least, by the photo-inducible circadian phase. We have seen in the greenfinch that a mechanism has evolved in which the photo-inducible phase for LH release is apparently entirely separate from that promoting FSH release (Fig. 9), occurring much earlier in the 24 h cycle. In the house sparrow, however, Murton, Lofts & Orr (1970) have been able to show that the pattern is quite different. By subjecting winter birds to an identical experimental set-up as used for the greenfinch investigation, these workers were able to demonstrate that in this species a marked peak of LH release in relation to a light interruption early in the

24 h cycle does not occur (Fig. 10); there is a greater overlap of LH and FSH release. These differences correlate with differences in the ecology and ethology of the two species.

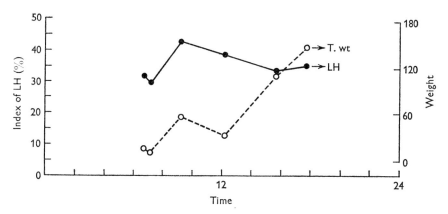

Fig. 10. Testicular development and plasma LH levels of *Passer* exposed to the same asymmetric skeleton photoperiods as shown in Fig. 9. A significantly higher peak of LH does not occur in birds with undeveloped testes, cf. Fig. 9. From Murton, Lofts & Orr (1970).

PHOTO-REFRACTORINESS

We have so far discussed only the gonadal growth phase of the annual reproductive cycle. In many wild species, however, the period of reproduction is terminated abruptly by a spontaneous rapid regression of the gonads at a time when the bird is experiencing normally stimulatory photoperiods. Birds caught during this phase of their annual cycle cannot be photo-stimulated experimentally and are said to be photo-refractory. A characteristic period of post-nuptial refractoriness is not an inevitable part of all avian breeding cycles, however, and has been shown to be absent in some photoperiodic species such as the bobwhite quail, *Colinus virginianus* (Kirkpatrick, 1959), and the wood-pigeon (Lofts, Murton & Westwood, 1967b). The development of photo-refractoriness may be regarded as an adaptive mechanism preventing the production of young at times when stimulatory photoperiods still occur but other environmental needs for successful breeding, such as an adequate food supply, are becoming reduced. In many species it also provides a period for the post-nuptial moult and autumnal hyperphagia prior to the winter migration. In the case of the wood-pigeon this is unnecessary since it is a resident species which can take advantage of the autumnal grain harvest to provide for late hatched young, but a photo-refractory state is an essential component of the closely related turtle dove, *Streptopelia turtur*, which migrates to central Africa for the winter and therefore requires a 'safety device' to curtail its breeding season and allow sufficient time for the necessary pre-migratory changes to occur (Lofts, Murton & Westwood 1967c).

The duration of photo-refractoriness varies from species to species and generally in many temperate zone birds may last for some 2–4 months, so that by the time the birds regain their photo-sensitivity it is usually late autumn and declining daylengths inhibit any gametogenetic recovery. The termination of photo-refractoriness in a number of species has been experi-mentally advanced by artificially exposing them to a period of short days (see review by Lofts & Murton, 1968), and it is only by exposure to a species-specific period of short photoperiods that these animals regain their capacity to respond to stimulatory daylengths. Conversely, continued exposure to long days prolongs it in many, but not all species. Murton, Lofts & Westwood (1970) have recently demonstrated in house sparrows that this part of the annual cycle might also have a circadian-based mecha-nism determining its duration. Thus, when groups of photo-refractory birds were maintained for 35 days under a daily photoperiod of 6 h plus an additional 1 h light pulse given as an interruption of the dark period at 8, 12 and 16 h from 'dawn', respectively, only the birds that had received a light interruption at 8 h from 'dawn' responded when trans-ferred to a 16 L–8 D light schedule and underwent a gonadal recrudescence; whereas the other two groups remained photo-refractory and their gonads remained fully regressed and spermatogenetically inactive (Plate 3 a–d). These results make it clear that the total quantity of light received by the bird is unimportant in determining the duration of photo-refractoriness since all groups were exposed to the same daily ration of light (i.e. 6 h + 1 h). Hamner (1968) similarly claims a circadian component timing the refractory period of the house finch, and distinguishes between the absolute refrac-tory period and the relative refractory period (Farner & Follett, 1966). The former marks the period during which even continuous light fails to evoke a spermatogenetic response, and in this finch lasts for about a month. In the latter period, the bird gradually recovers its photo-sensitivity and will show a differential response to a light stimulus, depending on when it is applied during this gradual recovery phase, so that finches exposed to an 18 L–6 D for several weeks during the beginning of this phase in late September show a far slower rate of testicular response than if the same experiment is performed in December.

The physiological basis for photo-refractoriness still remains one of the unresolved problems of reproductive endocrinology, and the gonad, pituitary and hypothalamus have all in turn been suggested as the locus of the blockage. The early experiment of Benoit et al. (1950) which demonstrated that compensatory hypertrophy of the residual testis did not occur in hemicastrated ducks when this was carried out during the refractory phase, but did when performed during the non-refractory period, indicated a cessation of gonadotrophin output during the refractory phase and suggested the hypothalamo–pituitary axis and not the pituitary–gonad axis as the primary site of the mechanism. This has been supported too by the results of a recent series of experiments by Farner (personal communication) which have shown that the gonadotrophin content of the pituitaries of castrated white-crowned sparrows become depleted at the same time as intact control birds entering the refractory stage. At the present time, however, it would still seem premature to exclude completely the gonads from any influential role in determining whether a photo-refractory state persists. When wild birds enter their post-nuptial refractory phase a proliferation of new Leydig cells occur (Marshall, 1949, 1955) so that the interstitium becomes very conspicuous and extensive. These cells are at first without any lipoidal inclusions. When the birds emerge from their refractory state the interstitial condition changes in its histological appearance, becoming more condensed and more lipoidal. This natural difference in the appearance of the interstitial tissue of the refractory bird and the non-refractory bird has been observed in a number of species (Lofts & Murton 1968; Lofts, 1970) and has also been duplicated by photo-experimentation. For example, the interstitial condition of wild mallards, *Anas platyrhynchos*, caught during their refractory phase in August and maintained in a protracted refractory condition by exposure to a 16 L–8 D photoperiod to the end of December was as extensive as in the natural refractory condition, whereas that of a parallel group, in which the refractory phase had been terminated by placing the birds under a short 8 L–16 D light in regime, was much more condensed and lipoidal (Plate 3e, f) although spermatogenetically they were identical (Lofts & Coombs, 1965). In *Carpodacus* too an accumulation of interstitial lipids takes place at the termination of photorefractoriness (Hamner, 1968). On the basis of this and other cytological data Lofts (1970) has argued that the gonad cannot be excluded from any consideration of the endocrinological mechanisms underlying the maintenance of a photo-refractory state. A positive conclusion about possible gonadal feedback during this period can only be reached however, when we eventually measure the circulating testosterone levels of the photo-refractory bird.

REFERENCES

ABLANALP, H. (1966). Selection for egg number in chicken and quail populations held under diverse lighting. *Proc. 13th World's Poult. Cong.* pp. 70–4.

ASCHOFF, J. (1955). Annual periodicity in the reproduction of warm-blooded animals. *Studium gen.* **8**, 742–76.

ASSENMACHER, I. (1958). Recherches sur le contrôle hypothalamique de la fonction gonadotrope préhypophysaire chez le canard. *Archs Anat. microsc. Morph. exp.* **47**, 448–572.

BAGSHAWE, K. D., ORR, A. H. & GODDEN, J. (1968). Cross-reaction in radioimmunoassay between human chorionic gonadotrophin and plasma from various species. *J. Endocr.* **42**, 513–18.

BAGSHAWE, K. D., WILDE, C. E. & ORR, A. H. (1966). Radioimmunoassay for human chorionic gonadotrophin and luteinizing hormone. *Lancet*, i, 1118–21.

BAKER, J. R. (1938). The evolution of breeding seasons. In *Evolution*, pp. 161–77 (ed. G. R. de Beer). Oxford University Press.

BARTHOLOMEW, G. A. (1949). The effect of light intensity and day length on reproduction in the English sparrow. *Bull. Mus. Comp. Zool. Harv.* **101**, 432–76.

BAYLÉ, J. D. & ASSENMACHER, I. (1967). Contrôle hypothalamo–hypophysaire du fonctionnement thyroidien chez la caille. *C. r. hebd. Séanc. Acad. Sci., Paris* **264**, 125–8.

BECK, S. D. (1968). *Insect Photoperiodism.* New York: Academic Press.

BENOIT, J. & ASSENMACHER, I. (1959). The control by visible radiations of the gonadotropic activity of the duck hypophysis. *Recent Prog. Horm. Res.* **15**, 143–64.

BENOIT, J., ASSENMACHER, I. & BRARD, E. (1956a). Apparition et maintien de cycles sexuels non saisonniers chez le Canard domestique placé pendant plus de trois ans a l'obscurité totale. *J. Physiol., Paris* **48**, 388–391.

BENOIT, J., ASSENMACHER, I. & BRARD, E. (1956b). Etude de l'évolution testiculaire du Canard domestique soumis tres jeune à un éclairement artificiel permanent pendant deux ans. *C. r. hebd. Séanc. Acad. Sci., Paris* **242**, 3113–15.

BENOIT, J., ASSENMACHER, I. & WALTER, F. X. (1950). Résponses du mécanisme gonado-stimulant à l'eclairement artificiel et de la préhypophyse aux castrations bilatérale et unilatérale, chez le canard domestique mâle, au cours de la période de régression testiculaire saisonnière. *C. r. Séanc. Soc. Biol.* **144**, 573–7.

BÜNNING, E. (1936). Die endogene Tagesrhythmik als Grundlage der photoperiodischen Reaktion. *Ber. dt. bot. Grs.* **54**, 590–607.

BÜNNING, E. (1960). Circadian rhythms and time-measurement in photoperiodism. *Cold Spring Harb. Symp. quant. Biol.* **25**, 249–56.

BÜNNING, E. (1963). *Die Physiologische Uhr.* Berlin: Springer-Verlag.

BÜNSOW, R. C. (1960). The circadian rhythm of photoperiodic responsiveness in *Kalanchoe. Cold Spring Harb. Symp. quant. Biol.* **25**, 257–60.

CHOWERS, I. & McCANN, S. M. (1965). Content of luteinizing hormone-releasing factor and luteinizing hormone during the estrous cycle and after changes in gonadal steroid titres. *Endocrinology* **76**, 700–8.

DOLNIK, V. R. (1963). Kolichestvennoe issledovanie zakonomernostei vesennovo rosta semenikov u neskolkikh vidov vyurkovykh ptits (Fringillidae). *Dokl. Akad. Nauk SSSR*, **149**, 191–3.

ELEFTHERIOU, B. E., ZOLOVICK, A. J. & NORMAN, R. L. (1967). Effects of amygdaloid lesions on plasma and pituitary levels of luteinizing hormone in the male deermouse. *J. Endocr.* **38**, 469–74.

FARNER, D. S. (1961). Comparative physiology: photoperiodicity. *Ann. Rev. Physiol.* **23**, 71–96.

FARNER, D. S. (1964). The photoperiodic control of reproductive cycles in birds. *Am. Scient.* **52**, 137–56.

FARNER, D. S. (1965). Circadian systems in the photoperiodic responses of vertebrates. In *Circadian Clocks*, pp. 357–69 (ed. J. Aschoff). Amsterdam: North-Holland.

FARNER, D. S. & FOLLETT, B. K. (1966). Light and other environmental factors affecting avian reproduction. *J. Anim. Sci.* **25**, 90–118.

FARNER, D. S., FOLLETT, B. K., KING, J. R. & MORTON, M. L. (1966). A quantitative examination of ovarian growth in the white-crowned sparrow. *Biol. Bull. mar. biol. Lab., Woods Hole* **130**, 67–75.

FARNER, D. S. & WILSON, A. C. (1957). A quantitative examination of testicular growth in the white-crowned sparrow. *Biol. Bull. mar. biol. Lab., Woods Hole* **113**, 254–67.

FOLLETT, B. K. (1969). Gonadotrophin releasing activity in the quail hypothalamus. Seminar on *Hypothalamic and Endocrine Functions in Birds*, Tokyo, pp. 33–4.

FOLLETT, B. K. & FARNER, D. S. (1966*a*). The effects of the daily photoperiod on gonadal growth, neurohypophysial hormone content and neurosecretion in the hypothalamo–hypophysial system of the Japanese quail (*Coturnix coturnix japonica*). *Gen. comp. Endocr.* **7**, 111–24.

FOLLETT, B. K. & FARNER, D. S. (1966*b*). Pituitary gonadotrophins in the Japanese quail (*Coturnix coturnix japonica*) during photoperiodically induced gonadal growth. *Gen comp. Endocr.* **7**, 125–31.

FOLLETT, B. K. & SHARP, P. J. (1969). Circadian rhythmicity in photoperiodically induced gonadotrophin release and gonadal growth in the quail. *Nature, Lond.* **223**, 968–71.

FURR, B. J. A. & CUNNINGHAM, F. J. (1970). The biological assay of chicken pituitary gonadotrophins. *Brit. Poult. Sci.* (in the Press).

GOGAN. F. (1968). Sensibilité hypothalamique à la testostérone chez le canard. *Gen. comp. Endocr.* **11**, 316–27.

GOGAN, F., KORDON, C. & BENOIT, J. (1963). Retentissement de lésions de l'éminence médiane sur la gonadostimulation du canard. *C. r. Séanc. Soc. Biol.* **157**, 2133–6.

GRABER, J. W., FRANKEL, A. I. & NALBANDOV, A. V. (1967). Hypothalamic center influencing the release of LH in the cockerel. *Gen. comp. Endocr.* **9**, 187–92.

HAMNER, K. C. & TAKIMOTO, A. (1964). Circadian rhythms and plant photoperiodism. *Am. Nat.* **98**, 295–323.

HAMNER, W. H. (1963). Diurnal rhythm and photoperiodism in testicular recrudescence of the house finch. *Science, N.Y.* **142**, 1294–5.

HAMNER, W. H. (1964). Circadian control of photoperiodism in the house finch demonstrated by interrupted night experiments. *Nature, Lond.* **203**, 1400–1.

HAMNER, W. H. (1966). Photoperiodic control of the annual testicular cycle in the house finch. *Carpodacus mexicanus. Gen. comp. Endocr.* **7**, 224–33.

HAMNER, W. H. (1968). The photorefractory period of the house finch. *Ecology* **49**, 211–27.

JACKSON, G. K. & NALBANDOV, A. V. (1969). Luteinizing hormone releasing activity in the chicken hypothalamus. *Endocrinology* **84**, 1262–6.

JALLAGEAS, M. & ATTAL, J. (1968). Dosage par chromatographie en phase gazeuse de la testostérone plasmatique non conjuguée chez le Canard, la Caille, le pigeon. *C. r. hebd. Séanc. Acad. Sci., Paris* **267**, 341–3.

KANEMATSU, S., SONODA, T., KII, M. & KATO, Y. (1966). Effects of hypothalamic lesions on the gonad in the hen. *Jap. J. vet. Sci.* **28** (Suppl.), 451.

KING, J. R., FOLLETT, B. K., FARNER, D. S., MORTON, M. L. (1966). Annual gonadal cycles and pituitary gonadotropins in *Zonotrichia leucophrys gambelii*,
Condor **68**, 476–87.

KIRKPATRICK, C. M. (1959). Interrupted dark period: tests for refractoriness in
bobwhite quail hens. In *Photoperiodism and Related Phenomena in Plants and
Animals*, pp. 751–8 (ed. R. B. Withrow). Washington, D.C., American Association for the Advancement of Science.

KOBAYASHI, H. & FARNER, D. S. (1966). Evidence of a negative feedback on photoperiodically induced gonadal development in the white-crowned sparrow,
Zonotrichia leucophrys gambellii. *Gen. comp. Endocr.* **6**, 443–52.

LACK, D. (1950). The breeding seasons of European birds. *Ibis* **92**, 288–316.

LACK, D. (1967). Interrelations in breeding adaptations as shown by marine
birds. *Int. Orn. Congr.* **14**, 3–42.

LEHRMAN, D. S. (1961). Hormonal regulation of parental behaviour in birds and
infrahuman mammals. In *Sex and Internal Secretions*, vol. II, 1268–382 (ed.
W. C. Young). Baltimore: Williams and Wilkins Co.

LEPKOVSKY, S. & YASUDA, M. (1966). Hypothalamic lesions, growth and body
composition of male chickens. *Poult. Sci.* **45**, 582–8.

LOFTS, B. (1962). The effects of exogenous androgen on the testicular cycle of the
weaver-finch, *Quelea quelea*. *Gen. comp. Endocr.* **2**, 394–406.

LOFTS, B. (1970). Cytology of the gonads and feed-back mechanisms with respect
to photo-sexual relationships in male birds. In *La Photoregulation de la Reproduction chez les Oiseaux et les Mammifères*. Paris: C.N.R.S. (In the Press.)

LOFTS, B. & COOMBS, C. J. F. (1965). Photoperiodism and the testicular refractory
period in the mallard. *J. Zool., Lond.* **146**, 44–54.

LOFTS, B. & MURTON, R. K. (1966). The role of weather, food and biological factors in timing the sexual cycle of woodpigeons. *Br. Birds* **59**, 261–80.

LOFTS, B. & MURTON, R. K. (1968). Photoperiodic and physiological adaptations
regulating avian breeding cycles and their ecological significance. *J. Zool.,
Lond.* **155**, 327–94.

LOFTS, B., MURTON, R. K. & WESTWOOD, N. J. (1966). Gonadal cycles and the
evolution of breeding seasons in British Columbidae. *J. Zool. Lond.* **150**,
249–72.

LOFTS, B., MURTON, R. K. & WESTWOOD, N. J. (1967*a*). Interspecific differences
in photosensitivity between three closely related species of pigeons. *J. Zool.,
Lond.* **151**, 17–25.

LOFTS, B., MURTON, R. K. & WESTWOOD, N. J. (1967*b*). Photo-responses of the wood-
pigeon (*Columba palumbus*) in relation to the breeding season. *Ibis* **109**, 338–51.

LOFTS, B., MURTON, R. K. & WESTWOOD, N. J. (1967*c*). The experimental demonstration of a post-nuptial refractory period in the turtle dove, *Streptopelia
turtur*. *Ibis* **109**, 352–8.

MA, R. C. S. & NALBANDOV, A. V. (1963). Hormonal activity of the autotransplanted adenohypophysis. In *Advances in Neuroendocrinology*, pp. 306–12.
University of Illinois Press.

MARSHALL, A. J. (1949). On the function of the interstitium of the testis; the sexual
cycle of a wild bird (*Fulmarus glacialis* L.). *Q. Jl Microsc. Sci.* **90**, 265–80.

MARSHALL, A. J. (1955). Reproduction in birds: the male. *Mem. Soc. Endocr.* **4**,
75–88.

MARSHALL, A. J. (1959). Internal and environmental control of breeding. *Ibis* **101**,
456–78.

MARSHALL, A. J. (1970). Environmental factors other than light involved in the
control of sexual cycles in birds and mammals. In *La Photoregulation de la
Reproduction chez les Oiseaux et les Mammifères*. Paris: C.N.R.S. (In the Press.)

MENAKER, M. (1965). Circadian rhythms and photoperiodism in *Passer domesticus*. In *Circadian Clocks*, pp. 385–95 (ed. J. Aschoff). Amsterdam: North Holland.

MENAKER, M. & ESKIN, A. (1967). Circadian clock in photoperiodic time measurement: a test of the Bünning hypothesis. *Science, N.Y.* **157**, 1182–5.

MIDDLETON, J. (1965). Testicular responses of house sparrows and white-crowned sparrows to short daily photoperiods with low intensities of light. *Physiol. Zoöl.* **38**, 255–66.

MURTON, R. K. & ISAACON, A. J. (1962). The functional basis of some behaviour in the woodpigeon *Columba palumbus*. *Ibis* **104**, 503–21.

MURTON, R. K., BAGSHAWE, K. D. & LOFTS, B. (1969). The circadian basis of specific gonadotrophin release in relation to avian spermatogenesis. *J. Endocr.* **45**, 311–12.

MURTON, R. K., LOFTS, B. & ORR, A. H. (1970). The significance of circadian based photosensitivity in the house sparrow, *Passer domesticus*. *Ibis* (In the Press).

MURTON, R. K., LOFTS, B. & WESTWOOD, N. J. (1970). Manipulation of photorefractoriness in the house sparrow (*Passer domesticus*) by circadian light regimes. *Gen. comp. Endocr.* **14**. (In the Press).

MURTON, R. K., LOFTS, B. & WESTWOOD, N. J. (1970). The circadian basis of photo-periodically controlled spermatogenesis in the greenfinch *Chloris. chloris*. *J. Zool.* (In the Press).

PHILLIPS, R. E. (1964). 'Wildness' in the mallard duck: Effect of brain lesions and stimulation on 'escape behaviour' and reproduction. *J. comp. Neurol.* **122**, 139–55.

PHILLIPS, R. E. & TIENHOVEN, A. VAN (1960). Endocrine factors involved in the failure of pintail ducks *Anas acuta* to reproduce in captivity. *J. Endocr.* **21**, 253–61.

PITTENDRIGH, C. S. (1966). The circadian oscillation in *Drosophila pseudoobscura* pupae; a model for the photoperiodic clock. *Z. Pflanzenphysiol.* **54**, 275–307.

PITTENDRIGH, C. S. & MINIS, D. H. (1964). The entrainment of circadian oscillations by light and their role as photoperiodic clocks. *Am. Nat.* **98**, 261–94.

POLIKARPOVA, E. (1940). Influence of external factors upon the development of the sexual gland of the sparrow. *Dokl. akad. Nauk SSSR* **26**, 91–5.

RAMIREZ, V. D. & SAWYER, C. H. (1965). Fluctuations in hypothalamic LH–RF (luteinizing hormone–releasing factor) during the rat estrous cycle. *Endocrinology* **76**, 282–9.

SAWYER, C. H. (1967). Some endocrine aspects of forebrain inhibition. *Brain Res.* **6**, 48–59.

SHARP, P. J. & FOLLETT, B. K. (1969). The effect of hypothalamic lesions on gonadotrophin release in Japanese quail (*Coturnix coturnix japonica*). *Neuroendocrinology* **5**, 205-219.

STETSON, M. H. (1969). The role of the median eminence in control of photoperiodically induced testicular growth in the white-crowned sparrow, *Zonotrichia leucophrys gambelii*. *Z. Zellforsch. mikrosk. Anat.* **93**, 369–94.

TIXIER-VIDAL, A., FOLLETT, B. K. & FARNER, D. S. (1968). The anterior pituitary of the Japanese quail, *Coturnix coturnix japonica*. The cytological effects of photoperiodic stimulation. *Z. Zellforsch. mikrosk. Anat.* **92**, 610–35.

UEMURA, H. (1964). Effects of gonadectomy and sex steroids on the acid phosphatase activity of the hypothalamo–hypophysial system in the bird, *Emberiza rustica latifascia*. *Endocr. jap.* **11**, 185–203.

WILDE, C. E., ORR, A. H. & BAGSHAWE, K. D. (1967). A sensitive radio-immunoassay for human chorionic gonadotrophin and luteinizing hormone. *J. Endocr.* **37**, 23–35.

WILSON, F. E. (1967). The tubero–infundibular neuron system: a component of the photoperiodic control mechanism of the white-crowned sparrow, *Zonotrichia leucophrys gambelii. Z. Zellforsch. mikrosk. Anat.* **82**, 1–24.

WOLFSON, A. (1966). Environmental and neuroendocrine regulation of annual gonadal cycles and migratory behaviour in birds. *Recent Prog. Horm. Res.* **22**, 177–244.

DISCUSSION

DE WILDE: Professor Lofts, I think your paper was of unusual interest because it emphasizes so much some things I wanted to say in my talk. We share similar points of view and in the insect work, of course, a very similar approach has been followed. We have, however, obtained rather diverse results. It has been shown in some cases that it is highly probable that circadian sensitivity is involved in photoperiodic induction, especially in cases where a light break of only a few milliseconds flash at a fixed moment in the dark period suffices to break the induction of diapause. But in the Colorado beetle, on the other hand, we have never succeeded in doing this. It seems in insects there are two types of photoperiodic induction mechanisms. That sounds very strange because it seems such a basic thing. It seems that these responses have occurred already in the stage of evolution where there were, perhaps, no birds or insects. On the other hand, diapause has, very probably, been evolved in different insect species independently and it may well be that the basic mechanism may also be very different. We have tried to find out whether the level of juvenile hormone in the 24 h cycle would change, in fact, in the same way as you have done. In our experiments on the Colorado beetle the titre of juvenile hormone has never been found to vary. In fact, the level seems to be regulated in a sort of steady-state situation during a long period of reproduction and also during diapause. We have not been able to find any oscillations over 24 h.

LOFTS: I regard the photoperiodic mechanisms as having been subjected to selective pressures the same as other regulating components of the reproductive cycle and this has resulted in variations in the patterns to be found in different species of animals. This has become very evident in the data of these two species, the greenfinch and the house-sparrow. There is some ecological background information which Dr Murton has that correlates very nicely with the data on the greenfinch. This is a species of bird which undergoes a very intense, very active, territorial behaviour at the very beginning of its seasonal cycle, and of course territorial behaviour has been associated with high androgen titres. It may be that the sudden peak of LH right at the beginning of this particular phase, which when translated into terms of the annual cycle, occurring early in the season, could be a basic selective feature stimulating the bird to undergo this very intensive territorial behaviour which is a prominent part of its established breeding pattern.

QUAY: I wonder if we can examine a little more closely this very interesting circadian rhythm in *photosensitivity*. There are certain potentially very important implications here. It would seem to be implied that this rhythm in nocturnal photosensitivity is not as prone to phase shifting, induced by the environmental photoperiod, as rapidly as certain other circadian rhythms which have light as a *zeitgeber*, such as rhythms in locomotor activity or other behaviours. Is it in fact true that the rhythm in photosensitivity is more stable in terms of phase-shifting in response to changes in timing of environmental illumination?

LOFTS: It is not isolated and you can get a phase-shifting of the photoinducible phase with changing photoperiods.

QUAY: How long a time? In an experimental design one wants to know whether these nocturnal light flashes will set up a new cycle length or not.

LOFTS: Dr Follett has some evidence on this in the quail, where he did get a phase-shifting which he measured by a phase-shifting in the peak of testicular growth when he changed the primary photoperiod in these light-interruption experiments. Perhaps he would like to come into the discussion at this point.

FOLLETT: I don't really think Dr Quay that the rhythm of photosensitivity in our birds is different from any other circadian rhythm, since it shows very similar properties. If one exposes quail to an asymmetric skeleton using 6 h of light as the primary photoperiod and interrupting the night with 15 min flashes, the rhythm seems to be located somewhere between 11 and 15 h after the main onset of the photoperiod. If 3 h of light are used as the primary photoperiod the rhythm is advanced, while if the photoperiod is increased to 9 h then the rhythm is phase delayed.

QUAY: I was not questioning the circadian nature of your rhythm, I was trying to see to what extent it differed in its phase-shifting response to environmental light; that is, whether it was more or less sensitive to *zeitgeber* activity of environmental light. Thank you very much for your added information.

MURTON: Further to the points raised by Dr Quay, the experiments described by Professor Lofts took place over a 3-month period—from the time when subjects were initially captured until they were sacrificed. We have no data on how stable the circadian oscillation would remain over longer periods, especially if, as with natural conditions, the onset of dawn and the total photoperiod was changing. We have no reason to think that phase-shifting would not occur over a longer period.

It is, of course, a feature of most wild-bird populations that they do show far more temporal and spatial stability than causal observations would suggest. For example, migrant species move within relatively restricted limits, which means that they experience a limited range of photoperiods. Furthermore, birds accidentally shifted to the wrong geographical location show inappropriate breeding responses, presumably because any circadian-based mechanisms are disturbed. In a similar manner the daily routines of birds seem remarkably stereotyped and stable; for instance, wood-pigeons feed their young with crop milk—which depends on prolactin secretion—twice per day (at around 10.00 h and 17.00 h) and the exact times show little daily variation for any particular pair. In other words, birds live within a predictable range of circumstances which would not preclude them being dependent, for much reproductive behaviour at least, on circadian-based rhythms, accepting that phase-shifting and re-cycling phenomena must be involved.

FENWICK: It would be most interesting to find out if this night-time sensitivity exists throughout the entire 12 months of the year. Let me add a comment. In goldfish exposure to various photoperiods throughout the year—photoperiods ranging from 1 h of light per 24 h right through to continuous light—results in gonadal stimulation for a very limited duration of time during the year. This sensitive phase coincides with the time of year when the gonads were normally stimulated in nature. If the night-time sensitivity that you are finding doesn't persist throughout the whole year, but is only effective at that time of year when the animals are normally being stimulated under natural conditions, I wonder if it's all that important talking about this light-sensitive period as the timing mechanism since its appearance coincides with the natural daylength. So how is it timing?

LOFTS: The answer to that of course is that these birds were caught in the winter time when the gonads were spermatogenetically inactive and the light-sensitive phase was still present. The location of this phase within the 24 h cycle was such that in winter-time, when daylengths may be of only 8 h duration, this phase coincides with the night. The hypothesis is, as far as time measurement is concerned, that as daylengths then increase in spring and lengthen to 10 or 12 h the photoperiod coincides more and more with the light-sensitive phase and there is a stimulation of gonadotrophin secretion. You can always photostimulate testes which are already coming up into breeding conditions with light, but by night-interruption technique you also indicate a light-sensitive phase when the gonads are normally fully regressed. A 7 h light period is not normally stimulatory, but if you break it up into 6 h and give the other hour at a specific time of night time, although the birds are being exposed to the same total amount of light in the 24 h period, they can be stimulated into breeding conditions because the light pulse coincides with their photo-sensitive phase.

FENWICK: Thank you, that answers the question, but the point that I did want to make was that goldfish, at this one particular time of year, and only this time of year, show a relationship between the size of the gonads and the amount of light per day. It was almost a linear relationship from 1 h of light to 24 h of light. So the possibility existed that in the fish there is a light-sensitive phase at a particular time of year.

DESCRIPTION OF PLATES

PLATE 1

(a) Testis of feral pigeon (C. *livia*) after 3 weeks administration of testosterone (1 mg/day). The seminiferous tubules are greatly regressed and spermatogenetic activity retarded. The number of germ cells in the germinal epithelium is much reduced.

(b) Gelatine section of the same testis as shown in (a). Testicular regression is accompanied by a build-up of lipoidal material within the seminiferous tubules.

(c) Testis of greenfinch in late December. There are only spermatogonia in the germinal epithelium and the interstitial cells are small with spindle-shaped nuclei.

(d) Gelatine section of the same testis as in (c) stained to show the lipoidal condition of the interstitial cells (cf. Plate 2d).

PLATE 2

(a) Testis of a greenfinch kept on an asymmetric skeleton photoperiod (6L+1L:17D) 17 h from 'dawn'. The bird has been stimulated into full spermatogeneiss.

(b) Testis of a greenfinch on an asymmetric skeleton photoperiod (6L+1L:17D) with a light pulse 7·5 h from 'dawn'. There is no spermatogenetic activity but there has been considerable development of the interstitium (cf. Plate 1c).

(c) A high power of the testis of specimen seen in (b), showing the proliferation of the interstitial cells.

(d) Gelatine section of same specimen as (b) and (c) to show the lipoidal condition of the interstitial tissue, cf. control specimen in Plate 1 (d).

PLATE 3

(a) Testis of photo-refractory house-sparrow, showing the completely regressed seminiferous tubules with a spermatogenetically inactive germinal epithelium.

(b) Gelatine section of photo-refractory control sparrow, showing the densely lipoidal seminiferous tubules.

(c) Testis of sparrow caught during the photorefractory period and submitted to long (16L/8D) photoperiods after being kept for 35 days under an asymmetric skeleton photo-period (6L+1L:17D) with the light pulse given 8 h from 'dawn'. The photo-refractory phase has been broken and the birds have been stimulated into full breeding condition.

(d) Testis of refractory sparrow submitted to the same photoperiodic schedule as the bird in (c) except that the light pulse was given 15 h after 'dawn'. In this case photo-refractoriness remains and the tubules are still regressed although a marked clearance of tubule lipids has occurred (cf. b).

(e) Testis of photorefractory mallard which has been maintained in this condition by exposure to a 16L/8D photoperiod. The interstitial cells are less condensed and have more rounded nuclei than birds that are no longer refractory (cf. f).

(f) Testis of a mallard caught during the refractory period and maintained on an 8L/16D photoperiod until the refractory state has terminated. The interstitial cells are much more condensed and the nuclei more spindle-shaped than during the refractory condition. (cf. e).

THE PHYSIOLOGICAL AND MORPHOLOGICAL RESPONSE OF MAMMALS TO CHANGES IN THEIR SODIUM STATUS

BY B. A. SCOGGINS, J. R. BLAIR-WEST,
J. P. COGHLAN, D. A. DENTON, K. MYERS,
J. F. NELSON, ELSPETH ORCHARD
AND R. D. WRIGHT

Sodium-depleted environments occur in many of the arid regions of the earth. As the sodium content of rain water decreases with increasing distance from the sea, many arid and mountain areas have a short supply of sodium and chloride (Hutton, 1958). The water table of arid areas is well below the surface and sodium and its soluble salts are leached from the soil. Plants grown in this type of area consequently have a low concentration of sodium. Analyses of grasses and foliage of trees eaten by sheep and cattle in areas of Central Australia showed sodium concentrations of 1–3 m-equiv/kg dry wt, compared with the 100–350 m-equiv/kg dry wt of grass from coastal regions (Denton, Goding, Sabine & Wright, 1961). Conversely, in these same areas bore, pan and spring water may be high in sodium content. In alpine grasslands which are covered by snow for the winter months of each year, the melting snow in springtime leaches sodium from the top soil. This was characteristic of the Snowy Plains area investigated in this study.

Observations on different mammalian species, in particular ruminant herbivores, in the natural environment have shown that many exhibit an avid appetite for salt. In the Altai mountains, stags, deer, chamois and other herbivores have been reported to have licked out caves in the slaty rocks of high salt content (Bunge, 1873). Similarly, in the game reserves of Africa, Hemingway (1966) described how kudu and other types of buck are attracted to salty pools and salt licks. Mountain gorillas have also occasionally been observed to eat soil which on analysis was found to have a high salt content (Schaller, 1965). Herds of reindeer in the Arctic have been reported to drink sea water (Bell, 1963) and the shoe hare (*Lepus americanus*) populations in Alaska have been seen gnawing wood which had been immersed in salt water (Bell, personal communication). During spring or summer in the Rocky Mountains, Rooseveldt elk and deer, bighorn sheep, goats, moose, caribou, mule deer and white tailed deer all seek out natural dry or wet licks or are attracted to salt blocks

(Dalke, Beeman, Kindel, Robel & Williams, 1965; Cowan & Brink, 1949). Analyses of the licks showed them to contain a number of trace metals and considerable amounts of sodium. In Montana, wild ruminants were only attracted to a fraction of the total available licks—to those licks which contained large amounts of sodium bicarbonate and sodium sulphate (Knight & Mudge, 1967). In domestic animals appetite for salt has been long recognized (Lehmann, 1854); Giambattista Basile (1632) in one of the tales of the Pentameron remarks on the gluttony of sheep for salt. In sheep made sodium deficient by salivary loss from a permanent unilateral parotid fistula it has been shown that they selectively graze sodium-rich plants relative to the intake of control animals (Arnold, 1964). Cattle have been observed to seek out and lick the saddle cloths of horses (Loeb, personal communication). In south-eastern and central Australia kangaroos are attracted to the salt licks put out for cattle and sheep. Studies on cattle eating pasture low in sodium by Bott, Denton, Goding & Sabine (1964) showed that they excreted urine virtually free of sodium and had a salivary sodium/potassium ratio indicative of sodium deficiency and increased aldosterone secretion.

Laboratory investigations in the Eutherian mammals—sheep (Blair-West, Coghlan, Denton, Goding, Wintour & Wright, 1963) dog (Davis, 1967) and man (Gocke, Gerten, Sharwood & Laragh, 1969) have shown that when depletion of total body sodium stores occurs the activity of the renin–angiotensin–aldosterone system is increased and renal conservation of sodium occurs. In marsupials, studies on the American opossum, *Didelphis virginiana*, have shown that aldosterone secretion rises following infusion of extracts of *Didelphis* kidney and that the juxtaglomerular index of granulation in the kidneys is increased following diuretic induced sodium depletion (Johnston, Davis & Hartroft, 1967). More recently in the possum, *Trichosurus vulpecula*, it has been demonstrated that plasma renin concentration and activity was elevated following diuretic induced sodium depletion (Reid & MacDonald, 1969). Further, these animals exhibited a salt appetite and had increased juxtaglomerular cell granulation.

This paper reviews our investigations in Australia on the physiological and morphological response of both native marsupial and introduced species of animals to a sodium-deficient environment under natural conditions. These observations have been published in a preliminary form (Blair-West, Coghlan, Denton, Nelson, Orchard, Scoggins, Wright, Myers & Junqueira, 1968).

Native and introduced species of mammals have been studied in four different areas of south-eastern Australia which differ in their sodium content of soil and plants.

A. Alpine: Snowy Plains and Tantangara: a region in the Snowy Mountains of New South Wales, an area snow-covered for 3–4 months of the year and low in sodium.

B. Grassland: Canberra: temperate grassland close to Canberra in the Australian Capital Territory, an area of relatively low sodium status.

C. Sea Coast: Welshpool: temperate grassland close to the ocean coast in Victoria, an area of high sodium status.

Fig. 1. Map of Australia to show localities of areas under investigation. (*A*) Alpine area at Tantangara, (*B*) grassland area at Canberra, (*C*) sea coast at Welshpool, (*D*) desert at Calindary.

D. Desert: Calindary: arid desert area in inland New South Wales, an area of salt pan and high sodium status.

The exact location of these different areas is shown in Fig. 1.

The indigenous animal species studied were two marsupial phalangeroids —the eastern grey kangaroo (*Macropus giganteus*) and the common wombat (*Vombatus hirsutus* Perry). The introduced species were the wild rabbit (*Oryctolagus cuniculus*), sheep (*Ovis aries*), fox (*Vulpes fulva*) and cattle (*Bovis taurus*). With the exception of some rabbits from Tantangara and Canberra which were captured and kept in cages overnight prior to sacrifice, all animals were located at night by spotlight and shot through the head without chase. Blood samples were obtained by cardiac puncture within 2–3 min of death. Immediately after collection the blood was haemolysed with distilled water and then deep frozen until analysis. In eutherian mammals it is known that the onset of adrenal response to ACTH (adrenocorticotrophic hormone) infused into the adrenal artery occurs after 3–4 min (Urquhart, 1966; Blair-West, Coghlan, Denton, Goding, Orchard, Scoggins, Wintour & Wright, 1967). Minimal or no changes in peripheral blood corticosteroid concentration would have resulted with this procedure of sample collection. The small volumes of blood obtained from the rabbits made it necessary to pool samples from different rabbits and this was done according to sex and incidence of pregnancy and/or lactation. Urine specimens for electrolyte determination were obtained by bladder puncture. Sections of the left adrenal gland, left kidney, parotid and submandibular glands were placed in 10 % formal saline for histology. The entire right adrenal was put into buffered 10 % formalin. A section of the left kidney was also put into Helley's solution. The remainder of the left kidney and the right kidney were deep frozen for renal renin analysis. With the sheep, specimens of peripheral blood were also collected from the heart immediately after the animals were shot through the head whilst grazing quietly. In the sheep in which steroid estimates were made by adrenal vein cannulation, the animals were rounded up 24 h before-hand, and were collected individually from the yard, anaesthetized, and the left adrenal vein and parotid duct cannulated. The salt appetite of the rabbits in the different areas was investigated by allowing them access to salt-impregnated pegs. Soft-wood pegs impregnated with either sodium bicarbonate, sodium chloride, potassium chloride, magnesium chloride or distilled water were firmly driven into the ground in the region of the rabbit warrens. Access to the pegs by cattle and sheep was prevented by enclosing the peg area with wire netting which was open sufficiently at the bottom to allow access by rabbits. Pegs were weighed at the beginning of the experiment and at the end of 9 days. Electrolyte analysis of the pegs showed that the outer wood contained up to 1800 m-equiv/kg of sodium whereas the core contained about 20 m-equiv/kg. The resulting salt content following impregnation differed with the different salt solutions. The concentrations of Na and K in whole blood, plasma, urine, faeces and grass samples were measured by a Technicon autoanalyser.

Aldosterone, cortisol and corticosterone in peripheral blood samples were measured using a double isotope derivative dilution procedure (Coghlan & Scoggins, 1967a). Specificity was established by multiple derivative formation and further chromatographic steps as has been previously described for sheep blood (Coghlan & Scoggins, 1967a). Although blood from adrenalectomized animals was not available, the non-specific blank of the method for aldosterone in these species is probably similar to that reported for human plasma and sheep blood (Coghlan & Scoggins, 1967 b) as concentrations of less than 1 ng/100 ml blood have been found in the sodium-replete rabbit, wombat and kangaroo.

The adrenal, kidney, parotid and submandibular sections were stained with Mayer's haemalum and PAS (periodic acid–Schiff) and the kidney sections which had been fixed in Helley's solution were stained with Bowies. The right adrenal was cleaned of fat, weighed and sectioned. Sections of 15 μ were mounted and stained with haematoxylin and a 50:50 solution of Sudan III and IV. The stained sections were magnified 25 times, photographed and areas of the various zones measured by planimetry.

Pastures in each of the regions where animals were obtained contained the following major plant species:

A. Alpine: *Danthonia, Poa, Themeda, Carex, Sphagnum, Trifolium, Festuca, Rumex, Hypochaeris, Bromus* and *Hordeum*.

B. Grassland: the same as for alpine but with a predominance of *Trifolium* and *Lolium*.

C. Sea Coast: *Holcus, Poa, Lolium, Stipa, Acaena, Selaginella* and *Lobelia*.

D. Desert: *Atriplex, Themeda, Panicum, Marsilea, Eragrostis, Muehlenbeckia* and *Astrebla*.

Table 1. *Sodium and potassium concentration in mixed grass and soil samples*

Area	Season	Mixed grass (m-equiv/kg dry wt)		Soil (m-equiv/kg dry wt)	
		Na$^+$	K$^+$	Na$^+$	K$^+$
Alpine	Spring	0·2	457	0·8	10·0
	Summer	1·5	308	—	—
	Autumn	1·8	165	0·2	7·0
	Winter	2·7	69	<0·1	16·0
Grassland	Spring	7·5	289	0·8	4·3
Desert	Summer	200	110	3·6	11·0
Sea Coast	Summer	189	242	7·8	5·2
	Winter	149	284	2·7	5·5

The sodium and potassium content of typical mixed samples of these various plant species are shown on Table 1. The sodium content of the grasses and soil from the two areas of high sodium status was much higher than that for either the alpine or grassland regions. The lowest sodium levels were seen in the samples taken in late spring and early summer from Tantangara. One sample of grass taken in the spring from heavily timbered country close to where kangaroos were found had a much higher sodium content (8·5 m-equiv/kg dry wt) than other samples from this region. Seed husks, seeds and stalks obtained from *Stylidium graminiforum* in the autumn at Tantangara contained between 20–175 m-equiv sodium/ kg dry wt. This finding was unexpected and is being investigated. In the desert area analysis of *Atriplex* showed a sodium content of 2,200 m-equiv/ kg dry wt but the rabbits were grazing on grass similar in composition to that shown on Table 1.

Table 2. *Urinary electrolyte concentration in rabbits*

		Urinary (m-equiv/l.)	
Area	Season	Na⁺	K⁺
Alpine	Spring	0·6 ± 0·4 (19)	214 ± 77 (18)
	Summer	0·5 ± 0·1 (18)	270 ± 99 (18)
	Autumn	2·6 ± 2·9 (12)	466 ± 134 (6)
	Winter	6·4 ± 1·2 (5)	390 ± 96 (5)
Grassland	Spring	18·0 ± 14·6 (12)	219 ± 42 (12)
Desert	Summer	139·0 ± 97 (15)	319 ± 124 (15)

Table 3. *Mean plasma and urinary* $[Na^+]$ *and* $[K^+]$ *range (m-equiv/l.) and number of samples for wombats and kangaroos from alpine area and sea coast*

	Alpine (spring)	Sea Coast (summer–winter)
	Kangaroo	
Plasma Na⁺	138 (136–140) 7	140 (131–150) 5
Plasma K⁺	8·8 (7·7–10·0)	8·7 (5·9–10·0)
Urinary Na⁺	0·4 (0·2–0·8) 8	268 (79–496) 5
Urinary K⁺	336 (48–665)	130 (45–275)
	Wombat	
Plasma Na⁺	134 (128–141) 9	131 (127–134) 2
Plasma K⁺	8·5 (5·1–9·5)	6·4 (6·2–6·5)
Urinary Na⁺	0·3 (0·1–0·8) 13	23 (2–50) 4
Urinary K⁺	134 (17–262)	108 (74–165)

Urinary $[Na^+]$ and $[K^+]$ in animals from the different areas are shown on Tables 2 and 3. Kangaroos, wombats and rabbits from the alpine region excreted urine virtually free of sodium. Although the urinary $[Na^+]$ increased slightly in rabbits from this region during autumn and winter, values were still much less than those found in Canberra or in the desert. Samples taken from six sheep in the Snowy Mountains during spring also had urine free of sodium (mean 0·4 m-equiv/l.). Foxes shot in this same area had a urinary $[Na^+]$ ranging from 6–47 m-equiv/l., suggesting that these animals were not as sodium depleted as the other species. With the exception of the wombats all animals obtained from areas rich in sodium had a high urinary sodium excretion (Tables 2, 3). Urinary potassium excretion was high and mean values exceeded 100 m-equiv/l.

in all the areas studied (Tables 2, 3). Urinary [K⁺] in foxes and sheep was 187 and 211 m-equiv/l. respectively.

Although problems in the collection and storage of whole blood and plasma samples in the field has made collection of samples for electrolyte determinations difficult, a few unhaemolysed plasma samples have been obtained in wombats and kangaroo (Table 3). Considerable variation was seen between animals and the mean results show little difference in either [Na⁺] or [K⁺] between the two areas.

Electrolytes in hard and soft faeces were investigated in rabbits from the alpine and desert regions. The soft faeces of the alpine rabbits contained little sodium (5 m-equiv/kg dry wt) and on redigestion there was further sodium reabsorption so that the hard faeces were virtually free of sodium (2 m-equiv/kg dry wt). There was also a large reabsorption of potassium from the soft faeces. In the desert rabbits there was considerable sodium in both soft (42 m-equiv/kg dry wt) and hard faeces (28 m-equiv/kg dry wt).

In an earlier study from our laboratory (Bott *et al.* 1964) samples of parotid saliva and urine were collected from Aberdeen Angus and Hereford cattle which had been grazing on low sodium mountain pastures (2 m-equiv/kg wet wt) or on river flats irrigated periodically with water of very low sodium content producing a low sodium pasture (10–12 m-equiv/kg wet wt). The results from this area were compared with similar animals fed on irrigated pasture but with access to sodium supplements of 1,500 m-equiv/day or with animals that had been fed on pasture containing 84 m-equiv/kg wet wt. Fig. 2 shows the results for the various groups of animals. The animals on the low sodium diet had a markedly reduced salivary sodium/potassium ratio and low urine [Na⁺] compared with the control animals which excreted considerable amounts of sodium. The changes in salivary and urinary electrolytes in the sodium-deficient animals were consistent with an increased production of aldosterone.

The mean peripheral blood concentrations of aldosterone, cortisol and corticosterone for rabbits obtained from the sodium-deficient alpine region in spring, summer and winter and from Canberra and the desert region are shown on Fig. 3. Blood aldosterone levels were markedly elevated in the animals from the sodium-deficient area compared with either the intermediate Canberra region or the high-sodium desert area. The mean values for males, non-pregnant and pregnant and/or lactating females are shown on Table 4. Similar trends are seen in the three groups of animals. The highest values were obtained in the pregnant and/or lactating females from the sodium-deficient area. During the spring in both the Tantangara and Canberra areas all the female rabbits obtained were pregnant. The mean blood levels of cortisol and corticosterone (Fig. 4) were elevated in the spring Tantangara and Canberra rabbits. In these two groups of animals which were obtained in the same season the elevated levels

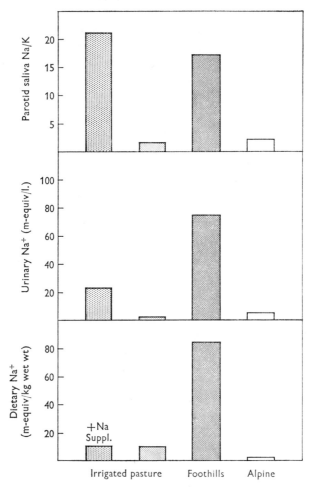

Fig. 2. Dietary, urinary and parotid salivary electrolytes in Aberdeen Angus cattle grazing on irrigated pasture with (▨) and without (▨) sodium supplement and Hereford cattle in foothills (▨) and mountain grassland (□).

probably resulted from the additional handling procedures that a number of these rabbits received; they were captured and caged overnight prior to sampling. This procedure was subsequently abandoned. Other animals which were shot at the same time in the same area had corticosterone and cortisol levels similar to those found in the other regions.

Blood corticosteroid levels in kangaroos from the alpine and sea-coast

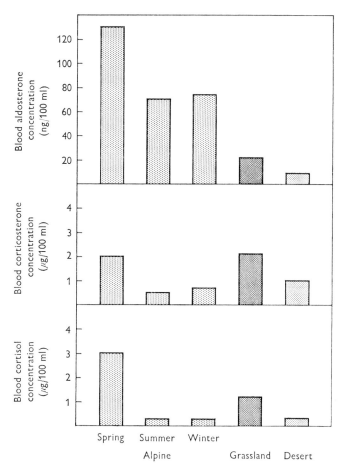

Fig. 3. Blood aldosterone, cortisol and corticosterone levels in rabbits from the different areas under investigation.

Table 4. *Peripheral blood aldosterone levels of rabbits ng/100 ml (mean and number of animals)*

		Male	Female	
Region	Season		Non-pregnant	Pregnant and/or lactating
Alpine	Spring	56 (5)	—	187 (10)
	Summer	65 (8)	42 (3)	158 (1)
	Winter	70 (6)	79 (5)	—
Grassland	Spring	20 (8)	—	23 (6)
Desert	Summer	11 (7)	7 (3)	3 (1)

areas are shown on Fig. 4. Aldosterone levels were markedly elevated in
sodium-depleted kangaroos of both sexes. The high level of 13·6 ng/100 ml
seen in one sodium-replete female was in an animal with a 60-day-old
pouch young. One of the deplete females (11·0 ng/100 ml) had a foetus

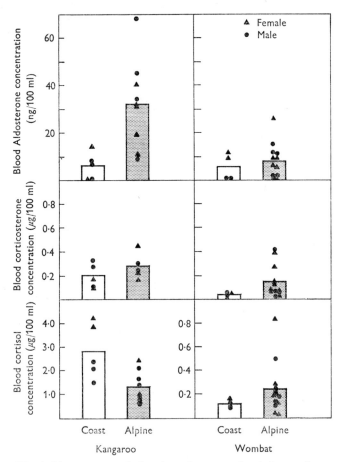

Fig. 4. Blood aldosterone, cortisol and corticosterone levels in wombats and kangaroos
from the alpine and sea coast areas.

in the pouch. With the exception of two female sodium-replete kangaroos,
cortisol and corticosterone levels were similar in both groups of animals.

Fig. 4 also shows the blood corticosteroid levels in wombats from the
same two areas. A number of animals of both sexes from the sodium-
deficient region with low urinary Na^+ had blood aldosterone levels similar
to those seen in the replete animals and the means of the two groups are
probably not significantly different. The two female wombats from the

Victorian sea coast (Welshpool) area had pouch young, 3 and 12 weeks old respectively.

Blood cortisol and corticosterone levels were similar in animals of both sexes and from the two areas. Cortisol levels in this species were much lower than those seen in the kangaroo.

Blood corticosteroids in three male foxes from the alpine Tantangara region are shown on Table 5. Aldosterone levels were high in this small sample even though these animals had some sodium in the urine. Samples from foxes from areas high in sodium have yet to be obtained. The higher steroid levels seen in fox 3 may have resulted from this animal being chased prior to being shot.

Table 5. *Peripheral blood corticosteroids in foxes and sheep from the alpine region—Tantangara (spring)*

Species	Aldosterone (ng/100 ml)	Corticosterone (μg/100 ml)	Cortisol (μg/100 ml)
Fox 1	17·1	0·09	0·91
Fox 2	25·7	0·50	1·28
Fox 3	107·9	0·58	3·84
Sheep 1	77·6	0·06	0·22
Sheep 2	36·0	0·03	0·68
Sheep 3	108·4	0·04	0·17

The peripheral blood aldosterone level in three sheep (wethers) from Tantangara were also very high compared with the normal sodium-replete level of about 1 ng/100 ml. Aldosterone secretion rates were measured by cannulation of the adrenal vein in other anaesthetized sheep from this area. In three animals, urine free of sodium and low salivary sodium/potassium ratio was found and high aldosterone secretion rates (14·0, 15·6 and 19·8 μg/hr) were observed. Four other animals were apparently sodium-replete as aldosterone secretion rates were about 2 μg/h, as seen in normal sodium-replete animals in the laboratory. The basis of this difference in a small sample of animals which had spent the winter and early spring in the snow region has yet to be determined. The possibility of selective feeding requires study.

The mean renin concentrations found in the renal cortices of rabbits from the three regions and during different seasons for the Tantangara animals are shown on Fig. 5. Although there was considerable variation between rabbits (Blair-West *et al.* 1968), the levels found in the alpine animals in spring and summer were much higher than levels found from other regions. The highest values were seen in pregnant female rabbits. The renal renin content of samples taken in the autumn were surprisingly

low considering the urinary sodium excretion of these animals was still
low. Renal renin concentrations of sheep from the sodium-deficient area
(mean $12·9 \pm 3·4$ U/g, $N = 7$) were also elevated above the levels found in
the laboratory for normal sodium-replete animals (mean $8·0 \pm 5·7$, $N = 25$).

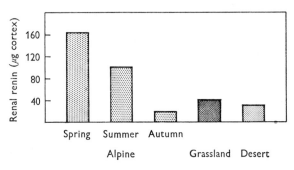

Fig. 5. Renin content of renal cortex of rabbits from different areas.

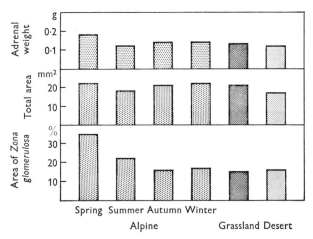

Fig. 6. Mean adrenal weight, total adrenal area and area of zona glomerulosa as a
percentage total area in rabbits from the alpine, grassland and desert areas.

In other sheep from the alpine region but with a reasonably high urinary
sodium excretion the renal renin concentration was normal. Cortices of
kangaroo and wombat kidney were incubated in the same way as the
sheep and rabbit samples against a prepared sheep substrate (Blair-West
et al. 1967). The concentration found in all animals was less than 1 U/g of
cortex.

Fig. 6 shows the weight and total area of the adrenal gland, and
the area of the zona glomerulosa as a percentage of the total area for the

sodium-deficient rabbits from the alpine and Canberra regions. This figure also shows the seasonal variation in these parameters for animals from the alpine Tantangara area. At the time of most severe sodium deficiency, spring in Tantangara, the rabbit adrenal glands were heavier, had a larger area and a significantly increased proportion of zona glomerulosa when compared with the other rabbits. Similar, although not as marked, changes were seen in kangaroo and wombat adrenals when the alpine and sea-coast animals were compared (Table 6).

Table 6. *Mean values for adrenal weight, total area and area of zona glomerulosa as percentage of total area for wombat and kangaroo*

Species	Area	Season	No. of animals	Adrenal wt (g)	Total area (mm²)	Area glomer- ulosa (% total)
Wombat	Alpine	Spring	6	0·73	47·9	27·6
	Sea coast	Summer	4	0·37	38·4	23·3
Kangaroo	Alpine	Spring	4	1·12	65·9	24·5
	Sea coast	Summer	8	0·72	45·0	15·0

Plate 1 shows histological sections of sodium-deplete and sodium-replete kangaroo and wombat adrenal glands from the alpine and sea-coast regions. The zona glomerulosa is wider in the sodium-deficient animals. This is particularly obvious in the section of the alpine kangaroo. The wombat adrenal also has a prominent capsule.

The morphology of the kidneys of the sodium-deficient wombats and kangaroos was compared with that of kidneys from animals obtained from the sea coast. Histological differences were observed particularly in the wombats (Plate 2). There was evidence of hypertrophy of the JGA- (juxta glomerular apparatus) macula densa region in the alpine animals (Plate 2).

Striking structural changes were seen in the salivary glands* of the sodium-depleted animals from Tantangara when compared with the animals from the sodium-rich sea-coast area. The marked increase in the duct system of the parotid gland of the sodium-deficient kangaroo is shown in Table 7. The height of parotid duct cells was $19·5 \pm 3·2 \mu$ for sodium-deficient and $14·0 \pm 3·0 \mu$ ($P < 0·01$) for sodium-replete animals. The vascular pattern of the deficient animals was different from that usually seen and was characterized by abundant blood vessels around the striated ducts. Plate 3 shows the loops of capillaries penetrating the outer wall of the striated ducts and running into the epithelial zone. Similar but not as substantial changes were seen in the wombat. Submandibular

* These studies were done in collaboration with Dr C. L. Junqueira, Faculty of Medicine University of São Paulo, Brazil.

glands from the same species from the alpine region had four times the area as striated ducts compared with sodium-replete animals.

The parotid gland of sodium-deficient rabbits also had an increase in the volume of striated ducts—5·9 % as compared with 1·5 % in the replete animals. In submandibular glands, striated ducts were only seen in the sodium-deficient rabbits. The epithelium of the parotid was increased in height 18·3 μ in deplete animals, compared with the 10·9 μ seen in replete animals.

Table 7. *Constitution of* Macropus giganteus (*grey kangaroo*) *parotid gland expressed in percentage of total volume*

Area	Season	Serous cell	Striated duct	Excretory duct
Alpine	Spring	86·3	5·3	3·1
		79·6	12·4	3·6
		83·2	9·5	4·8
Sea coast	Spring	95·4	0	1·5
		94·3	0·3	0·6
		92·7	0·2	1·3

Electron micrographs of rabbit parotid glands from both regions are shown on Plate 4. In sodium deficiency the density of the acinar granules is increased, the basal cell membranes are profusely infolded and the striated ducts have much higher cells. The infoldings of the basal-cell membranes are limiting portions of cell cytoplasm that are packed with mitochondria with very little cytoplasm between them. The conspicuous intercellular clear channels are suggestive of intercellular channels involved in ion and fluid transport. In the sodium-replete animals the striated ducts had short prismatic or cubic cells with very few infoldings of the basal cell membrane. Mitchondria were sparse and separated by cytoplasm with loosely arranged cristae and no intracellular channels.

The appetite for sodium, potassium and magnesium salts in rabbits in the different regions is shown in Fig. 7. In the spring and summer in the alpine Tantangara region, the rabbits devoured and in some cases completely ate out of the ground the pegs impregnated with sodium chloride and sodium bicarbonate. Some gnawing of the potassium chloride pegs occurred but the magnesium chloride and distilled-water pegs were hardly touched. In the late autumn the rabbits showed little salt appetite. The animals from Canberra or the desert showed no salt appetite. Plate 5, taken with a telescopic lens, shows wild rabbits at the pegs and their dominant interest in the pegs impregnated with sodium.

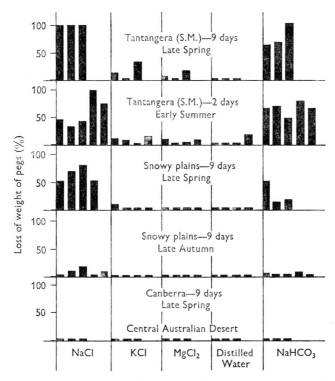

Fig. 7. The percentage loss of weight of soft wooden pegs impregnated with various salt solutions. Pegs were placed in vicinity of warrens for 9 days. The upper four panels record seasonal observations from two different sites in the alpine area and the bottom panel results from the grassland and desert regions. SM, Snowy Mountains.

DISCUSSION

Equivocal findings in early studies (Britton & Silvette, 1937) on the effect of adrenalectomy upon the electrolyte metabolism of the opossum (*Didelphis virginiana*) led Hartmann, Smith & Lewis (1943) to conclude that the adrenal in the marsupial species played a role of lesser importance in electrolyte metabolism than it did in the cat, dog and man. More recent studies, however, both *in vitro* (Brownell, Beck & Besch, 1967) and *in vivo* by direct adrenal vein cannulation (Johnston *et al.* 1967) have shown that aldosterone is produced by the opossum adrenal. Aldosterone has been measured in the peripheral blood of the wombat (*Vombatus hirsutus*) (Coghlan & Scoggins, 1967*b*); in the kangaroo' aldosterone has been identified in the adrenal venous blood of *Megaleia rufus* and *Macropus canguru major* (Weiss & McDonald, 1967) and in the peripheral blood of *Macropus giganteus* (Coghlan & Scoggins, 1967*b*). Using a method with a sensitivity of about o·1 μg Weiss & McDonald (1966a) failed to find aldosterone in adrenal venous plasma of the possum (*Trichosurus vulpecula*). However, adrenalectomy studies in both the possum (Reid & McDonald, 1968) and the quokka (*Setonyx brachyurus*) (Buttle, Kirk & Waring, 1952) suggest that a mineralocorticoid is produced by these two species. Aldosterone has also been identified in the peripheral blood of the tiger cat (*Dasyurus maculatus*), pademelon wallaby (*Thylogale billardieri*) and red-necked wallaby (*Wallabia rufogrisea frutica*) (unpublished observations). The studies on blood aldosterone levels in both sodium-deplete and sodium-repleted kangaroos confirm our earlier observations that sodium-deficiency increases aldosterone production in this species (Coghlan & Scoggins, 1967b; Blair-West et al. 1968). In the wombat, although elevated aldosterone levels were found in some cases, the mean levels for animals from the differing areas are not significantly different. Further studies on the response to sodium depletion in this species are in progress; in particular, a close examination of the dietary habits of the sea-coast animals which showed a relatively low urinary Na^+ excretion (mean 23 m-equiv/l.).

Although aldosterone secretion is increased during pregnancy in man it is not known whether this occurs in marsupials. Some of the animals in this study had young in the pouch; although it is not possible to draw definite conclusions, in the sodium-replete animals of both species the highest aldosterone levels seen were in the pregnant animals.

The question whether the renin–angiotensin–aldosterone system shows similar correlations with change in sodium status in marsupials as have been observed in eutherian mammals is of great interest. Due to the difficulties of obtaining plasma samples in the field our preliminary observations in the kangaroo and wombat from the sodium-deficient areas were made on renal renin for which low values were obtained (Blair-West et al. 1967). Plasma samples from kangaroos, wallabies and wombats have been incubated with sheep substrate (renin concentration assay) and the production of pressor material demonstrated. In a recent report by Reid & McDonald (1969) on the possum (*Trichosurus vulpecula*), a large concentration of renin in the renal cortex was shown when incubation was with sheep renin substrate.

Evidence that marsupials have a renin–angiotensin system which responds to sodium deficiency has been obtained from studies on the possum (Reid & McDonald, 1969) and American opossum (Johnston et al. 1967). In diuretic-induced sodium depletion, plasma renin concentration in the possum rose from 46 to 182 U/ml and plasma renin activity from 2·0 to 5·9 ng/h/ml. In the American opossum, dexamethasone-treated or hypophysectomized, animals infused with an opossum kidney extract showed a significant increase in adrenal production of aldosterone, cortisol and corticosterone. In relation to the question of possible species differences in the renin–angiotensin system it can be noted that in a eutherian mammal, the laboratory rat, the physiological role of this system in the control of aldosterone is also uncertain (Palmore, Marieb & Mulrow, 1969).

A further attempt in wombats to assess the renin–angiotensin system was made by measurement of angiotensin II concentration. The radio-immunoassay procedure of Catt, Cain & Coghlan (1967) was used. Low values were found compared with the concentration usually found in sodium-deficient mammals. Although the antibody used is equally reactive with the valine-5 and isoleucine-5 forms of angiotensin II (Catt *et al.* 1967) care must be taken in interpreting these low values as negative findings, as the sequence of marsupial angiotensin II is not known. The low angiotensin II values are none the less consistent with the low renin activity and low blood aldosterone levels seen in these animals. In other wombats from the low-sodium area some increase in aldosterone levels was found. In all the animals from this region the urinary sodium excretion was very low and consistent with the morphological changes in the adrenal, parotid and kidney. Similar morphological changes occurred in the kangaroos where aldosterone levels were markedly elevated. Wombats from the high sodium sea-coast area had only about 20 m-equiv/l. of sodium in their urine compared with 268 m-equiv/l. in urine of kangaroos from the same area. The wombats are a burrowing animal with dietary habits different from those of the other animals studied. They eat roots of plants and trees, and consequently probably have a much lower sodium intake even in the sea-coast area. Further studies are in progress to investigate the response and sensitivity of the renin–angiotensin–aldosterone system to changes in sodium status in this species.

The plasma [K^+] in both wombats and kangaroos from both the alpine and coastal areas was much higher than the level found in eutherian mammals. Although it is not easy to collect samples for analysis, care was taken to ensure that unhaemolysed plasma was obtained, and that blood was centrifuged in the field within a short time (30 min to 1 h) of collection. Similar high [K^+] levels have also been seen in samples taken from two species of wallaby (*Wallabia rufogrisea frutica* and *Thylogale billardieri*). Other recent studies on Australian marsupials have reported similar high values. Kinnear, Purohit & Main (1968) reported values of 6·2 and 7·7 m-equiv/l. as being in the normal range and Dawson & Denny (1969) found levels ranging from 6·5 to 10·9 m-equiv/l. in the kangaroos *Megaleia rufa* and *Macropus robustus* grazing throughout the year in an arid environment. They reported a mean value of 7·1 m-equiv/l. in animals sampled in the laboratory. In the American opossum levels in the range 4·0–5·0 m-equiv/l. have been reported by Smith, Lewis & Hartmann (1943) and Britton & Silvette (1937). The high plasma [K⁺] may be interesting in relation to the control of the renin and aldosterone systems in marsupials in view of the known direct effect of increased plasma [K⁺] on aldosterone secretion (Blair-West *et al.* 1963) and the inhibition of renin release by potassium loading (Veyrat, Brunner, Grandchamp & Muller, 1967).

The rabbits from the sodium-deficient Tantangara region showed a marked increase in all the parameters of the system measured. This was particularly true in the spring and summer, when the levels of available dietary sodium were at their lowest. Renal renin levels were 3–5 times higher during this time than the levels seen in animals shot in the same area in the autumn, or in either the grassland or desert area. Blood aldosterone levels were also very high in the Tantangara rabbits. During the spring and summer all the female rabbits that were obtained were either pregnant and/or lactating; the results in females, therefore, may reflect an additional effect of the reproductive process on sodium deficiency.

Morphological changes in the rabbit adrenals showed that in the severely depleted animal the zona glomerulosa was grossly enlarged and occupied about 35% of the total area of the cortex. In more extensive studies on the rabbit Myers (1967) has seen adrenals in which half the cortex is occupied by the zona glomerulosa, the glomerular zone apparently invading the zona fasciculata. Preceding this extension of the glomerular zone into the zona fasciculata the lipid content of the inner zone is reduced and numerous small staining nuclei with mitotic figures appear along the interface of the two zones (Myers, 1967). Accessory adrenal tissue and cortical nodules have also been observed in rabbits. These appear to contain both glomerulosa and fasciculata cells and have been induced by challenging sodium-depleted rabbits with ACTH (Myers, 1967).

The high blood levels of aldosterone found in the foxes from the sodium-deficient area were unexpected as it was thought that carnivorous animals would be less liable to sodium deficiency as they feed on herbivores which, although depleted, have a high tissue content of sodium. However, these animals had only moderate amounts of sodium in their urine (6–47 m-equiv/l.) and had a large zona glomerulosa of the adrenal cortex. Other observations in the field have shown that rather than living entirely on a diet of rabbits and other small animals, foxes tend to be omnivorous. Further studies are in progress to examine larger numbers of this species in both the coastal high sodium and the alpine sodium-deficient areas.

Sheep from the alpine area with very low urinary [Na⁺] showed increased activity of the renin–angiotensin–aldosterone system similar to that seen in laboratory experiments with sheep with a parotid fistula (Blair-West et al. 1967). In other sheep, some of which had considerable sodium in their urine, aldosterone and renal renin levels were similar to those found in the sodium replete animal. It is possible that these latter animals were showing some selectivity in their grazing behaviour similar to that reported for sheep with a parotid fistula by Arnold (1964).

A number of studies on adrenal venous blood have shown that cortisol is the major glucocorticoid produced by the adrenal in the American opossum (Johnston et al. 1967), kangaroo (Weiss & McDonald, 1967), brush-tailed possum (Chester Jones, Vinson, Jarrett & Sharman, 1964; Weiss & McDonald, 1966a) and wombat (Weiss & McDonald, 1966b). Corticosterone has also been identified in smaller amounts in all these species. In the present studies cortisol and corticosterone levels were measured in peripheral blood and confirm the observations on the presence of these steroids in adrenal venous blood. However, in the wombat cortisol and corticosterone blood levels are similar (means of 0·18 and 0·10 μg/ 100 ml respectively). In contrast, cortisol is the major component of the kangaroo blood. The corticosteroid levels in both species are much lower than those previously reported for peripheral blood in these species (Weiss & McDonald, 1966b, 1967). It is of interest that in the monotreme, Tachyglossus aculeatus, corticosterone is the major steroid in adrenal venous (Weiss & McDonald, 1965) and peripheral blood (Coghlan & Scoggins, unpublished observations).

The rabbit secretes corticosterone rather than cortisol but is said to reverse this pattern under the influence of ACTH (Kass, Hechter, Macchi & Mou, 1954; Krum & Glenn, 1965). In the rabbits shot prior to sampling, corticosterone was the major steroid. Other animals, captured, kept overnight, and presumably stressed by this procedure, had high blood levels of both cortisol and corticosterone. In contrast to the ACTH studies reported by Krum & Glenn (1965) corticosterone secretion did not appear to be reduced at the expense of the increased cortisol production in the stressed rabbits.

The cortisol and corticosterone levels reported in the fox are the first measurements made of peripheral blood steroids in this species. The peripheral blood corticosteroids in the sodium-depleted sheep were similar to those found in trained sheep in the laboratory (Coghlan & Scoggins, 1967a).

The observations on the blood glucocorticoid levels in the various animal species studied in their natural environment confirms laboratory experiments on the lack of any obligatory change in cortisol or corticosterone production in sodium depletion.

It has been possible to extend our observations on the adaptation of various marsupial and animal species to sodium-deficient environments to include man. Dietetic studies have shown that the primitive mountain tribes of New Guinea have an almost exclusively vegetarian diet and that they consume 200–400 times more potassium than sodium (Oomen, Spoon, Heesterman, Ruinard, Luyken & Slump, 1961). In collaboration with Professor W. V. Macfarlane we have recently carried out an investigation on these natives. In an isolated highland area at Koinambe the staple diet is sago, taro and the introduced sweet potato; the mean urinary [Na$^+$] and [K$^+$] were 27 and 224 m-equiv/l. respectively. Urinary, salivary and sweat electrolytes were consistent with values seen during sodium deficiency. The mean blood aldosterone level was 23·0 ng/100 ml—a value markedly elevated above that of natives from a village (Pari) with access to foods containing higher levels of salt (mean 9·9 ng/100 ml). In this latter group mean urinary Na$^+$ and K$^+$ were 54 and 212 m-equiv/l. respectively. Both groups had potassium excretion much higher than that of other natives with a more European type of diet. In Madang, a coastal area, urinary sodium excretion was high (156 m-equiv/l.) and blood aldosterone levels were of the order found in European subjects on a similar sodium intake.

Plasma renin activity measured by Dr S. L. Skinner, (Skinner, 1967) was high in all groups compared with values found in Europeans and did not correlate well with either sodium intake, sodium excretion or blood aldosterone levels. The sweat and salivary sodium to potassium ratio was also low in the four groups studied although urinary sodium excretion was high (> 150 m-equiv/l.) in Madang and Goroke natives. These latter two groups also had a much lower potassium intake. Observations made in New Guinea show a good correlation between sodium status and aldosterone blood levels. As these people come in contact more and more with Europeans, salt consumption increases and sweat and urinary sodium rises while blood aldosterone levels fall. There is some evidence that these changes are associated with an increase in systemic blood pressure. Systolic blood pressure was approximately 20 mm Hg higher in the Madang group than that in the group at Koinambe. It is not easy to reconcile the plasma renin activities with the sodium intake and further studies will be required to determine the significance of these findings.

Denton (1965), in a review on the evolutionary aspects of the emergence of aldosterone secretion and salt appetite, proposed that the endocrine control of the digestive secretory mechanism is an adaptation with considerable survival value in ruminant animals. In herbivorous animals the volume of saliva secreted each day is large and the rumen content of sodium represents a significant part of the total body sodium pool. The rate of salivation and the rumen pool size of sodium are altered by the composition of the food (Wilson & Tribe, 1963). Thus the rapid change from lush to dehydrated pasture which occurs in arid areas following the change from the monsoonal rainy season to intense continental heat may impose a further stress on an animal eating pasture already low in sodium. Starvation followed by a period of active feeding may similarly result in tissue depletion of sodium. In cattle, faecal loss of sodium is also a significant route of sodium wastage. In this species a change of diet from a moderate to low sodium content reduced loss of sodium in faeces from 750 to 196 m-equiv/day (Renkema, Senshu, Gallard & Brouwer, 1962). Exposure of animals to an environment which varies in sodium status may also impair growth and development of these grazing animals.

Recently, Morris & Gartner (personal communication) have shown that young cattle fed a low sodium dietary regime of all grain plus protein supplements had a growth rate only 65 % of that seen in animals whose diet was supplemented with sodium. The animals without sodium supplementation had marked increases in zona glomerulosa area and a salivary sodium to potassium ratio indicative of severe sodium deficiency. The period of most severe sodium deficiency in the alpine area, the spring and early summer, coincides with the breeding season of the rabbit and other species. The sodium requirements of lactation, particularly in the rabbits in this area, imposes a further profound stress on the mechanisms regulating sodium homeostasis.

Population density may be an important determinant of the ability of a species to adapt to an environment. Christian (personal communication) has proposed that a high population density increases the size and activity of the zona fasciculata and reticularis and that the high glucocorticoid secretion causes both natriuresis and kaliuresis. This effect may thus prevent an adequate response to sodium deficiency. In wild rabbits (Blair-West et al. 1967) an increased density from 35 to 200 animals to the acre resulted in a marked reduction in the zona glomerulosa and an increase in the zona fasciculata, a condition which may severely aggravate the effect of sodium depletion.

In summary, these studies have shown that in the Southern Alps of Australia—a severely sodium-deficient area—the herbivorous animals, both indigenous marsupials and introduced rabbits, sheep and cattle, respond by activating physiological mechanisms to conserve sodium. It is evident that strong selection pressures operate favouring those animals with effective conservatory mechanisms. The herbivores may be expected to be exposed to this ecological stress to an extent much greater than the carnivores which feed upon them. Additional factors such as population density, disease and the reproductive process may well impose further demands on the mechanisms regulating sodium homeostasis.

The authors wish to thank C. L. Junqueira, Faculty of Medicine, University of São Paulo, Brazil, for his collaboration in these studies. We are also grateful to Professor W. V. Macfarlane, Department of Animal Physiology, University of Adelaide, Adelaide, and to Dr S. L. Skinner, Department of Physiology, University of Melbourne for allowing us to use unpublished observations from our joint New Guinea studies. This work was supported by U.S. Public Health Service Research Grants HE 06284 and AM 08701, The National Health and Medical Research Council of Australia, the Rural Credits Fund of the Reserve Bank of Australia, the Wool Industry Research Fund of Australia, the Anti-Cancer Council of Victoria, and the Ingram Trust.

Angiotensin levels were determined in collaboration with Dr K. Catt and M. D. Cain, Department of Medicine, Monash University, Victoria.

Plates 1, 3, 4 and 5 and Fig. 7 have been reproduced from Blair-West et al. (1967), Nature, Lond. 217, 922.

REFERENCES

ARNOLD, G. W. (1964). Some principles in the investigation of selective grazing. Proc. Aust. Soc. Animal Prod. 5, 258.

BASILE, G. (1632). The Pentameron. London: Hamlyn.

BELL, F. R. (1963). The variations in taste thresholds of ruminants associated with sodium depletion. In Olfaction and Taste (ed. Y. Zotterman). Oxford: Pergamon Press.

BLAIR-WEST, J. R., COGHLAN, J. P., DENTON, D. A., GODING, J. R., WINTOUR, M. & WRIGHT, R. D. (1963). The control of aldosterone secretion. Rec. Progr. Horm. Res. 19, 311.

BLAIR-WEST, J. R., COGHLAN, J. P., DENTON, D. A., GODING, J. R., ORCHARD, E., SCOGGINS, B. A., WINTOUR, M. & WRIGHT, R. D. (1967). Mechanisms regulating aldosterone secretion during Na deficiency. Proc. IIIrd Int. Congr. Nephrol. 1, 201.

BLAIR-WEST, J. R., COGHLAN, J. P., DENTON, D. A., NELSON, J. F., ORCHARD, E., SCOGGINS, B. A., WRIGHT, R. D., MYERS, K. & JUNQUIERA, C. L. (1968). Physiological, morphological and behavioural adaptation to a sodium deficient environment by wild native Australian and introduced species of animals. Nature, Lond. 217, 922.

BOTT, E., DENTON, D. A., GODING, J. R. & SABINE, J. R. (1964). Sodium deficiency and corticosteroid secretion in cattle. Nature, Lond. 202, 461.

BRITTON, S. W. & SILVETTE, H. (1937). Further observations on sodium chloride balance in the adrenalectomised opossum. Am. J. Physiol. 118, 21.

BROWNELL, K. A., BECK, R. R. & BESCH, P. K. (1967). Steroid production by the normal opossum (Didelphis virgiana) adrenal in vitro. Gen. comp. Endocr. 9, 214.

BUNGE, G. (1873). On the significance of common salt and the behaviour of potassium salts in the human organism. *Ger. Zeits. Biol.* **9**, 104.

BUTTLE, J. M., KIRK, R. L. & WARING, H. (1952). The effects of complete adrenal-ectomy in the wallaby (*Setonyx brachydrus*). *J. Endocr.* **8**, 281.

CATT, K. J., CAIN, M. D. & COGHLAN, J. P. (1967). Measurement of angiotensin II in blood. *Lancet* ii, 1005.

CHESTER-JONES, I., VINSON, G. P., JARRETT, I. G. & SHARMAN, G. B. (1964). Steroid components in the adrenal venous blood of *Trichosurus vulpecula* Kerr. *J. Endocr.* **29**, 149.

COGHLAN, J. P. & SCOGGINS, B. A. (1967a). The measurement of aldosterone in peripheral blood of man and sheep. *J. Clin. Endocr. Metab.* **27**, 1470.

COGHLAN, J. P. & SCOGGINS, B. A. (1967b). The measurement of aldosterone, cortisol and corticosterone in the blood of the wombat (*Vombatus hirsutus* Perry) and the kangaroo (*Macropus giganteus*). *J. Endocr.* **39**, 445.

COWAN, I. McT. & BRINK, V. C. (1949). Natural game licks in the rocky mountain national parks of Canada. *J. Mamm.* **30**, 379.

DALKE, P. D., BEEMAN, R. D., KINDEL, F. J., ROBEL, R. J. & WILLIAMS, T. R. (1965). Use of salt by elks in Idaho. *J. Wildl. Mgmt* **29**, 319.

DAVIS, J. O. (1967). The regulation of aldosterone secretion. In *The Adrenal Cortex* (ed. A. Eisenstein). Boston: Little, Brown and Co.

DAWSON, T. J. & DENNY, M. J. S. (1969). Seasonal variation in the plasma and urine electrolytes of the arid zone kangaroos *Megaleia rufa* and *Macropus robustus*. *Aust. J. Zool.* (In the Press.)

DENTON, D. A. (1965). Evolutionary aspects of the emergence of aldosterone secretion and salt appetite. *Physiol. Rev.* **45**, 254.

DENTON, D. A., GODING, J. R., SABINE, J. R. & WRIGHT, R. D. (1961). In *Salinity Problems in the Arid Zones*, p. 193. Paris: UNESCO.

GOCKE, D. J., GERTEN, J., SHERWOOD, L. M. & LARAGH, J. H. (1969). Physiological and pathological variations of plasma angiotensin II in man. *Circ. Res.* **24**, (Suppl. 1), 131–148.

HARTMAN, F. A., SMITH, D. E. & LEWIS, L. A. (1943). Adrenal functions in the opossum. *Endocrinology* **32**, 340.

HEMINGWAY, E. (1966). *Green Hills of Africa*. London: Penguin.

HUTTON, J. T. (1958). The chemistry of rainwater with particular reference to conditions in south-eastern Australia. In *Climatology and Micro-climatology*, p. 285. Paris: UNESCO.

JOHNSTON, C. I., DAVIS, J. O. & HARTROFT, P. M. (1967). Renin–angiotensin system, adrenal steroids and sodium depletion in a primitive mammal, the American opossum. *Endocrinology* **81**, 633.

KASS, E. H., HECHTER, O., MACCHI, I. A. & MOU, T. W. (1954). Changes in patterns of secretion of corticosteroids in rabbits after prolonged treatment with ACTH. *Proc. Soc. exp. Biol. Med.* **85**, 583.

KINNEAR, J. E., PUROHIT, K. G. & MAIN, A. R. (1968). The ability of the Tammar Wallaby (*Macropus eugenii marsupialia*) to drink sea water. *Comp. Biochem. Physiol.* **25**, 761.

KNIGHT, R. R. & MUDGE, M. R. (1967). Characteristics of some natural licks in the sun river area, Montana. *J. Wildl. Mgmt* **31**, 293.

KRUM, A. A. & GLENN, R. E. (1965). Adrenal steroid secretion in rabbits following prolonged ACTH administration. *Proc. Soc. exp. Biol. Med.* **118**, 255.

LEHMAN, C. G. (1854). *Physiological Chemistry*. London: Harrison and Sons.

MYERS, K. (1967). Morphological changes in the adrenal glands of wild rabbits. *Nature, Lond.* **213**, 147.

OOMEN, H. A. P. C., SPOON, W., HEESTERMAN, J. E., RUINARD, J., LUYKEN, R. & SLUMP, P. (1961). The sweet potato as the staff of life of the highland Papuan. *Trop. Geogr. Med.* **13,** 55.

PALMORE, W. P., MARIEB, N. J. & MULROW, P. J. (1969). Stimulation of aldosterone secretion by sodium depletion in nephrectomised rats. *Endocrinology* **84,** 1342.

REID, I. A. & McDONALD, I. R. (1968). Bilateral adrenalectomy and steroid replacement in the marsupial *Trichosurus vulpecula*. *Biochem Physiol.* **26,** 613.

REID, I. A. & McDONALD, I. R. (1969). The renin–angiotensin system in a marsupial, *Trichosurus vulpecula*. *J. Endocr.* **44,** 231.

RENKEMA, J. A., SENSITU, T., GALLARD, B. D. E. & BROUWER, E. (1962). Regulation of sodium excretion and retention by the intestine in cows. *Nature, Lond.* **195,** 389.

SCHALLER, G. (1965). *The Year of the Gorilla*. London: Collins.

SKINNER, S. L., (1967). Improved assay methods for renin 'concentration' and 'activity' in human plasma. Methods using selective denaturation of renin substrate. *Circulation Res.* **20,** 391–402.

SMITH, D. E., LEWIS, A. & HARTMAN, F. A. (1943). Sodium retention in the opossum. *Endocrinology* **32,** 437.

URQUHART, J. (1966). A comparison of the adrenal actions of ACTH and of cyclic 3′5′ AMP. *Proc. 48th Meeting Endocr. Soc.* p. 62.

VEYRAT, R., BRUNNER, H. R., GRANDCHAMP, A. & MULLER, A. F. (1967). Inhibition of renin by potassium in man. *Acta Endocr.* (Suppl.), **119,** 86.

WEISS, M. & McDONALD, I. R. (1965). Corticosteroid secretion in the monotreme *Tachyglossus aculeatus*. *J. Endocr.* **33,** 203.

WEISS, M. & McDONALD, I. R. (1966). Corticosteroid secretion in the Australian phalanger (*Trichosurus vulpecula*). *Gen. comp. Endocr.* **7,** 345.

WEISS, M. & McDONALD, I. R. (1966*b*). Adrenocortical secretion in the wombat, *Vombatus hirsutus* Perry. *J. Endocr.* **35,** 207.

WEISS, M. & McDONALD, I. R. (1967). Corticosteroid secretion in kangaroos (*Macropus canguru major* and *Megaleia rufus*). *J. Endocr.* **39,** 251.

WILSON, A. D. & TRIBE, D. E. (1963). The effect of diet on the secretion of parotid saliva by sheep. 1. The daily secretion of saliva by caged sheep. *Aust. J. agric. Res.* **14,** 670.

DISCUSSION

PHILLIPS: Dr Scoggins, may I congratulate you on a very fine presentation. I think you will be interested in an observation which Dr Neil Holmes and I made about ten years ago on one specimen of the Californian bobtail deer, which was donated to us by Dr Cowan in Vancouver. We found very surprisingly when we looked at the adrenal that the zona glomerulosa was about 30% of the cortex and we had no explanation for it at that time, but it fits in very nicely with what you've been talking about.

Might I ask if you have any evidence that the availability of salt in the environment does, in fact, dictate population density levels in any of these wild species.

SCOGGINS: Our studies of salt appetite have primarily been centred around those situations which the C.S.I.R.O. Division of Wild Life have

PLATE I

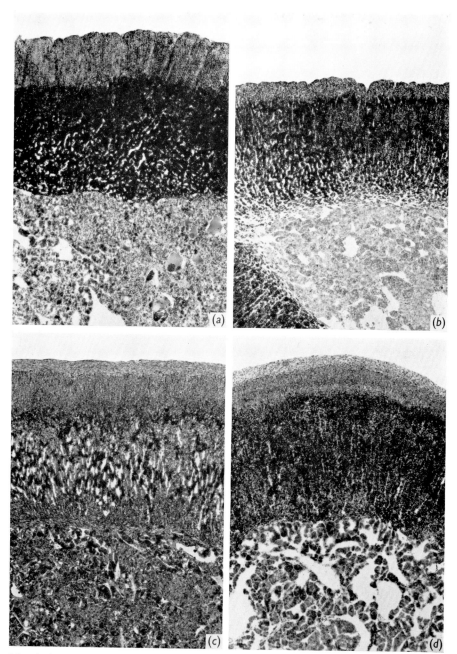

Morphology of the adrenal cortex of sodium-deficient (*a*) and sodium-replete (*b*) kangaroo (*Macropus giganteus*) and of the sodium-deficient (*c*) and sodium-replete (*d*) wombat (*Vombatus hirsutus*). The area of the zona glomerulosa is wider in the sodium-deficient animals. ×27.

(*facing p.* 600)

PLATE 2

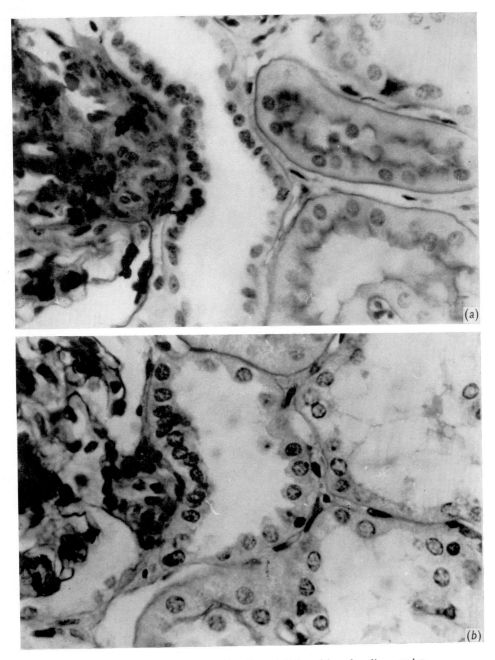

Morphology of renal cortex of sodium-deficient (*a*) and sodium-replete
(*b*) wombat (*Vombatus hirsutus*).

PLATE 3

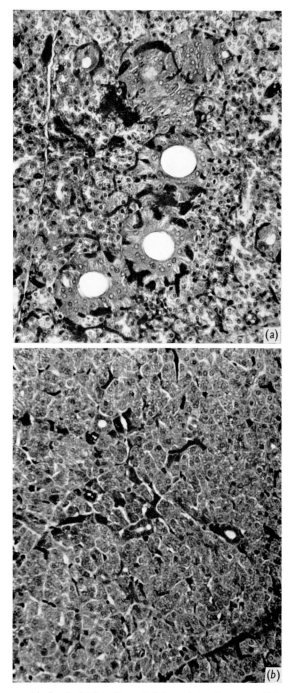

Section of the parotid gland of (a) sodium-deficient alpine kangaroo (*Macropus giganteus* compared with that of (b) sodium-replete sea-coast animal. × 60.

PLATE 4

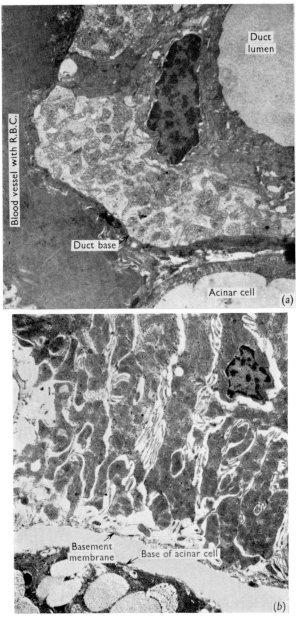

Electron micrograph of the parotid gland of a sodium-replete rabbit (*Oryctolagus cuniculus*) (*a*) compared with that of a sodium-deficient alpine wild rabbit (*b*). × 10,000.

PLATE 5

A photograph showing wild rabbits in the sodium-deficient alpine area attacking pegs impregnated with sodium bicarbonate and sodium chloride (label detached). This photograph was taken by Mr E. Slater, C.S.I.R.O. Wildlife Division, Canberra.

been observing for other reasons, and they have been studying population changes. I think this is something that will increase in the future, and we are starting to look at other geographical areas, in fact, where we have facilities to make such observations. But there is certainly good evidence that the drive for sodium is extremely strong; in fact, some of the pegs we put in were eaten out the same evening—completely eaten out of the ground. I think that certainly the animals do look for sodium and they have a strong drive to find it. Past this, we haven't, at this stage, extended our observations.

FERGUSON: It seems to me, Dr Scoggins, that a comparison of daily sodium and potassium losses in urine, saliva, and sweat would more readily be made if we had some knowledge of the total output of each ion per day, rather than the concentration in each of these fluids. Do you know whether the production of saliva, sweat and urine is the same in these groups of people?

SCOGGINS: You are talking primarily about the native groups in New Guinea. I think that Professor Macfarlane has shown that the water turnover in these groups is, in fact, the same. I think you can realize that when you're doing this in the field—the work that I've described this morning has all been done in the field—the problems of collecting 24 h urine collections from people who are frightened of even the sight of you is rather difficult. I think that he does an admirable job to collect the blood samples, which, I believe, he trades for a polaroid photograph of themselves.

IDLER: I would just like to make an observation on an aspect of Dr Scoggins and his group's work that I know he hasn't had time to cover in his presentation, and that is their methodology. I think this has been a very major contribution to this field; you can now look at aldosterone levels and have some confidence in them. In this respect I would like to mention that we have spent quite a bit of time applying these methods to fish plasma and they work very well, except that there is a need to carry out a preliminary chromatography prior to the acetylation step in our animals; otherwise, the recovery of [^{14}C]aldosterone is extremely low through procedures. With this modification the procedures worked extremely well for the identification and also for the quantification of aldosterone in the blood of herring. Salmon, cod and alewife all gave values not significantly different from blanks and confirm the reliability of the method.

I would also like to ask Dr Scoggins a question. I seem to recall that Dr Weiss reported in Milan, and possibly in a subsequent publication, that Reichstein's S was the principal steroid in kangaroo plasma. I was wondering whether you might like to comment, since I notice you found cortisol at levels between 1·5 and 3·5 μg/100 ml.

SCOGGINS: When we did these studies we weren't aware of what she had being doing; she has subsequently published it. I'll answer your question first. We haven't yet measured S in any of the animals, but are doing this. We have primarily been interested in measuring aldosterone, although we have the capability to measure large numbers of other steroids. We are more concerned, in fact, with getting data that we can interpret physiologically. We use the cortisol and corticosterone as an index of ACTH and we measure the aldosterone, being the mineralocorticoid of primary importance in the physiological situation that we are looking at. The other thing is that all our data are in fact radically different from hers. The levels that we have here are a fraction of the ones that she has reported in the same species. I think that the technique that we used to obtain the samples is relatively good in terms of getting unstressed animals. They are killed with one shot; the blood is taken very quickly into blood packs and frozen in dry ice. We feel this is a very good technique compared with that of getting these wild animals into a laboratory and having them under anaesthesia in a very stressed state.

IDLER: From a physiological point of view it was rather interesting that Weiss's work suggested that 11-hydroxylation was not taking place in marsupials. I was wondering if you have noticed anything that might have been Reichstein's S.

SCOGGINS: It would have to be there in enormous concentration to see it. We aren't looking specifically for anything that we don't really want.

HOLMES: I would like to comment on the observation that, not only do some species actively seek salt, but also that others appear to regulate their intake. Dr Fletcher and I recently reported that the duck progressively increased the intake of Na^+ when the Na^+ concentration of the drinking water was increased. When the Na^+ concentration of the drinking water exceeded 200 mM sodium chloride, however, the daily intake of Na^+ remained constant even though the birds became increasingly dehydrated when the volume of ingested saline was reduced. These data suggest therefore that, in addition to seeking Na^+, certain vertebrate species may also possess a mechanism whereby the intake of Na^+ is regulated (G. L. Fletcher & W. N. Holmes (1968), *J. exp. Biol.* **49**, 325–39).

SUMMING UP OF CONFERENCE

By PROFESSOR H. A. BERN

Yesterday evening we were regaled with a most remarkable spectrum of after-dinner speeches. It would seem most appropriate at this juncture to institute a new tradition, of which I would be strongly in favour, namely the after-speech dinner. You avid auditors eminently deserve such a reward for your patient attention to what I, necessarily the most constant auditor of us all, consider to have been a particularly valuable series of presentations. Nevertheless, I fear that you are to receive no fine dinner at my hands, but only a verbal *smörgåsbord*. In a topic as broad as 'Hormones and the Environment,' there is no single common theme that can possibly emerge. In addition, as Dr John Ball remarked to me, there have been some papers devoid of hormones and other papers devoid of the environment; I don't think there have been any papers devoid of both. When I consented to make these concluding remarks, inspired by the great good humour of Professor Phillips and Dr Benson as to the exhilaratingly joyous consequences of agreeing to their request, I was aware of the honour being paid me. As the Symposium has proceeded, I have become somewhat less conscious of the honour, but certainly fully cognizant of the onus. Whereas it ill-behoves me to stand before you tongue-tied (a natural garrulity is, in any case, difficult to repress completely), the one gift I can bring is the commendable one of brevity. As you well know, all lengthy speeches begin with such a statement. Nevertheless, there are a few points I would like to call to your attention which have come to mind as we have proceeded in our deliberations.

We have had it brought home rather forcefully that the very environment that we are discussing is in part a *milieu* of chemical stimuli, acting upon organisms and parts of organisms, much in the same way as hormones act within organisms. The intraspecific pheromones and the interspecific ectohormones, discussed by Dr Barth, among insects, remind us that the problems involved in understanding the mechanisms of hormone action are repeated again with these chemical agents acting through the environment to co-ordinate responses between organisms of the same or different species. Indeed, Dr Barth's reference to Professor Lüscher's continuing analysis of termite colonies indicates clearly that the old term 'super-organism', used to describe colonies of insects, is hardly irrelevant and, indeed, may aid conceptually in our understanding of mutualistic and community relations among animals and plants, where ectohormones integrating the activities of organisms can be analogized with hormones integrating the activities of organs. The discussion of mammalian skin glands by Dr Strauss gave us additional information relevant to the aerial broth in which we live. The biotic world is naturally contaminated with the chemical effusions of its component parts. It was fascinating—and not a little disturbing—to consider some of man's plans to add his own selection of chemical pollutants of hormonal and ecto-hormonal significance to this already complex chemical environment.

Another aspect of chemical regulation becomes apparent when one remembers that the *milieu interne* is the environment of cells and organs which do not happen to be exposed to the *milieu externe*. When searching for factors that control the function of endocrine glands and the response of target organs, we should not forget that the same ionic changes in the environment that may directly affect gill cells, for example, as Dr Maetz has discussed, when reflected internally may have the power to affect directly endocrine cells (as possible examples: prolactin cells, zona glomer-ulosa cells, juxtaglomerular cells, neurosecretory cells—or at least their endings). Such direct responses of internal cells, may I say primary or primitive in nature, may be more involved in regulation than we are accustomed to consider. One value of elaborate diagrammatic schemes delineating control over organ function, such as Professor Holmes's treatment of the avian salt gland, is that they remind us of the possible importance of the primary stimuli. The diet is obviously part of the environment, and Dr Scoggins's paper also points to responses to primary stimuli occurring possibly in mammals including man.

It was good to have Professor Heller's meaningful discussion of neural factors in the control of water balance. The endocrinologist is often commendably more cognizant of the influence of non-endocrine control systems upon his target organs, or target processes, than the physiologist, who is not endocrinologically disposed, seems aware of endocrine control systems. The history of our knowledge of salt-gland physiology in birds, to refer to these organs again, is especially revealing in this regard. The students of nervous control of the salt glands have acted as if the endocrine system had not yet been invented. The students of endocrine control are some-what less recalcitrant *vis-à-vis* the nervous system, as was indicated by Professor Phillips in his comment during the discussion.

In fact, there is little doubt that endocrinologists have finally reached a level of understanding of the real meaning of neuroendocrine integration which does not slight the nervous system: Dr Tindal's exquisite analysis of the neural background operative in oxytocin release in the hitherto well-known, but not well understood, suckling reflex; Professor Martini's dissection of the ways in which hormones feed back upon the brain to direct its secretion of hypophysiotrophic hormones; Professor Lofts's pleasantly comprehensible discussion of photoperiodic control of gonadal function in birds; Professor Lederis's description of a part of the piscine nervous system which acts in the same fashion as a 'straight' endocrine gland secreting into the blood; Professor Quay's attempts to define and separate the receptor and effector functions of the pineal complex. All reflect the sophistication of modern vertebrate endocrinology which has inextricably become neuroendocrinology. This appreciation of the fundamental endocrine-regulatory contribution of the nervous system is rather old hat for invertebrate endocrinologists, as was indicated in Professor de Wilde's discussion of the control of insect diapause. I am increasingly impressed, and occasionally more than a little depressed, by the complexity of the problems raised by the interaction of environmental cycles of various temporal characteristics with endogenous rhythms.

Dr Tindal's valuable preoccupation with one of the more noticeable skin glands should raise some interest in possible parallel mechanisms controlling the responses of less prominent integumentary structures. The skin men among us—Professor Ebling, Dr Strauss, Dr Maderson— could well ask whether plucking a hair, or even scraping a scale, may not have neuroendocrine repercussions in a minor key reminiscent of those seen after suckling a teat.

Professor Etkin's survey of amphibian metamorphosis stimulated an awareness of some interesting complexities that are involved in considering effects of environmental factors. And I am sure that Dr Mordue and his colleagues and Professor de Wilde would agree that the same argument may apply to insects. In situations where there is a mixture of cell populations—larval and adult—in amphibians and insects, environmental effects, mediated by the endocrine system, may be manifested in a manner dependent upon which cell population has a future, so as to speak. Thus, in early larval stages one might expect different repercussions as the result of activation or suppression of endocrine organs from what might occur at later stages. Environmental stimuli may similarly have different effects during the several stages of the sexual transformation in fishes, as discussed by Dr Reinboth. This point is also pertinent, I think, to Professor Bellamy's discussion of the ageing phenomenon. Laboratory mammals subjected to certain exogenous hormonal stimuli during the perinatal period are known to manifest profound effects in later life. Subjection to environmental stimuli affecting hormone production at this period might be expected to have similar profound consequences, again possibly evident only at a considerably later time. The same kind of stimulation in later postnatal life may have only minimal or transitory effects. Such 'critical periods' may occur in the life-history of other animals as well. Indeed, one of the facets of Dr Milner's study of starvation effects on human infants is the early age at which endocrine involvement is seen, with the possibility that permanent changes may occur, to be noted possibly decades later in those individuals that survive.

From Professor Dantzler's and Dr Maetz's reports we are reminded of the need for a deep knowledge of the background of the phenomenon the endocrinologist assumes he is coping with. One can no longer speak meaningfully of hormones and osmoregulation unless one can think realistically at the cellular and subcellular levels, as Dr Maetz does so skilfully, and at the level of whole-organism physiology, as was so thoroughly exemplified by Professor Dantzler regarding desert vertebrates. It is in this area, among others, where one can well praise the Chester Jones school. Dr Chan's presentation was characterized by an admirable ability to present concretely the changes that occur simultaneously in a series of features as a result of manipulations of endocrine organs related to osmoregulatory capacity. Several considerations of osmoregulatory phenomena have made it evident that we must think also in terms of exchanges occurring at a variety of surfaces. Dr Olivereau made this clear in relation to prolactin in teleost fishes, Professors Heller and Dantzler in relation to water movement in vertebrates generally, Dr Mordue in relation to water balance in insects.

A last point before I conclude concerns a topic which I would much enjoy seeing elaborated. This has to do not with hormones and the environment, but rather with the endocrinologist and his environment. I have already referred to this in relation to the modern endocrinologist's need for a fundamental understanding of the phenomenon he is studying. The renal physiologist must know his renal physiology; the endocrinologist interested in vasopressin physiology must know not only the endocrinology of neurohypophysial factors, but also his renal physiology. The variety of phenomena available to the endocrinologist for study is virtually unlimited. That is why there are so many of us. Inherent in the topics of this Symposium has been the theme of comparative endocrinology: endocrinology oriented toward understanding how animals of all kinds adapt to the challenges presented by the environment in which they find themselves, and how endocrine organs and endocrine mechanisms have evolved and have contributed to the resolution of these adaptational problems. I commend Professor Lofts in particular for his use of the language appropriate to this orientation.

Comparative endocrinology has more to suggest regarding the ontogeny of endocrine mechanisms, and the nature of endocrine mechanisms during ontogeny, than may be fully appreciated, I believe. This is not a plea for a neo-Haeckelian approach. However, I suggest that information regarding both the chemistry and function of hormones in lower vertebrates may point to the role they subserve in the early development of higher vertebrates, including man. Prolactin has been discussed by Professor Etkin in regard to its developmental influence in amphibians and by Dr Olivereau in regard to its osmoregulatory influence in teleosts. One can ask some obvious questions for which no answers yet exist, as to the possible contributions of this hormone in the aquatic environment that is imposed upon all amniotes during their embryonic and foetal lives.

As Professor Chester Jones pointed out yesterday evening, comparative endocrinology has not always been respectable. It is to him, as a real pioneer of our field, not only in the laboratory and in the classroom, and at his desk writing *The Adrenal Cortex*, but also formally at scientific sessions and informally at the bar, that we owe much of our present stature. Comparative endocrinology recognizes that that which is general in regard to endocrine biology will emerge when we know enough about the varieties of patterns of hormonal mediation occurring in organisms of all kinds. All animals, and plants too, are fair game, including the human species, as Dr Scoggins's and Dr Milner's able discussions of nutritional–hormonal interrelations so clearly demonstrated, in the obtainment of the general picture we seek.

I would recommend to my good friend and respected colleague, Ian Chester Jones, that he might consider expanding on the subject of the comparative endocrinologist and his environment at some early date. We would all profit from such an historical analysis, and he could well take considerable pride in the fact that an important part of his account would have to be autobiographical.

INDEX

acetazolamide, and urate clearance in
  Natrix, 181
acetylcholine
  response of urotensin assay systems to,
    468–9, 470
  and salt gland in aquatic birds, 107
  and water metabolism, 453, 454
acetylcholinesterase in brain, 453
  after pinealectomy, 428–30
acetylserotonin methyl transferase, in
    retina and pineal of amphibia, 444
Acheta, juvenile hormone in, 508
acid-base balance, maintained by gills,
    11–12
acid phosphatases, of hypothalamus, after
    castration, 551
acidosis, and urate clearance in *Natrix*,
    181–2
Acronyta rumicis (Noctuid moth), geo-
    graphic races of, 487, 490
actinomycin D, and sodium pump of gill,
    20–1, 22
adaptations
  ageing and capacity for, 310–14
  to external salinity: in gill, 20–4, 25; in
    intestine, 4–9, 24
  proximate, 545–6
adenohypophysis, *see* pituitary, anterior
Adoxophyes reticulana (fruit-tree leaf-
    roller), diapause in, 495, 496, 498
adrenal cortex
  age-related changes in, 324
  effects of ACTH on, 23, 101–4
  and hair growth, 235–6
  in malnutrition, 192–3
  and osmoregulation in teleosts, 31–55
  pituitary control of function of, 9, 93–8
  see also adrenocorticotrophin, corti-
    costeroids
adrenalectomy
  compared with removal of corpuscles of
    Stannius, 44
  effects of starvation after, at different
    ages, 325–6
  hair colour after, 230
  hair growth after, 226
  in marsupials, 592
  and osmoregulation in teleosts, 38–9,
    40, 41, 42, 72
  plasma renin content after, 101
  sebaceous glands after, 354
adrenaline
  and blood flow in gills, 43
  responses of urotensin assay systems to,
    470
  and water metabolism, 450, 451, 455

adrenals, weight of
  after blocking of ACTH secretion,
    408
  in sodium-depleted herbivores,
    588–9
  in young rats isolated and not handled,
    437
adrenergic mechanisms, in water meta-
    bolism, 450–2, 454–6
adrenocorticosteroids, *see* corticosteroids
adrenocorticotrophin (ACTH)
  age and response to, 318, 323, 325
  cells of anterior pituitary producing,
    58–9, 64, 65
  effects of, on adrenal explants *in vitro*,
    101–4
  and hair colour, 230
  and hair growth, 226
  hypothalamic releasing factor for (CRF),
    144, 415–16
  and lactation, 239, 242
  and osmoregulation in teleosts, 9, 23,
    24, 38, 39, 42, 43, 72, 74
  and plasma renin, 99
  receptors for, in median eminence,
    412
  release of, from pituitaries *in vitro*,
    89–93
  and salt-gland secretion in aquatic birds,
    88
  and sebaceous glands, 354, 355
  sexual maturity and response to, in
    salmonids, 45
  and skin-shedding in snakes, 268
adult diapause, 498
Aedes, temperature and sexual differentia-
    tion in, 529
ageing, and endocrine responses to
    environment, 303–39
aggregation pheromone of *Blatella*, 383
Albula vulpes, urophysis of, 465
albumin of blood, in malnutrition, 196
aldosterone
  age and response to, 326; and secretion
    of, 321, 324
  in marsupials, 591, 592
  methods for determining, 580, 601
  and osmoregulation in teleosts, 34, 35,
    37, 39, 40, 42, 46, 54–5
  in plasma after hypophysectomy in duck,
    97–8
  renin-angiotensin system and, 98, 578,
    593
  in sodium depletion, 583–4, 585, 586–7,
    591, 596
  and water metabolism, 449, 456

39   [609]   B M E 18

Lophortyx spp. (quails), kidney function
in, 162, 163
luteinizing hormone (LH)
amount of, in pituitary: after implantations in median eminence, 409,
410, 413; at different ages, 323
of hosts, and ovarian regression in
rabbit fleas, 394
photoperiod and secretion of, in birds,
550, 562, 563–5
receptors for, in median eminence, 412–13
and territorial behaviour in birds,
572–3
luteinizing-hormone releasing factor
(LH-RF), 410
in birds, 559, 561
site of synthesis of, 414, 415
Lymantria dispar, diapause in, 498

Macrodytes, allatectomy of, 534
Macropodus opercularis, artificial sex-
inversion in, 531
Macropus canguru major (kangaroo),
aldosterone in blood of, 591
Macropus giganteus (kangaroo), sodium-
depleted, 582–3, 589–90 blood
corticosteroids of, 584, 586, 591, 592
Macropus robustus, plasma potassium in,
593
malnutrition, and endocrine system in
humans, 191–212
Malpighian tubes in locusts, regulation of
excretion by, 125, 126, 127
mammals
control and function of skin glands in,
341–71
control of moulting in, 215–37
kidney function in, 158–61, 177
nitrogenous excretory products, of 169,
179
pineal in, 427–8
sites of water intake by, 456, 457
sodium-depleted, 577–602
mammary gland
environmental stimuli and, 239–58
innervation of, 242, 251–2
marasmus, 192, 193, 194, 195–6
marking behaviour, of terrestrial mammals
403
marsupials
lactation in, 240
sensory pathways in, 249
sodium depletion in, 591–4
median eminence
effects of extracts of, on release of
ACTH from pituitary tissue *in vitro*,
90–3
receptors for hormones in, 408–13
stores of releasing factors in, 414–18

in tadpoles at different stages of
metamorphosis, 146, Pl. 2
Megaleia rufus (kangaroo)
aldosterone in blood of, 591
plasma potassium in, 593
Megoura vitiae (aphid), photoperiodic
response in, 493
melanophore-stimulating hormone (MSH),
60, 72
and hair colour, 230
and lactation, 239
pineal and, 232, 424
Melanoplus, effect of oat diet on development of, 124
Melanoplus differentialis (grasshopper),
oxygen uptake in diapause of, 499
melatonin
photoperiod and secretion of, 232
pineal and secretion of, 424
site of synthesis of, 444–5
in teleosts, 425
memory, in conditional release of
hormones, 248
metabolic rate
age and, 307, 312
malnutrition and, 194–5
not affected by thyroxin in amphibia,
150–1, 154
metamorphosis in amphibia, endocrine
mechanism of, 137–55
metaraminal (α-adrenergic), and water
metabolism, 450, 451, 455
Metatetranychus ulmi (fruit-tree red
spider), diapause in, 491
methyl farnesoate dihydrochloride,
synthetic analogue of juvenile
hormone, 395, 398, 404
17α-methyl-β-nortestosterone (anti-
androgen), and sebaceous glands, 347
methyl testosterone
after hypophysectomy of *Fundulus*, 535
and sex determination in *Hemihaplochromis*, 543
metopirone, adrenocortical blocking
agent, 43
mice
at different ages, 305, 322–3
hair growth in, 217, 222, 224, 226, 232
Microtus agrestis, hair growth in, 236
Microtus montanus, hair growth in, 220
milk-ejection reflex, 239, 241, 245–8
Mimas tiliae, diapause in, 503–4, 507
mink, hair growth in, 217, 218, 225, 226
minnow, in deionized water, 66
Misgurnus fossilis, in deionized water, 66
mitochondria, 172, 291
loss of, with ageing, 307, 308
steroid-hydroxylating enzymes in 35,
37, 290